Food Bioactives and Nutraceuticals

Syed Amir Ashraf • Mohd Adnan
Editors

Food Bioactives and Nutraceuticals

Dietary Medicine for Human Health and Well-Beings

Springer

Editors
Syed Amir Ashraf
Department of Clinical Nutrition
University of Ha'il, College of Applied Medical Sciences
Ha'il, Saudi Arabia

Mohd Adnan
Department of Biology
University of Ha'il, College of Science
Ha'il, Saudi Arabia

ISBN 978-981-97-9722-6 ISBN 978-981-97-9723-3 (eBook)
https://doi.org/10.1007/978-981-97-9723-3

This Springer imprint is published by the registered company Springer Nature Singapore Pte Ltd.
The registered company address is: 152 Beach Road, #21-01/04 Gateway East, Singapore 189721, Singapore

This book is dedicated with full beliefs to all the educators, mentors, teachers, and guides who loves to teach and have been beacons of light for their students throughout their lives. To every student, whom we entrust with the future and who truly enjoy studying.

Preface

A new food health paradigm that has emerged in recent years has increased the focus on the benefits of diet. Furthermore, modern-day population has embraced a new lifestyle as a result of the rapid urbanization, which has altered basic food eating patterns globally. As a result, consumption of junk food has risen dramatically, raising concerns about a number of chronic diseases like diabetes, obesity, cardiovascular disease, and several disorders linked to an unbalanced diet and nutrition. Thus, the potential of nutraceuticals as balanced nutrition has been recognized due to the significance of natural compounds, such as food bioactives with their function as a health promoter for therapeutic management of various metabolic diseases. Nutraceuticals have the potential to improve and maintain human health without posing any risks due to their natural occurrence.

The most recent scientific data on many aspects of functional foods and nutraceuticals are presented in this book. Nutraceuticals from plant, animal, and microbial sources are covered, along with the role that dietary bioactives play in the prevention and therapeutic management of a range of chronic and metabolic diseases. Nutraceuticals as intelligent foods are also discussed. For the future requirements of human wellness and well-being, this book presents excellent prospects for food scientists and the food industry to develop innovative foods or food components. They fall under the following categories: prebiotics, probiotics, polyunsaturated fatty acids, antioxidants, dietary fiber, vitamins, and spices. The editors and authors of this book hope to provide the most recent findings and reviews on the different bioactive ingredients found in traditional foods, as well as their nutraceutical value for leading a healthy lifestyle. Additionally, the potential therapeutic application of food bioactives as a new generation nutraceuticals from novel sources is detailed and discussed.

The book will prove to be a valuable reference book for/in food scientists, food analysts, food technologists, new product development research, food and pharma industry professionals, food ingredient suppliers, food processing industries, nutritionists, dieticians, nutraceuticals, health and wellness, and regulatory agencies. It will also be helpful for PhD students with research interests in related subjects, as well as for students seeking undergraduate and postgraduate degrees in food science, nutrition biology, clinical nutrition and dietetics, biotechnology, food bioengineering, and food technology, among other fields.

Ha'il, Saudi Arabia

Syed Amir Ashraf
Mohd Adnan

Acknowledgments

Editors are eternally grateful to their parents and families. None of this would have been possible without their everlasting love, encouragement, and endless support.

Editors are graciously thankful to all the authors of this book; without them this book would not exist. Your participation, contribution, and faith in this book motivated us to finish it.

What we have accomplished together will make a difference

Contents

Chapter 1
Food Bioactives as a Novel Source for Nutraceuticals

Syed Amir Ashraf, Humera Banu, Abd Elmoneim O. Elkhalifa, Abdul Haque, Md Faruque Ahmad, and Danish Mahmood

Abstract Food provides a variety of nutrients that are vital to human health, as well as chemical components present in food with bioactive attributes believed to be crucial for both preventing illness and health promotion. Food bioactive compounds have garnered attention recently due to its significant therapeutic values, and there is discussion on how to utilize these, when making recommendations for public health as nutraceuticals. Food bioactives are naturally occurring biologically active chemical components derived from various food sources (plants, animal and microbial), which possess significant therapeutic properties against various ailments. Food or food products contains an abundant amount of bioactive compounds such as (alkaloids, glycosides, phenol, tannins, carotenoids), which is considered for its therapeutic efficacy against various clinical disorders such as cardiovascular issues, diabetes, inflammatory disease, cancer, arthritis and immune-modulators. Therefore, food or food bioactive can play a very significant role as nutraceuticals for the management of various therapeutic illness. A nutraceutical is a "food or food bioactive obtained from plant or animal with significant health benefits" mostly employed to prevent or cure diseases. In this chapter, efforts were made to bring altogether the possible wide range of various food bioactive compounds present in food and its

S. A. Ashraf (✉) · H. Banu · A. E. O. Elkhalifa
Department of Clinical Nutrition, College of Applied Medical Sciences, University of Ha'il, Ha'il, Saudi Arabia
e-mail: ao.abdalla@uoh.edu.sa

A. Haque
Department of Post-Harvest Engineering and Technology, Faculty of Agricultural Sciences, Aligarh Muslim University, Aligarh, India

M. F. Ahmad
Department of Clinical Nutrition, College of Nursing and Health Sciences, Jazan University, Jazan, Saudi Arabia
e-mail: mfahmad@jazanu.edu.sa

D. Mahmood
Department of Pharmacology and Toxicology, College of Pharmacy, Qassim University, Buraydah, Saudi Arabia

S. Amir Ashraf, M. Adnan (eds.), *Food Bioactives and Nutraceuticals*,
https://doi.org/10.1007/978-981-97-9723-3_1

possible application as nutraceuticals. Additionally, this chapter will also draw the interest of nutritionists, food scientists, pharmaceutical industries and food businesses in order to improve the application of bioactive molecules in nutraceutical industry and facilitate their commercialization in order to support and promote wellness, wellbeing, and a healthy lifestyle.

Keywords Functional food · Carotenoids · Dietary fiber · Flavonoids · Cancer · Diabetes

1.1 Introduction

Food bioactive compounds are naturally occurring chemicals found as secondary metabolites in plants, animals, or microbes that provide health benefits such as anti-obesity, antioxidants, anti-diabetic, anti-inflammatory, cardiovascular-protective, anti-cancer and many more. In recent years, food bioactives are being explored as nutraceuticals due to growing understanding of nutrition, health, and illness (Sorrenti et al. 2023; Teodoro 2019). Nutraceuticals provide a wide range of potential benefits including a good track record in the prevention and treatment of complex illnesses. Nutraceuticals are the formulation of nutrient/nutrients, food bioactive compounds, which helps in prevention and treatment of several diseases. Nutraceuticals are characterized as 'specially designed preparations', formulated with the aim of fulfilling specific dietary requirements including preventive health care approach. Foods or a part of foods that are beneficial in providing various health benefits including the treatment and/or prevention of the disease considered as nutraceuticals (Puri et al. 2022). When used in conjunction with conventional drug therapy, nutraceuticals with a well-established safety profile and efficacious safety profile may be a viable therapeutic alternative for the prevention of diabetes, hypertensive diseases and other chronic disease as presented in Fig. 1.1. According to Puri et al. (2022), the combination of a novel nutraceutical and non-steroidal anti-inflammatory medicines (NSAIDs) has demonstrated potential as a treatment for osteoarthritis, hence enhancing the product's safety and efficacy for commercial use (Puri et al. 2022).

This article examines current evidence-based research on bioactive substances in foods and their relationship for the management of human health. It investigates the evidence for various bioactive substances in our diet and their potential advantages in the fight against obesity, diabetes, cardiovascular disease, and cancer. Some examples of food bioactive compounds are alkaloids, glycosides, polyphenols, dietary fiber, antioxidants, carotenoids, omega-3 fatty acids etc. Beyond the nutritional components, functional foods, dietary supplements, and nutraceuticals include bioactive chemicals that have positive effects on health. The majority of what are known as "bioactive" compounds are phenolic compounds, which are found in vegetables, fruits, grains, spices, and foods and beverages that are derived from them, such as tea, olive oil, fruit juices, wine, chocolate, and beer. Terpenoids

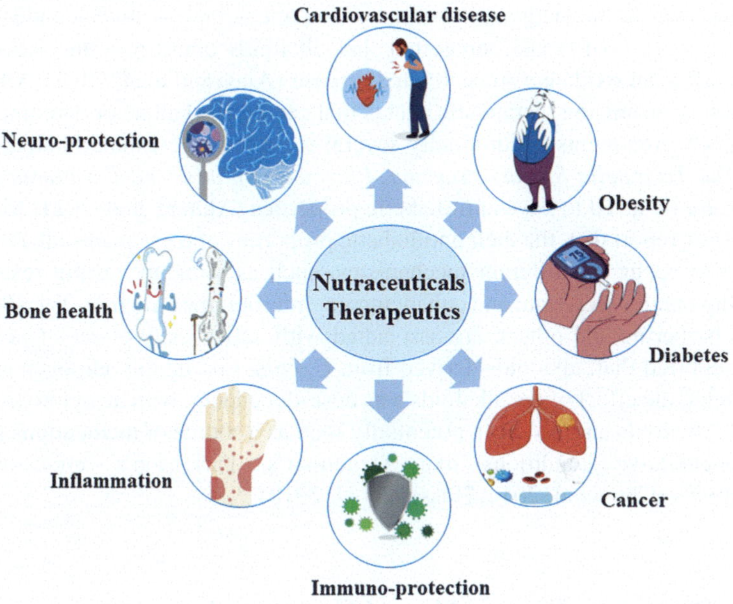

Fig. 1.1 Pictorial representation of food bioactive as nutraceuticals and its therapeutic application

(carotenoids and phytosterols), polyphenols, alkaloids, and other nitrogen-containing constituents (glucosinolates) are among these compounds (Sorrenti et al. 2023; Samtiya and Aluko 2021; Elkhalifa et al. 2021). Various food bioactive and its potential were put forward further to understand its possible role as nutraceuticals.

1.2 Food Bioactives and Their Possible Applications as Nutraceuticals

1.2.1 Alkaloids

As a naturally occurring bioactive compound, alkaloids contain a nitrogen atom in its ring structure that contributes to several compounds with neutral or mildly acidic properties. Bio-heterocyclic alkaloids are the main secondary metabolites recognized to have medicinal properties (Teodoro 2019). Secondary metabolism (specialized metabolism) produces a large group of natural compounds known as alkaloids. Generally found in plants but also in animals, insects, marine invertebrates and some microorganisms (Bribi 2018). Their precursors (mainly amino acids) are derivatives of metabolic pathways such as glycolysis. They can be found in mitochondria, vesicles, chloroplasts, and vacuoles. A number of plant alkaloids have been shown to be anticancer agents in some drug-resistant cancers, which is of particular interest to medicinal chemistry and medicine. A study conducted by Alasvand et al. (2019) also suggested that alkaloids could prevent and/or treat inflammation, an oxidative stress-related disease (Alasvand et al. 2019). Alkaloids are currently promising therapeutic agents that can be used alone or combined with other cancer treatments. Additionally several reports suggest that, alkaloids from plants like *Trigonella foenum-graecum, Rhizoma coptidis, and Ervatamia microphylla* were believed to have anti-diabetic properties (Ahmad et al. 2021; Ma et al. 2019). They report that, the their antidiabetic properties of these plants alkaloids are believed to occur via different mechanisms, such as reducing insulin resistance, increasing insulin secretion, and enhancing gut microbiome structure. The alkaloids such as berberine and others, are associated with anti-obesity effects. Choi et al. (2014) reported that, alkaloids derived from *C. chinensis* inhibit adipogenesis in 3 T3-L1 cells; the effect of the alkaloids was dose-dependent, with no cytotoxic effect evident. Alkaloids has also been potentially seen as a source of medications to treat neurodegenerative conditions like Alzheimer's, Parkinson's, epilepsy, and Huntington's (Choi et al. 2014; Hussain et al. 2018).

1.2.2 Phenolics, Flavonoids and Carotenoids

According to Singh et al. (2015), phenols and carotenoids are thought to be the most significant bioactive or phytochemical substances that can improve human health. The majority of fruits and vegetables with orange and yellow hues are rich in lipophilic molecules called carotenoids (Elkhalifa et al. 2021; Ahmed et al. 2014; Singh et al. 2015). For the food business, in particular, these substances are highly helpful as food additives that promote health and as colors. Zeaxanthin, lutein, and β-cryptoxanthin, for instance, have been shown to protect against sunburn-related illnesses, delay the onset of cataracts, and lower the risk of cardiovascular disease (Benke and Benke 2014; Weikel et al. 2014). Furthermore, because carotenoids have strong antioxidant properties that have been demonstrated, there is a good chance that they will lower the risk of developing certain chronic diseases (Gürbüz et al. 2018). Natural antioxidants called polyphenols are primarily found in foods and beverages, drink mixes, spices, fruits, vegetables, and cereals. They are also found in medicinal plants. Some groups of polyphenols include anthocyanins, flavonoids, and phenolic acids. Biological properties such as anti-inflammatory, anti-aging, and anti-cancer effects are believed to be due to presence of natural antioxidants, particularly those in the form of carotenoids and polyphenols (Xu et al. 2017). Biogenic amine formation has been verified by decarboxylation of amino acids or amination of aldehydes and ketones. Dopamine, serotonin, and noradrenaline are examples of biogenic amines that have been found in banana peels and pulp (Bennett et al. 2010; Singh et al. 2016). Because they are high in naturally occurring antioxidants like vitamin C (ascorbic acid) and vitamin E (tocopherol), it is advised to consume fruits and vegetables. Modulating immune function, lowering

DNA damage, and enhancing lipid peroxidation are just a few of the health-promoting qualities of antioxidants. The two best foods for dietary catechins are red wine and tea. Aglycones or gallic acid esterification are the typical forms of catechins. Additionally, the most significant flavones are luteolin and apigenin. Red pepper and celery comprise the majority of the human diet. Edible plants like plums, eggplants, and berries have blue, red, and purple hues due to the presence of anthocyanins like cyanidin, delphinidin, and pelargonidin. Isoflavonoids that are mostly found in legumes include genistein and daidzein. These compounds are most abundant in soybeans and products derived from soybeans (Erlund 2004; Liggins et al. 2000). D'Andrea (2015) state that apples, onions, and tea are just a few of the foods that contain quercetin, a flavonoid of the flavonol type. Fruits such as grapes and citrus fruits are the main sources of naringenin, a naturally occurring flavanone that has been shown to support insulin signaling and regulate brain function (D'Andrea 2015; Ghofrani et al. 2015). Regarded as the final members of the flavonoid group, proanthocyanidins are also known as condensed tannins. These polymers, or oligomers, are connected by interflavan bonds and are called catechol (Fvan-3-ol). Plant-based foods and drinks, such as red wine, grains, and fruits, are rich in them (Pereira-Caro et al. 2020). Sulfur compounds found in plants are another class of bioactive substances (OSCs) (Moreno-Ortega and Ordóñez 2021). According to recent research, the most significant onion thiosulfinates are isoalliin (organic) (Moreno-Ortega et al. 2020) Garlic, a member of the Allium genus, has been used traditionally around the globe as a medicinal plant. Health-promoting qualities like anti-cancer and anti-diabetic properties have been found in earlier research. According to reports, the presence of bioactive compounds like phenolic compounds and organosulfur components is what causes these health benefits (Moreno-Ortega and Pereira-Caro 2020). Virgin olive oil is well-known and well-liked in the Mediterranean diet for its flavor and nutritional content. The phenolic compounds found in virgin olive oil, such as hydroxytyrosol, tyrosol, flavonoids, vanillic acid, and p-coumaric acid, are primarily responsible for their health benefits(Quintero-Flórez et al. 2018). Because phenolic components can donate electrons or hydrogen atoms, they are regarded as important antioxidants or free radical inhibitors (Singh et al. 2016). According to Singh et al. (2015) and Kim et al. (2016a), antioxidant-rich fruits and vegetables may reduce the symptoms of heart disease, arthritis, neurodegenerative diseases, cancer, atherosclerosis, and brain dysfunction (Singh et al. 2015; Kim et al. 2016a). Cheng et al. (2014) conducted research pointed out that Rutgers Scarlet Lettuce (RSL) aqueous extract rich in polyphenols may have applications as an antidiabetic agent. After 28 days of oral treatment, mice made obese on a high-fat diet showed improved glucose metabolism when given RSL extract (RSLE) at doses of 100 or 300 mg/kg (Cheng et al. 2014). Hidalgo-Liberona et al. (2021) conducted a recent clinical study involving 51 adults showed that polyphenols may regulate intestinal health and avert pathological or unhealthy conditions linked to decreased intestinal permeability (Hidalgo-Liberona et al. 2021).

1.2.3 Terpenes and Terpenoids

Terpenoids comprise a vast and varied class of naturally occurring organic compounds that are available in different combinations and modifications among all living things. According to Perveen et al. (2018), terpenoids are responsible for the yellow and red hues of tomatoes and sunflowers, as well as the scent of clove, eucalyptus, ginger, and cinnamon (Shagufta 2018). A growing number of terpenoid compounds with established medical uses have been registered as pharmaceuticals on the market as a result of the wide variety of terpenoid structures and activities stimulating interest in their commercial application (Ludwiczuk et al. 2017).

According to Álvarez-Martínez et al. (2021), terpenes have shown antimicrobial effects on bacteria that are both susceptible and resistant to antibiotics, despite their capacity to stimulate cell rupture and inhibit the synthesis of proteins and DNA (Álvarez-Martínez et al. 2021). In an investigation by Gallucci et al. (2009) revealed that terpenes including thymol, eugenol, geraniol, carvacrol, and carvone had antibacterial effects on Staphylococcus aureus. Furthermore, in aromatic and therapeutic plants, terpenoids are significant secondary metabolites that contribute significantly to disease resistance. Using monoterpenoids as an example, these antibacterial molecules interfere with the growth, reproduction, and metabolism of microbes (Gallucci et al. 2009; Burt 2004). According to Yang et al. (2020), terpenoids promote transdermal absorption, prevent and treat cardiovascular disease, and have hypoglycemic and anticancer, anti-inflammatory, antibacterial, antiviral, and antimalarial properties. Additionally, an isolated active ingredient from the stevia plant (*Stevia rebaudiana* (Bertoni) Hemsl.) is called diterpenoid stevioside (SVS) (Yang et al. 2020). Jeppesen et al. (2006) reported that SVS significantly reduced the hyperglycemia of the rats. The active component of ginsenoside, a traditional Chinese medicine, is found in ginseng. Ginsenoside-Rb1 (G-Rb1) has been demonstrated to be a promising anti-inflammatory agent through additional research. G-Rb1 has a potent anti-inflammatory effect by significantly reducing NF-B activation. (NF-B is also a regulator of TNF- production). Additionally, it was found that ginsenosides-Rb2 (G-Rb2) and ginsenosides-Rd (G-Rd) had neuroprotective qualities (Jeppesen et al. 2006; Kim et al. 2017). Terpenes are one of the most promising sources of novel antimicrobial drugs, with efficacy against viruses, bacteria, fungus, and protozoa (Hassan et al. 2015). Artemisinin is a sesquiterpene lactone chemical discovered in the 1970s in *Artemisia annua* Linn. Artemisinin's chemical structure was later modified to produce antimalarial medications such as artesunate, arteether, and artemether. These medications have little side effects and can successfully kill plasmodium during the red blood cell phase (Yang et al. 2020).

1.2.4 Saponins

It has been shown that saponins are amphiphilic molecules with carbohydrate and either triterpenoid or steroid moieties, with antifungicidal, antimicrobial, antiviral, anti-inflammatory, anticancer, antioxidant, and immunomodulatory properties. A variety of traditional and herbal medicines have long relied on saponins from natural sources (Juang and Liang 2020). Although saponins were believed to be antinutritional factors, research has revealed that some saponin stereoisomers may have stereospecific pharmacological activities such as anti-tumor, antibacterial, antioxidative, anti-inflammatory, antidiabetic, and neuroprotective properties. Furthermore, it influences immunological response, cardiovascular health, and ion channel current regulation in a conventional way. The substance has various noteworthy effects (Saleri et al. 2017; Ashour et al. 2019). Among these are adjuvant, cholesterol-binding, anti-inflammatory, and anticancer properties. In the intestinal tract, phytosterols, such as diosgenin and its derivatives, are well known for lowering blood cholesterol levels by limiting cholesterol absorption (Wang et al. 2012). A dioscin, protodioscin, and trillin phytochemical from Dioscorea nipponica rhizomes showed antihyperlipidemic activity by increasing SOD and lipid peroxidation (Wang et al. 2012; Moghimipour et al. 2015). Further studies have shown that saponins contain dioscoresides, which may be useful in treating and preventing cardiovascular disease (Moghimipour and Handali 2015; Parama et al. 2020). The pharmaceutical industry also uses saponins for semisynthesis of steroidal drugs, as reported by Ashour et al. (2019). As saponins have a number of health benefits and can be found in a wide range of foods, researchers are increasingly interested in studying them, including tea, grains, legumes, and many medicinal plants. In addition, there have been studies on the safe limits for saponins from different food sources(Ashour et al. 2019).

1.2.5 Dietary Fiber

According to Timm et al. (2023), dietary fiber is composed of cellulose, hemicellulose, pectin, and lignin. The components of plant cell walls that are most frequently recognized as dietary fibers are found there. A plant's response to damage causes it to secrete certain fibers, while other fibers are a component of the intracellular cement. Most of the polysaccharides found in human diets are non-cellulosic, with trace amounts of cellulose and lignin. Due to the fact that oligosaccharides are referred to as prebiotic fibers, they have recently been included in regulations and labels as dietary fiber (Timm et al. 2023). In general, cellulose content is higher in fruits and vegetables than in cereals. Fruits with edible seeds have the highest lignin content e.g. G. ripe veggies like carrots and other root vegetables, or in strawberries (Timm et al. 2023). The species, maturity, and constituents of a plant determine its fiber composition (i.e. E. stem, leaf, or root). Human food contains not only lignin and polysaccharides but also plant-based materials like cutin, waxes, and trace amounts of proteins and lipids that are indigestible in the small intestine and resemble dietary fiber. Since non-hydrolyzable starch and non-enzymatic browning products are frequently indigestible, they are categorized as dietary fiber. Free radical and reactive oxygen species control is linked to diets high in polyphenols (Gebicki and Nauser 2021). Glycemic management has been linked to plant sources that are high in bioactives and insoluble dietary fiber (IDF). According to Krawęcka et al., adding stinging nettle to durum wheat pasta in amounts of 1, 2, and 3 percent lowered the glycemic index (GI) (2021). Addition of 3% stinging nettle to the pasta resulted in the lowest glycemic index. Even though the GI index of wheat durum pasta was already low, it was further lowered. The percentage of bread that had ginseng insoluble dietary fiber (GSI) reduced as the amount increased, starting at 2%. Most likely as a result of a rise in IDF, the addition of GSI also increased the capacity for absorbing cholesterol. The fiber from mango pulp had a low GI index due to its balanced IDF to SDF ratio (Jiang et al. 2021; Sudha et al. 2015). Potential treatments for blood glucose control include GSI, stinging nettle, and fibers from mango pulp. These can be particularly helpful for those who already have type 2 diabetes or are at risk of getting it. There is an abundance of bioactive compounds in insoluble dietary fiber. It may be incorporated into food products or used as a nutritional supplement. While certain bioactive chemicals can be extracted, processed, and stored with care, most of them demonstrated an increase in bioactive content even when only a small amount of IDF was added (Timm et al. 2023). Furthermore, a summarized table (Table 1.1) is been presented to understand the food bioactive and its possible role as nutraceuticals with the potential mechanism of action against various disorders.

1.3 Therapeutic Approach of Food Bioactives as Nutraceuticals

1.3.1 Food Bioactives as a Source of Nutraceuticals for Obesity Management

Globally, obesity and its associated conditions constitute a significant health concern. Reducing food intake and increasing physical activity as forms of treatment have not always been successful, particularly when it comes to chronically ill senior patients. Nutraceuticals derived from nature and can be utilize in medicine for their physiological and nutritional benefits. It might show to be a different approach to weight loss and the reduction of metabolic disorders linked to obesity with the enhancement of health, delaying the aging process, preventing chronic diseases, lengthening life expectancy, or supporting the body's structure or function (Singab et al. 2024). Various food bioactive discussed further to explore the need of these bioactive as nutraceuticals for obesity management. Polyphenols, abundant

Table 1.1 Food bioactive compounds and their health benefits

Bioactive compound	Food source	Function	Mechanism of action	Impact on health	References
Catechins (epigallocatechin gallate)	Green tea	Suppresses appetite and ↑ se satiety Inhibits proliferation of 3T3-L1 adipocytes	The inhibition of the catechol-O-methyltransferase (COMT) ↑ se thermogenesis ↑se fat oxidation Lowers phosphorylated ERK1/2, cdk2, and cyclin D1 protein levels	Anti-obesity/ prevention and management of obesity	Stuby et al. (2019), Hursel et al. (2009), Huda et al. (2024), Dulloo et al. (1999)
Gingerol	Ginger	Reduce triglycerides, inhibit lipid accumulation ↓ se expression of adipogenesis-related transcription factors ↓se lipogenic enzymes ↓se lipid levels	Suppress differentiation of preadipocytes into adipocytes Inhibit pancreatic lipase and amylase activity	Anti-obesity/ prevention and management of obesity	Tzeng and Liu (2013), Li and Zhou (2015), Suk et al. (2015)

(continued)

Table 1.1 (continued)

Bioactive compound	Food source	Function	Mechanism of action	Impact on health	References
Resveratrol	Grapes, berries Peanuts	Regulate energy metabolism Inhibits adipogenesi, lipogenesis, lipolysis, thermogenesis, and fatty acid oxidation Enhance energy expenditure	Activates sirtuins enzymes, Inhibits preadipocyte differentiation, ↓**se** adipocyte proliferation, Induces adipocyte apoptosis Promotes activity of brown adipose tissue (BAT) and white adipose tissue (WAT)	Anti-obesity/ prevention and management of obesity	Ahn et al. (2007), Rayalam et al. (2008), Liu et al. (2018), Mercader et al. (2011), Zhou et al. (2019)
		Lowers cholesterol formation and LDL cholesterol levels Anti-inflammatory	Inhibits the enzyme 3-hydroxy-3-methylglutaryl coenzyme A reductase Stimulates expression of LDL receptors in the liver, Lowers C-reactive proteins and TNF-alpha Inhibits cyclooxygenase	Cardio protective	Capurso et al. (2023), Teimouri et al. (2022)
		Protects against DNA damage and mutations. Reduces mitogenic and antiapoptotic action.	Suppresses carcinogenic metabolic activation Lower plasma levels of 1 IGF-1 and IGFBP-3.	Anti-cancer	Patel et al. (2011)
Capsaicinoids	Capsicum	Appetite suppressants ↑se energy expenditure ↑se lipid oxidation	Release of catecholamines. Stimulation of the TRPV1 receptor	Anti-obesity/ prevention and management of obesity	Stuby et al. (2019), Whiting et al. (2012)

Quercetin	Cranberries	Suppress fat accumulation, Enhance fat breakdown	Activates the MAPK signaling pathway Down-regulated transcription factors of the adipogenesis pathway Inhibits the activity of adipogenesis-related enzymes Regulates the critical ERK and JNK pathways.	Anti-obesity/ prevention and management of obesity	Chen and Jiang (2016), Kowalska et al. (2014)
		Improves glucose metabolism and insulin sensitivity Reduces the severity of symptoms in subjects with type 2 diabetic neuropathy.	Stimulates the enzyme AMPK Inhibits G6pase, (enzyme in liver cells) ↑ the activity of GS, (enzyme in liver cells)	Anti-diabetic/ prevention and management of diabetes	Eid et al. (2015)
Anthocyanin	Berries	↑se glucose uptake Improves glucose-lowering and insulin sensitivity ↓se risk of diabetes and its complications	Activates pAMPK/AMPK, GLUT-4, and SGLUT-1, activates adiponectin-AMPK pathway, and GLP-1 upregulation.	Prevention and management of diabetes	Hameed and Galli (2020)

(continued)

Table 1.1 (continued)

Bioactive compound	Food source	Function	Mechanism of action	Impact on health	References
Oats Beta-glucan (OBG)	Oats	Reduce postprandial glucose level	Reduces starch gelatinization, digestion and alters food product.	Prevention and management of diabetes	Battilana et al. (2001)
		↓se total cholesterol ↓se LDL cholesterol	Inhibits cholesterol uptake. ↑bile acid excretion	Cardio protective	Gudbrandsen et al. (2005)
		Reduces risk of skin, epithelial lung cancer and colon cancer.	Reduces fecal bile acid levels. Produces SCFAs. Aids in Para cancerous apoptosis	Anti-cancer	Choromanska et al. (2018), Choromanska et al. (2015), Rui-Ling Shen et al. (2016)
Pectin	Cereals, legumes, fruits, and vegetables	↓se plasma cholesterol ↓se LDL cholesterol	Inhibits cholesterol uptake ↑se bile acid excretion and prevents reabsorption	Cardio protective	Cerda et al. (1988)
Lycopene	Red, pink, and orange fruit and vegetables: such as tomatoes, red bell pepper, carrots, apricots, melons, papayas, grapes, peaches, guava, watermelons, and cranberries	Lower blood pressure	Reduces oxidation of LDL Improves function and dilation of the endothelium in arteries Anti -inflammatory	Cardio protective	Khan et al. (2021)
		Protects against prostate, colorectal and breast cancers	Inhibits cell proliferation, gene expression, cell cycle regulation Induces apoptosis ↓ IGF-1R pathway	Anti-cancer	Agarwal and Rao (2000), Kelkel et al. (2011)
Capsaicin	Capsicum	Suppress human neoplasia/cell growth	Generates ROS species ↑apoptosis Activation of TRPV1 Prevents cell proliferation	Anti-cancer	Zheng et al. (2016)

Phytoestrogen		↑ antioxidant enzymes, SOD, CAT, and Gpx. Lowers incidence of estrogen-dependent malignancies, breast and prostate cancer.	Block aromatase enzyme Reduces estrogen production	Anti-cancer	Robb and Stuart (2011), Kampkötter et al. (2008), Surico et al. (2017), Torrens-Mas and Roca (2020), Gorzkiewicz and Bartosz (2021)
Organosulfur compounds Allicin Diallyl di sulfide (DADS)	Garlic	Antiproliferative and apoptotic effects in human cell lines	Induce cell cycle arrest through release of ROS. Elevation of exogenous oxidants Activates caspase-3 Degrades PARP and fragmentation of DNA	Anti-cancer	Merhi et al. (2008), Wu et al. (2005), Kwon et al. (2002)
β-Carotene	Carrot	Inhibit cellular proliferation Induce apoptosis and cellular arrest	Suppresses MAPK/ERK and PI3L/AKT pathways	Anti-cancer	Milani et al. (2017)
Curcumin	Turmeric	Suppresses angiogenesis and induction of apoptosis	Production of ROS Downregulation of NF-κB and COX-2 pathway Inhibits activating protein-1 (AP-1), activation of c-Jun N-terminal kinases (JNK) Activates caspase 3 and the release of cytochrome	Anti-cancer	Ismail and Othman (2019)

phytochemicals found in plant-based diets, are natural compounds found in fruits and colorful vegetables like green tea, berries, grapes, chocolate, nuts, and spices (Ramírez-Moreno and Arias-Rico 2022).Both animal and clinical studies have demonstrated that polyphenols have health benefits, particularly in obese individuals. Recent research has linked bioactive compounds to effects on cell functions in obesity, such as reducing inflammation (Wang et al. 2014), inhibiting adipogenesis and lipogenesis (Kowalska et al. 2014), inducing apoptosis (Wu et al. 2005), and regulating genes involved in adipogenesis, lipolysis, and fatty acid oxidation. In obese individuals excess adipose tissue produces inflammatory substances such as TNF-α and IL-6, which reduce adiponectin synthesis and increase IL-6 levels. The inflammatory profile is linked to excess reactive oxygen species, which may contribute to chronic systemic inflammation, cardio metabolic illnesses, and obesity-related consequences (Boqué et al. 2013). Recent research indicates that bioactive components like ginger, resveratrol, and flaxseed have promising effects on reducing inflammation and obesity-related issues with low risk. Catechins are reported to found in green tea, black tea, and red wine. Green tea is a rich source of different bioactive polyphenolics like flavonols, flavones, and flavan-3-ols. Green tea contains bioactive compounds catechin that inhibits catechol-O-methyltransferase (COMT), an enzyme that degrades norepinephrine, potentially increasing thermogenesis, fat oxidation, and altering appetite (Hursel et al. 2009; Dulloo et al. 1999). In another study by Hung et al. (2005) reported that green tea catechins inhibit the proliferation of 3T3-L1 adipocytes by lowering phosphorylated ERK1/2, cdk2, and cyclin D1 protein levels. Clinical studies have demonstrated the beneficial effects of catechins (270–1200 mg/day), including decrease in blood leptin levels, body weight, and absorption of fatty acids (Meydani and Hasan 2010; Sun et al. 2016). Gingerol found in ginger, a plant predominantly utilized in culinary and medical applications, is also renowned for its anti-obesity properties, owing to the active components 6-gingerol and 6-shogaol. These compounds reduce the differentiation of preadipocytes into adipocytes, reduce triglycerides, inhibit lipid accumulation, and decrease the expression of adipogenesis-related transcription factors and lipogenic enzymes. They also inhibit pancreatic lipase and amylase activity, resulting in reduced lipid levels (Tzeng and Liu 2013; Li and Zhou 2015; Suk et al. 2015).

Resveratrol, a bioactive compound found in grapes, berries, and peanuts, has been found to have anti-inflammatory, anti-cancer, and anti-aging benefits. It activates sirtuins enzymes, which regulate energy metabolism and cellular stress response. Resveratrol may also help prevent or treat obesity, which is a major risk factor for many chronic diseases. Resveratrol may counteract obesity by affecting several aspects of energy metabolism such as by inhibiting preadipocyte differentiation, decreasing adipocyte proliferation, inducing adipocyte apoptosis, inhibits adipogenesis (Ahn et al. 2007; Rayalam et al. 2008; Liu et al. 2018), lipogenesis, lipolysis, thermogenesis, and fatty acid oxidation (Mercader et al. 2011) as demonstrated by experimental animal models and clinical studies. Resveratrol, according to Zhou et al. (2019), can enhance energy expenditure by promoting the activity of brown adipose tissue (BAT) and white adipose tissue (WAT), two forms of fat tissue with distinct roles. BAT burns calories to generate heat, whereas WAT stores energy.

Resveratrol may stimulate BAT and cause WAT to brown more. In obese humans, 150 mg of RSVL daily dose, increased energy expenditure and lowered AT lipolysis and plasma fatty acids, replicating the effects of calorie restriction, according to a randomized, double-blind crossover research (Zhou et al. 2019; Timmers et al. 2011). Meanwhile, capsaicinoids a bioactive compound found in capsicum, has been linked to weight loss, inflammation reduction, and pain relief when applied topically. It has been discovered that capsaicinoids increase the effects of calorie restriction on weight reduction. They may inhibit appetite, satiety, and fat oxidation in addition to raising energy expenditure (Tremblay et al. 2016).

Quercetin is a very famous flavonoid present in various fruits like cranberries and nuts. It has been reported to suppress fat accumulation, enhance fat breakdown, and improve insulin sensitivity. The study by Kowalska et al. (2014), found that cranberries can inhibit adipogenesis, reduce pre-adipocyte viability and proliferation, and stimulate lipolysis in adipocytes. They also down-regulated transcription factors of the adipogenesis pathway, suggesting their potential for natural bioactive substances in adipose tissue mass modification(Kowalska et al. 2014). Another study by Chen et al. (2016) found that quercetin reduced adipogenesis and apoptosis The implication is that quercetin inhibits adipogenesis via activating the MAPK signaling pathway. At the same time, quercetin triggered apoptosis in mature adipocytes via regulating the critical ERK and JNK pathways (Chen and Jiang 2016).

1.3.2 Food Bioactives as a Source of Nutraceuticals for Diabetes Management

With fewer side effects than herbal medications, nutraceuticals provide an additional promising therapeutic option for diabetes. It is said that nutritional supplements possess strong pro-health, and curative properties. The purpose of this study is to discuss and highlight some of the most popular nutraceuticals for preventing diabetes. Quercetin, a natural flavonoid, has been found to have antidiabetic effects in animal studies, particularly beneficial for those with type 2 diabetes. Jeong et al. (2012) observed that mice given quercetin had reduced glucose levels but no insulin alterations. Furthermore, mice given 0.08% quercetin demonstrated improvements in plasma adiponectin, HDL cholesterol, total cholesterol, and triacylglycerols (Jeong et al. 2012). According to Lai et al. (2012) quercetin enhanced renal function in diabetic nephropathic rats by inhibiting upregulation of CTGF and TGF-β1. TGF-β1 and CTGF are key factors in end-stage renal disease, which is linked to diabetic nephropathy. In streptozotocin-induced diabetic Sprague-Dawley rats, quercetin decreased kidney-to-body weight ratio and reduced CTGF and TGF-β1 overexpression in renal tissues (Lai et al. 2012). Eid et al. (2015) investigated the effect of quercetin, on glucose homeostasis in muscle and liver cells. The study found that quercetin stimulated the enzyme AMPK and increased the transport and expression of GLUT4, a glucose transporter, in muscle cells. Quercetin also activated AMPK and inhibited G6pase, an enzyme that produces glucose, in liver cells. Quercetin also slightly increased the activity of GS, an enzyme that stores glucose as glycogen, in liver cells. This suggests quercetin could be a potential treatment for type 2 diabetes, as it improves glucose metabolism in both tissues. For humans, quercetin has been demonstrated to reduce the severity of numbness, jolting pain, and irritation in people with type 2 diabetic neuropathy (Eid et al. 2015; Valensi et al. 2005). However, human clinical investigations on quercetin and diabetes control are few and inconclusive. Quercetin has potential for treating diabetes and may be a useful adjuvant therapy for type 2 diabetes, but further studies and high-quality trials needed to confirm its efficacy and safety.

1.3.3 Food Bioactives as a Source of Nutraceuticals for Cardiovascular Diseases

The study of potential treatments to lower risk and decrease the progression of cardiovascular disease is centered on the concerning pace of increase in this condition. We provide an overview of the latest research on the cardio-protective impact of several food bioactive as nutraceuticals to discuss the possible processes by which they may impede the advancement of cardiovascular disease, and emphasize the advantages of consuming nutraceuticals. Dietary fiber, a type of carbohydrate found in plant-based foods like cereals, legumes, fruits, and vegetables, is not fully digested by the human body but can be fermented by the gut microbiota. Fiber has prebiotic, cholesterol-lowering, and glucose-regulating effects, potentially reducing inflammation, oxidative stress, and blood pressure, and has been shown to reduce CVD risk factors. Beneficial effects of fiber operate via a diverse range of mechanisms. Fibers are often present in vegetables, containing large amounts of bioactive compounds like pectins, β-glucan, resveratrol, polyphenols, and hydro-colloids. Cross-over studies showed that pectin or cellulose containing food reduced low-density lipoprotein cholesterol (LDL-C) in mildly hyper-cholesterolemic individuals (Brouns et al. 2012). A study by Cerda et al. (1988) indicated that a grapefruit pectin-supplemented with diet, without lifestyle adjustments, could considerably reduce plasma cholesterol. The study included 27 adult volunteers who were at medium to high risk of developing coronary heart disease as a result of elevated cholesterol levels. Oat grains contain β-glucan, (OBG), a dietary fiber with cholesterol-lowering and antidiabetic effects, along with other bioactive compounds like phenolic acids, tocols, sterols, avenacosides, and avenanthramides (Cerda et al. 1988; Paudel et al. 2021). OBG's cholesterol-lowering effect occurs through the formation of a viscous layer in the small intestine, inhibiting cholesterol uptake and increasing bile acid excretion by preventing reabsorption, which stimulates bile acid synthesis from cholesterol, reducing LDL cholesterol levels (Gudbrandsen et al. 2005). Resveratrol has been reported to have significant impact on human health, including the lipid profile, where it predominantly inhibits the enzyme

3-hydroxy-3-methylglutaryl coenzyme A reductase, hence lowering cholesterol formation. Additionally, resveratrol promotes the liver's production of LDL receptors, which lowers LDL cholesterol levels (Capurso et al. 2023).

1.3.4 Food Bioactives as a Source of Nutraceuticals for Cancer Prevention and Management

Food bioactive compounds obtained through diet or herbal sources are known as nutraceuticals. They have potential to induce apoptosis in cancer cells hence alleviating their use, in many circumstances or, in combination with chemotherapy. Over 25 years of research has identified capsaicin's multifaceted bioactivity, demonstrating its antitumor properties through cytotoxicity and immunological adjuvancy against 74 types of cancer (Luján-Méndez and Roldán-Padrón 2023). Capsaicin has gained attention for its potential in cancer prevention and treatment, with preclinical studies display it can reduce suppress human neoplasia by producing reactive oxygen species and growing apoptosis (Zheng et al. 2016). Liu et al. (2016) study examined the effects of capsaicin on U251 cells using MTT and LDH assays and immunofluorescence. Results showed reduced cell viability and elevated LC3II expression, suggesting capsaicin may cause apoptosis in U251 cells by inhibiting autophagy. Additionally, resveratrol protects against DNA damage and the subsequent early mutations, demonstrating detoxification action and suppressing carcinogenic metabolic activation (Patel et al. 2011; Liu et al. 2016). Phytoestrogens interact with ERs and stimulate the transcription of numerous target genes. This leads to an improvement in mitochondrial function and an increase in antioxidant enzyme levels, including glutathione peroxidase (Gpx), catalase (CAT), and superoxide dismutase (SOD) (Robb and Stuart 2011; Kampkötter et al. 2008; Surico et al. 2017). The G-protein-coupled estrogen receptor 1 (GPER/GPR30) is utilized by phytoestrogens, which have non-genomic effects (Molina et al. 2018; Rietjens et al. 2017; Cappelletti et al. 2006). Aromatase is an enzyme that transforms male hormones (androgens) into estrogens. Phytoestrogens block aromatase, which reduces estrogen production and lowers the incidence of estrogen-dependent malignancies as reported by Torrens-Mas and Roca (2020) (Torrens-Mas and Roca 2020). Resveratrol, a phytoestrogen found in grapes, can inhibit aromatase activity and slow the development of breast and prostate cancer cells (Gorzkiewicz and Bartosz 2021). Sulfur-containing organic compounds (OSCs) are abundantly present in Allium species and are believed to have shown anti-proliferative and apoptotic effects in human cell lines (Merhi et al. 2008). Diallyl di sulfide (DADS) induces apoptosis in HL-60 cells by upregulating the expression of the C3 botulinum toxin substrate 2 GTPase Ras-related. DNA breakage, PARP breakdown, and caspase-3 activation cause DADS-induced apoptosis (Kwon et al. 2002). On the other hands, high dietary fiber foods like oats contain biologically active compounds called beta-glucans (OBG), which have been found to reduce cancer risk (Martínez-Villaluenga and Peñas 2017). OBG has been found to have antitumor activity against skin epithelial lung cancer and colon cancer cells (Choromanska et al. 2018).

1.3.5 Food Bioactives as a Source of Nutraceuticals for Immuno-Protection and Inflammation

Flavonoids, which are produced by the secondary metabolism of plants and include about 6000 phenolic chemicals, can be included in a diet by eating a variety of edible plants. The various types of flavonoids are flavonols (like kaempferol, quercetin, and isoquercetin, etc.), flavones (like apigenin glycosides, luteolin and chrysin), isoflavones (like daidzein, glycitein and genistein), flavanones (like naringenin, eriodictyol, and hesperidin), flavanols, and anthocyanidins (like pelargonidin, cyanidin, and malvidin) (Pérez-Cano and Castell 2016). According to Martínez et al. (2019), flavonoids and their derivatives prevent a variety of transcriptional factors that affect immune cell activation, differentiation, proliferation, and production of regulatory T cells. Certain flavonoids have anti-inflammatory properties via blocking NF-κB and the NLRP3 inflammasome, suppressing the production of pro-inflammatory cytokines including IL-1β, IL-2, IL-6, TNF-α, and IL-17A, reducing reactive oxygen and nitrogen species, and downregulating chemokines (Martínez et al. 2019). Other bioactive compound such as luteolin has been found to immunosuppressive power especially the expression of proinflammatory cytokines, including TNF-α, IL-6, IFN-γ, IL-5 including CD4+CD25− T cell differentiation (Kang et al. 2010; Kao et al. 2011; Kim et al. 2019). Studies suggest that luteolin serves as a natural regulator of immune system inflammatory signaling processes (Hussain et al. 2023). Apigenin a flavonoid has numerous biological effects is apigenin as it not only directly kills tumor cells, but it also modifies the immune system to increase immune cells' ability to fight cancer. Enhancing the bioavailability of dietary plant-derived COX-2 and NF-κB inhibitors, like apigenin, may be beneficial for preventing inflammation-based tumors overexpressing COX-2 (colon, breast), as well as for suppressing inflammation in lupus and other Th17-mediated diseases like rheumatoid arthritis, Crohn's disease, and psoriasis (Kang et al. 2009). Naringenin, a flavonoid found in citrus fruits, has a very wide range of pharmacological activities. Recent studies reported that treatment with naringenin reduced tumor invasion by lowering the expression of metalloproteinases (do Nascimento et al. 2022). Another recent study improvement in the innate immune response, which was suppressed by porcine reproductive and respiratory syndrome virus infection. These findings indicate that naringenin could be development as a drug against immunosuppressive viral infectious disease infections (Yu et al. 2023).

Plants contain quercetin, a flavonoid chemical with a wide range of biological functions (Islam et al. 2021). In 2021, the finding of a study reported that quercetin at the dose of 50 mg/kg for 15 days orally lowered doxorubicin-induced toxicities on the bone marrow, spleen, and antibody-producing cells. The study advocated the use of quercetin for chemotherapy patients to minimize doxorubicin induced inflammation and immunotoxicity, which may help to prevent chemotherapy failure (Farag et al. 2021). A recent report suggested quercetin possesses estrogen-like effects besides being a powerful anti-inflammatory, antioxidant in nature. It was found to regulate the functions of maternal-fetal interface immune cells and could maintain the dynamic balance of maternal and fetal immunity by lowering cell toxicity, reducing excessive apoptosis of the tissue cells and inhibiting excessive inflammatory reactions (Fu et al. 2023). Genistein belongs to the subgroup of isoflavones in the flavonoid family. It is regarded as a phytoestrogen having similar chemical structure to mammalian estrogens, and is derived primarily from leguminous plants. The mechanism of genistein antiinflammatory and immunosuppression involve the inhibition of NF-κB activation, prostaglandins (PGF2α and PGE2) and reduced expression of COX2, suppression of proinflammatory cytokines, reduced expression of inducible nitric oxide synthase and nitric oxide production, and also potent inhibition of reactive oxygen species and free radical scavenging activities have been noted (Goh et al. 2022; Sharifi-Rad et al. 2021). Also, modulation of the cell cycle, apoptosis, angiogenesis, and metastasis have been reported with genistein (Tuli et al. 2019). Several studies have reported immunomodulatory and immunoprotective properties of catechin and its metabolites. In 2016, Kim et al. reported that tea catechins metabolites could effectively enhance immune activity and prevent cancer by enhancing CD4+ T cell activity as well as natural killer (NK) cell activity (Kim et al. 2016b). Anthocyanins are colored water-soluble pigments, belong to the phenolic group. They are responsible for the various colors of the fruits and vegetables. Blueberry anthocyanin enriched extracts has been reported to reduce cyclophosphamide-induced leukocyte infiltration and inflammatory cytokines expression, contain oxidative stress and cardiomyocyte apoptosis which were linked closely to its anti-inflammatory, antioxidant and anti-inflammatory properties (Liu et al. 2018). Recent studies have shown potent immune system modulatory properties of these anthocyanins. In a study involving cadmium chloride—induced immunosuppression in rat model, foods supplemented with anthocyanin have shown immune stimulatory activities. In 2021, a study involving anthocyanin fucoidan nanocomplex showed immunostimulatory activities of anthocyanin complex which stimulated the production of various cytokines and inducible NO synthase expression via NF-κB signaling in the serum and spleen of immunosuppressed mice (Han et al. 2021). In 2022, Chen et al. reported a study involving purple red bran anthocyanin from rice which was found to stimulate the cytokines production, reduce endotoxin and lipopolysaccharide binding protein in the serum, and upregulate the expression of tight junction proteins and NF-κB pathway proteins (Chen et al. 2022). Antiinflammatory and immunosuppressive activities of curcumin has been reported to through its modulatory role on various cells such as macrophages, dendritic cells, B cells, T cells, and natural killer cells to modify the body's defence capacity (Allegra et al. 2022; Moudgil and Venkatesha 2022). Several experimental and pharmacologic studies have demonstrated curcumin's as a strong anti-inflammatory agent (Fadus et al. 2017). In 2006, Jacksopn et al. reported that micromolar dose of curcumin poetntly prevented the expression of collagenase and stromelysin expression (Jackson et al. 2006). It has been found inhibits neutrophil activation, synoviocyte proliferation, prevent tumor growth, proliferation, invasion, metastasis and angiogenesis (Wang et al. 2020). In 2014, a study in an experimental model of arthritis reported that dietary curcumin was able to reduce inflammation following oral administration of curcumin (Nonose et al. 2014). Nonose et al. observed that antinflammatory activaties of curcumin was associated with the suppression of NF-κB activation in CD4+ T cells whose profliferation was also found to modulated by curcumin. Also, studies suggest that beneficial effect of curcumin was due to the increased activation of peroxisome proliferator-activated receptor-γ (Jacob et al. 2007).

1.3.6 Food Bioactives as a Source of Nutraceuticals for the Management of Bone Health

According to earlier studies, incorporating certain nutraceuticals—like lycopene, curcumin, lutein, anthocyanin, resveratrol, and isoflavones and consuming these food bioactive nutraceuticals can be effective ways to support bone health. Osteoporosis is a frequently occurring bone disorder presented as low bone strength, low bone mass and microstructural deterioration of bone tissue with a consequent increase in bone fragility. Postmenopausal women are most likely to experience it because of an imbalance in bone growth and loss caused by estrogen insufficiency. In 2009, Kim et al. reported that lutolin stimulated osteoclast formation and prevented bone loss in ovariectomized (OVX) animal model. The finding of the study suggested that these effects of luteolin on bone was linked to its inhibitory effects on the regulation of receptor activator of nuclear factor-κB ligand (RANKL)-induced osteoclast formation and prevent bone loss in OVX mice (Kim et al. 2011). Another recent study involving luteolin oral intake reduced maxillary alveolar bone resorption in animal model of periodontitis induced by *P. gingivalis* infection (Kariu et al. 2023). In 2014, in a study Jung reported that there was a protective effects of apigenin against oxidative stress induced bone cells loss and it was linked to the activation of the hydrogen peroxide-induced downregulation of genes for phosphatidylinositol 3′-kinase, protein kinase B2 and extracellular signal-related kinase 2 important for bone cell survival-related signaling pathways (Jung 2014). In an *in-vitro* study involving human osteoblast cells obtained from human jaw reported that treatment with apigenin was able to stimulate human oral osteoblasts growth and mineralization following incubation for 48 hours with 5 μM concentration of apigenin (D'Amico et al. 2022) A recent study has reported potential effects of apigenin on bone regeneration process (Lorusso and Scarano 2023).

A meta-analysis and systematic review of 10 studies reported that naringenin treatment signifcantly increased the bone mineral density and promoted osteogenesis in OVX rats (Xu et al. 2020). Another, systematic review of original articles involving naringin have found that it enhanced osteoblastogenesis and bone formation through various signalling pathways such as bone morphogenetic protein-2 (BMP-2)/p38 mitogen-activated protein kinase (MAPK)/Runx2/Osterix, SDF-1/CXCR4, and PI3K/Akt/c-Fos/c-Jun/AP-1/ Wnt/β-catenin, and VEGF/VEGFR signalling pathways (Gan et al. 2023; Li et al. 2024; Nor Muhamad and Ekeuku 2022). Also, it suppressed lipopolysaccharide (LPS) induced M1 like macrophage which proinflammatory in nature. Naringenin was remarkable in reversing the LPS-induced bone loss and healed calvarial defect. Naringenin has potential anabolic properties with regard to pathological bone loss and have protective effects against bone and cartilage disorders improving bone quality, reducing bone resorption, and bone mineral deposition (Gan et al. 2023; Oršolić et al. 2022). It could be incorporated into the diet through fruit intake or supplements to prevent bone loss. Additionally, animal studies have reported the protective effects of quercetin on bone. Quercetin has been reported to promote proliferation and osteogenesis of bone marrow cells. Quercetin acts through the activation of bone morphogenetic protein signaling pathway subsequently increase the downstream genes expression via estrogen receptor (Pang et al. 2018).

The findings from systematic review of literatures involving *in vivo* and *in vitro* studies reported bone protective effects of kaempferol. The bone protective mechanisms of kaempferol included the inhibition of adipogenesis, oxidative stress, inflammation, osteoclastic autophagy and osteoblastic apoptosis and activation of osteoblastic autophagy. Kaempferol mediated anti-osteoporotic effects via the control of estrogen receptor, BMP-2, NF-κB, MAPK and mammalian target of rapamycin (mTOR) signaling pathways (Wong and Chin 2019). In 2017, in a study oral intake of kaempferol for 8 week inhibited bone loss in ovariectomized rats (Nowak et al. 2017). Hence, studies suggest that kaempferol could be a drug which exhibit great potentials to be developed into a new drug for treating glucocorticoid-induced osteoporosis.

Genistein, belongs to natural isoflavones, rich in soybean and leguminous plants, exhibits estrogen-like activity in protecting against osteoporosis and osteoarthritis. In 2002, Morabito et al. reported the finding of a randomized double-blind placebo-controlled study where genistein exhibited positive effects on bone loss reducing bone resorption and increasing bone formation in postmenopausal women (Morabito et al. 2002). In an animal model of bilateral ovariectomy, genistein was found act via interaction with Parathyroid hormone (PTH)/ Parathyroid hormone receptor 1 (PTHR1), exhibited significant mineral restoring effect besides it acted as an estrogen. Also, genistein's osteoporosis preventive effects was found to be partly PTH/PTHR1-dependent. It possesses better tolerance and have more clinical benefits and few undesirable side effects; and hence, could be a good option for the prevention and treatment of post-menopausal osteoporosis (Marini et al. 2008; Miao et al. 2012).

Delphinidin, a major anthocyanidin compound belonging to the category of flavonoids; rich in blue berries. In recent past, their role as bone protectors have been greatly highlighted with literatures reporting that they protect against bone loss by controlling the differentiation of osteoblasts and osteoclasts (Mao et al. 2021). The extract containing delphinidin was found to stimulate the osteoblastic differentiation of MC3T3-E1 cells by increasing alkaline phosphatase activity, via increased mRNA expression of BMP2, Runx2, Osterix, Ocn, and matrix extracellular phosphoglycoprotein; and also inhibit bone resorption by suppressing the NF-κB pathway (Nagaoka et al. 2019). The findings from studies increasingly report that delphinidin was a strong inhibitor of bone loss in women who frequently encounter bone fracture and osteoporosis after menopause (Mao et al. 2021).

Cyanidin, cyanidin chloride and cyanidin-3-glucoside are anthocyanins which are abundant cherry fruits. In 2018, in an in vitro study, Cheng et al. reported that bone loss, hydroxyapatite resorption and RANKL-induced signal pathways was inhibited while, in their in vivo study they reported that OVX-induced bone resorption was protected by the treatment with cyanidin chloride. Their study findings potentially indicated that cyanidin chloride could be used therapeutically for bone disorders. It was found that cyanidin chloride mediated prevention of the bone loss via inhibition of RANKL-induced NF-κB activation, suppression of IκB-α degradation, and attenuation of the extracellular signal-regulated kinases phosphorylation. Also, In 2019 Kim et al. and in 2021, Hu et al. reported bone regenerative properties of cyanidin-3-glucoside (Kim et al. 2019; Hu et al. 2021). They found that cyanidin-3-glucoside regulated osteoblasts and osteoclasts growth via the ERK1/2 pathway (Kim et al. 2019; Hu et al. 2021; Cheng et al. 2018). A lipid-soluble carotenoid found in nature in both all-trans and cis forms is lycopene. Lycopene is abundant in tomatoes and tomato-based products. Lycopene intake has been shown to prevent bone loss. In one of the in vitro and pilot clinical studies, Russo et al. showed that intake of lycopene was linked to the activation of the WNT/β-catenin and ERK1/2 pathways, increased expression of RUNX2, alkaline phosphatase, COL1A and reduced expression of RANKL Saos-2, which contributed to the prevention of osteoporosis in women after their menopause (Russo et al. 2020). Intake of lycopene daily 30 mg or more has been found to reduce bone resorption markers in postmenopausal women (Walallawita et al. 2020). In 2021, Qi et al. reported that lycopene lowered diabetic induced bone loss and the results of the study indicated that nutritional intervention with lycopene could treat diabetic osteoporosis (Qi et al. 2021). In a recent preclinical study, Semeghini et al. reported that there was an increased in the total number of bone forming cells as wells as bone resorptive cells when compared to sham and ovariectomized groups following treatment with lycopene for 8 weeks, orally at dose of 10 mg/kg. The authors of the study concluded that lycopene could stimulate the bone cell metabolism and bone remodeling initiating osteoporosis (Semeghini et al. 2022). A recent study concluded that lycopene inhibited oxidative stress, cell senescence and the senescence-associated secretory phenotype, and hence, it could be beneficial in managing the senile osteoporosis (Wang et al. 2023).

Curcumin, a natural phenolic compound with various pharmacological and biological activities, including antioxidant, anti-apoptotic, and anti-inflammatory. In 2015, Chen et al. reported that curcumin was beneficial in the treatment of glucocorticoid-induced osteoporosis (Chen et al. 2016). In 2020, a study showed that the deleterious effects of diabetic osteoporosis was reversed by curcumin in a dose-dependent manner (He et al. 2020). In 2020, a study reported by Liang et al. reported that migration of osteoclast precursors was prevented by curcumin due to reduction in the expression of chemokines such as CCL3 which inhibited osteoclasts formation (Liang et al. 2020).

A review of literature involving evidence from preclinical studies have indicated that supplementation with resveratrol attenuated the bone loss resulting from different etiologies and pathologies such as in ovariectomized rodent models of rapid bone loss due to estrogen-deficiency; age-related bone loss; hind limb unloaded rat model, etc. (Tou 2015). A double-blind randomized placebo-controlled trial involving 192 type 2 diabetic outpatients receive resveratrol 500 mg/day for 6 months observed the bone protective effects of resveratrol (Bo et al. 2018). A recent meta-analysis of evidence from preclinical studies 15 animal experiments involving osteoporosis rats showed that there was an increase in bone mineral density in osteoporosis rats treated with resveratrol which was attributed to improvement in the bone microstructure and the regulation of calcium and phosphorus metabolism (Zhao et al. 2023).

1.3.7 Food Bioactives as a Nutraceuticals in Neuro-Protection

Numerous studies demonstrate the same or complimentary neuroprotective mechanisms of nutraceuticals, indicating the need for novel integrated strategies that combine pharmacological interventions to maximize their neuroprotective effects. The potential of several dietary bioactive ingredients as nutraceuticals for the management of neuroprotective impact is outlined below.

Quercetin has been extensively explored for its neuroprotective effects such as Alzheimer's disease (AD), Huntington's disease (HD), Amyloid β peptide, Parkinson's disease (PD) (Islam et al. 2021; Evans et al. 2022). Low doses of quercetin has been found prevent cell toxicity by oxidative stress in neurons (Costa et al. 2016). It has been found to suppress neuroinflammatory processes by reducing the expression of pro-inflammatory cytokines, such as NF-kB and iNOS, and stimulated neuronal regeneration. (Khan and Ullah 2019). Improvement in motor functions in acute spinal cord injury, synaptic plasticity including reduction in hippocampal neuronal cell death and α-synuclein fibrillization, and also reversal of AD have been noted with isoquercetin administration (Figueira et al. 2017). In the PD, quercetin was found to protect mitochondrial dysfunction, reduce the deaths of dopaminergic neurons besides improving the learning, memory, and cognitive functions (Anand David et al. 2016; Ay et al. 2017). The mechanism involved in Quercetin's anti-inflammatory, antioxidative and neuroprotective properties is believed to activate AMP-activated protein kinase, a cellular energy sensor that modulates inflammation and oxidative stress (Chiang et al. 2023). Also, quercetin was found to influence microRNA expression which could control inflammation, differentiation, proliferation, apoptosis and immune response. A recent study has linked neuroprotective effects of quercetin to the amelioration of ferroptosis, a newly discovered non-apoptotic and iron-dependent type of cell death (Mohan et al. 2024).

A study has found neuroprotective effects of kaempferol reducing neuropathic pain via the reduced activation of TLR4/NF-κB (Alexander et al. 2023) pathways. Kaempferol beneficial neuroprotective actions is exhibited though multiple mechanisms which include the modulation of several proinflammatory signaling pathways such as NF-kB, p38 p38MAPK, AKT, and β-catenin cascade. Some of the recent studies have reported the beneficial effects of kaempferol in spinal cord injury and cerebral ischemia induced brain injury (Li et al. 2023; Nguyen and Le 2024). Increasing evidence suggest the neuroprotective potential of genistein and daidzein in neurodegenerative disorders and other neuronal disorders. Both *in-vivo* and *in-vitro* investigation have reported neuroprotective effects and antioxidant activities of phytoestrogens genistein and daidzein (Valeri et al. 2012). These phytoestrogens mediate neuroprotective effects via their antioxidant properties and by interacting with the estrogen receptor (Gorzkiewicz and Bartosz 2021). An experimental study reported the neuroprotective activity of daidzein against MPTP-induced PD (Wu and Wang 2022). In atrazine induced PD model, soybean isoflavones containing majorly the genistein and daidzein have shown protection against mitochondrial dysfunction and neurotoxicity by activating mitophagy which is a process critical for maintaining proper cellular functions (Li et al. 2021). Epigallocatechin gallate, the main effective catechin, possesses activities such as antioxidant, antiinflammatory, and antiapoptotic effects including neuroprotective effects against neural injuries and neurodegenerative diseases (Khalatbary and Khademi 2020). Several human interventional and observational studies reported that catechin intake could benefit neurodegenerative disorders. These beneficial effects of catechins have been attributed to its anti-inflammatory and free radical scavenging and metal ions chelating activities which is mediated through prevention of excessive cytokines production and inflammatory pathways (Afzal et al. 2022).

1.4 Conclusion

Food bioactive compounds, found in fruits, vegetables, cereals, oilseeds, and oils, have potential to prevent and treat chronic diseases by modulating oxidative stress, inflammation, and gut microbiota. However, their consumption is often low due to poor dietary habits, food processing, and bioavailability. To improve public health and reduce chronic disease burden, there is a need to increase awareness of the health benefits of food bioactive compounds and promote their intake through dietary recommendations as nutraceuticals. Based upon the literature review, it showed that food bioactive have a tremendous scope for the development for nutraceuticals. As many food or food derived components has been scientifically proven for its efficacy against various clinical disorders. Moreover, further research warranted to assess the best practices for consuming bioactive food compounds as a nutraceuticals. Additionally, safety, toxicity, and amount to be consume for a particular ailment must be analyses. Additionally, establishing the safety and effectiveness of all bioactive substances used as nutraceuticals requires further research.

References

Afzal O, Dalhat MH, Altamimi ASA, Rasool R, Alzarea SI (2022) Green tea catechins attenuate neurodegenerative diseases and cognitive deficits. Molecules 27(21):7604

Agarwal S, Rao AV (2000) Tomato lycopene and its role in human health and chronic diseases. CMAJ 163(6):739–744

Ahmad MF, Ahmad FA, Ashraf SA, Saad HH, Wahab S, Khan MI et al (2021) An updated knowledge of Black seed (Nigella sativa Linn.): review of phytochemical constituents and pharmacological properties. J Herb Med 25:100404

Ahmed F, Fanning K, Netzel M, Turner W, Li Y, Schenk PM (2014) Profiling of carotenoids and antioxidant capacity of microalgae from subtropical coastal and brackish waters. Food Chem 165:300–306

Ahn J, Lee H, Kim S, Ha T (2007) Resveratrol inhibits TNF-α-induced changes of adipokines in 3T3-L1 adipocytes. Biochem Biophys Res Commun 364(4):972–977

Alasvand M, Assadollahi V, Ambra R, Hedayati E, Kooti W (2019) Antiangiogenic effect of alkaloids. Oxid Med Cell Longev 2019:9475908

Alexander C, Parsaee A, Vasefi M (2023) Polyherbal and multimodal treatments: kaempferol- and quercetin-rich herbs alleviate symptoms of Alzheimer's disease. Biology 12(11):1453

Allegra A, Mirabile G, Ettari R, Pioggia G (2022) The impact of curcumin on immune response: an immunomodulatory strategy to treat sepsis. Int J Mol Sci 23(23):14710

Álvarez-Martínez FJ, Barrajón-Catalán E, Herranz-López M, Micol V (2021) Antibacterial plant compounds, extracts and essential oils: an updated review on their effects and putative mechanisms of action. Phytomedicine 90:153626

Anand David AV, Arulmoli R, Parasuraman S (2016) Overviews of biological importance of quercetin: a bioactive flavonoid. Pharmacogn Rev 10(20):84–89

Ashour A, Abed El Aziz M, Melad A (2019) A review on saponins from medicinal plants: chemistry, isolation, and determination. J Nanomed Res 7:282–288

Ay M, Luo J, Langley M, Jin H, Anantharam V, Kanthasamy A et al (2017) Molecular mechanisms underlying protective effects of quercetin against mitochondrial dysfunction and progressive dopaminergic neurodegeneration in cell culture and MitoPark transgenic mouse models of Parkinson's disease. J Neurochem 141(5):766–782

Battilana P, Ornstein K, Minehira K, Schwarz JM, Acheson K, Schneiter P et al (2001) Mechanisms of action of beta-glucan in postprandial glucose metabolism in healthy men. Eur J Clin Nutr 55(5):327–333

Benke KK, Benke KE (2014) Experimental design issue for assessment of carotenoids lutein and zeaxanthin in age-related eye disease study 2 formulation for age-related macular degeneration. JAMA Ophthalmol 132(7):904

Bennett RN, Shiga TM, Hassimotto NM, Rosa EA, Lajolo FM, Cordenunsi BR (2010) Phenolics and antioxidant properties of fruit pulp and cell wall fractions of postharvest banana (Musa acuminata Juss.) cultivars. J Agric Food Chem 58(13):7991–8003

Bo S, Gambino R, Ponzo V, Cioffi I, Goitre I, Evangelista A et al (2018) Effects of resveratrol on bone health in type 2 diabetic patients. A double-blind randomized-controlled trial. Nutr Diabetes 8(1):51

Boqué N, de la Iglesia R, de la Garza AL, Milagro FI, Olivares M, Bañuelos O et al (2013) Prevention of diet-induced obesity by apple polyphenols in Wistar rats through regulation of adipocyte gene expression and DNA methylation patterns. Mol Nutr Food Res 57(8):1473–1478

Bribi N (2018) Pharmacological activity of Alkaloids: a review. Asian J Bot (1):1–6. https://doi.org/10.63019/ajb.v1i2.467

Brouns F, Theuwissen E, Adam A, Bell M, Berger A, Mensink RP (2012) Cholesterol-lowering properties of different pectin types in mildly hyper-cholesterolemic men and women. Eur J Clin Nutr 66(5):591–599

Burt S (2004) Essential oils: their antibacterial properties and potential applications in foods—a review. Int J Food Microbiol 94(3):223–253

Cappelletti V, Miodini P, Di Fronzo G, Daidone MG (2006) Modulation of estrogen receptor-β isoforms by phytoestrogens in breast cancer cells. Int J Oncol 28(5):1185–1191

Capurso C, Bellanti F, Lo Buglio A, Vendemiale G (2023) Cardioprotective effects of resveratrol in the Mediterranean diet: a short narrative review. Die Dermatol 2(2):174–190

Cerda JJ, Robbins FL, Burgin CW, Baumgartner TG, Rice RW (1988) The effects of grapefruit pectin on patients at risk for coronary heart disease without altering diet or lifestyle. Clin Cardiol 11(9):589–594

Chen S, Jiang H (2016) Therapeutic effects of quercetin on inflammation, obesity, and type 2. Diabetes 2016:9340637

Chen Z, Xue J, Shen T, Mu S, Fu Q (2016) Curcumin alleviates glucocorticoid-induced osteoporosis through the regulation of the Wnt signaling pathway. Int J Mol Med 37(2):329–338

Chen T, Shen M, Yu Q, Chen Y, Wen H, Lu H et al (2022) Purple red rice anthocyanins alleviate intestinal damage in cyclophosphamide-induced mice associated with modulation of intestinal barrier function and gut microbiota. Food Chem 397:133768

Cheng DM, Pogrebnyak N, Kuhn P, Poulev A, Waterman C, Rojas-Silva P et al (2014) Polyphenol-rich Rutgers Scarlet Lettuce improves glucose metabolism and liver lipid accumulation in diet-induced obese C57BL/6 mice. Nutrition 30(7–8 Suppl):S52–S58

Cheng J, Zhou L, Liu Q, Tickner J, Tan Z, Li X et al (2018) Cyanidin chloride inhibits ovariectomy-induced osteoporosis by suppressing RANKL-mediated osteoclastogenesis and associated signaling pathways. Int J Obesity 233(3):2502–2512

Chiang MC, Tsai TY, Wang CJ (2023) The potential benefits of quercetin for brain health: a review of anti-inflammatory and neuroprotective mechanisms. Int J Mol Sci 24(7):6328

Choi JS, Kim JH, Ali MY, Min BS, Kim GD, Jung HA (2014) Coptis chinensis alkaloids exert anti-adipogenic activity on 3T3-L1 adipocytes by downregulating C/EBP-α and PPAR-γ. Fitoterapia 98:199–208

Choromanska A, Kulbacka J, Rembialkowska N, Pilat J, Oledzki R, Harasym J et al (2015) Anticancer properties of low molecular weight oat beta-glucan—an in vitro study. Int J Biol Macromol 80:23–28

Choromanska A, Kulbacka J, Harasym J, Oledzki R, Szewczyk A, Saczko J (2018) High- and low-molecular weight oat beta-glucan reveals antitumor activity in human epithelial lung cancer. Pathol Oncol Res 24(3):583–592

Costa LG, Garrick JM, Roquè PJ, Pellacani C (2016) Mechanisms of neuroprotection by quercetin: counteracting oxidative stress and more. Oxidative Med Cell Longev 2016:2986796

D'Amico E, Pierfelice TV, Iezzi G, Di Pietro N, Lepore S, Lorusso F et al (2022) Apigenin promotes proliferation and mineralization of human osteoblasts and up-regulates osteogenic markers. Appl Sci 12(17):8510

D'Andrea G (2015) Quercetin: a flavonol with multifaceted therapeutic applications? Fitoterapia 106:256–271

do Nascimento RP, Dos Santos BL, Amparo JAO, Soares JRP, da Silva KC, Santana MR et al (2022) Neuroimmunomodulatory properties of flavonoids and derivates: a potential action as adjuvants for the treatment of glioblastoma. Pharmaceutics 14(1):116

Dulloo AG, Duret C, Rohrer D, Girardier L, Mensi N, Fathi M et al (1999) Efficacy of a green tea extract rich in catechin polyphenols and caffeine in increasing 24-h energy expenditure and fat oxidation in humans. Am J Clin Nutr 70(6):1040–1045

Eid HM, Nachar A, Thong F, Sweeney G, Haddad PS (2015) The molecular basis of the antidiabetic action of quercetin in cultured skeletal muscle cells and hepatocytes. Pharmacogn Mag 11(41):74–81

Elkhalifa AEO, Alshammari E, Adnan M (2021) Okra (Abelmoschus esculentus) as a potential dietary medicine with nutraceutical importance for sustainable health applications. Molecules 26(3):696

Erlund I (2004) Review of the flavonoids quercetin, hesperetin, and naringenin. Dietary sources, bioactivities, bioavailability, and epidemiology. Nutr Res 24(10):851–874

Evans JA, Mendonca P, Soliman KFA (2022) Neuroprotective effects and therapeutic potential of the citrus flavonoid hesperetin in neurodegenerative diseases. Nutrients 14(11):2228
Fadus MC, Lau C, Bikhchandani J, Lynch HT (2017) Curcumin: an age-old anti-inflammatory and anti-neoplastic agent. J Tradit Complement Med 7(3):339–346
Farag MR, Moselhy AAA, El-Mleeh A, Aljuaydi SH, Ismail TA (2021) Quercetin alleviates the immunotoxic impact mediated by oxidative stress and inflammation induced by doxorubicin exposure in rats. Antioxidants (Basel) 10(12):1906
Figueira I, Menezes R, Macedo D, Costa I, Dos Santos CN (2017) Polyphenols beyond barriers: a glimpse into the brain. Curr Neuropharmacol 15(4):562–594
Fu Z, Tian Y, Zhou X, Lan H, Wu S, Lou Y (2023) Effects of quercetin on immune regulation at the maternal-fetal interface. Zhejiang Da Xue Xue Bao Yi Xue Ban 52(1):68–76
Gallucci MN, Oliva M, Casero C, Dambolena J, Luna A, Zygadlo J et al (2009) Antimicrobial combined action of terpenes against the food-borne microorganisms Escherichia coli, Staphylococcus aureus and Bacillus cereus. Flavour Fragr J 24(6):348–354
Gan J, Deng X, Le Y, Lai J, Liao X (2023) The development of naringin for use against bone and cartilage disorders. Molecules 28(9):3716
Gebicki JM, Nauser T (2021) Fast antioxidant reaction of polyphenols and their metabolites. Antioxidants (Basel) 10(8):1297
Ghofrani S, Joghataei MT, Mohseni S, Baluchnejadmojarad T, Bagheri M, Khamse S et al (2015) Naringenin improves learning and memory in an Alzheimer's disease rat model: insights into the underlying mechanisms. Eur J Pharmacol 764:195–201
Goh YX, Jalil J, Lam KW, Husain K, Premakumar CM (2022) Genistein: a review on its anti-inflammatory properties. Front Pharmacol 13:820969
Gorzkiewicz J, Bartosz G (2021) The potential effects of phytoestrogens: the role in neuroprotection. Molecules 26(10):2954
Gudbrandsen OA, Wergedahl H, Liaset B, Espe M, Berge RK (2005) Dietary proteins with high isoflavone content or low methionine-glycine and lysine-arginine ratios are hypocholesterolaemic and lower the plasma homocysteine level in male Zucker fa/fa rats. Br J Nutr 94(3):321–330
Gürbüz N, Uluişik S, Frary A, Frary A, Doğanlar S (2018) Health benefits and bioactive compounds of eggplant. Food Chem 268:602–610
Hameed A, Galli M (2020) Select polyphenol-rich berry consumption to defer or deter diabetes and diabetes-related complications. Nutrients 12(9):2538
Han NR, Kim HJ, Lee JS, Kim HY, Moon PD, Kim HM (2021) The immune-enhancing effect of anthocyanin-fucoidan nanocomplex in RAW264.7 macrophages and cyclophosphamide-induced immunosuppressed mice. J Food Biochem 45(4):e13631
Hassan ST, Masarčíková R, Berchová K (2015) Bioactive natural products with anti-herpes simplex virus properties. J Pharm Pharmacol 67(10):1325–1336
He J, Yang X, Liu F, Li D, Zheng B, Abdullah AO et al (2020) The impact of curcumin on bone osteogenic promotion of MC3T3 cells under high glucose conditions and enhanced bone formation in diabetic mice. Coatings 10(3):258
Hidalgo-Liberona N, González-Domínguez R, Vegas E, Riso P, Del Bo' C, Bernardi S, Peron G, Guglielmetti S, Gargari G, Kroon PA, Cherubini A, & Andrés-Lacueva C (2021) Correction to "Increased Intestinal Permeability in Older Subjects Impacts the Beneficial Effects of Dietary Polyphenols by Modulating Their Bioavailability". J Agric Food Chem, 69(10), 3266. https://doi.org/10.1021/acs.jafc.1c01240
Hu B, Chen L, Chen Y, Zhang Z, Wang X, Zhou B (2021) Cyanidin-3-glucoside regulates osteoblast differentiation via the ERK1/2 signaling pathway. ACS Omega 6(7):4759–4766
Huda H-S-A, Majid NBA, Chen Y, Adnan M, Ashraf SA, Roszko M et al (2024) Exploring the ancient roots and modern global brews of tea and herbal beverages: a comprehensive review of origins, types, health benefits, market dynamics, and future trends. Food Sci Nutr 00:1–18. https://doi.org/10.1002/fsn3.4346
Hung PF, Wu BT, Chen HC, Chen YH, Chen CL, Wu MH, Liu HC, Lee MJ, Kao YH (2005) Antimitogenic effect of green tea (-)-epigallocatechin gallate on 3T3-L1 preadipocytes

depends on the ERK and Cdk2 pathways. Am J Phys Cell Phys 288(5):C1094–C1108. https://doi.org/10.1152/ajpcell.00569.2004

Hursel R, Viechtbauer W, Westerterp-Plantenga MS (2009) The effects of green tea on weight loss and weight maintenance: a meta-analysis. Int J Obes (Lond) 33(9):956–961

Hussain G, Rasul A, Anwar H, Aziz N, Razzaq A, Wei W et al (2018) Role of plant derived alkaloids and their mechanism in neurodegenerative disorders. Int J Biol Sci 14(3):341–357

Hussain MS, Gupta G, Goyal A, Thapa R, Almalki WH (2023) From nature to therapy: Luteolin's potential as an immune system modulator in inflammatory disorders. J Biochem Mol Toxicol 37(11):e23482

Islam MS, Quispe C, Hossain R, Islam MT, Al-Harrasi A, Al-Rawahi A et al (2021) Neuropharmacological effects of quercetin: a literature-based review. Front Pharmacol 12:665031

Ismail NI, Othman I (2019) Mechanism of apoptosis induced by curcumin in colorectal cancer. Int J Mol Sci 20(10):2454

Jackson JK, Higo T, Hunter WL, Burt HM (2006) The antioxidants curcumin and quercetin inhibit inflammatory processes associated with arthritis. Inflamm Res 55(4):168–175

Jacob A, Wu R, Zhou M, Wang P (2007) Mechanism of the anti-inflammatory effect of curcumin: PPAR-gamma activation. PPAR Res 2007:89369

Jeong SM, Kang MJ, Choi HN, Kim JH, Kim JI (2012) Quercetin ameliorates hyperglycemia and dyslipidemia and improves antioxidant status in type 2 diabetic db/db mice. Nutr Res Pract 6(3):201–207

Jeppesen PB, Dyrskog SE, Agger A, Gregersen S, Colombo M, Xiao J et al (2006) Can stevioside in combination with a soy-based dietary supplement be a new useful treatment of type 2 diabetes? An in vivo study in the diabetic goto-kakizaki rat. Rev Diabet Stud 3(4):189–199

Jiang G, Feng X, Wu Z, Li S, Bai X, Zhao C et al (2021) Development of wheat bread added with insoluble dietary fiber from ginseng residue and effects on physiochemical properties, in vitro adsorption capacities and starch digestibility. LWT 149:111855

Juang YP, Liang PH (2020) Biological and pharmacological effects of synthetic saponins. Molecules 25(21):4974

Jung WW (2014) Protective effect of apigenin against oxidative stress-induced damage in osteoblastic cells. Int J Mol Med 33(5):1327–1334

Kampkötter A, Chovolou Y, Kulawik A, Röhrdanz E, Weber N, Proksch P et al (2008) Isoflavone daidzein possesses no antioxidant activities in cell-free assays but induces the antioxidant enzyme catalase. Nutr Res 28(9):620–628

Kang HK, Ecklund D, Liu M, Datta SK (2009) Apigenin, a non-mutagenic dietary flavonoid, suppresses lupus by inhibiting autoantigen presentation for expansion of autoreactive Th1 and Th17 cells. Arthritis Res Ther 11(2):R59

Kang OH, Choi JG, Lee JH, Kwon DY (2010) Luteolin isolated from the flowers of Lonicera japonica suppresses inflammatory mediator release by blocking NF-kappaB and MAPKs activation pathways in HMC-1 cells. Molecules 15(1):385–398

Kao TK, Ou YC, Lin SY, Pan HC, Song PJ, Raung SL et al (2011) Luteolin inhibits cytokine expression in endotoxin/cytokine-stimulated microglia. J Nutr Biochem 22(7):612–624

Kariu T, Hamada N, Lakshmyya K (2023) Luteolin inhibits Porphyromonas gingivalis growth and alleviates alveolar bone destruction in experimental murine periodontitis. Biosci Biotechnol Biochem 88(1):37–43

Kelkel M, Schumacher M, Dicato M, Diederich M (2011) Antioxidant and anti-proliferative properties of lycopene. Free Radic Res 45(8):925–940

Khalatbary AR, Khademi E (2020) The green tea polyphenolic catechin epigallocatechin gallate and neuroprotection. Nutr Neurosci 23(4):281–294

Khan H, Ullah H (2019) Neuroprotective effects of quercetin in Alzheimer's disease. Biomolecules 10(1):59

Khan UM, Sevindik M, Zarrabi A (2021) Lycopene: food sources, biological activities, and human health benefits. Oxid Med Cell Longev 2021:2713511

Kim TH, Jung JW, Ha BG, Hong JM, Park EK, Kim HJ et al (2011) The effects of luteolin on osteoclast differentiation, function in vitro and ovariectomy-induced bone loss. J Nutr Biochem 22(1):8–15
Kim MJ, Moon Y, Tou J, Mou B, Waterland N (2016a) Nutritional value, bioactive compounds and health benefits of lettuce (Lactuca sativa L.). J Food Compos Anal 49:19–34
Kim YH, Won YS, Yang X, Kumazoe M, Yamashita S, Hara A et al (2016b) Green tea catechin metabolites exert immunoregulatory effects on CD4(+) T cell and natural killer cell activities. J Agric Food Chem 64(18):3591–3597
Kim JH, Yi YS, Kim MY, Cho JY (2017) Role of ginsenosides, the main active components of Panax ginseng, in inflammatory responses and diseases. J Ginseng Res 41(4):435–443
Kim JM, Yang YS, Park KH (2019) The ERK MAPK pathway is essential for skeletal development and homeostasis. Int J Mol Sci 20(8):1803
Kowalska K, Olejnik A, Rychlik J, Grajek W (2014) Cranberries (Oxycoccus quadripetalus) inhibit adipogenesis and lipogenesis in 3T3-L1 cells. Food Chem 148:246–252
Krawęcka A, Sobota A, Pankiewicz U, Zielińska E, Zarzycki P (2021) Stinging nettle (Urtica Dioica l.) as a functional component in durum wheat pasta production: impact on chemical composition, in vitro glycemic index, and quality properties. Molecules 26:6909
Kwon KB, Yoo SJ, Ryu DG, Yang JY, Rho HW, Kim JS et al (2002) Induction of apoptosis by diallyl disulfide through activation of caspase-3 in human leukemia HL-60 cells. Biochem Pharmacol 63(1):41–47
Lai PB, Zhang L, Yang LY (2012) Quercetin ameliorates diabetic nephropathy by reducing the expressions of transforming growth factor-β1 and connective tissue growth factor in streptozotocin-induced diabetic rats. Ren Fail 34(1):83–87
Li C, Zhou L (2015) Inhibitory effect 6-gingerol on adipogenesis through activation of the Wnt/β--catenin signaling pathway in 3T3-L1 adipocytes. Toxicol In Vitro 30(1 Pt B):394–401
Li P, Yao LY, Jiang YJ, Wang DD, Wang T, Wu YP et al (2021) Soybean isoflavones protect SH-SY5Y neurons from atrazine-induced toxicity by activating mitophagy through stimulation of the BEX2/BNIP3/NIX pathway. Ecotoxicol Environ Saf 227:112886
Li E, Yan K, Zhang R, Zou P, Li S, Ma Q et al (2023) Kaempferol protects against apoptosis in PC12 cells exposed to hydrogen peroxide by activating Akt1. Nat Prod Commun 18:1934578X2311704
Li X, Zhou X, Huang Z, Chen K, Jiang X, Lai R et al (2024) Study on the mechanism of naringin in promoting bone differentiation: in vitro and in vivo study. Heliyon 10(2):e24906
Liang Z, Xue Y, Wang T, Xie Q, Lin J, Wang Y (2020) Curcumin inhibits the migration of osteoclast precursors and osteoclastogenesis by repressing CCL3 production. BMC Complement Med Ther 20(1):234
Liggins J, Bluck LJ, Runswick S, Atkinson C, Coward WA, Bingham SA (2000) Daidzein and genistein contents of vegetables. Br J Nutr 84(5):717–725
Liu YP, Dong FX, Chai X, Zhu S, Zhang BL, Gao DS (2016) Role of autophagy in capsaicin-induced apoptosis in U251 glioma cells. Cell Mol Neurobiol 36(5):737–743
Liu X, Zhao H, Jin Q, You W, Cheng H, Liu Y et al (2018) Resveratrol induces apoptosis and inhibits adipogenesis by stimulating the SIRT1-AMPKα-FOXO1 signalling pathway in bovine intramuscular adipocytes. Mol Cell Biochem 439(1–2):213–223
Lorusso F, Scarano A (2023) Effectiveness of apigenin, resveratrol, and curcumin as adjuvant nutraceuticals for calvarial bone defect healing: an in vitro and histological study on rats. Nutrients 15(5):1235
Ludwiczuk A, Skalicka-Woźniak K, Georgiev MI (eds) (2017) Chapter 11—Terpenoids. In: Pharmacognosy fundamentals, applications and strategies. Elsevier, pp 233–266. https://doi.org/10.1016/B978-0-12-802104-0.00011-1
Luján-Méndez F, Roldán-Padrón O (2023) Capsaicinoids and their effects on cancer: the "double-edged sword" postulate from the molecular scale. Cells 12(21):2573

Ma H, He K, Zhu J, Li X, Ye X (2019) The anti-hyperglycemia effects of Rhizoma Coptidis alkaloids: a systematic review of modern pharmacological studies of the traditional herbal medicine. Fitoterapia 134:210–220
Mao W, Huang G, Chen H, Xu L, Qin S, Li A (2021) Research progress of the role of anthocyanins on bone regeneration. Front Pharmacol 12:773660
Marini H, Bitto A, Altavilla D, Burnett BP, Polito F, Di Stefano V et al (2008) Breast safety and efficacy of genistein aglycone for postmenopausal bone loss: a follow-up study. J Clin Endocrinol Metab 93(12):4787–4796
Martínez G, Mijares MR, De Sanctis JB (2019) Effects of flavonoids and its derivatives on immune cell responses. Recent Patents Inflamm Allergy Drug Discov 13(2):84–104
Martínez-Villaluenga C, Peñas E (2017) Health benefits of oat: current evidence and molecular mechanisms. Curr Opin Food Sci 14:26–31
Mercader J, Palou A, Bonet ML (2011) Resveratrol enhances fatty acid oxidation capacity and reduces resistin and retinol-binding protein 4 expression in white adipocytes. J Nutr Biochem 22(9):828–834
Merhi F, Auger J, Rendu F, Bauvois B (2008) Allium compounds, dipropyl and dimethyl thiosulfinates as antiproliferative and differentiating agents of human acute myeloid leukemia cell lines. Biologics 2(4):885–895
Meydani M, Hasan ST (2010) Dietary polyphenols and obesity. Nutrients 2(7):737–751
Miao Q, Li JG, Miao S, Hu N, Zhang J, Zhang S et al (2012) The bone-protective effect of genistein in the animal model of bilateral ovariectomy: roles of phytoestrogens and PTH/PTHR1 against post-menopausal osteoporosis. Int J Mol Sci 13(1):56–70
Milani A, Basirnejad M, Shahbazi S, Bolhassani A (2017) Carotenoids: biochemistry, pharmacology and treatment. Br J Pharmacol 174(11):1290–1324
Moghimipour E, Handali S (2015) Saponin: properties, methods of evaluation and applications. Annu Res Rev Biol 5:207–220
Moghimipour E, Kooshapour H, Parkhideh S, Handali S (2015) In-vitro evaluation of complex forming affinity of total saponins extracted from Ziziphus spina-christi and Quillaja saponaria with cholesterol. Res J Pharm, Biol Chem Sci 6:619–624
Mohan S, Alhazmi HA, Hassani R, Khuwaja G, Maheshkumar VP, Aldahish A et al (2024) Role of ferroptosis pathways in neuroinflammation and neurological disorders: from pathogenesis to treatment. Heliyon 10(3):e24786
Molina L, Bustamante FA, Bhoola KD, Figueroa CD (2018) Possible role of phytoestrogens in breast cancer via GPER-1/GPR30 signaling. Clin Sci (Lond) 132(24):2583–2598
Morabito N, Crisafulli A, Vergara C, Gaudio A, Lasco A, Frisina N et al (2002) Effects of genistein and hormone-replacement therapy on bone loss in early postmenopausal women: a randomized double-blind placebo-controlled study. J Bone Miner Res 17(10):1904–1912
Moreno-Ortega A, Ordóñez JL (2021) Changes in the organosulfur and polyphenol compound profiles of black and fresh onion during simulated gastrointestinal digestion. Foods 10(2):337
Moreno-Ortega A, Pereira-Caro G (2020) Bioaccessibility of bioactive compounds of 'fresh garlic' and 'black garlic' through in vitro gastrointestinal digestion. Foods 9(11):1582
Moreno-Ortega A, Pereira-Caro G, Luis Ordóñez J, Manuel Muñoz-Redondo J, Moreno-Rojas R, Pérez-Aparicio J et al (2020) Changes in the antioxidant activity and metabolite profile of three onion varieties during the elaboration of 'black onion'. Food Chem 311:125958
Moudgil KD, Venkatesha SH (2022) The anti-inflammatory and immunomodulatory activities of natural products to control autoimmune inflammation. Int J Mol Sci 24(1):95
Nagaoka M, Maeda T, Chatani M, Handa K, Yamakawa T, Kiyohara S et al (2019) A delphinidin-enriched maqui berry extract improves bone metabolism and protects against bone loss in osteopenic mouse models. Antioxidants (Basel) 8(9):386
Nguyen LTT, Le XT (2024) Kaempferol-3-O-(2″-O-galloyl-β-D-glucopyranoside): a novel neuroprotective agent from Diospryros kaki against cerebral ischemia-induced brain injury. J Nat Med 78(2):312–327

Nonose N, Pereira JA, Machado PR, Rodrigues MR, Sato DT, Martinez CA (2014) Oral administration of curcumin (Curcuma longa) can attenuate the neutrophil inflammatory response in zymosan-induced arthritis in rats. Acta Cir Bras 29(11):727–734

Nor Muhamad ML, Ekeuku SO (2022) A scoping review of the skeletal effects of naringenin. Nutrients 14(22):4851

Nowak B, Matuszewska A, Nikodem A, Filipiak J, Landwójtowicz M, Sadanowicz E et al (2017) Oral administration of kaempferol inhibits bone loss in rat model of ovariectomy-induced osteopenia. Pharmacol Rep 69(5):1113–1119

Oršolić N, Nemrava J, Jeleč Ž (2022) Antioxidative and anti-inflammatory activities of chrysin and naringenin in a drug-induced bone loss model in rats. Int J Mol Sci 23(5):2872

Pang XG, Cong Y, Bao NR (2018) Quercetin stimulates bone marrow mesenchymal stem cell differentiation through an Estrogen receptor-mediated pathway. Biomed Res Int 2018:4178021

Parama D, Boruah M, Yachna K, Rana V, Banik K, Harsha C et al (2020) Diosgenin, a steroidal saponin, and its analogs: effective therapies against different chronic diseases. Life Sci 260:118182

Patel KR, Scott E, Brown VA, Gescher AJ, Steward WP, Brown K (2011) Clinical trials of resveratrol. Ann N Y Acad Sci 1215:161–169

Paudel D, Dhungana B, Caffe M, Krishnan P (2021) A review of health-beneficial properties of oats. Foods 10(11):2591

Pereira-Caro G, Gaillet S, Ordóñez JL, Mena P, Bresciani L, Bindon KA et al (2020) Bioavailability of red wine and grape seed proanthocyanidins in rats. Food Funct 11(5):3986–4001

Pérez-Cano FJ, Castell M (2016) Flavonoids, inflammation and immune system. Nutrients 8(10):659

Perveen S (2018) Introductory chapter: terpenes and terpenoids. IntechOpen. https://doi.org/10.5772/intechopen.79683

Puri V, Nagpal M, Singh I (2022) A comprehensive review on nutraceuticals: therapy support and formulation challenges. Nutrients 14(21):4637

Qi SS, Shao ML, Sun Z, Chen SM, Hu YJ, Li XS et al (2021) Chondroitin sulfate alleviates diabetic osteoporosis and repairs bone microstructure via anti-oxidation, anti-inflammation, and regulating bone metabolism. Front Endocrinol (Lausanne) 12:759843

Quintero-Flórez A, Pereira-Caro G, Sánchez-Quezada C, Moreno-Rojas JM, Gaforio JJ, Jimenez A et al (2018) Effect of olive cultivar on bioaccessibility and antioxidant activity of phenolic fraction of virgin olive oil. Eur J Nutr 57(5):1925–1946

Ramírez-Moreno E, Arias-Rico J (2022) Role of bioactive compounds in obesity: metabolic mechanism focused on inflammation. Foods 11(9):1232

Rayalam S, Yang JY, Ambati S, Della-Fera MA, Baile CA (2008) Resveratrol induces apoptosis and inhibits adipogenesis in 3T3-L1 adipocytes. Phytother Res 22(10):1367–1371

Rietjens I, Louisse J, Beekmann K (2017) The potential health effects of dietary phytoestrogens. Br J Pharmacol 174(11):1263–1280

Robb EL, Stuart JA (2011) Resveratrol interacts with estrogen receptor-β to inhibit cell replicative growth and enhance stress resistance by upregulating mitochondrial superoxide dismutase. Free Radic Biol Med 50(7):821–831

Rui-Ling Shen ZW, Dong J-L, Xiang Q-S, Liu Y-Q (2016) Effects of oat soluble and insoluble β-glucan on 1,2-dimethylhydrazine-induced early colon carcinogenesis in mice. Food Agric Immunol 27(5):657–666

Russo C, Ferro Y, Maurotti S, Salvati MA, Mazza E, Pujia R et al (2020) Lycopene and bone: an in vitro investigation and a pilot prospective clinical study. J Transl Med 18:43. https://doi.org/10.1186/s12967-020-02238-7

Saleri FD, Chen G, Li X, Guo M (2017) Comparative analysis of saponins from different phytolaccaceae species and their antiproliferative activities. Molecules 22(7):1077

Samtiya M, Aluko RE (2021) Potential health benefits of plant food-derived bioactive components: an overview. Foods 10(4):839

Semeghini MS, Scalize PH, Coelho MC, Fernandes RR, Pitol DL, Tavares MS et al (2022) Lycopene prevents bone loss in ovariectomized rats and increases the number of osteocytes and osteoblasts. J Anat 241(3):729–740
Shagufta P (2018) Introductory chapter: terpenes and terpenoids. In: Shagufta P, Areej A-T (eds) Terpenes and terpenoids. IntechOpen, Rijeka, p Ch. 1
Sharifi-Rad J, Quispe C, Imran M, Rauf A (2021) Genistein: an integrative overview of its mode of action, pharmacological properties, and health benefits. Oxid Med Cell Longev 2021:3268136
Singab ANB, Elhawary EA, Elkhawas YA, Fawzy IM, Moussa AY, Mostafa NM (2024) Role of nutraceuticals in obesity management: a mechanism and prospective supported by molecular docking studies. J Med Food 27(2):176–197
Singh JP, Kaur A, Shevkani K, Singh N (2015) Influence of jambolan (yzygium cumini) and xanthan gum incorporation on the physicochemical, antioxidant and sensory properties of gluten-free eggless rice muffins. Int J Food Sci Technol 50(5):1190–1197
Singh B, Singh JP, Kaur A, Singh N (2016) Bioactive compounds in banana and their associated health benefits - a review. Food Chem 206:1–11
Sorrenti V, Burò I, Consoli V (2023) Recent advances in health benefits of bioactive compounds from food wastes and by-products: biochemical aspects. Int J Mol Sci 24(3):2019
Stuby J, Gravestock I, Wolfram E, Pichierri G, Steurer J, Burgstaller JM (2019) Appetite-suppressing and satiety-increasing bioactive phytochemicals: a systematic review. Nutrients 11(9):2238
Sudha ML, Indumathi K, Sumanth MS, Rajarathnam S, Shashirekha MN (2015) Mango pulp fibre waste: characterization and utilization as a bakery product ingredient. J Food Meas Charact 9:382–388
Suk S, Seo S, Yu J, Yang H, Jeong E, Jang Y et al (2015) A bioactive constituent of ginger, 6-shogaol, prevents adipogenesis and stimulates lipolysis in 3T3-L1 adipocytes. J Food Biochem 40(1)
Sun NN, Wu TY, Chau CF (2016) Natural dietary and herbal products in anti-obesity treatment. Molecules 21(10):1351
Surico D, Ercoli A, Farruggio S, Raina G, Filippini D, Mary D et al (2017) Modulation of oxidative stress by 17 β-estradiol and genistein in human hepatic cell lines in vitro. Cell Physiol Biochem 42(3):1051–1062
Teimouri M, Homayouni-Tabrizi M, Rajabian A, Amiri H, Hosseini H (2022) Anti-inflammatory effects of resveratrol in patients with cardiovascular disease: a systematic review and meta-analysis of randomized controlled trials. Complement Ther Med 70:102863
Teodoro AJ (2019) Bioactive compounds of food: their role in the prevention and treatment of diseases. Oxidative Med Cell Longev 2019:3765986
Timm M, Offringa LC, Van Klinken BJ (2023) Beyond insoluble dietary fiber: bioactive compounds in plant foods. Nutrients 15(19):4138
Timmers S, Konings E, Bilet L, Houtkooper RH, van de Weijer T, Goossens GH et al (2011) Calorie restriction-like effects of 30 days of resveratrol supplementation on energy metabolism and metabolic profile in obese humans. Cell Metab 14(5):612–622
Torrens-Mas M, Roca P (2020) Phytoestrogens for cancer prevention and treatment. Biology 9(12):427
Tou JC (2015) Resveratrol supplementation affects bone acquisition and osteoporosis: pre-clinical evidence toward translational diet therapy. Biochim Biophys Acta 1852(6):1186–1194
Tremblay A, Arguin H, Panahi S (2016) Capsaicinoids: a spicy solution to the management of obesity? Int J Obes 40(8):1198–1204
Tuli HS, Tuorkey MJ, Thakral F, Sak K, Kumar M, Sharma AK et al (2019) Molecular mechanisms of action of genistein in cancer: recent advances. Front Pharmacol 10:1336
Tzeng TF, Liu IM (2013) 6-gingerol prevents adipogenesis and the accumulation of cytoplasmic lipid droplets in 3T3-L1 cells. Phytomedicine 20(6):481–487
Valensi P, Le Devehat C, Richard JL, Farez C, Khodabandehlou T, Rosenbloom RA et al (2005) A multicenter, double-blind, safety study of QR-333 for the treatment of symptomatic diabetic peripheral neuropathy. A preliminary report. J Diabetes Complicat 19(5):247–253

Valeri A, Fiorenzani P, Rossi R, Aloisi AM, Valoti M, Pessina F (2012) The soy phytoestrogens genistein and daidzein as neuroprotective agents against anoxia-glucopenia and reperfusion damage in rat urinary bladder. Pharmacol Res 66:309–316. https://doi.org/10.1016/j.phrs.2012.06.007

Walallawita US, Wolber FM, Ziv-Gal A (2020) Potential role of lycopene in the prevention of postmenopausal bone loss: evidence from molecular to clinical studies. Int J Mol Sci 21(19):7119

Wang T, Choi RC, Li J, Bi CW, Ran W, Chen X et al (2012) Trillin, a steroidal saponin isolated from the rhizomes of Dioscorea nipponica, exerts protective effects against hyperlipidemia and oxidative stress. J Ethnopharmacol 139(1):214–220

Wang S, Moustaid-Moussa N, Chen L, Mo H, Shastri A, Su R et al (2014) Novel insights of dietary polyphenols and obesity. J Nutr Biochem 25(1):1–18

Wang Y, Lu J, Jiang B, Guo J (2020) The roles of curcumin in regulating the tumor immunosuppressive microenvironment (review). Oncol Lett 19(4):3059–3070

Wang S, Heng K, Song X, Zhai J, Zhang H, Geng Q (2023) Lycopene improves bone quality in SAMP6 mice by inhibiting oxidative stress, cellular senescence, and the SASP. Mol Nutr Food Res 67(24):e2300330

Weikel KA, Garber C, Baburins A, Taylor A (2014) Nutritional modulation of cataract. Nutr Rev 72(1):30–47

Whiting S, Derbyshire E, Tiwari BK (2012) Capsaicinoids and capsinoids. A potential role for weight management? A systematic review of the evidence. Appetite 59(2):341–348

Wong SK, Chin KY (2019) The osteoprotective effects of kaempferol: the evidence from in vivo and in vitro studies. Drug Des Devel Ther 13:3497–3514

Wu Q, Wang M (2022) Daidzein exerts neuroprotective activity against MPTP-induced Parkinson's disease in experimental mice and lipopolysaccharide-induced BV2 microglial cells. J Biochem Mol Toxicol 36(2):e22949

Wu X, Kassie F, Mersch-Sundermann V (2005) Induction of apoptosis in tumor cells by naturally occurring sulfur-containing compounds. Mutat Res 589(2):81–102

Xu DP, Li Y, Meng X, Zhou T, Zhou Y, Zheng J et al (2017) Natural antioxidants in foods and medicinal plants: extraction, assessment and resources. Int J Mol Sci 18(1):96

Xu Z, Shi L, Wang Y, Zhang J, Huang L, Zhang C et al (2020) Pathological findings of COVID-19 associated with acute respiratory distress syndrome. Lancet Respir Med 8(4):420–422

Yang W, Chen X, Li Y, Guo S, Wang Z, Yu X (2020) Advances in pharmacological activities of terpenoids. Nat Prod Commun 15:1934578X2090355

Yu J, Shi H, Song K, Yang Y, Li X, Peng L et al (2023) Naringenin improves innate immune suppression after PRRSV infection by reactivating the RIG-I-MAVS signaling pathway, promoting the production of IFN-I. Viruses 15(11):2172

Zhao J, Zhou G, Yang J, Pan J, Sha B, Luo M et al (2023) Effects of resveratrol in an animal model of osteoporosis: a meta-analysis of preclinical evidence. Front Nutr 10:1234756

Zheng L, Chen J, Ma Z, Liu W, Yang F, Yang Z et al (2016) Capsaicin enhances anti-proliferation efficacy of pirarubicin via activating TRPV1 and inhibiting PCNA nuclear translocation in 5637 cells. Mol Med Rep 13(1):881–887

Zhou L, Xiao X, Zhang Q, Zheng J, Deng M (2019) Deciphering the anti-obesity benefits of resveratrol: the "gut microbiota-adipose tissue" axis. Front Endocrinol 10:413

Chapter 2
Future of Nutraceuticals as a Safe and Intelligent Food

Erandi Escamilla-García, Katiushka Arevalo-Niño, Jorge Alberto García-Fajardo, Daniel Sánchez-Romo, Hamlet Avilés-Arnaut, Ana Laura De la Garza, Said Emmanuel Flores-Niño, Luis Galán-Wong, and Benito Pereyra-Alférez

Abstract Economic, political, cultural, and educational factors and increased population density have led to development and advances in science and technology over time. However, every action and progress had an impact. This chapter proposes to address the topic of nutraceuticals from two perspectives. First, we want to present a general overview of how technological advances, industrialization, globalization, and climate change have altered the quality of the natural foods produced, their consumption, the lifestyle and the consumer's health. It has led to new challenges in creating and consuming safe food and the need for the population to live in a new way. In the background, the chapter seeks to promote the consumption of nutraceuticals by emphasizing their classification, the mechanism of action it has in the body, making known some conditions or disorders that are studied, using nutraceuticals as part of the treatment, as well as the different food groups where these elements are. It also seeks to publicize the most current techniques in extracting, producing, and purifying nutraceutical compounds. Within the natural sources of nutraceuticals such as plants, fruits, and vegetables, naturally, fermented foods, biomass, and algae will be exposed to sources that benefit health.

E. Escamilla-García (✉) · K. Arevalo-Niño · D. Sánchez-Romo · H. Avilés-Arnaut · S. E. Flores-Niño · L. Galán-Wong · B. Pereyra-Alférez
Universidad Autónoma de Nuevo León, Facultad de Ciencias Biológicas, Instituto de Biotecnología, San Nicolas de los Garza, Nuevo León, Mexico
e-mail: erandi.escamillagrc@uanl.edu.mx; katiushka.arevalonn@uanl.edu.mx; hamlet.avilesarn@uanl.edu.mx; luis.galanwn@uanl.edu.mx; benito.pereyraal@uanl.edu.mx

J. A. García-Fajardo
Subsede Noreste, Centro de Investigación y Asistencia en Tecnología y Diseño del Estado de Jalisco, Parque de Investigación e Innovación Tecnológica, Apodaca, Mexico
e-mail: jgarcia@ciatej.mx

A. L. De la Garza
Universidad Autónoma de Nuevo León, Facultad de Salud Pública y Nutrición, Centro de Investigación en Nutrición y Salud Pública, Monterrey, Nuevo León, Mexico
e-mail: ana.dlgarzah@uanl.edu.mx

S. Amir Ashraf, M. Adnan (eds.), *Food Bioactives and Nutraceuticals*,
https://doi.org/10.1007/978-981-97-9723-3_2

Keywords Safe food · Nutrition · Health · Nutraceuticals · Bioactive compounds · Sustainability

2.1 Introduction

Recent data from the World Health Organization (Organization WH 2024) report more than "200 diseases in the world, caused by food contaminated by microorganisms, parasites, and chemical substances. Around the world, an estimated 600 million – almost 1 in 10 people – fall ill after eating contaminated food each year, resulting in 420 000 deaths and the loss of 33 million healthy life years". Moreover 70% of deaths globally are because of the development of Non-Communicable Diseases (NCDs) (Raiten et al. 2015; Organisation PAH 2020). Different causes are at the origin, such as changes in food production and consumption systems linked to the issue of globalization and all that this entails (technological advances, ultra-processed foods, climate change). The One Health approach (Sinclair 2019) supports global health security by improving coordination, collaboration, and communication at the human-animal-environment interface to address shared health threats such as zoonotic diseases, antimicrobial resistance, and food safety. Then, WHO aims to enhance the capacity to prevent, detect and respond to public health threats associated with unsafe food at the global and country levels. Implementing the WHO Global Strategy for Food Safety (2022–2030) is underway (Nations 2015), monitoring the global burden of foodborne diseases and supporting countries in burden estimates. The eminent global crisis related to food safety and security, nutrition, and its impact on health has launched efforts by the health sector, food industry, governments, scientists, and society, in general, to implement natural food alternatives and develop another biological origin that ensures the well-being of humans and animals—case, particularly of functional foods rich in bioactive components, essential oils, and so-called nutraceuticals.

2.2 Challenges in the Production and Consumption of Safe Food

2.2.1 Globalization on Food Systems

Globalization emerged at the end of the twentieth century as a complex phenomenon encompassing various areas of humanity, including politics, economics, and food. Globalization originated as an evolution of social activities due to technological development and communication, facilitating interculturalism through products, services and international connections (Lang 1999). However, global goes beyond international, so all processes must act in real-time to be global. This phenomenon has led to the migration of people and the convergence of cultures and societies (Flores 2016). Furthermore, it has led to the dissemination of culture, encouraging the adoption of cultural forms, influences, and practices from other parts of the world (Giménez 2002). Regarding food, the influence of globalization is reflected in the supply of foods available from all continents, proposing them as universal and accessible options in any season of the year (Pere 2018).

This gastronomic phenomenon is possible due to the population's adaptation to other foods or eating styles (Gayathri et al. 2017). In recent years, there has been a change in eating habits, where different dietary patterns have gained popularity in the Western world due to concerns for human health and sustainability (Craig et al. 2021). Understanding the impact of globalization on food preferences, dietary patterns and the association with health may be an area of research of interest to both industry and academia. That led to knowing dietary trends and evaluating whether these patterns are related to a good state of health and the population's demand. For example, various studies have established that globalization has transformed towards an industrialized food environment with a predominance of ultra-processed foods characterized by high energy density and low nutritional quality (de la Garza et al. 2022). Together with the urbanization and commercialization of the food industry, this has generated the adoption of a "Western" dietary pattern (DiNicolantonio and O'Keefe 2021).

Conversely, there is a great interest in developing safe, healthy and sustainable foods and nutraceuticals that can maintain and promote health (Weiskirchen and Weiskirchen 2016). The dissemination and use of nutraceuticals as dietary solutions for health problems has developed thanks to the commercialization and entry of local markets abroad; i.e., tea, an aromatic drink prepared by infusing the leaves of the *Camellia sinensis* plant, originates from China. It is currently the most consumed drink around the world after water (Peñarrieta et al. 2014). *Camellia sinensis* plant has different bioactive compounds such as polyphenols, methylxanthines, among others, with a beneficial effect on health (Peñarrieta et al. 2014), which made it possible to be used as a dietary supplement worldwide. Although considering the impact of globalization on food, dietary patterns, and health is complex, it can help to think that food is not only to satisfy vital human needs but also to promote health and social and cultural identity.

2.2.2 *Food Safety*

2.2.2.1 Contamination by Microorganisms

Food safety is a fundamental characteristic that any food must comply with and maintain to avoid alterations not only in the physical and nutritional aspects of food but also in the health status of those who consume it (Barnes et al. 2022; Organization WH 2022). When the international food safety regulations and standards established in the Codex Alimentarius (FAO/WHO 2020a) are not complied with, the protection of the food chain that includes all the hygienic conditions necessary to produce safe and healthy food is broken, i.e. the protection of food from primary production to the final consumer is lost (FAO/WHO 2020a). An example of this is the care of good animal feed, which is essential for the health and well-being of the animal itself, but also to guarantee the production of safe and quality products of animal origin for safe consumption. Another important point lies in the development of bacteria, viruses, parasites, and fungi in food that put food safety and the health of people and animals at risk, due to the highly toxic substances they produce. An example of this are the aflotoxins produced by fungi such as *Aspergillus flavus* and *Aspergillus parasiticus* (Shabeer et al. 2022), which are responsible for contaminating cereal crops during harvest or storage and are persistent in hot places. At the same time, they are responsible for causing chronic and acute effects on people and animals. Over time, different molecular, analytical, and sensor techniques and methods (Shabeer et al. 2022), have been developed for the detection of these mycotoxins and prevent agricultural losses and health damage. On the other hand, and a parallel problem, is the indiscriminate use of antimicrobial and antifungal chemical agents and resistance, which for years has represented a severe global problem for animal and human health. In addition to the above-mentioned impacts on animal and plant products, shelf life, food security and an economic imbalance in the agricultural sector are severely affected (FAO/WHO 2020b). In the face of the growing concern to maintain food security in the world, it has been a great challenge to have a more critical control than the one that has used for decades, by taking advantage of the development of new technologies and alternatives based on science and biotechnology. This, to prevent or eliminate the development of microbials and their metabolites such as toxins that can intoxicate the consumer and even cause death. In scientifically based biotechnological-developments that work and continuous improvement is sought to make the cost-benefit and its application in the field more efficient, it is through biological control. In the case of mycotoxins, using bacteria (*Bacillus* spp, *Pseudomonas* spp, *Lactobacillus* spp.) (Akhtar et al. 2023) yeasts and fungi (nontoxogenic *Aspergillus* spp, *Trichoderma* spp, *Penicillium* spp, various yeasts), producers of bioactive compounds with antagonistic activity in aflatoxigenic molds (Ren et al. 2020). The fact is that more and more research based on science and biotechnology is being found in the search for alternatives for preserving food that ensure safety and, at the same time, consumers' health. We found food technologies ranging from drying and dehydration (Venkitasamy et al. 2017; Liu et al. 2023a) to the production and use of microorganisms' bacteriocins as food preservatives (Castellano et al. 2023). The application of these biomolecules is an alternative to replace chemical preservatives of food bio conservation and biocontrol.

2.2.2.2 Contamination by Chemical Substances

Food contamination can arise from multiple origins, including agriculture and industrial pollution via air, water, and soil. Hence, it is crucial to regulate chemical packaging, pesticides, and antibiotics in the food industry to avoid undesired

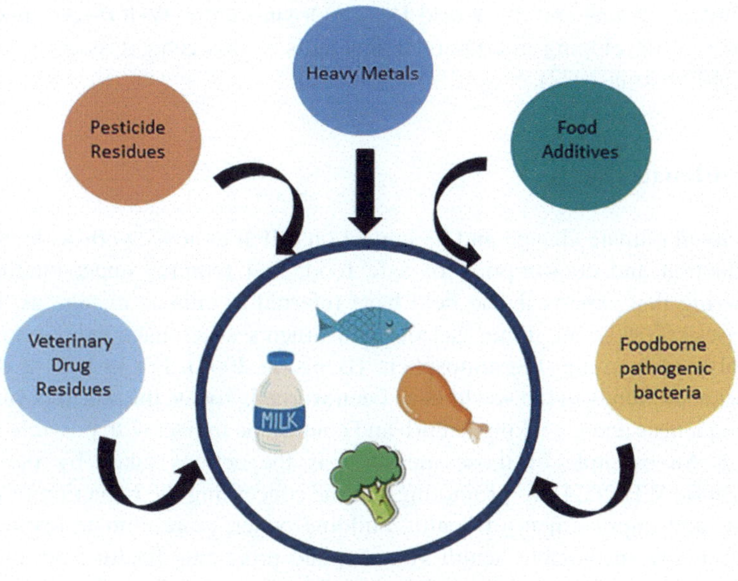

Fig. 2.1 Food contamination by chemical substances

foodstuff residues (Lebelo et al. 2021). As illustrated in Fig. 2.1, contamination by chemical substances, particularly residues of pesticides, veterinary drugs, and illegal additives, is a significant concern in food safety and environmental health. Research has underscored the presence of multiple residues in fruits and vegetables where specific samples are tainted with different pesticide residues (Dubey and Sharma 2022).

In general, substances found in human food correspond to one of the six following classifications: (1) Residues, (2) Unavoidable Contaminants, (3) Prohibited Substances, (4) Supplements, (5) Food Additives, and (6) Cosmetic Additives (Keener 2022). The market increase in the demand for top-quality animal protein underscores the need to safeguard animals from infectious diseases, which means it requires increasing the use of veterinary therapeutics. Nonetheless, the primary challenge lies in the emergence of bacteria or parasites that are resistant to treatment (Akhtar et al. 2023).

Typically, food additives perform the following functions: (1) to improve and maintain the nutritional value of food; (2) to extend the shelf life of food; (3) to make food processing more convenient for healthy beings; (4) to enhance the flavor and grace the color of food; (5) to ensure the safety of microorganisms; (6) to keep the continuity and unity of food quality (Wu 2021). Artificial synthetic food additives have gradually replaced natural food additives, and many problems related to food additives involve the abuse of food additives or even toxic additives (Wu et al. 2022). Continuous efforts are underway to enforce strict restrictions on the amounts of residues present in food products. The Food and Agriculture Organization (FAO) of the United Nations and the World Health Organization (WHO) recognized the importance of developing international standards in this context as early as 1960 (Dubey and Sharma 2022).

2.2.2.3 Climate change

Talking about climate change and its impact on different areas worldwide, such as the production and consumption of safe food, first requires understanding and remembering that experts in the field have referred to climate change as "greenhouse gases" (GHGs) are minor gaseous constituents which have radiation properties capable of warming the atmosphere (Beniston 2002). For at least a decade, climate change trends in the world have been warned. Today, the world is suffering the consequences and is seeking to curb and control the impact with possible human measures. An example of those measures is the actions taken by the WHO (Organization WH 2024), as promoting actions concerning the reduction of carbon emissions and improvement of health, building better, more climate-resilient and environmentally sustainable health systems, and protecting health from the wide range of impacts of climate change. To address food security from the point of view of climate change is to speak of a chain made up of linked links; it means an increase in atmospheric CO_2 increases diseases in plants, in crop yields, in the cost of plants as food, in the nutrition security and health of those who consume it (Przyrembel 2022; Raza and Bebber 2022). And not least, it limits their accessibility. Reducing emissions of GHGs through better food and sustainable diets can result in substantial gains for health, mainly through reduced air pollution. The agri-food sector offers critical solutions for development and is central to hunger and poverty eradication. They are in addition to Hunger Zero (Goal 2) from the Sustainable Development Goals, which are 17 interconnected goals established by United Nations Organization to achieve by 2030 (Nations U 2015).

2.2.3 *Bioeconomy as a sustainable and economic way*

Bioeconomics was initially introduced in the latter part of the twentieth century by an American economist of Romanian descent (Georgescu-Roegen 1971). He contended that economic processes are subject to the thermodynamic law of entropy, advocating for the minimization of scarce raw material consumption to preserve existing reserves for future generations (Skydan et al. 2022). The FAO defines bioeconomy as utilizing biological resources, processes, and principles based on knowledge to sustainably generate goods and services across various economic sectors, as seen in Fig. 2.2. Proponents of the bioeconomy highlight the prospects of innovations to produce food, materials, and other products more efficiently and, therefore, sustainably (Biber-Freudenberger et al. 2020). They argue that bioeconomy or eco-industrial development has the potential to provide solutions for some

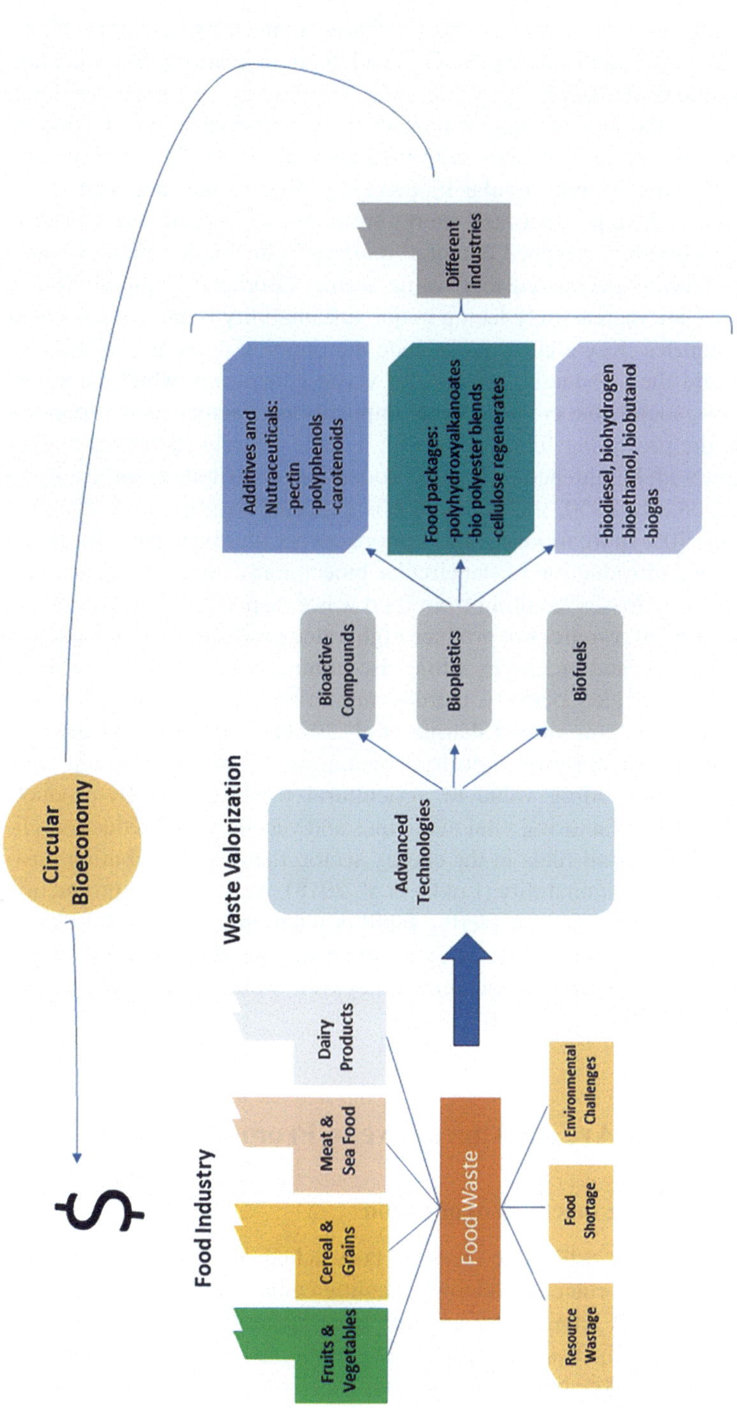

Fig. 2.2 Food waste valorization (Circular Bioeconomy)

of the most pressing sustainability challenges including combating hunger (SDG 2), providing clean and affordable energy (SDG 7) and the fight against climate change (SDG 13) (D'Amico et al. 2022).

The circularity of the bioeconomy is related to the sustainable valorization of bioresources. Organic waste is an asset with the potential to significantly drive the shift towards a sustainable and circular bioeconomy. Recent advancements have produced new value-added products and energy from various organic waste materials. However, considerable untapped potential remains in the exploitation of waste and by-products within specialized high-value sectors (Borhane Mahjoub 2020). Today's economies are increasingly facing major sustainability issues in developed or developing countries. Key challenges include the ongoing depletion of primary fossil resources and the substantial increase in waste generation, which have far-reaching economic, social, and environmental implications. Factors such as population growth and the prevailing linear economic model, wherein increased market demand correlates with heightened industrial production, contribute significantly to the current situation. By 2050, the population is expected to grow by more than 50%, which means 70% more food and 100% more energy will be needed (Borhane Mahjoub 2020). The introduction of the circular bioeconomy concept offers a viable solution to mitigate losses resulting from food waste disposal. It involves utilizing renewable biological resources to produce high-value products such as bioactive compounds, bioplastics, and bioenergy while also aiming to conserve the value of resources for more extended periods, thereby minimizing waste generation and reducing greenhouse gas emissions (Sharma et al. 2021). Over recent years, biotechnology has been transforming industrial operations by aiding in large-scale agricultural production, adding value to agricultural products and by-products through processing, manufacturing vital medicines and vaccines, and reducing reliance on petroleum and fossil fuels in the energy sector, thereby contributing positively to environmental sustainability (Lokko et al. 2018). At least, food production becomes increasingly pressing, considering rapid population growth and the challenges posed by climate change. It is essential to ensure an adequate food supply and adopt sustainable resource management strategies to mitigate adverse environmental effects (Tyczewska et al. 2018).

2.3 Nutraceuticals Are an Alternative to Promoting Health

(a) **Classification and the mechanism of action**

Over the last 20 years, rapid lifestyle changes, busy schedules, and sedentary jobs have underscored the importance of adopting healthy eating habits tailored to individual needs. The public's interest in using diet to mitigate the risk of chronic illnesses beyond just meeting basic nutritional requirements is growing. In today's modern society, ailments like obesity, diabetes, hypertension, stress, allergies, cardiovascular issues, and cancer have become prevalent concerns (Nwanodi 2017;

Garza-Juárez et al. 2023). Stephen L. Defelice, the founder and chairman of the Foundation of Innovation Medicine, coined the term in 1989 by blending "nutrition" and "pharmaceutical." Nutraceuticals, derived from food sources, are purported to offer additional health benefits beyond essential nutrition (Puri et al. 2022). Nutraceuticals encompass natural bioactive or chemical compounds that act as nutritional supplements but also possess health-promoting, disease-preventing, or curative attributes, as shown in Fig. 2.3. The reasons behind the increasing popularity of nutraceuticals are as follows (Puri et al. 2022):

1. A growing segment of consumers is concerned about the escalating costs of healthcare.
2. Individuals dissatisfied with pharmaceutical products for health enhancement are turning to nutraceuticals to improve their well-being and prevent chronic diseases.
3. Healthcare providers recognize that our highly processed food supply, cultivated using chemical fertilizers, pesticides, herbicides, and often genetically modified seeds, lacks the essential nutrients necessary for optimal health.
4. Those who prioritize prevention over treatment.
5. Individuals suffering from chronic conditions who have found no relief from conventional medical treatments.
6. Patients with limited options.

The classification and definition of nutraceuticals tend to overlap due to the similarity among their chemical constituents and functions in delivering health benefits. Some of the most common ways of classifying nutraceuticals are on chemical nature, mechanism of action and food sources (Tsiaka et al. 2022). Classification based on origin source (Nwanodi 2017; Bonciu 2020) is below:

1. Plants: lycopene, ascorbic acid, cellulose, allicin, gallic acid, quercetin, lutein, β-Glucans, daidzein, minerals
2. Animals: Lecithin, calcium, coenzyme Q10, conjugated linoleic acid, selenium, and zinc.
3. Microorganisms: *Bifidobacterium bifidum, Lactobacillus acidophilus, Saccharomyces boulardii, Streptococcus salivarius.*

Prebiotics and probiotics are considered nutraceutical products due to their health benefits. A probiotic is composed of living microorganisms provided as dietary supplements. When ingested in sufficient quantities, these microorganisms confer health benefits by improving the balance of intestinal microbes (Damián et al. 2022; Siciliano et al. 2021). Familiar sources of probiotics include yoghurt, cultured buttermilk, and cheese. Most probiotic products on the market contain lactic acid bacteria (LAB), such as *Lactobacilli, Streptococci*, and *Bifidobacteria,* typically found in the healthy human gastrointestinal microbiota (Siciliano et al. 2021). Prebiotics are dietary components related to the gut microbiota, potentially enhancing metabolic function, promoting mineral absorption, bolstering the immune response, and fortifying the integrity of the intestinal mucosal barrier (Damián et al. 2022). Among the most recognized prebiotics are fructo-oligosaccharides (found in fruits and

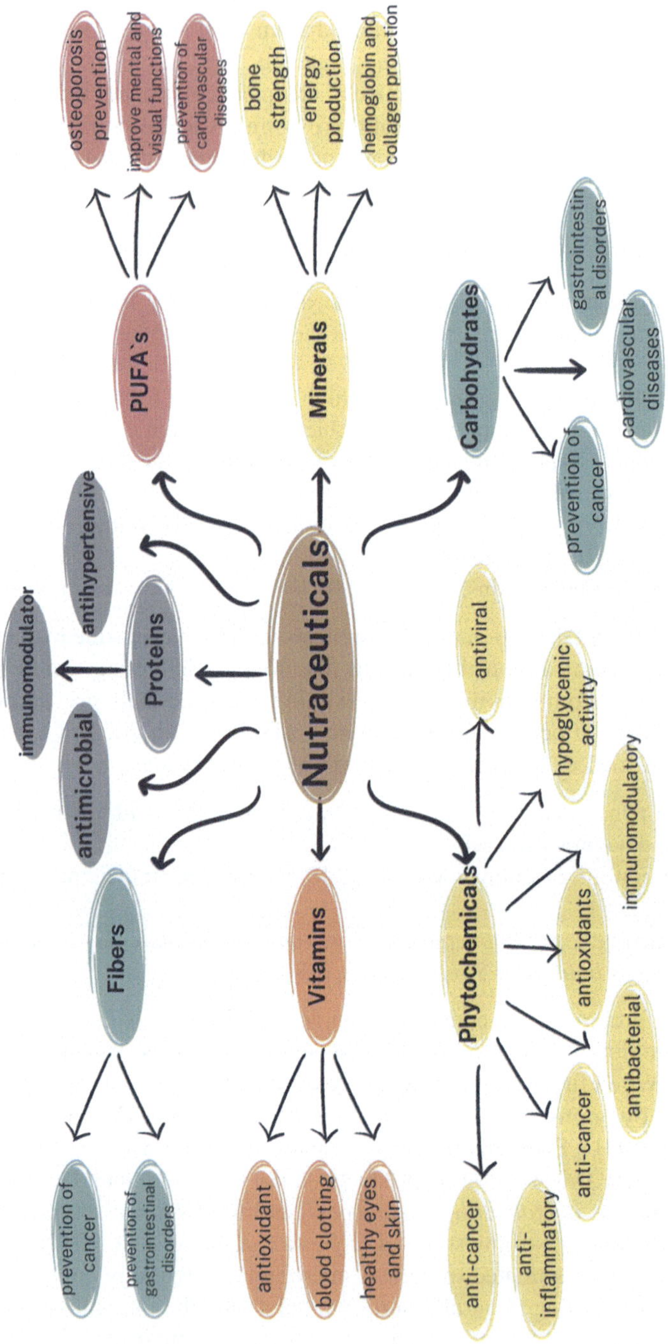

Fig. 2.3 Nutraceuticals based on the chemical nature

vegetables), inulin (derived from chicory root and *Agave tequilana*), and galacto-oligosaccharides (found in milk) (Durazzo et al. 2020).

(b) **Diseases as a focus of nutraceutical treatment**

Nutraceuticals are considered nutritious components displaying an essential biological activity to maintain optimal health or fitness (Puri et al. 2022). Today, several health disorders are controlled with dietary or herbal products, contributing to human welfare. These metabolites may be an isolated nutrient from fruits, vegetables, whole grains or genetically engineered foods or beverages (Bergamin et al. 2019; Granato et al. 2020). Among these highlighted polyphenols, fatty acids, vitamins, flavonoids, tannins, and thiocarbamates have been used to treat stomach pain and diabetes until cancer and cardiovascular and neurological disorders (Wu 2021; Cheng et al. 2022; Wu et al. 2021). Bioactive ingredients can be in two classes: (i) Dietary and (ii) Herbal. The first class comprises carbohydrates, proteins, lipids, vitamins, mushrooms, and probiotics. The second class comprises anthraquinones, alkaloids, tannins, carotenoids, flavonoids, and bitters (Sachdeva et al. 2020). Research made last decade demonstrated that the consumption of carbohydrates such as inulin and fructose combined with proteins promote the growth of beneficial microorganisms like *Bifidobacterium lactis* and others, which help to diminish the progress and abundance of harmful microbes in our gut (Sachdeva et al. 2020; Hayes and Tiwari 2015).

In a special section are the vitamins used to reduce diverse ill types. For Vitamin D, corporal levels are lower worldwide despite its central role in autoimmune thyroid diseases and some cancers. Vitamin D regulates the genetic expression of many genes and could have activities in diverse pathways, such as in autoimmune diseases, cancers, metabolic syndromes, cardiovascular disease, and infection, as well as all-cause mortality (Puri et al. 2022; Kim 2017). Recent scientific data suggested that a diet based on vitamin C reduce stomach cancer and that vitamins B-complex might to prevent a putative Alzheimer disease (Puri et al. 2022). Probiotic agents are beneficial microorganisms, mainly bacterial and yeast species. Commercial brands include members of *Lactobacillus*, *Bifidobacterium*, and *Saccharomyces species*. Among the benefits of their intake, we can see an enhancement in immune function, a shortening of the gastrointestinal infectious, improved handling of women's bacterial vaginosis, and suppressively reduced dental caries too (Balthazar et al. 2022). On the other hand, the herbal class groups many plant types used in folk medicine worldwide for many years. A recent study reported that an isothiocyanate, recognized as 4-(α-L-rhamnosyloxy) benzyl isothiocyanate, obtained from *Moringa oleifera*, possesses exciting properties for the control of diabetes, neurologic diseases, ulcerative colitis, and several cancer types (Wu 2021). Illnesses provoked by viruses like influenza and COVID-19 are in the sight of alternative medicine. For example, the wild watermelon (*Citrullus lanatus* var. *citroides*) is an excellent alternative to control respiratory symptoms caused by the influenza virus. This plant juice can attack the entry and replication but also viral growth. Polyphenols, flavonoids, and prenylated compounds from wild watermelon juice have antiviral properties. These native plants provide helpful information for designing alternative therapies (Puri et al. 2022; Morimoto and Isegawa 2023).

(c) **Sustainable diets**

Throughout history, the study of food has primarily focused on analyzing the nutrients in food and their significance for human health (Johnston et al. 2014). Nowadays, diets also consider food production's impacts on the environment (Luzzani 2022). If we talk about the goals of sustainable diets, we can say that these are to promote healthy and sustainable eating habits. It means environmental and human health, resource usage, and the processes involved, focusing on production, processing, distribution, preparation, consumption, and food handling activities (Araújo et al. 2023; James-Martin et al. 2022; Meybeck and Gitz 2017). Figure 2.4 represents that understanding how food production and processing impact the environment is essential, considering indicators such as greenhouse gas (GHG) emissions, water withdrawal, and land use (Jennings et al. 2023). The global food chain accounts for about 30% of all greenhouse gas emissions worldwide, making it a significant source of emissions (Clark et al. 2020). That percentage includes four main components: the supply chain accounts for 18%, which includes processing, packaging, transportation, and sales; livestock and fisheries contribute 31% of food emissions; while crop production accounts for 27%, with 21% for human food and 6% for livestock feed crops, land use, which includes use for livestock (16%) and human food (8%), contributes 24% of GHG emissions (Ritchie 2023).

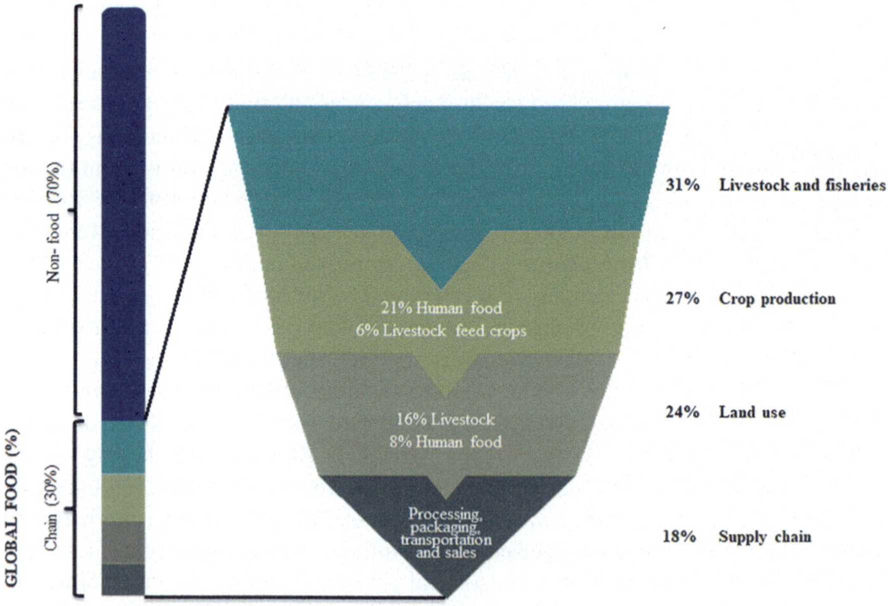

Fig. 2.4 Sustainable diets

Since carbon dioxide (CO_2) makes up most of anthropogenic GHG emissions, attempts to reduce emissions usually centre on this gas. Considering all greenhouse gases is crucial when creating plans to reduce emissions. However, based on their ability to cause global warming over 100 years, other GHGs, including nitrous oxide (N_2O) and methane (CH_4) have a higher capacity to absorb heat than CO_2 (Liu et al. 2023b). It's crucial to remember that the sort of food production determines how much emissions are produced while talking about the emissions caused by our food. Accurate reporting of these numbers is vital. Production of cereals emits 2.19 kg CO_2-equivalent; the output of pulses emits 0.94 kg CO_2-equivalent, and production of beef emits 32.28 kg CO_2-equivalent (Zhang et al. 2023). Climate change represents a significant global threat. It is essential to promote a sustainable diet as part of the transition to a lower-carbon future (Arrieta and González 2018). Maintaining current dietary patterns worldwide could lead to significant additional warming, How do you approach a sustainable diet? Healthier and environmentally beneficial actions include replacing meat with eggs, legumes or cheeses and reducing meat consumption by 20%. These sustainable diet options are more affordable, have a lower impact, and are environmentally healthier (Guy et al. 2024).

(d) **Initiatives technology trends**

At the global level, functional food sales have had a spectacular increase, especially in developed countries such as the United States, Japan, and the European Union. Recent economic estimations revealed an average annual growth rate of 8.5% (Bogue et al. 2017). This sustainable growth may be supported only through science and technology applications, which is the reason industrial processes are applied to traditional food products since digestive care (yoghurt), cardiac fitness (cereals), and drinks (sports) are subject to innovative technologies to enhance their beneficial properties (Puri et al. 2022). Today, plants contribute with thousands of plant natural products (PNPs) displaying some importance in dietary or human health. Draw attention that almost all these compounds fall into just three classes of products: (i) isoprenoids (terpenoids): sterols and carotenoids; (ii) phenylpropanoids and phenylpropanoid-derived compounds such as flavonoids, stilbenes and lignans draw attention; and (iii) alkaloids. Many molecules of medical and or alimentary importance, such as artemisinin (antimalarial drug) and paclitaxel (anticancer), are synthesized at lower concentrations in plants (Marienhagen and Bott 2013). Then genetic engineering is recognized as a powerful technology that must produce many metabolites. Thanks to it, insulin is now available at almost any pharmaceutical store. Since 1973, when the first recombinant microorganism (*Escherichia coli*) was reported, many industrial-relevant ones, such as *Corynebacterium glutamicum*, *Bacillus subtilis*, or the yeast *Saccharomyces cerevisiae* have been available. The above genetic systems allow the design of intricate metabolic pathways to get important metabolites from animals as plants.

Innovative ways include the design and development of formulations and the generation of new products obtained through recombinant technology using bacterium, yeast, or plants. These organisms could yield whether their molecules are essential in the human diet, for example, biotechnological (GMOs) corn and soybean (Dorchester 2022). Process innovation involving a significant bioavailability of active molecules present in food and past decades focused on physical and mechanical systems like hydrostatic pressure, ultrasounds, and microwaves. In general, the above process, searching for better preservation of natural elements (i.e. Vitamins, proteins, hormones), reduces foodborne microbial proliferation (Granato et al. 2020).

Some biotechnological products are below: (i) Polyunsatured fatty acids such as Omega-3 and eicosapentaenoic acid (EPA; C20:5n−3) are produced in significantly lower quantities in fish oil, now they have been produced for the yeast *Yarrowia lipolytica*; (ii) Polyphenols; as resveratrol antioxidant produced for grapes, is synthesized for *Escherichia coli* as *S. cerevisiae*; (iii) Carotenoids (Lycopene and Astaxanthin), produced by *S. cerevisiae*; and (iv) Amino acids as tryptophan, alanine, glutamic acid and phenylalanine in *Corynebacterium glutamicum* and *E. coli* (Yuan and Alper 2019).

2.4 Nutraceutical Sources: Elements of the Present, Essentials for the Future

(a) **Plants**

A nutraceutical food of plant origin presents excellent protein, fibre, lipids, vitamins, and minerals sources. Moreover, their demands are increasing due to the presence of bioactive components such as phenol, flavonoid, terpenes, polysaccharides and reported to possess antimicrobial, anticancer, anti-hyperlipidemia, antiproliferative, antidiabetic, antioxidant, anti-inflammatory, immunostimulant activity and hypocholesterolemic properties (Devi et al. 2004; Gutierrez-Del-Rio et al. 2018). Due to these beneficial health effects, they act as a potential source for the formulation of nutraceutical food products.

Phytochemicals are active plant chemicals accumulated in different parts, such as roots, stems, flowers, leaves, fruits, and seeds, which protect plants from diseases and damage. The most important class of bioactive compounds found in plants are polyphenols; the potent antioxidant properties of most polyphenols are mainly attributed to their hydroxyl group and have preventive effects against the development of cancer, cardiovascular diseases, neural degeneration, diabetes, and obesity (Mani et al. 2021). For this reason, polyphenol-containing products have been used as food supplements and nutraceuticals over the last few years. Some of the most prevalent plants used as a source of polyphenols are tea, coffee, cocoa, wine, beer, nuts, fruits, vegetables, and aromatic plants and species (Taroncher et al. 2021).

Herbal bioactive sources include anthraquinones, alkaloids, tannins, carotenoids, flavonoids, and dietary supplements such as carbohydrates, proteins, lipids, vitamins, probiotics, and prebiotics (Sachdeva et al. 2020). Root vegetable side streams (potato, carrot, sugar beet and beetroot) are commonly consumed root vegetables in human daily diet, but they also have high value-added compounds for nutraceutical and pharmaceutical purposes (Ying Zhou and Yang 2023). Cannabis compounds are essential in pharmaceuticals for the treatment of depression, anxiety, insomnia, epilepsy, and seizures or as ingredients in healthy functional foods (i.e. oil, tinctures, cookies, ice cream, chocolates, butter, brownies, cooking oil, jellies, beverages, capsules, pills) (Suárez-Jacobo et al. 2023; Mesa et al. 2023).

Furthermore, the need for innovative extraction and purification of components reveals the importance of developing new nutraceutical compositions. Different emerging techniques have the aim of improving not only extraction but also reducing operating costs and waste recovery with new applications. Therefore, technologies such as supercritical extraction, cold press extraction, fractional and molecular distillation, ultrasound-assisted extraction, microwave-assisted extraction, pressurized liquid extraction and pulsed electric field are some of the innovative processes used today.

(b) **Fruits and vegetables**

Fruits generally contain a significant source of phytochemicals, carbohydrates, vitamins, amino acids, peptides, and antioxidants that are of interest to be nutraceuticals (Mesa et al. 2023).

Studies have suggested that the bioactive compounds within prickly pear (Opuntia) may influence glucose metabolism by improving insulin sensitivity, reducing insulin resistance, and modulating gut microbiota composition. From traditional Chinese medicine, new indications and formulations for nutraceutical use were found for some plants, such as monk fruit as a sugar substitute and specific extracts of ginseng, which are increasingly popular (Williamson et al. 2020).

Market trends in the United States demonstrate the vital role of fruits as potential nutraceuticals and have reported rises in 2017 of nutraceuticals prepared with fruits such as cranberry (*Vaccinium macrocarpon*), garcinia (*Garcinia cambogia*), coconut oil (*Cocos nucifera*), ginseng (*Panax ginseng*), guarana (*Paullina cupana*), acai berry (*Euterpe oleracea*), green coffee (*Coffea arabica*) and cherry fruit (*Prunus* spp.) (Williamson et al. 2020). In addition, health professionals, pharmacists and nutritionists are showing interest in more advanced research disciplines on the medicinal properties of fruits and vegetables. There is a trend towards consuming nutraceutical foods of plant origin to prevent and reduce diseases. Therefore, in-depth studies are required to conduct further research on long-term consumer clinical trials, cosmeceutical properties, and benefits to treat diseases and make the most of the potential of bioactive components placed in food.

(c) **Naturally fermented foods**

In the field of industrial biotechnology there are liquid, semi-solid and solid fermentation, state. Foods exposed to this process, also known as functional foods, have aroused the scientific community as they have multiple health benefits (Damián et al. 2022), because of their bioactive compounds as carotenes, ascorbic acid, lycopene, rivoflavin (Mathur et al. 2020). There are different cultures in the world that have adopted the fermentation method, which they have adapted into traditional gastronomy, so it is to be expected that carbon sources vary in each region of the world. Different cultures have adopted the fermentation method, which they have adapted into traditional gastronomy, so it is important to note that carbon sources vary in each region. Foods such as cereals, legumes, milk, and vegetables are some of the most common sources of carbon, and due to their composition, are degraded by microorganisms generally recognized as safe (GRAS). Among those that abound are *Lactobacillus* spp. and *Bifidobacterium* spp., mainly. In northern India (Das et al. 2016), regional foods such as Kimona are derived from soybean fermentation. It has cholesterol-lowering and antioxidant properties. On the Asian continent, cooking pork is a common feature due to its high protein contain. "Sour meat" is an example of fermented food made with salt, rice, and some spices to season the meat. One of its characteristics is the abundant presence of probiotics and a high number of free amino acids, depending on the temperature at which it is fermented (Zhao et al. 2023). In the west, specifically in Mexico, is tepache, a fermented drink made of pineapple peel, piloncillo (brown sugar) and water. Reports show that consuming tepache reduces the risk of gastrointestinal diseases thanks to the mesentericin and dextran produced in fermentation (Ojeda-Linares et al. 2017). Another biotechnology-based product that is constantly analyzed is beer, which has been shown to have beneficial health capabilities due to the presence of alcohol, improving cardiovascular health and insulin sensitivity in the blood, clarifying that these studies were a moderate alcohol consumption. Interestingly, beer byproducts such as spent hops seem to have antithrombotic activity (Lordan et al. 2019). All this food with functional properties is called nutraceutical (Garza-Juárez et al. 2023). Regarding nutraceuticals, the global food industry estimates a growth of at least $7 Billion by 2026 due to their importance as immune enhancers, cancer prevention, diabetes, and a regulator of chronic inflammation disorders, according to data collected (Sachdeva et al. 2020).

(d) **Biomass**

Due to global population growth, there is an increasing interest in addressing the shortage of animal and plant proteins in the diet. Proteins play essential roles in human cellular activities (Grossmann and Weiss 2021). An emerging alternative source is using microorganisms, such as bacteria, archaea, fungi, and some algae, as the primary producers of single-cell protein (SCP) (Molfetta et al. 2022). Fungi can grow on a wide range of substrates, have a significantly high protein content, between 44% and 45% of dry weight and contain some B-complex vitamins, making them one of the most widely used organisms (Bajić et al. 2023).

The production of SPCs refers to the extraction of proteins from the edible biomass produced by pure or mixed cultures of microorganisms. In addition, microorganisms have a series of properties:

- High protein content and nutritional value.
- Shorter production times.
- Adaptability to grow in various substrates.
- Production under fully controlled conditions less land use.

Table 2.1 Production of SCP by different yeast species

Yeast	Renewable feedstock	Fermentation mode	Biomass (g/L)	References
C. utilis	Fruit wastes from orange peel	Batch, flasks	15.71	Carranza-Méndez et al. (2022)
K. marxianus	Cheese whey	Batch	4.7 (mono culture), 6.1 (mixed culture)	Karim et al. (2020)
Y. lipolytica	Glycerol	Batch, bioreactor	19.7	Jach and Serefko (2018)
P. pastoris	Metanol	Bioreactor	63.37	Meng et al. (2023)
S. cerevisiae	Date juice	Batch bioreactor	42.8	Koukoumaki et al. (2024)
S. cerevisiae	Fruit wastes, from Mango, Prickly Custard, Apple, Papaya, Pineapple, Banana, Mangosteen, Cashew apple, Cacao, Jackfruit, Pomegranate	Batch, flasks	0.4	Koukoumaki et al. (2024)
P. kudriavzevii	Biogas slurry (derived from chicken manery).	Bottles	6.8	Koukoumaki et al. (2024)

Therefore, the potential use of SCP offers a more significant advantage due to the variety of products and attractive alternatives for use as a protein source (Onyeaka et al. 2022; Pereira et al. 2022). Numerous microorganisms are of biotechnological interest because of the processes to obtain SCP (see Table 2.1). Examples include *Candida utilis (*Carranza-Méndez et al. 2022*)*, *Kluyveromyces marxianus (*Karim et al. 2020*)*, *Yarrowia lipolytica (*James-Martin et al. 2022*;* Lopes et al. 2018*;* Sanya and Onesime 2022*)*, *Pichia pastoris (*Meng et al. 2023*)*, *Saccharomyces cerevisiae* and *Picchia kudriavzevii (*Koukoumaki et al. 2024*)*, due to its chemical composition, nutrients (vitamin B and folic acid), and its amino acid profile, in which lysine and threonine (essential amino acids) stand out, align with FAO standards (Nasseri et al. 2011). Crucial considerations for selecting products for SCP production should consider some essential criteria such as the availability of low-cost substrates for microorganism growth, accurate estimation of culture conditions for efficient substrate utilization, and efficiency in solving challenges related to transportation costs and pre-treatment of waste products (Spalvins et al. 2018).

(e) **Algae**

Despite its rich biochemical diversity, Algae's potential remains largely untapped in the nutraceutical realm. Algae, encompassing a wide array of photosynthetic organisms from microscopic diatoms to macroscopic seaweeds, offer a treasure trove of bioactive compounds (including polysaccharides, fatty acids, proteins and peptides, polyphenols, diterpenes, steroids, antioxidants, and alkaloids), with promising health benefits. This chapter delves into the current landscape of algae-derived nutraceuticals and outlines their indispensable role in shaping the future of preventive healthcare (García-Perez et al. 2023).

Among the most recent and extensively studied molecules sourced from algae are Omega-3 fatty acids, particularly eicosapentaenoic acid (EPA) and docosahexaenoic acid (DHA). These essential fatty acids, crucial for cardiovascular health and cognitive function, are primarily extracted from species like *Nannochloropsis* and *Schizochytrium*. Their abundance in certain algae species underscores the potential of algae as sustainable alternatives to traditional marine sources (Jesionowska et al. 2023). Pigments with antioxidant activity, such as β-carotene from *Dunaliella salina* and phycocyanin from *Arthrospira platensis*, help neutralize free radicals in the body, reducing oxidative stress and protecting cells from damage. Its inclusion in the diet can promote cardiovascular health, strengthen the immune system, and reduce the risk of chronic diseases (Kusmayadi et al. 2021). Fucoxanthin, a carotenoid found in brown seaweeds such as *Undaria pinnatifida* and *Sargassum*, has garnered attention for its anti-obesity and anti-inflammatory properties. Its role in promoting thermogenesis and inhibiting adipogenesis highlights its potential as a natural remedy for metabolic disorders (Kumari and Garima 2023). Astaxanthin, another potent carotenoid from microalgae like *Haematococcus pluvialis*, has gained popularity for its antioxidant power, surpassing even renowned antioxidants like vitamin E and beta-carotene. Its applications span skin health to sports nutrition, offering a natural solution to combat oxidative stress and enhance endurance (Cao et al. 2023).

Moreover, algae-derived polysaccharides, such as fucoidan from brown algae and carrageenan from red algae, exhibit a myriad of biological activities, including immunomodulation, anticoagulant properties, and potential anticancer effects (Sanniyasi et al. 2023). Their structural complexity and diverse functionalities make them invaluable assets in functional foods and dietary supplements. As we venture into the future, algae hold immense promise as sustainable sources of novel nutraceuticals. However, concerted efforts are warranted to overcome existing hurdles in scalability, cost-effectiveness, regulatory frameworks to realize the transformative potential of algae-derived nutraceuticals on a global scale, the sustainable and green extraction of high-value compounds, and finally, exploring the biological properties of these compounds using *in vitro*, *ex vivo*, and *in vivo* models (Tiwari and García-Vaquero 2022). In conclusion, algae represent a veritable treasure trove of bioactive compounds with far-reaching implications for human health and well-being. Embracing algae as nutraceutical sources not only offers sustainable solutions but also underscores the imperative of harnessing nature's bounty for the betterment of society.

2.5 Conclusion

The future of nutraceuticals is auspicious due to their ability to prevent a comprehensive list of diseases. In addition, products on the market facilitate access to this type of therapy to improve health. At the same time, the high interest of the scientific community supports the acceptance of these by consumers, generating acceptance and opening alternatives for the innovation and development of functional foods with natural ingredients and easy access to make at home. The challenges in producing and consuming safe food are important issues requiring urgent, permanent actions, such as increasing investments in public policies and research on food bioactives and nutraceuticals worldwide. To have and maintain good health is to understand that we must invest in our future life. People who move more and maintain a proper diet without excesses live longer. It must be transmitted in the family and at school to children from a very early age as part of a comprehensive education, and this is something like a pedagogy of longevity.

References

Akhtar M, Nosheen A, Keyani R, Yasmin H, Naz R, Mumtaz S et al (2023) Biocontrol of Rhizoctonia solani in basmati rice by the application of *Lactobacillus* and *Weissella* spp. Sci Rep 13(1):13855. https://doi.org/10.1038/s41598-023-41058-9

Araújo RG, Chavez-Santoscoy RA, Parra-Saldívar R, Melchor-Martínez EM, Iqbal HMN (2023) Agro-food systems and environment: sustaining the unsustainable. Curr Opin Environ Sci Health 34. https://doi.org/10.1016/j.coesh.2022.100413

Arrieta EM, González AD (2018) Impact of current, National Dietary Guidelines and alternative diets on greenhouse gas emissions in Argentina. Food Policy 79:58–56. https://doi.org/10.1016/j.foodpol.2018.05.003

Bajić B, Vučurović D, Vasić Đ, Jevtić-Mučibabić R, Dodić S (2023) Biotechnological production of sustainable microbial proteins from agro-industrial residues and by-products. Food Secur 12(1). https://doi.org/10.3390/foods12010107

Balthazar CF, Guimarães JF, Coutinho NM, Pimentel TC, Ranadheera CS, Santillo A et al (2022) The future of functional food: emerging technologies application on prebiotics, probiotics and postbiotics. Compr Rev Food Sci Food Saf 21(3):2560–2586. https://doi.org/10.1111/1541-4337.12962

Barnes J, Whiley H, Ross K, Smith J (2022) Defining food safety inspection. Int J Environ Res Public Health 19(2). https://doi.org/10.3390/ijerph19020789

Beniston M (2002) Climatic change: possible impacts on human health. Swiss Med Wkly 132(25–26):332–337. https://doi.org/10.4414/smw.2002.10041

Bergamin A, Mantzioris E, Cross G, Deo P, Garg S, Hill AM (2019) Nutraceuticals: reviewing their role in chronic disease prevention and management. Pharmaceut Med 33(4):291–309. https://doi.org/10.1007/s40290-019-00289-w

Biber-Freudenberger L, Ergeneman C, Förster JJ, Dietz T, Börner J (2020) Bioeconomy futures: expectation patterns of scientists and practitioners on the sustainability of bio-based transformation. Sustain Dev 28(5):1220–1235. https://doi.org/10.1002/sd.2072

Bogue J, Collins O, Troy AJ (2017) Market analysis and concept development of functional foods. In: Debasis Bagchi SN (ed) Developing new functional food and nutraceutical products. Academic, pp 29–45. https://doi.org/10.1016/B978-0-12-802780-6.00002-X

Bonciu E (2020) Aspects of the involvement of biotechnology in functional food and nutraceuticals. Agric Food Sci LXIII(2):261–266

Borhane Mahjoub ED (2020) Chances and challenges of an organic waste–based bioeconomy. Curr Opin Green Sustain Chem 25. https://doi.org/10.1016/j.cogsc.2020.100388

Cao Y, Yang L, Qiao X, Xue C, Xu J (2023) Dietary astaxanthin: an excellent carotenoid with multiple health benefits. Crit Rev Food Sci Nutr 63(18):3019–3045. https://doi.org/10.1080/10408398.2021.1983766

Carranza-Méndez RC, Chávez-González ML, Sepúlveda-Torre L, Aguilar CN, Govea-Salas M, Ramos-González R (2022) Production of single cell protein from orange peel residues by Candida utilis. Biocat Agric Biotechnol 40. https://doi.org/10.1016/j.bcab.2022.102298

Castellano P, Melian C, Burgos C, Vignolo G (2023) Bioprotective cultures and bacteriocins as food preservatives. Adv Food Nutr Res 106:275–315. https://doi.org/10.1016/j.fm.2023.104275

Cheng Y, Liu J, Ling Z (2022) Short-chain fatty acids-producing probiotics: a novel source of psychobiotics. Crit Rev Food Sci Nutr 62(28):7929–7959. https://doi.org/10.1080/10408398.2021.1920884

Clark MA, Domingo NGG, Colgan K, Thakrar SK, Tilman D, Lynch J et al (2020) Global food system emissions could preclude achieving the 1.5 degrees and 2 degrees C climate change targets. Science 370(6517):705–708. https://doi.org/10.1126/science.aba7357

Craig WJ, Mangels AR, Fresán U, Marsh K, Miles FL, Saunders AV et al (2021) The safe and effective use of plant-based diets with guidelines for health professionals. Nutrients 13(11). https://doi.org/10.3390/nu13114144

D'Amico G, Szopik-Depczyńska K, Beltramo R, D'Adamo I, Ioppolo G (2022) Smart and sustainable bioeconomy platform: a new approach towards sustainability. Sustain For 14(1):466. https://doi.org/10.3390/su14010466

Damián MR, Cortes-Perez NG, Quintana ET, Ortiz-Moreno A, Garfias Noguez C, Cruceño-Casarrubias CE et al (2022) Functional foods, nutraceuticals and probiotics: a focus on human health. Microorganisms 10(5). https://doi.org/10.3390/microorganisms10051065

Das G, Patra JK, Singdevsachan SK, Gouda S, Shin HS (2016) Diversity of traditional and fermented foods of the Seven Sister states of India and their nutritional and nutraceutical potential: a review. Front Life Sci 9(4):292–312. https://doi.org/10.1080/21553769.2016.1249032

de la Garza AL, Martínez-Tamez AM, Mellado-Negrete A, Arjonilla-Becerra S, Peña-Vázquez GI, Marín-Obispo LM et al (2022) Characterization of the cafeteria diet as simulation of the human western diet and its impact on the lipidomic profile and gut microbiota in obese rats. Nutrients 15(1):86. https://doi.org/10.3390/nu15010086

Devi PVIJ, Narzary P, Sharma D, Sultana F (2004) Bioactive compounds, nutraceutical values and its application in food product development of oyster mushroom. J Future Foods 4(4):335–342. https://doi.org/10.1016/j.jfutfo.2023.11.005

DiNicolantonio JJ, O'Keefe J (2021) The importance of maintaining a low Omega-6/Omega-3 ratio for reducing the risk of autoimmune diseases, asthma, and allergies. Mo Med 118(5):453–459. https://pubmed.ncbi.nlm.nih.gov/34658440/

Dorchester (2022) https://wwwpgeconomics.co.uk/

Dubey JK, Sharma A (2022) Pesticide residues and international regulations. In: Chakrabarti SK, Sharma S, Shah MA (eds) Sustainable management of potato pests and diseases. Springer, Singapore. https://doi.org/10.1007/978-981-16-7695-6_14

Durazzo A, Lucarini M, Santini A (2020) Nutraceuticals in human health. Food Secur 9(3). https://doi.org/10.3390/foods9030370

FAO/WHO (2020a) Codex Alimentarius International Food Standards

FAO/WHO (2020b) Report of the 51st session of the Codex Committee on food hygiene. Codex Alimentarius Commision

Flores MV (2016) La Globalización como Fenómeno Político, Económico y Social. Orbis Revista Científica Ciencias Humanas 12(34) https://www.redalyc.org/articulo.oa?id=70946593002

García-Perez P, Cassani L, García-Oliveira P, Xiao J, Simal-Gandara J, Prieto MA et al (2023) Algal nutraceuticals: a perspective on metabolic diversity, current food applications, and prospects in the field of metabolomics. Food Chem:409. https://doi.org/10.1016/j.foodchem.2022.135295

Garza-Juárez A, Pérez-Carrillo E, Arredondo-Espinoza EU, Islas JF, Benítez-Chao DF, Escamilla-García E (2023) Nutraceuticals and their contribution to preventing noncommunicable diseases. Food Secur 12(17). https://doi.org/10.3390/foods12173262

Gayathri RV, Ruchi V, Mohan. (2017) Impact of nutrition transition and resulting morbidities on economic and human development. Curr Diabetes Rev 13(5):452–460. https://doi.org/10.2174/1573399812666160901095534

Georgescu-Roegen N (1971) The entropy law and the economic process. Harvard University Press. https://doi.org/10.4159/harvard.9780674281653

Giménez G (2002) Globalización y cultura. Estudios Sociológicos XX(1):23–46. https://www.redalyc.org/pdf/598/59805802.pdf

Granato D, Barba FJ, Bursać Kovačević D, Lorenzo JM, Cruz AG, Putnik P (2020) Functional foods: product development, technological trends, efficacy testing, and safety. Annu Rev Food Sci Technol 11:93–118. https://doi.org/10.1146/annurev-food-032519-051708

Grossmann L, Weiss J (2021) Alternative protein sources as technofunctional food ingredients. Annu Rev Food Sci Technol 12:93–117. https://doi.org/10.1146/annurev-food-062520-093642

Gutierrez-Del-Rio I, Fernández J, Lombo F (2018) Plant nutraceuticals as antimicrobial agents in food preservation: terpenoids, polyphenols and thiols. Int J Antimicrob Agents 52(3):309–315. https://doi.org/10.1016/j.ijantimicag.2018.04.024

Guy DJ, Bray J, Appleton KM (2024) Select dietary changes towards sustainability: impacts on dietary profiles, environmental footprint, and cost. Appetite 194:107194. https://doi.org/10.1016/j.appet.2023.107194

Hayes M, Tiwari BK (2015) Bioactive carbohydrates and peptides in foods: an overview of sources, downstream processing steps and associated bioactivities. Int J Mol Sci 16(9):22485–22508. https://doi.org/10.3390/ijms160922485

Jach ME, Serefko A (2018) Nutritional yeast biomass: characterization and application. In: Holban AM, Grumezescu AM (eds) Handbook of food bioengineering, diet, microbiome and health. Academic, pp 237–270. https://doi.org/10.1016/B978-0-12-811440-7.00009-0

James-Martin G, Baird DL, Hendrie GA, Bogard J, Anastasiou K, Brooker PG et al (2022) Environmental sustainability in national food-based dietary guidelines: a global review. Lancet Planet Health 6(12):e977–e986. https://doi.org/10.1016/S2542-5196(22)00246-7

Jennings R, Henderson AD, Phelps A, Janda KM, van den Berg AE (2023) Five U.S. dietary patterns and their relationship to land use, water use, and greenhouse gas emissions: implications for future food security. Nutrients 15(1). https://doi.org/10.3390/nu15010215

Jesionowska M, Ovadia J, Hockemeyer K, Clews KC, Xu Y (2023) EPA and DHA in microalgae: health benefits, biosynthesis, and metabolic engineering advances. J Am Oil Chem Soc 100(11):831–842. https://doi.org/10.1002/aocs.12718

Johnston JL, Fanzo JC, Cogill B (2014) Understanding sustainable diets: a descriptive analysis of the determinants and processes that influence diets and their impact on health, food security, and environmental sustainability. Adv Nutr 5(4):418–429. https://doi.org/10.3945/an.113.005553

Karim A, Gerliani N, Aider M (2020) *Kluyveromyces marxianus*: an emerging yeast cell factory for applications in food and biotechnology. Int J Food Microbiol 333:108818. https://doi.org/10.1016/j.ijfoodmicro.2020.108818

Keener L (2022) Food safety and regulatory survey of food additives and other substances in human food. In: Martinović A, Oh S, Lelieveld H (eds) Ensuring global food safety. Academic, pp 259–273. https://doi.org/10.1016/B978-0-12-816011-4.00005-7

Kim D (2017) The role of vitamin D in thyroid diseases. Int J Mol Sci 18(9). https://doi.org/10.3390/ijms18091949

Koukoumaki DI, Tsouko E, Papanikolaou S, Ioannou Z, Diamantopoulou P, Sarris D (2024) Recent advances in the production of single cell protein from renewable resources and applications. Carbon Resour Convers 7(2). https://doi.org/10.1016/j.crcon.2023.07.004

Kumari A, Garima BN (2023) A comprehensive review on algal nutraceuticals as prospective therapeutic agent for different diseases. 3 Biotech 13(2):44. https://doi.org/10.1007/s13205-022-03454-2

Kusmayadi A, Leong YK, Yen HW, Huang CY, Chang JS (2021) Microalgae as sustainable food and feed sources for animals and humans - biotechnological and environmental aspects. Chemosphere 271:129800. https://doi.org/10.1016/j.chemosphere.2021.129800

Lang T (1999) Diet, health and globalization: five key questions. Proc Nutr Soc 58(2):335–343. https://doi.org/10.1017/s0029665199000452

Lebelo K, Malebo N, Mochane MJ, Masinde M (2021) Chemical contamination pathways and the food safety implications along the various stages of food production: a review. Int J Environ Res Public Health 18(11). https://doi.org/10.3390/ijerph18115795

Liu S, Qiu Y, Ji K, Ozturk S, Erdoğdu F, Qin W et al (2023a) Effect of oil exposure stages on the heat resistance of Salmonella enterica serovar Enteritidis phage type 30 in peanut flour. Food Microbiol 113:104275. https://doi.org/10.1016/j.fm.2023.104275

Liu G, Zhang F, Deng X (2023b) Half of the greenhouse gas emissions from China's food system occur during food production. Commun Earth Environ 4. https://doi.org/10.1038/s43247-023-00809-2

Lokko Y, Heijde M, Schebesta K, Scholtès P, Van Montagu M, Giacca M (2018) Biotechnology and the bioeconomy-towards inclusive and sustainable industrial development. New Biotechnol 40(Pt A):5–10. https://doi.org/10.1016/j.nbt.2017.06.005

Lopes M, Gomes AS, Silva CM, Belo I (2018) Microbial lipids and added value metabolites production by *Yarrowia lipolytica* from pork lard. J Biotechnol 265:76–85. https://doi.org/10.1016/j.jbiotec.2017.11.007

Lordan R, O'Keeffe E, Tsoupras A, Zabetakis I (2019) Total, neutral, and polar lipids of brewing ingredients, by-products and beer: evaluation of antithrombotic activities. Food Secur 8(5). https://doi.org/10.3390/foods8050171

Luzzani G (2022) The sustainability of diets: current understanding and shortcomings. Curr Opin Environ Sci Health 30. https://doi.org/10.1016/j.coesh.2022.100398

Mani JS, Johnson JB, Hosking H, Ashwath N, Walsh KB, Neilsen PM et al (2021) Antioxidative and therapeutic potential of selected Australian plants: a review. J Ethnopharmacol 268:113580. https://doi.org/10.1016/j.jep.2020.113580

Marienhagen J, Bott M (2013) Metabolic engineering of microorganisms for the synthesis of plant natural products. J Biotechnol 163(2):166–178. https://doi.org/10.1016/j.jbiotec.2012.06.001

Mathur H, Beresford TP, Cotter PD (2020) Health benefits of lactic acid bacteria (LAB) fermentates. Nutrients 12(6). https://doi.org/10.3390/nu12061679

Meng J, Liu S, Gao L, Hong K, Liu S, Wu X (2023) Economical production of *Pichia pastoris* single cell protein from methanol at industrial pilot scale. Microb Cell Factories 22(1):198. https://doi.org/10.1186/s12934-023-02198-9

Mesa NC, Alves IA, Vilela FMP, e Silva DM, Forero LAP, Novoa DMA et al (2023) Fruits as nutraceuticals: a review of the main fruits included in nutraceutical patents. Food Res Int 170:113013. https://doi.org/10.1016/j.foodres.2023.113013

Meybeck A, Gitz V (2017) Sustainable diets within sustainable food systems. Proc Nutr Soc 76(1):1–11. https://doi.org/10.1017/S0029665116000653

Molfetta M, Morais EG, Barreira L, Bruno GL, Porcelli F, Dugat-Bony E et al (2022) Protein sources alternative to meat: state of the art and involvement of fermentation. Food Secur 11(14). https://doi.org/10.3390/foods11142065

Morimoto R, Isegawa Y (2023) Anti-influenza virus activity of Citrullus lanatus var. citroides as a functional food: a review. Food Secur 12(20). https://doi.org/10.3390/foods12203866

Nasseri AT, Resoul-Amini S, Morowvat MH, Ghasemi Y (2011) Single cell protein: production and process. Am J Food Technol 6:103–116. https://doi.org/10.3923/ajft.2011.103.116

Nations U (2015) Transforming our world: the 2030 Agenda forSustainable Development. A/RES/70/1 - General Assemble. https://sdgs.un.org/publications/transforming-our-world-2030-agenda-sustainable-development-17981

Nwanodi O (2017) Nutraceuticals: curative integrative cancer treatment. Altern Integr Med 6(2). https://doi.org/10.4172/2327-5162.1000240

Ojeda-Linares C, Álvarez-Ríos GD, Figueredo-Urbina CJ, Islas LA, Lappe-Oliveras P, Nabhan GP et al (2017) Traditional fermented beverages from Mexico as a potential probiotic source. Ann Microbiol 67:577–586. https://doi.org/10.3390/foods10102390

Onyeaka H, Anumudu CK, Okpe C, Okafor A, Ihenetu F, Miri T et al (2022) Single cell protein for foods and feeds: a review of trends. Open Microbiol J 16. https://doi.org/10.2174/18742858-v16-e2206160

Organisation PAH (2020) Rapid Assessment of Service Delivery for NCDs during the COVID-19 Pandemic in the Americas. 8. https://www.who.int/publications/m/item/rapid-assessment-of-service-delivery-for-ncds-during-the-covid-19-pandemic

Organization WH (2022) Newsroom Food Safety. Available from: https://www.who.int/news-room/fact-sheets/detail/food-safety

Organization WH (2024) Food safety. Available from: https://www.who.int/health-topics/food-safety#tab=tab_1

Peñarrieta M, Tejeda L, Mollinedo P, VIlla J, Bravo J (2014) Phenolic compounds in food. Characterization and analysis, vol 1. CRC Press, Boca Raton, p 450. https://doi.org/10.1201/9781315120157

Pere CE (2018) Cocina-alimentación-futuro. [Cooking-feeding-future]. Nutr Hosp 35:49–51. https://doi.org/10.20960/nh.2125

Pereira AG, Fraga-Corral M, García-Oliveira P, Otero P, Soria-López A, Cassani L et al (2022) Single-cell proteins obtained by circular economy intended as a feed ingredient in aquaculture. Food Secur 11(18). https://doi.org/10.3390/foods11182831

Przyrembel H (2022) 1.4.3 Food safety. World Rev Nutr Diet 124:94–100. https://doi.org/10.1159/000516730

Puri V, Nagpal M, Singh I, Singh M, Dhingra GA, Huanbutta K, Dheer D, Sharma A, Sangnim TA (2022) A comprehensive review on nutraceuticals: therapy support and formulation challenges. Nutrients 14(21). https://doi.org/10.3390/nu14214637

Raiten DJ, Sakr Ashour FA, Ross AC, Meydani SN, Dawson HD, Stephensen CB et al (2015) Inflammation and nutritional science for programs/policies and interpretation of research evidence (INSPIRE). J Nutr 145(5):1039S–1108S. https://doi.org/10.3945/jn.114.194571

Raza MM, Bebber DP (2022) Climate change and plant pathogens. Curr Opin Microbiol 70:102233. https://doi.org/10.1016/j.mib.2022.102233

Ren X, Zhang Q, Zhang W, Mao J, Li P (2020) Control of aflatoxigenic molds by antagonistic microorganisms: inhibitory behaviors, bioactive compounds, related mechanisms, and influencing factors. Toxins (Basel) 12(1). https://doi.org/10.3390/toxins12010024

Ritchie HRM. Food production is responsible for one-quarter of the world's greenhouse gas emissions. Our World in Data 2023. https://ourworldindata.org/food-ghg-emissions

Sachdeva V, Roy A, Bharadvaja N (2020) Current prospects of nutraceuticals: a review. Curr Pharm Biotechnol 21(10):884–896. https://doi.org/10.2174/1389201021666200130113441

Sanniyasi E, Gopal RK, Damodharan R, Arumugam A, Sampath Kumar M, Senthilkumar N et al (2023) In vitro anticancer potential of laminarin and fucoidan from Brown seaweeds. Sci Rep 13(1):14452. https://doi.org/10.1038/s41598-023-41327-7

Sanya DRA, Onesime D (2022) New roles for *Yarrowia lipolytica* in molecules synthesis and biocontrol. Appl Microbiol Biotechnol 106(22):7397–7416. https://doi.org/10.1007/s00253-022-12227-z

Shabeer S, Asad S, Jamal A, Ali A (2022) Aflatoxin contamination, its impact and management strategies: an updated review. Toxins (Basel) 14(5). https://doi.org/10.3390/toxins14050307

Sharma P, Vivek KG, Ranjna S, Sunita V, Sang HK, Jonathan WC, Wong. (2021) Sustainable processing of food waste for production of bio-based products for circular bioeconomy. Bioresour Technol 325:124684. https://doi.org/10.1016/j.biortech.2021.124684

Siciliano RA, Reale A, Mazzeo MF, Morandi S, Silvetti T, Brasca M (2021) Paraprobiotics: a new perspective for functional foods and nutraceuticals. Nutrients 13(4). https://doi.org/10.3390/nu13041225

Sinclair JR (2019) Importance of a One Health approach in advancing global health security and the Sustainable Development Goals. Rev Sci Tech 38(1):145–154. https://doi.org/10.20506/rst.38.1.2949

Skydan OV, Yaremova M, Tarasovych LV, Dankevych VY, Kutsmus NM (2022) Possibilities of developing sustainable world by introducing bioeconomy: global perspective. Problemy Ekorozwoju 17(1):162–170. https://doi.org/10.35784/pe.2022.1.15

Spalvins K, Zihare L, Blumberga D (2018) Single cell protein production from waste biomass: comparison of various industrial by-products. Energy Procedia 147:409–418. https://doi.org/10.1016/j.egypro.2018.07.111

Suárez-Jacobo Á, Díaz Pacheco A, Bonales-Alatorre E, Castillo-Herrera GA, García-Fajardo JA (2023) Cannabis extraction technologies: impact of research and value addition in Latin America. Molecules 28(7). https://doi.org/10.3390/molecules28072895

Taroncher M, Vila-Donat P, Tolosa J, Ruiz MJ, Rodríguez-Carrasco Y (2021) Biological activity and toxicity of plant nutraceuticals: an overview. Curr Opin Food Sci 42:113–118. https://doi.org/10.1016/j.cofs.2021.05.008

Tiwari BK, García-Vaquero M (2022) Nutraceuticals from algae: current view and prospects from a research perspective. Mar Drugs 20(11). https://doi.org/10.3390/md20110671

Tsiaka T, Kritsi E, Tsiantas K, Christodoulou P, Sinanoglou VJ, Zoumpoulakis P (2022) Design and development of novel nutraceuticals: current trends and methodologies. Forum Nutr 2(2):71–90. https://doi.org/10.3390/nutraceuticals2020006

Tyczewska A, Woźniak E, Gracz J, Kuczyński J, Twardowski T (2018) Towards food security: current state and future prospects of agrobiotechnology. Trends Biotechnol 36(12):1219–1229. https://doi.org/10.1016/j.tibtech.2018.07.008

Venkitasamy C, Brandl MT, Wang B, McHugh TH, Zhang R, Pan Z (2017) Drying and decontamination of raw pistachios with sequential infrared drying, tempering and hot air drying. Int J Food Microbiol 246:85–91. https://doi.org/10.1016/j.ijfoodmicro.2017.02.005

Weiskirchen S, Weiskirchen R (2016) Resveratrol: how much wine do you have to drink to stay healthy? Adv Nutr 7(4):706–718. https://doi.org/10.3945/an.115.011627

Williamson EM, Liu X, Izzo AA (2020) Trends in use, pharmacology, and clinical applications of emerging herbal nutraceuticals. Br J Pharmacol 177(6):1227–1240. https://doi.org/10.1111/bph.14943

Wu L (2021) Analysis of food additives. In: Galanakis CM (ed) Innovative food analysis. Academic, pp 157–180. https://doi.org/10.1016/B978-0-12-819493-5.00007-8

Wu YY, Xu YM, Lau ATY (2021) Anti-cancer and medicinal potentials of moringa isothiocyanate. Molecules 26(24). https://doi.org/10.3390/molecules26247512

Wu L, Zhang C, Long Y, Chen Q, Zhang W, Liu G (2022) Food additives: from functions to analytical methods. Crit Rev Food Sci Nutr 62(30):8497–8517. https://doi.org/10.1080/10408398.2021.1929823

Ying Zhou YT, Yang B (2023) Root vegetable side streams as sources of functional ingredients for food, nutraceutical and pharmaceutical applications: the current status and future prospects. Trends Food Sci Technol 137:1–16. https://doi.org/10.1016/j.tifs.2023.05.006

Yuan SF, Alper HS (2019) Metabolic engineering of microbial cell factories for production of nutraceuticals. Microb Cell Factories 18(1):46. https://doi.org/10.1186/s12934-019-1096-y

Zhang M, Li H, Chen S, Liu Y, Li S (2023) Interrogating greenhouse gas emissions of different dietary structures by using a new food equivalent incorporated in life cycle assessment method. Environ Impact Assessment Rev 103. https://doi.org/10.1016/j.eiar.2023.107212

Zhao XFJ, Laghi L, Deng J, Dao X, Tang J, Ji L et al (2023) Characterization of flavor profile of "Nanx Wudl" sour meat fermented from goose and pork using gas chromatography–ion mobility spectrometryl- (GC–IMS) combined with electronic nose and tongue. Food Secur 12(11). https://doi.org/10.3390/foods12112194

Chapter 3
Production and Development of Nutraceuticals Using Sustainable Food System

Sonam Prabhakar, Pawan Prabhakar, Ashok Nimesh, and Mamoni Banerjee

Abstract The manufacturing and advancement of nutraceuticals within a sustainable food system entails the formulation of health-enhancing commodities derived from organic food sources, simultaneously accounting for the comprehensive social, economic, entrepreneurial and environmental ramifications of the entire procedure. Nutraceuticals refer to food products or constituents that offer health advantages surpassing fundamental nutritional value, encompassing essential elements like vitamins, minerals, antioxidants, or bioactive compounds. The sustainable food system endeavors to address the interconnected challenges of environmental degradation, food insecurity, and social inequities associated with the production and consumption of nutraceutical-based food. Its overarching goal is to facilitate the sustainable production, distribution, and consumption of such food, thereby fostering economic prosperity, social well-being, and environmental preservation. This approach seeks to ensure the availability of nutritious food for both current and future generations while mitigating the adverse impacts on the environment and rectifying social injustices. This chapter focuses on the distinct health advantages associated with the incorporation of sustainability principles into all stages of nutraceutical production. This approach enables the development of products that not only enhance health but also contribute to a food system that is more ecologically sustainable and socially responsible.

S. Prabhakar (✉) · A. Nimesh
Department of Anthropology and Tribal Studies, Central University of Jharkhand, Ranchi, India

P. Prabhakar (✉)
Bio-Research Laboratory, Rajendra Mishra School of Engineering Entrepreneurship, Indian Institute of Technology Kharagpur, Kharagpur, West Bengal, India

School of Business, Woxsen University, Hyderabad, Telangana, India

M. Banerjee
Bio-Research Laboratory, Rajendra Mishra School of Engineering Entrepreneurship, Indian Institute of Technology Kharagpur, Kharagpur, West Bengal, India
e-mail: mamoni@see.iitkgp.ac.in

S. Amir Ashraf, M. Adnan (eds.), *Food Bioactives and Nutraceuticals*,
https://doi.org/10.1007/978-981-97-9723-3_3

Keywords Nutraceuticals · Organic food · Phytochemicals · Antioxidants · Vitamins · Minerals · Food security · Sustainable food system

3.1 Introduction

Sustainability refers to the practice of fulfilling our current requirements while ensuring that future generations may also fulfil their own needs without any compromise. Aside from natural resources, we also require social and economic resources and sustainability encompasses more than simply environmentalism and conservation of biodiversity (Thomsen 2013; Wu et al. 2023). The concept of sustainability includes considerations for both social equality and economic growth among the people in any country and around the globe. Sustainability is often defined by the interconnection of three overarching dimensions: environment, economy, and society (Mensah 2019; Guang-Wen et al. 2023). It takes into account the needs of both current and future generations and presently, it is critical to evaluate the carbon emissions in the environment caused by the substances that are being consumed by the animals and human and results in global climate change. Therefore, a sustainable food system is an integral objective of sustainability which is defined as a system that ensures sufficient nutrition and food security for everyone, while preserving the financial, social, and environmental resources needed to continue providing food security and nutrition for future generations without causing any negative impact (Garnett 2014; Herrero et al. 2020; Wu et al. 2023). Sustainable food systems are initiated by implementing sustainable agriculture methods, establishing more sustainable food distribution networks, promoting the adoption of sustainable diets, and minimising food waste at every stage of the system (Allen et al. 2019). The global discourse on the viability of food systems is getting more acknowledgment within the international community. The sustainable food systems are considered to be for achieving the 17 sustainable development goals (SDGs). Presently, these systems lack efficiency and sustainability. By the year 2022, around 9.2% of the global population, or over 735 million individuals, were experiencing chronic hunger, which is 122 million higher than the figure in the year 2019 (UNICEF 2020; World Health Organization 2023; Guang-Wen et al. 2023). Furthermore, these anthropologic activities are significant contributors to the climate catastrophe. They are responsible for one-third of greenhouse gas emissions and account for 70% of global freshwater demand. The prevalence of recurring food crises and the on going global environmental changes have elevated food security and sustainability to the forefront of policy priorities (Aleksandrowicz et al. 2016; Malhi et al. 2021). The requirements for developing a sustainable food system vary depending on the environmental, social, and economic condition of each nation. Therefore, the metrics used to measure progress towards sustainability must accurately represent the individual conditions of each country. In recent years, there has been an immense rise in the popularity of nutraceuticals. Nutraceuticals refer to a diverse variety of food-derived products that offer health advantages beyond basic nutrition (Brown et al. 2021). Sustainable food systems are intentionally developed to be ecologically responsible, socially fair, and commercially feasible. Their focus lies on optimising resource use, minimising waste generation, conserving biodiversity, and promoting local economies. The essential elements comprise of organic farming, agroecology, permaculture, and sustainable aquaculture (Velten et al. 2015). These systems strive to reduce the environmental impact of food production while still ensuring that people have access to healthy and reasonably priced food. The growth of the nutraceuticals and functional foods market is fuelled by the growing consumer knowledge regarding the health and nutritional advantages of these foods in improving health and avoiding diseases (Béné et al. 2019; Wezel et al. 2020). Simultaneously, there is an increasing demand for sustainability in the production of food, propelled by environmental considerations and the necessity for enhanced resource utilisation. The convergence of these developments is resulting in the development of nutraceuticals derived from sustainable food systems, presenting a potential opportunity for advancements in the field of health and wellbeing (Garnett and Godfray 2012; Nasri et al. 2014). Nutraceuticals are derived from the sustainable food system by using the health benefit characteristics of plants, animals, and microbes that are grown by developing a natural environment. Now the consumers are much aware of the benefits of maintaining a nutritious diet and adopting a healthy lifestyle. They have broader selection of food and drink options on the market to meet their health and wellbeing requirements. Advancements in nutraceuticals and functional foods have substantiated that the bioactive constituents included in the diet have a significant therapeutic impact on human disease. Furthermore, there is a significant focus on food scientists in the identification and production of foods with enhanced bioactivity, eventually leading to improved health concern (Wichienchot and Ishak 2017; Cassani and Gomez-Zavaglia 2022). The relationship between nutraceutical development and sustainable food systems shows a great potential in tackling the two-fold concerns of human health and environmental sustainability. Nutraceuticals offer many therapeutic advantages in areas such as cough and cold, anti-arthritis, digestion, sleep problems, cancer therapy, depression, diabetes management, cholesterol control, blood pressure regulation, and pain relief (Williamson et al. 2020; Sachdeva et al. 2020; Bommakanti et al. 2023). The nutraceutical research and development is now concentrated on exploring the potential of various nutraceuticals in the pharma and healthcare industry. Nutraceuticals, produced from food sources, are acknowledged for their capacity to prevent chronic illnesses and enhance general well-being. Nevertheless, it is imperative to analyse their progress from the perspective of sustainability, including ecological responsibility, financial feasibility, and societal fairness. The development of nutraceuticals within the context of sustainable food systems signifies a hopeful integration of health, nutrition, and environmental conservation (Kadiri et al. 2016; Benković 2019; Jain et al. 2022). The necessity for food with additional health advantages is also shown by trends in population demographics and socio-economic shifts. The rise in life expectancy, leading to a larger elderly population and a demand for better quality of life, along with rising healthcare expenses, have prompted governments, researchers, healthcare professionals, and the food industry to explore more efficient ways of managing these developments. To achieve sustainability in the development of nutraceuticals, it is important to reduce the environmental effect while ensuring efficiency and scalability. This entails choosing raw materials that have little impact on the environment and utilising industrial methods that reduce carbon emissions. The plant-based ingredients, which often have less resource need and emission compared to animal-derived components, are becoming more popular (Singh et al. 2019; Dominguez et al. 2022). Furthermore, the use of by-products derived from food production, such as fruit peels, seeds, and pulp, not only improves the sustainability of nutraceuticals but also mitigates food waste. Methods such as green extraction techniques, which employ solvents that are ecologically benign and need less energy, are vital in this context. Nutraceuticals can play a role in promoting health, preventing chronic diseases, delaying the aging process, and supporting the structure or function of the body. (Miladinov 2020; Elkhalifa et al. 2021). Therefore, nutraceuticals from sustainable foods offer supplementary health benefits beyond basic nutrition, frequently used for prevention and treatment of chronic as well as acute human disorders. The sustainable food systems aimed to produce these nutraceuticals that preserves resources, minimises environmental consequences, and promotes economic and social welfare. This paper provides an in-depth analysis of the status of nutraceutical development in sustainable food systems, including important trends, problems, and potential future directions.

3.2 Nutraceuticals for Human Health

A nutraceutical refers to a non-toxic dietary supplement derived from food extracts that has been scientifically demonstrated to provide health benefits for both the treatment and prevention of diseases. The growing fascination in nutraceuticals is driven by consumers being informed about epidemiological research that suggest a particular diet or dietary component is linked to a reduced risk of a specific illness (Santini and Novellino 2017; Taroncher et al. 2021). Nutraceuticals encompass a wide range of goods, including isolated nutrients, dietary supplements, genetically altered food, herbal remedies, and processed items including cereals, soups and drinks (Kapoor et al. 2020). Nutraceuticals encompass a wide variety of products, including dietary supplements, functional foods, medicinal foods, and more. Nutraceuticals encompass individual or combined pro- and pre-biotic food substances and food for specific medicinal purposes. Food supplements consist of individual or combined minerals, vitamins, amino acid supplements, nutritional foods, and herbal products. The inclusion of nutraceuticals in one's regular diet may help avoid pathological illnesses by reducing or eliminating the need for pharmaceuticals in those who are suitable for non-pharmacological therapy options (Da Costa 2017; Dudeja and Gupta 2017). Some foods, such as spices and herbs, have the ability to reduce the disease progression and promote better health conditions. Nevertheless, there may be a lack of scientific evidence supporting the safety, effectiveness, and impact on health and/or pathological conditions of these items (Dini 2018). In the pharmaceutical sector, conducting clinical trials *in vivo* or *in vitro* is a compulsory need to validate the effects of a drug or therapeutics for nutraceuticals there previously existed no mechanism for verifying the impact of for preventing or curing illnesses (Prabhakar and Mamoni 2021; Prabhakar et al., 2023a). Recently, there has been scientific testing and validation of nutraceuticals due to increasing awareness of health-related concerns and side effects. People are now recognising the direct and indirect role that nutraceuticals plays in promoting good health and avoiding any disease (Gupta et al. 2021).

3.3 Nutraceuticals: Classification and Their Origin

3.3.1 Dietary Supplements

Dietary supplements are consumable goods, including vitamins, minerals, and herbs that assert their ability to enhance overall well-being. The use of dietary supplements has grown in popularity as a method to promote health, optimise performance, and address nutritional deficiencies (Bailey et al. 2013). Dietary supplements encompass a wide range of goods designed to be consumed in order to fulfil vital nutritional needs. These products are offered in many formats such as tablets, capsules, gelatin capsules, soft gels, liquids, chewable preparations, and powders (Nori et al. 2023). They can include specific elements or a mixture of minerals, herbs, vitamins, fatty acids, amino acids, and other nutritional ingredients. Supplementing one's diet with necessary vitamins and minerals is crucial when nutritional needs are not adequately fulfilled by food solely. Dietary supplements are extensively utilised and have the ability to enhance health when carefully directed towards individuals who require them. Insufficient nourishment and deficits in essential nutrients are widespread issues that have a negative impact on worldwide health (Bronzato and Durante 2018; Binns et al. 2018). Enhancing the quality of food component is crucial for tackling these problems. However, those who are at risk of nutrient deficiencies might also benefit from taking dietary supplements or consuming fortified foods to satisfy their nutritional needs. In developing countries, it is beneficial to provide vitamin and iron supplements to address deficiencies in women of reproductive age, infants, and children. Similarly, folic acid supplementation is important for women of reproductive age and during pregnancy (Preedy et al. 2013; Olson et al. 2021). Vitamin A serves as a preventive measure against xerophthalmia and has the ability to reverse some instances when there is a gradual decline in eye health, finally resulting in permanent and irreversible blindness. Insufficient levels of Vitamin A also heighten vulnerability to, and mortality rates caused by, viral illnesses. Deficiency arises in communities who mostly consume starchy foods, with a basic diet that lacks animal fats and has limited intake of vividly coloured fruits or vegetables. Additionally, infants and children can benefit from vitamin D supplementation, while adults aged $\geq$65 years can benefit from calcium and vitamin D supplements to promote bone health (Lentjes 2019; Ahsan et al. 2023). Vegetables, complete grains, and vitamins are abundant in dietary phytochemicals. Dietary supplements are ingested either in their whole or as individual active components. The phytochemicals are quite varied and encompass carotenoids, phenolics, alkaloids, organosulfur compounds, and nitrogen-containing compounds (Prabhakar and Banerjee 2020). Niacin, often known as vitamin B3, effectively treats pellagra, a life-threatening condition linked to a diet mostly consisting of maize. The condition is distinguished diarrhoea, skin irritation, dementia, and finally death (Awuchi et al. 2020). Amino acids link together via peptide bonds to form polypeptides and proteins. An amino acid consists of an amino group ($-NH_2$), a carboxyl group (–COOH), and a distinctive side chain (R group) that possesses the features and qualities of the amino acid. There exist a total of 20 standard amino acids that are encoded by genes, and they may be categorised into two groups: essential and non-essential. Amino acids are essential for several biological functions, such as protein synthesis, enzyme activity, and the control of metabolic pathways (Damodaran and Parkin 2017). Additionally, they function as precursors for the production of hormones and neurotransmitters. The body cannot synthesise essential amino acids, including leucine, isoleucine, valine, lysine, methionine, phenylalanine, threonine, tryptophan, and histidine, and hence they must be acquired from the food (Dietzen 2018). Tryptophan may be taken from many foods such as cheese, chicken, egg whites, fish, milk, sunflower seeds, peanuts, pumpkin seeds, sesame seeds, and soybeans. It is also found in legumes, red meat, fish, cheese, milk, yoghurt, legumes, brewer's yeast (Alamgir 2018). Glutamine, on the other hand, is gained from protein-rich diets. Essential fatty acids (EFAs) are a category of polyunsaturated fats that are vital for several physiological processes but cannot be produced by the human body. Consequently, they must be acquired from dietary intake. They may be categorised into two main groups: Omega-3 Fatty Acids and omega-6 fatty acids. Essential fatty acids (EFAs), such as omega-3 fatty acids, are derived from plant oils including flaxseed, soybean, and canola oils, as well as from fish and other seafood. Omega-6 fatty acids are derived from vegetable oils such as sunflower, safflower, and maize oil, as well as animal items like meat and eggs. The equilibrium between omega-3 and omega-6 fatty acids is crucial in diet. (Cruzat et al. 2018; Newsholme et al. 2023; Khan et al. 2023).

3.3.2 Functional Foods

Functional foods are innovative food products that are designed to contain certain compounds or living microorganisms that may have potential health benefits or the ability to prevent diseases. Functional foods may also be defined as processed food that offers both medical or health advantages and a decrease in the risk of disease (Cencic and Chingwaru 2010). Therefore, functional food lacks a universally accepted definition, but one distinguishing characteristic is its capacity to prevent diseases while also providing nutritional and health-enhancing advantages. Individually, they are not regarded as preventive or therapeutic, and they are generally not necessary for a balanced diet. Functional foods encompass a diverse range of food types and are defined as fortified, enriched, or enhanced foods. These foods offer health benefits beyond basic nutrients (vitamins and minerals) when consumed in a sufficient amount as part of a balanced diet on a regular basis (Arshad et al. 2021; Khalid et al. 2022). A functional food is a type of food that may be either natural or modified by technical or biotechnological methods by adding or removing certain components. It is required to show its benefits in quantities that are typically ingested in a regular diet (Ballini et al. 2023). However, currently there is no special regulation in place to regulate them, save from ordinary food legislation. The US Food and Drugs Administration (FDA) in the USA lacks a regulatory framework that acknowledges functional foods as a separate category and does not offer a legal definition for functional foods (Wallace et al. 2010; Alamgir 2018). The implementation of regulations for the use of nutritional claims for foodstuffs established standardised norms across the European Union for the use of health and nutritional claims on food products. The primary goal of this regulation is to guarantee that food claims are supported by scientific data and are easily understood by consumers. The European Food Safety Authority (EFSA) distinguishes functional foods from the Food and Drug Administration (FDA) by defining them as "foods that have a positive impact on one or more specific bodily functions, going beyond basic nutritional effects, and contributing to improved health and well-being or a reduced risk of disease." Foods that are rich in nutrients, such as fruits, vegetables, nuts, whole grains, legumes, and seafood, are sometimes referred to as functional foods (Gilsenan 2011; Kušar et al. 2021). Furthermore, food scientists have developed functional foods, such as calcium-enriched orange juice, vitamin D-fortified milk, and fibre-enhanced cereals, which offer additional nutritional benefits. Other innovative functional food like probiotics are living microorganisms, usually bacteria or yeast that offers health advantages when ingested in sufficient quantities. They are commonly known as "beneficial" or "amiable" bacteria due to their role in preserving a harmonious equilibrium in the gut microbiome (Kechagia et al. 2013; Plaza-Diaz et al. 2019). Bacterial species such as *Lactobacillus* can be found in yoghurt and other fermented foods. Several kinds of bacteria have been found to be beneficial in treating diarrhoea and aiding those with lactose intolerance, a condition characterised by the inability to digest lactose, the sugar present in milk (Ayivi et al. 2020). Functional foods often include substantial amounts of bioactive chemicals, which are a characteristic feature. In general, these meals include of dietary fibre, omega-3 polyunsaturated fatty acids, carotenoids, as well as probiotics and prebiotics. They also contain foods that can decrease cholesterol (Damián et al. 2022). Probiotics are typically deemed safe for the majority of individuals, however, those with weakened immune systems or severe diseases should seek advice from a healthcare professional before to their usage. The following table provides a concise overview of several functional foods, their constituents and their impacts on human health (Table 3.1).

Table 3.1 Functional foods, their constituents and their impacts on health

S. No	Functional food	Components	Function	References
1.	Fortified bread, milk based drinks, fortified pasta and noodles,	Antioxidants	Helps in scavenging free radicals and reducing chances	Zeb (2020), Poljsak et al. (2021), Rahaman et al. (2023)
2.	Cereal, bread, oatmeal and whole grain oat, fruit jam and jelly, sausages, fish and fish products, whole grain oat, cereal β-glucan added to fruit drinks, yogurt and fermented milk containing *lactobacillus* bacteria	Dietary fibres	Reduces blood glucose and cholesterol level, relieve from constipation,	Yegin et al. (2020), Jahan et al. (2020)
3.	Eggs and eggs products; canned, corn; cornmeal	Carotenoids like lutein and zeaxanthin	Lutein can interact with free radicals and safeguard low-density lipoproteins from oxidative reactions. These factors are linked to the likelihood of developing atherosclerosis and ischaemic heart disease. Carotenoids, particularly lutein (at least 4 mg per day), are crucial in preventing age-related vision decline.	Honda (2020), Meléndez-Martínez et al. (2021), Crupi et al. (2023)
4.	Fermented dairy products like yogurt, cheese, vegetables juice, fermented meat and fruits	Prebiotics as well as Probiotics like fructans, inulin and essential polysaccharides like starch	Regulates immunity and improve immune system, treat gastric disorders like diarrhoea, irritable bowel diseases	Goel and Kumar (2021), Zoumpopoulou et al. (2018), Cunningham et al. (2021), Manzoor et al. (2022), Haque et al. (2024)

(continued)

Table 3.1 (continued)

S. No	Functional food	Components	Function	References
5.	Yogurt, cream cheese, margarine, milk-based drinks, yogurt-drinks, meat products, rye bread, soy or rice-based drinks, salads, mayonnaise, chocolates	Stanols and sterols	Lowering thrombocyte level,	Rysz et al. (2017), Vilahur et al. (2019), Barkas et al. (2023), Pederiva et al. (2024)
6.	Soy-milk, tofu, rye flaxseed-based bread, wholewheat cereal	Estrogen	Regulates menopause systems and lowers blood glucose levels, improves insulin sensitization and glucose tolerance	Swathi Krishna et al. (2022), Giampieri et al. (2022), Myrtsi et al. (2023)

3.3.3 Medicinal Foods

Medical foods are specifically designed and intended to manage the dietary demands of individuals with a particular ailment that require unique nutritional requirements that cannot be fulfilled by a regular diet alone. They serve as a interconnection between nutrition and medicine. These products are intended to be ingested as diet and include biologically active substances that can improve both physical and mental health. These meals can either be naturally present or intentionally created with additional bioactive components extracted from plants and used for their therapeutic attributes. These extracts are intricate mixture of multiple chemicals with diverse structures and origins. They have undergone extensive use and assessment over an extended period (Abuajah et al. 2015; Holmes et al. 2021).

3.4 Phytochemicals: Main Constituents of Nutraceuticals and Its Classifications

Phytochemicals are essential constituents of a nutritious diet, providing a variety of health advantages that surpass basic nutritional needs. By include a range of plant-based foods that are high in phytochemicals in their daily diet, individuals may harness the healing properties of these compounds to prevent and control chronic illnesses, boost immune function, and promote overall well-being. The ongoing study into phytochemicals is revealing their growing importance in the fields of nutrition and preventive medicine (Gupta et al. 2024). Phytocompounds are examined for their possible health advantages, such as their ability to reduce inflammation, act as antioxidants, prevent cancer, and protect the heart. They are a fundamental component of traditional medicine and are used as nutritional supplements (Prabhakar and Mamoni 2021).

3.4.1 Polyphenols

Polyphenols are a heterogeneous group of naturally existing compounds that are present in plants. Polyphenols are recognised for their antioxidant capabilities and have a vital function in safeguarding the body against many illnesses and enhancing general health. These compounds, which consist of one or more aromatic rings with different hydroxyl groups attached, are the most common secondary metabolites found in plants (Prabhakar et al. 2024). Because of their capacity to transfer electrons and hydrogen atoms to free radicals, so inhibiting oxidation, they serve as exceptional antioxidants and may be found in a range of phenolic structures, ranging from simple to complex. The formation of the parent structure occurs when a hydroxy group is substituted at the "ortho," "meta," or "para" positions of an aromatic ring. L-phenylalanine and l-tyrosine are essential components to produce phenolic acids in plants (Verma and Thakur 2021). This synthesis largely occurs through the shikimate/phenylpropanoid pathway. Phenolic acids, flavonoids, and tannins are the main phenolic components in our diet. These chemicals have a high correlation with antioxidant capacity. Phenolic compounds possess anti-inflammatory, anticancer, antibacterial, antidiabetic, and neuroprotective effects. Fruits contain a variety of phenolic compounds, with the five most prevalent being phenolic acids, stilbenes, lignans, flavonoids, and tannins or proanthocyanidins (Zagoskina et al. 2023). Polyphenols boost the body's response to insulin, help control blood sugar levels, and enhance lipid profiles, which makes them advantageous for the management and prevention of type 2 diabetes and metabolic disorders. Polyphenols promote digestion, decrease gut inflammation, and improve nutritional absorption via regulating gut flora (Kaur et al. 2021). Moreover, the phenolic acids, which include derivatives of benzoic and cinnamic acid, together with flavonoids, are the predominant phenolic compounds found in the diet account for high majority of all dietary phenolic compounds, respectively (Shahidi and Ambigaipalan 2015).

3.4.2 Terpenoids

Terpenoids, or isoprenoids, are a vast and varied group of organic compounds that exist naturally and are formed from terpenes. They are constructed using isoprene units, which are five-carbon building blocks, organised in different configurations. Terpenoids are classified into monoterpenoids, sesquiterpenoids, and diterpenoids (Panigrahy et al. 2021). Monoterpenoids are organic compounds that consist of two isoprene units, which means they have a total of 10 carbon atoms. Some examples of these compounds include menthol, limonene, and geraniol. While Sesquiterpenoids are consist of three isoprene units, which means they have a total of 15 carbon atoms. Some examples of these compounds include farnesol and humulene. Diterpenoids consist of four isoprene units, which amount to a total of 20 carbon atoms (Maurya et al. 2021). Triterpenoids are organic compounds that consist of six isoprene units, which mean they have a total of 30 carbon atoms. Some examples of substances are squalene and sterols such as cholesterol. Tetraterpenoids are organic compounds that consist of eight isoprene units, which amounts to a total of 40 carbon atoms. Terpenoids are primary constituents of essential oils and are accountable for their fragrant characteristics (Lund et al. 2019). Vitamins A (retinol) and E (tocopherol) are terpenoids that play crucial roles in eyesight, immunological function, and antioxidant protection. Terpenoids serve as precursors to both steroids and steroid hormones. Terpenoid pathways are responsible for the production of cholesterol, an essential element of cell membranes, as well as hormones such as oestrogen and testosterone (Tetali 2019).

3.4.3 Saponins

Saponins are a heterogeneous collection of plant chemicals that possess soap-like characteristics. Glycosides are compounds composed of a sugar molecule attached to a non-sugar component, usually a triterpenoid or steroid. Saponins are present in several plant species, such as beans, legumes, herbs, and some grains. Saponins have diverse biological and pharmacological properties (Oakenfull and Sidhu 2023). The principal active components in many traditional medicines are chosen according to their expression of diverse structural variations on sugar chains and aglycones. The aglycone moiety of saponins is linked to one or more carbohydrate chains. The aglycone or sapogenin unit is composed of either a more abundant triterpene unit or a sterol unit. In both steroid and triterpenoid saponins, the carbohydrate side-chain is commonly attached to the third carbon of sapogenin (Moses et al. 2014). The portion of the molecule that consists of a sugar, which can be either a single sugar or a complicated sugar chain, is known as glycoside component. The non-sugar component can either be a steroid, known as steroidal saponins, or a triterpenoid, referred to as triterpenoid saponins (Abdelrahman et al. 2020).

3.4.4 Alkaloids

Alkaloids are often produced from amino acids such lysine, tyrosine, tryptophan, and ornithine. The biosynthesis of alkaloid is a series of enzyme processes that transform these amino acids into the intricate structures that are distinctive of alkaloids. Alkaloids are a diverse group of naturally occurring organic compounds that contain nitrogen atoms. These compounds are primarily produced by plants, although they can also be found in some animals and fungi (Bhambhani et al. 2021).

Alkaloids have significant physiological effects on humans and animals, making them important in both traditional and modern medicine. Lately, there has been significant interest in the potential of nutraceuticals, which are dietary components that offer medicinal or health advantages. Alkaloids like caffeine and berberine possess notable antioxidant properties, aiding in the neutralisation of detrimental free radicals and the mitigation of oxidative stress. Substances such as capsaicin and curcumin possess potent anti-inflammatory characteristics, which are advantageous in the treatment of chronic illnesses associated with inflammation. Specific alkaloids, such as synephrine and caffeine, can increase metabolic rate and facilitate the breakdown of fat, so assisting in the control of weight and promoting metabolic well-being (Hussain et al. 2018; Heinrich et al. 2021). Alkaloids like theobromine and resveratrol have demonstrated the ability to enhance cardiovascular health by increasing blood circulation, decreasing blood pressure, and reducing cholesterol levels (Bøhn et al. 2012; Rojas et al. 2022). Huperzine A and caffeine has neuroprotective properties, which may help in the prevention and treatment of neurodegenerative disorders such as Alzheimer's and Parkinson's disease (Lichota et al. 2019). Alkaloids, due to their wide range of bioactivities and positive effects on health, show significant potential as nutraceuticals. Including them in dietary supplements and functional foods can promote several areas of health, including as metabolic and cardiovascular health, cognitive function, and antibacterial defence. Supplements containing alkaloids can deliver potent amounts of advantageous substances, promoting several areas of well-being, including vitality, cognitive abilities, and metabolic well-being (AlAli et al. 2021). Alkaloid-fortified foods offer additional health advantages beyond their fundamental nutritional value. For instance, energy beverages that include caffeine or chocolates that are fortified with theobromine. Nutraceuticals that include cardioprotective alkaloids such as theobromine and resveratrol have the ability to promote heart health and prevent cardiovascular illnesses. Further investigation and advancement in this domain will augment our comprehension of the potential of alkaloids, resulting in groundbreaking products that boost health and well-being (Bommakanti et al. 2023).

3.4.5 Carotenoids

Carotenoids are a group of pigments that are naturally present in plants, algae, and photosynthetic microorganisms. These chemicals are accountable for the crimson, amber, and tangerine hues of several fruits and vegetables. Carotenoids have significance not only for the well-being of plants but also for the human body, providing many health advantages. This makes them desirable as both dietary elements and nutraceuticals (Maoka 2020). Carotenoids can counteract the harmful effects of free radicals, safeguarding cells from oxidative harm and diminishing the likelihood of developing chronic ailments like heart disease and cancer. Lutein and zeaxanthin are mostly found in the macula of the eye and play a crucial role in safeguarding against age-related macular degeneration (AMD) and cataracts. They achieve this by filtering out detrimental blue light and offering antioxidant defence. Beta-carotene derived from carrots, lycopene obtained from tomatoes, and lutein extracted from spinach is among the significant carotenoids. In addition, they can shield the skin from harm caused by UV radiation and enhance both the health and look of the skin (Crupi et al. 2023; Rasmus and Kozłowska 2023).

3.4.6 *Flavonoids*

Flavonoids are a class of polyphenolic chemicals that are synthesised by plants as secondary metabolites. Flavonoids are abundant in several fruits, vegetables, and other food crops. They exhibit beneficial biochemical effects on several illnesses such as cardiovascular disease and atherosclerosis, in addition to other bioactivities. Flavonoid chemicals are often present in flowers, leaves, and seeds of plants. Plant flavonoids, which are secondary metabolites, have crucial functions in regulating plant responses to both biological and non-biological environmental stimuli (Banwo et al. 2021; Prabhakar et al. 2023a). They contribute to the pigmentation of fruits and flowers to lure animals, and shield plants from living threats like herbivores, bacteria, fungus, as well as non-living threats like ultraviolet (UV) light exposure. Flavonoids are plant compounds that are classified as secondary metabolites. They have a specific structure called a benzo-γ-pyrone skeleton. Flavonoids are synthesised by plants through many processes, including the phenylpropanoid system, the shikimate pathway, and the flavonoid pathway. Flavonoids are the primary bioactive compounds found in plants. Phenolic compounds, like usual, have the ability to behave as powerful substances that have antioxidant and metal chelating properties. In addition, they have long been acknowledged to contain anti-inflammatory, antiallergic, hepatoprotective, antithrombotic, antiviral, and anticarcinogenic properties (Bautista-Hernández et al. 2022; Prabhakar and Banerjee 2020).

3.5 Sustainable Food Systems

The global and local food systems are being confronted with pressing and intricate difficulties. The combination of population expansion, rising affluence, and urbanisation has resulted in an increase in the worldwide demand for food. The objective of food systems companies should be to ensure food security for all, while minimising the ecological footprint of intensive food production. Ensuring sufficient dietary energy for the world's population is one aspect of food security, while ensuring appropriate nutrients and calories at a reasonable price is another concern (Calicioglu et al. 2019; Abu Hatab et al. 2019). A sustainable food system should further advance livelihoods and food cultures while ensuring fairness, justice, and equity. The global population is steadily growing. By 2050, the global population is projected to reach 9.3 billion. If we continue consuming at our current rate, the worldwide need for animal-based protein would rise to 1250 million tonnes annually. However, new research indicates that over 690 million individuals worldwide, accounting for 8.9 percent of the global population, were expected to be undernourished in 2019. Furthermore, this number is projected to surpass 840 million by the year 2030 (Serraj et al. 2019; Fanzo et al. 2022). Food systems extend beyond individuals' food access and dietary choices. A food system may be defined as a collection of activities and their resulting outcomes. The concept of healthy diets from sustainable food systems is also influenced by the passage of time. Agricultural techniques, commodities trading, food processing, and retail undergo continuous evolution. For instance, agricultural practices are not fixed and undergo changes, which might lead to measurable improvements over time. The dimensions of sustainable food systems, including health/nutrition, economy, society, and the environment, are interconnected within a system (Drewnowski et al. 2020; Ambikapathi et al. 2022). A diet that is rich in vegetables, fruits, whole grains, legumes, nuts, and seeds, and low in animal-based foods is more beneficial for health and has a smaller negative impact on the environment in terms of greenhouse gas emissions, energy, land, and water usage compared to the typical diet. Even when considering the inclusion of energy, land, and water usage in these calculations, it still does not provide a complete evaluation of the overall environmental effect of a certain eating pattern. The study primarily focuses on the environmental consequences of food production and has mainly been carried out in high-income countries. There are still important deficiencies in research, collaboration, and integration, particularly in relation to the social and economic aspects of sustainability (Springmann et al. 2018; Polyak et al. 2023).

3.5.1 Millets

Millets, a group of diminutive-seeded cereals, are grown worldwide, with a particular focus on Asia and Africa. Millets are renowned for their high nutritional value and their capacity to be produced in an environmentally sustainable manner. As a result, they have gained recognition as a viable source for nutraceuticals. Nutraceuticals are products derived from food sources that provide health and medical advantages, such as the prevention and treatment of diseases. Millets are a rich source of carbohydrates, which can be converted into energy. In addition, they possess substantial quantities of proteins and dietary fibres, which are crucial for maintaining digestive well-being (Saini et al. 2021; Porwal et al. 2023). Millets include a plethora of essential vitamins and minerals, including B-vitamins (niacin, B6, folic acid), iron, magnesium, phosphorus, potassium, and zinc. These micronutrients are essential for a variety of metabolic processes and overall well-being. Millets are rich in bioactive substances such as phenolic acids, flavonoids, tannins, and lignans. These plant-based compounds have antioxidant, anti-inflammatory, and antibacterial activities, which contribute to their potential as nutraceuticals. Millets' high fibre content aids in reducing cholesterol levels, thereby decreasing the likelihood of cardiovascular illnesses (Saleh et al. 2013; Ambati and Sucharitha 2019). Moreover, the inclusion of magnesium contributes to the maintenance of cardiovascular well-being through the regulation of blood pressure. Millets include a high amount of dietary fibre, which facilitates proper digestion and helps avoid the occurrence of constipation. The fibre also functions as a prebiotic, promoting the proliferation of advantageous intestinal microorganisms. Millets include a low-fat content and a high fibre content, rendering them a suitable meal choice for the purpose of weight management. They induce satiety and decrease total caloric consumption. Millets are naturally gluten-free, making them suitable for individuals with celiac disease or gluten sensitivity (Jacob et al. 2024).

3.5.2 Marine Algae

Algae possess inherent characteristics that render them an appealing and environmentally friendly source of sustenance. They contain a high amount of easily digestible proteins, lipids, carbohydrates, vitamins, and minerals. Algae include important bioactive compounds such as phycocyanin, allophycocyanin, astaxanthin, lutein, zeaxanthin, canthaxanthin, β-carotene, and oleic acid. These components significantly enhance immunity and contribute to general nutritional well-being. Moreover, the growing popularity of polysaccharides, proteins, and polyphenols derived from algae is attributed to their potential nutritional benefits and effectiveness in treating many degenerative diseases (Babich et al. 2022). Algae-derived nutraceuticals are becoming increasingly important in the field of dietary supplements, mainly due to the concept of algal prebiotics and their ability to impact gut flora. Their cohesive, thickening, and stabilizing characteristics have facilitated the creation of goods like agar, alginate, and carrageenan. Furthermore, algae are utilised in the food business as dietary supplements and as an additive to enhance the functionality of food. Algae are included into meat products, including pasty, steaks, frankfurters, and sausages, as well as fish, fish products, and oils, in order to enhance their quality. Algae-enriched cereal-based products, such as pasta, wheat, and bread, belong to another category of items (Boukid and Castellari 2022; Kumari et al. 2023). Algae can be utilised in the production of fermented functional food due to their unique characteristics. Primarily, dairy products such as cheese, cream, milk desserts, yoghurt, cottage cheese, and processed cheese are examples of fermented items that incorporate algae. The combination of fermented products containing a significant amount of lactic acid bacteria with algae that have biologically active metabolites of natural origin enables the creation of products that are rich in nutrients and establishes a whole new category of fermented food (Ścieszka and Klewicka 2019).

3.5.3 Fungal Source

Fungi are crucial for nutraceutical development and multiple fungal species are highly nutritious and serve as a valuable reservoir of bioactive substances that possess both nutritional and pharmacological effects (Niego et al. 2021). Currently, mushrooms have garnered significant attention from scientists who are actively exploring novel medicinal possibilities. Mushrooms are increasingly being incorporated into our daily diet because of its nutritional worth, low fat level, and high protein content (Sonam et al. 2021a, b). In addition to their nutritional benefits, fungi possess antiviral, antibacterial, anticancer, immune-boosting, and therapeutic properties. Researchers have discovered numerous kinds of fungi that serve as highly abundant sources of nutraceuticals. Certain species, such as Chinese Cordyceps and Ganoderma species, possess traditional therapeutic properties (Ślusarczyk et al. 2021). Mushrooms are a form of fungi that contain bioactive compounds with various functional qualities, such as antioxidants, anticancer chemicals, antibiotics, immunomodulators, hypocholesterolemics, and hypoglycemics. Additionally, this encompasses Cordyceps, a type of mushroom that has medicinal attributes and contains several bioactive compounds. Cordyceps is also utilised in traditional medicine, namely for its cordycepin content. Cordycepin is a promising bioactive compound with the ability to combat diseases such as cancer, cardiovascular disease, and diabetes, which gives it significant importance as a mushroom (Chugh et al. 2022). Fungi such as Ganoderma have demonstrated efficacy in preventing a range of kidney illnesses and are commonly employed in traditional medicine. Recently, Ganoderma has exhibited activity by containing many bioactive chemicals that elicit distinct biological reactions, as observed through mass spectrometry and spectroscopic studies. Studies have identified several nutraceuticals produced from mushrooms that can alleviate neurological illnesses such as mental impairment (dementia) and ataxia. Beta-glucan is a prevalent constituent of nutraceuticals and can be found in several types of fungi. Cell walls of certain plants, such as oats and barley, contain a naturally occurring polysaccharide known as beta-glucan. Beta-glucans also exhibit anticancer effects by inducing apoptosis in cancer cells. The fungicidal capabilities of certain fungi are also beneficial for using fungal substances in many nutraceutical sectors (Bhambri et al. 2022; Ekiz et al. 2023).

3.5.4 Single-Cell Protein

Single-cell protein is a rich source of dietary fibres, minerals, amino acids, vitamins, and other essential nutrients. It gives immediate energy and is a useful supplement for undernourished children. Additionally, it provides nourishment to the skin and eyes and inhibits the buildup of high-density lipids, thereby diminishing obesity. Additionally, it aids in the management of obesity by reducing blood sugar levels.

Single-cell proteins are a valuable feedstock for cattle, fish, and birds due to their high protein content, making them an excellent and acceptable protein supply (Bratosin et al. 2021; Salazar-López et al. 2022). Single cell proteins (SCPs) are obtained by extracting cells from microorganisms that have a high protein content. These cells can be dried and/or the proteins can be purified. SCPs possess very desirable characteristics as a dietary supplement for humans, including a rich protein content with a diverse range of amino acids, minimal fat content, and a higher ratio of protein to carbohydrates compared to forages. SCPs consist of various vital nutrients, such as vitamins (including thiamine, riboflavin, pyridoxine, nicotinic acid, pantothenic acid, folic acid, biotin, cyanocobalamin, ascorbic acid, β-carotene, and α-tocopherol), necessary amino acids (such as lysine and methionine), minerals, nucleic acids, and lipids. Thus far, SCPs have been utilised for a diverse array of applications, including the manufacturing of food (such as scent carriers, vitamin carriers, and emulsifying acids) and feed for various animals (including pigs, poultry, cattle, and fish), as well as in the paper and lead industry (Bajpai 2017; Chama 2019). Single Cell Protein (SCP) is not affected by changes in the environment or climate, and can be generated consistently throughout the year due to the continuous availability of microorganisms. The longevity of protein preparation is contingent upon the characteristics of the protein and the storage conditions employed, and can range from a few days to over a year. Each protein has unique optimal storage conditions (Bratosin et al. 2021; Pereira et al. 2022).

3.5.5 Meat Alternatives

There is a consensus that shifting from meat-heavy diets to diets that are predominantly plant-based is crucial for mitigating the negative environmental effects of the food system and for enhancing human health and animal welfare. Nevertheless, our combined meat consumption continues to grow and is expected to continue increasing in the next 10 years. To address the expected increase in global meat consumption, it is contended that a significant decrease in average meat consumption levels, particularly in affluent societies, is of utmost importance. Meat products have a significant impact on civilization by providing proteins, vital amino acids, and microelements (Van der Weele et al. 2019; Lee et al. 2020). Meat and animal-derived products have a significant influence on the sustainability of diets, which raises the question of whether future diets should priorities minimal environmental implications. The primary issue associated with meat intake is not the act of consuming meat itself, but rather the significant problem of excessive consumption. The consumption of meat, particularly processed red meat, has been strongly linked to the risk of developing cancer and metabolic illnesses. The health hazards connected with consuming meat products are obviously influenced by the kind of meat and processing (red versus white, more processed versus less processed), as well as human lifestyle. There is a consensus that shifting from diets that heavily rely on meat to diets that are predominantly plant-based is crucial for mitigating the negative environmental effects of the food system and for enhancing human health and animal well-being (Smetana et al. 2015; Michel et al. 2021).

Nevertheless, our combined meat consumption continues to grow and is expected to continue increasing in the next decade. Presently, individuals residing in Western nations are ingesting meat at a pace that is five times more than it was two decades ago, surpassing consumption rates in underdeveloped countries by a factor of eight. The excessive rates of consumption and production not only result in obesity, hypertension, and heightened carcinogenic hazards but also contribute to escalated environmental consequences. Consumers are increasingly looking for alternatives to meat that provide the same flavour but without the harmful effects on the environment and health, to improve the overall health and sustainability of the food system (González et al. 2020). Plants continue to be the primary source of biomass utilised as a replacement for meat. Plant-based alternatives use key components such as soybeans, wheat, peas, and lupine. Raw grains must undergo processing in order to enhance the availability of nutrients and be regarded as viable alternatives to meat. Meat is a highly nutritious diet rich in protein; nevertheless, the process of converting feed and fodder into animal protein may not be environmentally sustainable due to the resources required and the limited availability of natural resources (Santo et al. 2020). Presently, there are several meat production farming methods, where the effectiveness of producing a unit of output is primarily influenced by factors like as food, breeds, management, and technology. Crop growth requires a reduced amount of resources per unit, making these goods a promising possibility for sustainable development and addressing the growing food need. Therefore, in industrialized nations that do not depend on animal breeding for basic survival, the adoption of meat substitutes might have positive environmental impacts in terms of preserving biodiversity, minimizing land and water usage, and decreasing greenhouse gas (GHG) emissions. To be able to get approval from non-vegetarian customers, plant-based foods must closely mimic the texture, flavour, look, scent, and taste of genuine animal products. Nevertheless, the extensive array of foreign components employed to imitate the sensory characteristics of meat results in distinct nutritional profiles for these items in comparison to animal foods. Consequently, despite the similarity of PBMAs to meat in terms of sensory perception, they cannot be regarded as a suitable substitute for animal products from a nutritional standpoint. Therefore, more research is required to not only assess the nutritional value of newly introduced plant-based meat products in the market but also to examine the effects of this substitution on human health indicators. Furthermore, it is necessary to implement comprehensive nutritional education initiatives that enhance consumers' understanding and consciousness regarding the distinctions between animal-derived and plant-derived goods (Smetana et al. 2015; Hemathilake and Gunathilake 2022).

3.6 Developing Nutraceuticals from Sustainable Food System: Problem, Challenges and Future Directions

Nutraceuticals offer a promising approach to enhancing health and preventing disease. They can be sourced from a wide array of foods, supplements, and natural products. By incorporating these into a balanced diet and healthy lifestyle, individuals can potentially improve their overall well-being and quality of life. The transition towards plant-based nutraceuticals is a noteworthy phenomenon, motivated by the advantages it offers in terms of health and sustainability. Plants provide a plentiful supply of bioactive substances, such as polyphenols, flavonoids, and carotenoids, that have been associated with numerous health advantages, including anti-inflammatory, antioxidant, and anticancer effects (Ruby et al. 2021). Plant cultivation frequently exhibits a reduced environmental footprint in comparison to animal-derived sources. Functional foods, also known as fortified foods, are experiencing a rise in popularity. These meals are enriched with extra nutrients or bioactive substances. These meals not only supply vital nutrients but also provide distinct health advantages, such as enhancing gastrointestinal health or enhancing the immune system. Functional foods are created by adding nutraceutical elements, such as probiotics, prebiotics, vitamins, and minerals, to common food products. This trend is in line with the concepts of sustainable food systems as it improves the nutritional composition of frequently consumed foods. The progress in biotechnology and genetics is vital in the advancement of nutraceuticals (Alcorta et al. 2021). Methods such as genetic modification and genome editing, such as CRISPR-Cas9, are currently employed to improve the nutritional composition and accessibility of plants. Genetically modified crops can be manipulated to generate increased quantities of nutrients or bioactive chemicals. These advancements not only enhance the effectiveness of nutraceuticals but also promote sustainability by boosting crop productivity and resistance to environmental pressures. Consumers are progressively seeking greater transparency regarding the origin and manufacturing processes of nutraceuticals (Smetana et al. 2023). Sustainable sourcing techniques guarantee the acquisition of materials from sources that do not exhaust natural resources or cause harm to the environment. Blockchain technology enables traceability solutions that empower consumers to authenticate the origins and sustainability credentials of nutraceutical items. This phenomenon is compelling corporations to embrace more environmentally friendly methods and to furnish comprehensive information regarding their supply chains. Sustainability in nutraceutical development necessitates minimizing environmental impact while maintaining efficiency and scalability. This involves selecting raw materials that are environmentally benign and employing production processes that lower carbon footprints. Plant-based ingredients, often more sustainable than animal-derived components, are increasingly favored due to their lower resource requirements and emissions (Paliwal et al. 2020).

Moreover, utilizing by-products from food production, such as fruit peels, seeds, and pulp, not only enhances the sustainability of nutraceuticals but also reduces food waste. Techniques such as green extraction methods, which use environmentally friendly solvents and lower energy inputs, are crucial in this regard. For nutraceuticals to be sustainable, their production must be economically viable. This involves optimizing agricultural practices to increase yield and reduce costs. Innovations such as precision farming, which utilizes technology to optimize field-level management about crop farming, and biotechnology, which enhances crop resilience and nutritional content, are pivotal (Sha et al. 2023; Prabhakar et al. 2023c, d). Nutraceuticals should be accessible to all socioeconomic groups, which involves tackling disparities in availability and affordability. Educational initiatives are necessary to raise awareness about the benefits of nutraceuticals and to promote their use, particularly in underprivileged communities (Anggraeni et al. 2022). Moreover, engaging communities in the development process can lead to products that are culturally acceptable and tailored to local health needs. The regulatory landscape for nutraceuticals is complex and varies significantly across regions. In many countries, nutraceuticals are regulated as dietary supplements rather than pharmaceuticals, which can lead to inconsistencies in quality and efficacy. Ensuring that nutraceutical products meet high safety and efficacy standards while navigating different regulatory frameworks is a significant challenge. Harmonizing regulations globally and establishing clear guidelines for the development and marketing of nutraceuticals are necessary steps forward. Developing new nutraceutical products involves substantial research and development (R&D) costs. This includes the identification and extraction of bioactive compounds, clinical trials to establish health benefits, and the development of sustainable production methods (Dwyer et al. 2018; Prabhakar et al. 2023a, b). Despite the growing interest in nutraceuticals, consumer awareness and acceptance can be barriers to their widespread adoption. Misconceptions about the efficacy and safety of nutraceuticals, as well as a lack of understanding about their benefits, can hinder market growth. Comprehensive consumer education campaigns and transparent communication about the science behind nutraceuticals are essential to build trust and acceptance. Sustainable nutraceutical development requires robust and resilient supply chains. This involves sourcing high-quality raw materials, ensuring fair trade practices, and maintaining the traceability of ingredients. Supply chain disruptions, such as those caused by geopolitical tensions or climate change, can pose significant risks. Developing diversified and resilient supply chains, supported by technology and innovation, is crucial for the sustainable development of nutraceuticals (Siddiqui and Moghadasian 2020; Lordan 2021). Adopting circular economy models in nutraceutical development can significantly enhance sustainability. This involves designing production processes that minimize waste and maximize the reuse and recycling of materials. For example, biowaste from nutraceutical production can be converted into bioenergy or used as animal feed, creating a closed-loop system. Implementing circular economy principles can reduce environmental impact, lower costs, and improve the overall sustainability of nutraceuticals. Personalised nutrition is the future of nutraceuticals, as it customizes dietary advice and nutraceutical interventions based on an individual's genetic, metabolic, and lifestyle profiles. The progress in genomes, metabolomics, and microbiomics is facilitating the creation of personalised nutraceutical products that can target specific health issues and enhance individual health results. This strategy not only improves the effectiveness of nutraceuticals but also supports sustainable practices by encouraging focused and efficient utilisation of resources (Gonçalves and Maximo 2023; Sgroi 2023).

3.7 Conclusion

Nutraceutical development from sustainable food systems represents a multifaceted and dynamic field that holds significant potential for improving human health and promoting environmental sustainability. By integrating environmental, economic, and social considerations into the development process, nutraceuticals can contribute to a healthier and more sustainable future. Sustainable food systems aim to produce these nutraceuticals in a way that conserves resources, reduces environmental impact, and supports economic and social well-being. Developing nutraceuticals within sustainable food systems is a multifaceted endeavor that requires a balance of ecological, economic, and social considerations. By adhering to sustainable practices in sourcing, production, and marketing, the nutraceutical industry can contribute to a healthier planet and healthier people, ensuring long-term benefits for both. This approach not only addresses the immediate need for effective health supplements but also supports the overarching goal of sustainability in agriculture and food production. Nutraceuticals from sustainable food systems represent a promising approach to enhancing health and preventing disease keeping sustainable growth. They bridge the gap between nutrition and medicine, offering a wide range of products that can support overall well-being.

Acknowledgement Authors are thankful to ministry of education, GoI for providing fellowship for the research work.

Conflict of Research Interests Authors declare no conflict of research interests.

References

Abdelrahman M, Jogaiah S, Abdelrahman M, Jogaiah S (2020) Metabolic and functional diversity of saponins. In: Bioactive molecules in plant defense: saponins, Springer Nature, Switzerland, pp 15–35

Abu Hatab A, Cavinato MER, Lagerkvist CJ (2019) Urbanization, livestock systems and food security in developing countries: a systematic review of the literature. Food Secur 11(2):279–299

Abuajah CI, Ogbonna AC, Osuji CM (2015) Functional components and medicinal properties of food: a review. J Food Sci Technol 52:2522–2529

Ahsan N, Imran M, Mohammed Y, Al Anouti F, Khan MI, Banerjee T et al (2023) Mechanistic insight into the role of vitamin D and zinc in modulating immunity against COVID-19: a view from an immunological standpoint. Biol Trace Elem Res 201(12):5546–5560

AlAli M, Alqubaisy M, Aljaafari MN, AlAli AO, Baqais L, Molouki A et al (2021) Nutraceuticals: transformation of conventional foods into health promoters/disease preventers and safety considerations. Molecules 26(9):2540

Alamgir ANM (2018) Vitamins, nutraceuticals, food additives, enzymes, anesthetic aids, and cosmetics. In: Therapeutic use of medicinal plants and their extracts: volume 2: phytochemistry and bioactive compounds, Springer Nature, Switzerland, pp 407–534

Alcorta A, Porta A, Tárrega A, Alvarez MD, Vaquero MP (2021) Foods for plant-based diets: challenges and innovations. Food Secur 10(2):293

Aleksandrowicz L, Green R, Joy EJ, Smith P, Haines A (2016) The impacts of dietary change on greenhouse gas emissions, land use, water use, and health: a systematic review. PLoS One 11(11):e0165797

Allen T, Prosperi P, Cogill B, Padilla M, Peri I (2019) A Delphi approach to develop sustainable food system metrics. Soc Indic Res 141:1307–1339

Ambati K, Sucharitha KV (2019) Millets-review on nutritional profiles and health benefits. Int J Recent Sci Res 10(7):33943–33948

Ambikapathi R, Schneider KR, Davis B, Herrero M, Winters P, Fanzo JC (2022) Global food systems transitions have enabled affordable diets but had less favourable outcomes for nutrition, environmental health, inclusion and equity. Nature Food 3(9):764–779

Anggraeni EW, Handayati Y, Novani S (2022) Improving local food systems through the coordination of agriculture supply chain actors. Sustain For 14(6):3281

Arshad MS, Khalid W, Ahmad RS, Khan MK, Ahmad MH, Safdar S et al (2021) Functional foods and human health: an overview. In: Functional foods phytochem health promoting potential, vol 3

Awuchi CG, Igwe VS, Amagwula IO (2020) Nutritional diseases and nutrient toxicities: a systematic review of the diets and nutrition for prevention and treatment. Int J Adv Acad Stud 6(1):1–46

Ayivi RD, Gyawali R, Krastanov A, Aljaloud SO, Worku M, Tahergorabi R et al (2020) Lactic acid bacteria: food safety and human health applications. Dairy 1(3):202–232

Babich O, Sukhikh S, Larina V, Kalashnikova O, Kashirskikh E, Prosekov A et al (2022) Algae: study of edible and biologically active fractions, their properties and applications. Plan Theory 11(6):780

Bailey RL, Gahche JJ, Miller PE, Thomas PR, Dwyer JT (2013) Why US adults use dietary supplements. JAMA Intern Med 173(5):355–361

Bajpai P (2017) Single cell protein production from lignocellulosic biomass. Springer, Singapore, pp 1–78

Ballini A, Charitos IA, Cantore S, Topi S, Bottalico L, Santacroce L (2023) About functional foods: the probiotics and prebiotics state of art. Antibiotics (Basel) 12(4):635

Banwo K, Olojede AO, Adesulu-Dahunsi AT, Verma DK, Thakur M, Tripathy S et al (2021) Functional importance of bioactive compounds of foods with potential health benefits: A review on recent trends. Food Biosci 43:101320

Barkas F, Bathrellou E, Nomikos T, Panagiotakos D, Liberopoulos E, Kontogianni MD (2023) Plant sterols and plant stanols in cholesterol management and cardiovascular prevention. Nutrients 15(13):2845

Bautista-Hernández I, Aguilar CN, Martínez-Ávila GC, Ilina A, Torres-León C, Verma DK, Chávez-González ML (2022) Phenolic compounds and antioxidant activity of Lippia graveolens Kunth residual leaves fermented by two filamentous fungal strains in solid-state process. Food Bioprod Process 136:24–35

Béné C, Oosterveer P, Lamotte L, Brouwer ID, de Haan S, Prager SD et al (2019) When food systems meet sustainability–current narratives and implications for actions. World Dev 113:116–130

Benković M (2019) Introduction to nutraceuticals and pharmaceuticals. In: Nutraceuticals and natural product pharmaceuticals. Academic, London, pp 1–31

Bhambhani S, Kondhare KR, Giri AP (2021) Diversity in chemical structures and biological properties of plant alkaloids. Molecules 26(11):3374

Bhambri A, Srivastava M, Mahale VG, Mahale S, Karn SK (2022) Mushrooms as potential sources of active metabolites and medicines. Front Microbiol 13:837266

Binns CW, Lee MK, Lee AH (2018) Problems and prospects: public health regulation of dietary supplements. Annu Rev Public Health 39(1):403–420

Bøhn SK, Ward NC, Hodgson JM, Croft KD (2012) Effects of tea and coffee on cardiovascular disease risk. Food Funct 3(6):575–591

Bommakanti V, Puthenparambil Ajikumar A, Sivi CM, Prakash G, Mundanat AS, Ahmad F et al (2023) An overview of herbal nutraceuticals, their extraction, formulation, therapeutic effects and potential toxicity. Separations 10(3):177

Boukid F, Castellari M (2022) Algae as nutritional and functional food sources. Food Secur 12(1):122

Bratosin BC, Darjan S, Vodnar DC (2021) Single cell protein: a potential substitute in human and animal nutrition. Sustain For 13(16):9284

Bronzato S, Durante A (2018) Dietary supplements and cardiovascular diseases. Int J Prev Med 9(1):80

Brown KA, Venkateshmurthy NS, Law C, Harris F, Kadiyala S, Shankar B et al (2021) Moving towards sustainable food systems: a review of Indian food policy budgets. Glob Food Sec 28:100462

Calicioglu O, Flammini A, Bracco S, Bellù L, Sims R (2019) The future challenges of food and agriculture: an integrated analysis of trends and solutions. Sustain For 11(1):222

Cassani L, Gomez-Zavaglia A (2022) Sustainable food systems in fruits and vegetables food supply chains. Front Nutr 9:829061

Cencic A, Chingwaru W (2010) The role of functional foods, nutraceuticals, and food supplements in intestinal health. Nutrients 2(6):611–625

Chama NT (2019) Production of single-cell protein from different substrates. Aust J Sci Technol 3:2208–6404

Chugh RM, Mittal P, Mp N, Arora T, Bhattacharya T, Chopra H et al (2022) Fungal mushrooms: a natural compound with therapeutic applications. Front Pharmacol 13:925387

Crupi P, Faienza MF, Naeem MY, Corbo F, Clodoveo ML, Muraglia M (2023) Overview of the potential beneficial effects of carotenoids on consumer health and well-being. Antioxidants (Basel) 12(5):1069

Cruzat V, Macedo Rogero M, Noel Keane K, Curi R, Newsholme P (2018) Glutamine: metabolism and immune function, supplementation and clinical translation. Nutrients 10(11):1564

Cunningham M, Azcarate-Peril MA, Barnard A, Benoit V, Grimaldi R, Guyonnet D et al (2021) Shaping the future of probiotics and prebiotics. Trends Microbiol 29(8):667–685

Da Costa JP (2017) A current look at nutraceuticals–key concepts and future prospects. Trends Food Sci Technol 62:68–78

Damián MR, Cortes-Perez NG, Quintana ET, Ortiz-Moreno A, Garfias Noguez C, Cruceño-Casarrubias CE et al (2022) Functional foods, nutraceuticals and probiotics: a focus on human health. Microorganisms 10(5):1065

Damodaran S, Parkin KL (2017) Amino acids, peptides, and proteins. In: Fennema's food chemistry. CRC Press, Boca Raton, pp 235–356

Dietzen DJ (2018) Amino acids, peptides, and proteins. In: Principles and applications of molecular diagnostics. Elsevier, Amsterdam, pp 345–380

Dini I (2018) Spices and herbs as therapeutic foods. In: Food quality: balancing health and disease. Academic, London, pp 433–469

Dominguez LJ, Veronese N, Baiamonte E, Guarrera M, Parisi A, Ruffolo C et al (2022) Healthy aging and dietary patterns. Nutrients 14(4):889

Drewnowski A, Finley J, Hess JM, Ingram J, Miller G, Peters C (2020) Toward healthy diets from sustainable food systems. Curr Dev Nutr 4(6):nzaa083

Dudeja P, Gupta RK (2017) Nutraceuticals. In: Food safety in the 21st century. Academic, London, pp 491–496

Dwyer JT, Coates PM, Smith MJ (2018) Dietary supplements: regulatory challenges and research resources. Nutrients 10(1):41

Ekiz E, Oz E, Abd El-Aty AM, Proestos C, Brennan C, Zeng M et al (2023) Exploring the potential medicinal benefits of Ganoderma lucidum: from metabolic disorders to coronavirus infections. Food Secur 12(7):1512

Elkhalifa AEO, Alshammari E, Adnan M, Alcantara JC, Awadelkareem AM, Eltoum NE, Mehmood K, Panda BP, Ashraf SA (2021) Okra (Abelmoschus esculentus) as a potential dietary medicine with nutraceutical importance for sustainable health applications. Molecules 26(3):696

Fanzo J, Rudie C, Sigman I, Grinspoon S, Benton TG, Brown ME et al (2022) Sustainable food systems and nutrition in the 21st century: a report from the 22nd annual Harvard Nutrition Obesity Symposium. Am J Clin Nutr 115(1):18–33

Garnett T (2014) Three perspectives on sustainable food security: efficiency, demand restraint, food system transformation. What role for life cycle assessment? J Clean Prod 73:10–18

Garnett T, Godfray C (2012) Sustainable intensification in agriculture. Navigating a course through competing food system priorities. Food climate research network and the Oxford Martin programme on the future of food, University of Oxford, UK, 51

Giampieri F, Godos J, Caruso G, Owczarek M, Jurek J, Castellano S et al (2022) Dietary phytoestrogen intake and cognitive status in southern Italian older adults. Biomol Ther 2022(12):760

Gilsenan MB (2011) Nutrition & health claims in the European Union: a regulatory overview. Trends Food Sci Technol 22(10):536–542

Goel G, Kumar A (eds) (2021) Advances in probiotics for sustainable food and medicine, vol 21. Springer, Singapore

Gonçalves MLM, Maximo GJ (2023) Circular economy in the food chain: production, processing and waste management. Circ Econ Sustain 3(3):1405–1423

González N, Marquès M, Nadal M, Domingo JL (2020) Meat consumption: which are the current global risks? A review of recent (2010–2020) evidences. Food Res Int 137:109341

Guang-Wen Z, Murshed M, Siddik AB, Alam MS, Balsalobre-Lorente D, Mahmood H (2023) Achieving the objectives of the 2030 sustainable development goals agenda: causalities between economic growth, environmental sustainability, financial development, and renewable energy consumption. Sustain Dev 31(2):680–697

Gupta RC, Lall R, Srivastava A (eds) (2021) Nutraceuticals: efficacy, safety and toxicity. Academic, London

Gupta V, Jamwal G, Rai GK, Gupta SK, Shukla RM, Dadrwal BK et al (2024) Biosynthesis of biomolecules from saffron as an industrial crop and their regulation, with emphasis on the chemistry, extraction methods, identification techniques, and potential applications in human health and food: a critical comprehensive review. Biocatal Agric Biotechnol:103260

Haque A, Ahmad S, Khan T, Alomrani SO, Adnan M, Kieliszek M, Ashraf SA (2024) Development of buffalo meat patties using pea peel powder as a value added ingredient: physicochemical, shelf stability, microstructural, sensory attributes analysis. J Food Meas Charact:1–20. https://doi.org/10.1007/s12011-023-03620-4

Heinrich M, Mah J, Amirkia V (2021) Alkaloids used as medicines: structural phytochemistry meets biodiversity—an update and forward look. Molecules 26(7):1836

Hemathilake DMKS, Gunathilake DMCC (2022) Agricultural productivity and food supply to meet increased demands. In: Future foods. Academic, London, pp 539–553

Herrero M, Thornton PK, Mason-D'Croz D, Palmer J, Benton TG, Bodirsky BL et al (2020) Innovation can accelerate the transition towards a sustainable food system. Nature Food 1(5):266–272

Holmes JL, Biella A, Morck T, Rostorfer J, Schneeman B (2021) Medical foods: science, regulation, and practical aspects. Summary of a workshop. Curr Dev Nutr 5:nzaa172

Honda M (2020) Nutraceutical and pharmaceutical applications of carotenoids. In: Pigments from microalgae handbook, pp 449–469, Springer Nature, Switzerland

Hussain G, Rasul A, Anwar H, Aziz N, Razzaq A, Wei W et al (2018) Role of plant derived alkaloids and their mechanism in neurodegenerative disorders. Int J Biol Sci 14(3):341

Jacob J, Krishnan V, Antony C, Bhavyasri M, Aruna C, Mishra K et al (2024) The nutrition and therapeutic potential of millets: an updated narrative review. Front Nutr 11:1346869

Jahan K, Qadri OS, Younis K (2020) Dietary fiber as a functional food. In: Functional food products and sustainable health, Springer Nature, Cham, Switzerland, pp 155–167
Jain S, Purohit A, Nema P, Vishwakarma H, Jain PK (2022) A brief review on nutraceuticals and its application. Asian J Dental Health Sci 2(1):7–13
Kadiri O, Olawoye B, Fawale OS, Adalumo OA (2016) Nutraceutical and antioxidant properties of the seeds, leaves and fruits of Carica papaya: potential relevance to humans diet, the food industry and the pharmaceutical industry-a review. Turkish J Agri-Food Tech 4(12):1039–1052
Kapoor N, Jamwal VL, Shukla MR, Gandhi SG (2020) The rise of nutraceuticals: overview and future. In: Biotechnology business-concept to delivery, Springer, Cham, Switzerland, pp 67–92
Kaur K, Asthir B, Verma DK (2021) Biosynthesis, bioavailability, and metabolism of plant polyphenols: biological activities and their potential benefits in human health. In: Phytochemicals in food and health. Apple Academic Press, Palm Bay, pp 231–255
Kechagia M, Basoulis D, Konstantopoulou S, Dimitriadi D, Gyftopoulou K, Skarmoutsou N, Fakiri EM (2013) Health benefits of probiotics: a review. ISRN Nutr 2013(1):481651
Khalid W, Ali A, Arshad MS, Afzal F, Akram R, Siddeeg A et al (2022) Nutrients and bioactive compounds of Sorghum bicolor L. used to prepare functional foods: a review on the efficacy against different chronic disorders. Int J Food Prop 25(1):1045–1062
Khan I, Hussain M, Jiang B, Zheng L, Pan Y, Hu J et al (2023) Omega-3 long-chain polyunsaturated fatty acids: metabolism and health implications. Prog Lipid Res 101255
Kumari A, Garima, Bharadvaja N (2023) A comprehensive review on algal nutraceuticals as prospective therapeutic agent for different diseases. 3 Biotech 13(2):44
Kušar A, Žmitek K, Lähteenmäki L, Raats MM, Pravst I (2021) Comparison of requirements for using health claims on foods in the European Union, the USA, Canada, and Australia/New Zealand. Compr Rev Food Sci Food Saf 20(2):1307–1332
Lee HJ, Yong HI, Kim M, Choi YS, Jo C (2020) Status of meat alternatives and their potential role in the future meat market—a review. Asian Australas J Anim Sci 33(10):1533
Lentjes MA (2019) The balance between food and dietary supplements in the general population. Proc Nutr Soc 78(1):97–109
Lichota A, Gwozdzinski L, Gwozdzinski K (2019) Therapeutic potential of natural compounds in inflammation and chronic venous insufficiency. Eur J Med Chem 176:68–91
Lordan R (2021) Dietary supplements and nutraceuticals market growth during the coronavirus pandemic–implications for consumers and regulatory oversight. PharmaNutrition 18:100282
Lund S, Hall R, Williams GJ (2019) An artificial pathway for isoprenoid biosynthesis decoupled from native hemiterpene metabolism. ACS Synth Biol 8(2):232–238
Malhi GS, Kaur M, Kaushik P (2021) Impact of climate change on agriculture and its mitigation strategies: a review. Sustain For 13(3):1318
Manzoor S, Wani SM, Mir SA, Rizwan D (2022) Role of probiotics and prebiotics in mitigation of different diseases. Nutrition 96:111602
Maoka T (2020) Carotenoids as natural functional pigments. J Nat Med 74(1):1–16
Maurya A, Mohan S, Verma SC (2021) Antidiabetic potential of naturally occurring sesquiterpenes: a review. Curr Top Med Chem 21(10):851–862
Meléndez-Martínez AJ, Böhm V, Borge GIA, Cano MP, Fikselová M, Gruskiene R et al (2021) Carotenoids: considerations for their use in functional foods, nutraceuticals, nutricosmetics, supplements, botanicals, and novel foods in the context of sustainability, circular economy, and climate change. Annu Rev Food Sci Technol 12(1):433–460
Mensah J (2019) Sustainable development: meaning, history, principles, pillars, and implications for human action: literature review. Cogent Soc Sci 5(1):1653531
Michel F, Hartmann C, Siegrist M (2021) Consumers' associations, perceptions and acceptance of meat and plant-based meat alternatives. Food Qual Prefer 87:104063
Miladinov G (2020) Socioeconomic development and life expectancy relationship: evidence from the EU accession candidate countries. Genus 76(1):2
Moses T, Papadopoulou KK, Osbourn A (2014) Metabolic and functional diversity of saponins, biosynthetic intermediates and semi-synthetic derivatives. Crit Rev Biochem Mol Biol 49(6):439–462

Myrtsi ED, Koulocheri SD, Haroutounian SA (2023) A novel method for the efficient simultaneous quantification of 67 phytoestrogens in plants and foodstuffs. Food Biosci 56:103357
Nasri H, Baradaran A, Shirzad H, Rafieian-Kopaei M (2014) New concepts in nutraceuticals as alternative for pharmaceuticals. Int J Prev Med 5(12):1487
Newsholme P, Diniz VLS, Dodd GT, Cruzat V (2023) Glutamine metabolism and optimal immune and CNS function. Proc Nutr Soc 82(1):22–31
Niego AG, Rapior S, Thongklang N, Raspé O, Jaidee W, Lumyong S, Hyde KD (2021) Macrofungi as a nutraceutical source: promising bioactive compounds and market value. J Fungi (Basel) 7(5):397
Nori LP, Sowmya P, Madhusree T, Raju KV (2023) Nutraceuticals: regulatory landscape and market dynamics in the Indian scenario. J Young Pharm 15(4):595–603
Oakenfull D, Sidhu GS (2023) Saponins. Toxicants of plant origin, 97–142
Olson R, Gavin-Smith B, Ferraboschi C, Kraemer K (2021) Food fortification: the advantages, disadvantages and lessons from sight and life programs. Nutrients 13(4):1118
Paliwal V, Chandra S, Sharma S (2020) Blockchain technology for sustainable supply chain management: a systematic literature review and a classification framework. Sustain For 12(18):7638
Panigrahy SK, Bhatt R, Kumar A (2021) Targeting type II diabetes with plant terpenes: the new and promising antidiabetic therapeutics. Biologia 76(1):241–254
Pederiva C, Biasucci G, Banderali G, Capra ME (2024) Plant sterols and stanols for Pediatric patients with increased cardiovascular risk. Children 11(1):129
Pereira AG, Fraga-Corral M, Garcia-Oliveira P, Otero P, Soria-Lopez A, Cassani L et al (2022) Single-cell proteins obtained by circular economy intended as a feed ingredient in aquaculture. Food Secur 11(18):2831
Plaza-Diaz J, Ruiz-Ojeda FJ, Gil-Campos M, Gil A (2019) Mechanisms of action of probiotics. Adv Nutr 10:S49–S66
Poljsak B, Kovač V, Milisav I (2021) Antioxidants, food processing and health. Antioxidants 10(3):433
Polyak E, Breitenbach Z, Frank E, Mate O, Figler M, Zsalig D et al (2023) Food and sustainability: is it a matter of choice? Sustain For 15(9):7191
Porwal A, Bhagwat G, Sawarkar J, Kamble P, Rode M (2023) An overview of millets-the nutri-cereals: its nutritional profile, potential health benefits and sustainable cultivation approach. Int J Sci Res Arch 10(1):841–859
Prabhakar P, Banerjee M (2020) Antidiabetic phytochemicals: a comprehensive review on opportunities and challenges in targeted therapy for herbal drug development. Int J Pharm Res 12(2):1673–1696
Prabhakar P, Mamoni B (2021) Technical problems, regulatory and market challenges in bringing herbal drug into mainstream of modern medicinal practices. Res J Biotechnol 16(3), 212–222
Prabhakar P, Mukherjee S, Kumar A, Kumar S, Verma DK, Dhara S et al (2023a) Optimization of MAE for Carica papaya phytochemicals, and its in silico, in vitro, and ex vivo evaluation: for functional food and drug applications. Food Biosci 54:102861, 1–24
Prabhakar S, Nimesh A, Prabhakar P (2023b) Startup ecosystem for rural development in India. In: Socio-economic disparities, vulnerable communities, and the future of work and entrepreneurship. IGI Global, Hershey, pp 26–50
Prabhakar P, Mukherjee S, Kumar A, Kumar S, Verma DK, Dhara S et al (2023c) Optimization of microwave-assisted extraction (MAE) of key phenolic compounds from pigeon pea (Cajanus cajan L.), their characterization, and measurement of their anti-diabetic and cytotoxic potential. J Food Meas Charact 17(6):5697–5720
Prabhakar S, Nimesh A, Prabhakar P (2023d) Assessment of start-up ecosystems for tribal women entrepreneurs in India. In: Diversity, equity, and inclusion efforts of businesses in rural areas. IGI Global, Hershey, Pennsylvania, USA, pp1–22
Prabhakar P, Verma DK, Banerjee M (2024) Advances in enzyme-assisted extraction technique for antidiabetic phytochemicals: an emerging approach for green and economical technologies. In: Enzymatic processes for food valorization. Academic Press, London, UK, pp-191–216
Preedy VR, Srirajaskanthan R, Patel VB (2013) Handbook of food fortification and health. From concepts to public health applications. Springer, Dordrecht

Rahaman MM, Hossain R, Herrera-Bravo J, Islam MT, Atolani O, Adeyemi OS et al (2023) Natural antioxidants from some fruits, seeds, foods, natural products, and associated health benefits: an update. Food Sci Nutr 11(4):1657–1670

Rasmus P, Kozłowska E (2023) Antioxidant and anti-inflammatory effects of carotenoids in mood disorders: an overview. Antioxidants (Basel) 12(3):676

Rojas R, Cabanillas B, Portales R, Candy R (2022) Bioactive compounds in native cocoa (Theobroma cacao L.). In: Native crops in Latin America. CRC Press, Boca Raton, pp 91–114

Ruby S, Prakash S, Kumar VP, Kumar TP, Prathab S (2021) A comprehensive review on nutraceuticals. Int J Pharm Sci Rev Res 68:136–148

Rysz J, Franczyk B, Olszewski R, Banach M, Gluba-Brzozka A (2017) The use of plant sterols and stanols as lipid-lowering agents in cardiovascular disease. Curr Pharm Des 23(17):2488–2495

Sachdeva V, Roy A, Bharadvaja N (2020) Current prospects of nutraceuticals: a review. Curr Pharm Biotechnol 21(10):884–896

Saini P, Islam M, Das R, Shekhar S, Sinha A, Prasad K (2023) Wheat bran as potential source of dietary fiber: prospects and challenges. J Food Compos Anal 116:105030

Salazar-López NJ, Barco-Mendoza GA, Zuñiga-Martínez BS, Domínguez-Avila JA, Robles-Sánchez RM, Ochoa MAV, González-Aguilar GA (2022) Single-cell protein production as a strategy to reincorporate food waste and agro by-products back into the processing chain. Bioengineering 9(11):623

Saleh AS, Zhang Q, Chen J, Shen Q (2013) Millet grains: nutritional quality, processing, and potential health benefits. Compr Rev Food Sci Food Saf 12(3):281–295

Santini A, Novellino E (2017) Nutraceuticals in hypercholesterolaemia: an overview. Br J Pharmacol 174(11):1450–1463

Santo RE, Kim BF, Goldman SE, Dutkiewicz J, Biehl EM, Bloem MW et al (2020) Considering plant-based meat substitutes and cell-based meats: a public health and food systems perspective. Front Sustain Food Syst 4:134

Ścieszka S, Klewicka E (2019) Algae in food: a general review. Crit Rev Food Sci Nutr 59(21):3538–3547

Serraj R, Krishnan L, Pingali P (2019) Agriculture and food systems to 2050: a synthesis. In: Agriculture & food systems to 2050, NJ, USA, pp 3–45

Sgroi F (2023) Circular economy and sustainable Agri-food systems. J Agric Food Res 14:100815

Sha SP, Modak D, Sarkar S, Roy SK, Sah SP, Ghatani K, Bhattacharjee S (2023) Fruit waste: a current perspective for the sustainable production of pharmacological, nutraceutical, and bioactive resources. Front Microbiol 14:1260071

Shahidi F, Ambigaipalan P (2015) Phenolics and polyphenolics in foods, beverages and spices: antioxidant activity and health effects–a review. J Funct Foods 18:820–897

Siddiqui RA, Moghadasian MH (2020) Nutraceuticals and nutrition supplements: challenges and opportunities. Nutrients 12(6):1593

Singh RB, Watanabe S, Li D, Nakamura T, Juneja LR, Takahashi T et al (2019) Diet and lifestyle guidelines and desirable levels of risk factors and protective factors for prevention of dementia-a scientific statement from Joint Symposium Of Jaas And Apcns. Biomed J Sci Tech Res 17(3):12844–12864

Ślusarczyk J, Adamska E, Czerwik-Marcinkowska J (2021) Fungi and algae as sources of medicinal and other biologically active compounds: a review. Nutrients 13(9):3178

Smetana S, Mathys A, Knoch A, Heinz V (2015) Meat alternatives: life cycle assessment of most known meat substitutes. Int J Life Cycle Assess 20:1254–1267

Smetana S, Ristic D, Pleissner D, Tuomisto HL, Parniakov O, Heinz V (2023) Meat substitutes: resource demands and environmental footprints. Resour Conserv Recycl 190:106831

Sonam, Shishir K, Prabhakar P (2021a) Constraint analysis in mushroom cultivation among women mushroom cultivators in the State of Bihar, India. Adv Manag 14(2):7–11

Sonam P, Kala S, Prabhakar P, Banerjee M (2021b) Post-training knowledge and attitude assessment of the women entrepreneurs towards mushroom cultivation. Int J Agric Environ Biotechnol 14(2):229–234

Springmann M, Wiebe K, Mason-D'Croz D, Sulser TB, Rayner M, Scarborough P (2018) Health and nutritional aspects of sustainable diet strategies and their association with environmental impacts: a global modelling analysis with country-level detail. Lancet Planet Health 2(10):e451–e461

Swathi Krishna S, Kuriakose BB, Lakshmi PK (2022) Effects of phytoestrogens on reproductive organ health. Arch Pharm Res 45(12):849–864

Taroncher M, Vila-Donat P, Tolosa J, Ruiz MJ, Rodriguez-Carrasco Y (2021) Biological activity and toxicity of plant nutraceuticals: an overview. Curr Opin Food Sci 42:113–118

Tetali SD (2019) Terpenes and isoprenoids: a wealth of compounds for global use. Planta 249:1–8

Thomsen C (2013) Sustainability (world commission on environment and development definition). In: Encyclopedia of corporate social responsibility, Springer, Cham, Switzerland, pp 2358–2363

UNICEF (2020) The state of food security and nutrition in the world 2020

Van der Weele C, Feindt P, van der Goot AJ, van Mierlo B, van Boekel M (2019) Meat alternatives: an integrative comparison. Trends Food Sci Technol 88:505–512

Velten S, Leventon J, Jager N, Newig J (2015) What is sustainable agriculture? A systematic review. Sustain For 7(6):7833–7865

Verma DK, Thakur M (eds) (2021) Phytochemicals in food and health: perspectives for research and technological development. CRC Press, Burlington

Vilahur G, Ben-Aicha S, Diaz-Riera E, Badimon L, Padró T (2019) Phytosterols and inflammation. Curr Med Chem 26(37):6724–6734

Wallace RB, Oria M, National Research Council (2010) The US Food and Drug Administration and imported food safety. In: Enhancing food safety: the role of the Food and Drug Administration. National Academies Press (US), Washington, DC

Wezel A, Herren BG, Kerr RB, Barrios E, Gonçalves ALR, Sinclair F (2020) Agroecological principles and elements and their implications for transitioning to sustainable food systems. A review. Agron Sustain Dev 40:1–13

Wichienchot S, Ishak WRW (2017) Sustainability of nutraceuticals and functional foods. In: Sustainability challenges in the Agrofood sector, pp 369–389

Williamson EM, Liu X, Izzo AA (2020) Trends in use, pharmacology, and clinical applications of emerging herbal nutraceuticals. Br J Pharmacol 177(6):1227–1240

World Health Organization (2023) The State of Food Security and Nutrition in the World 2023: urbanization, agrifood systems transformation and healthy diets across the rural–urban continuum, vol 2023. FAO, Rome

Wu L, Adebayo TS, Yue XG, Umut A (2023) The role of renewable energy consumption and financial development in environmental sustainability: implications for the Nordic countries. Int J Sustain Dev World Ecol 30(1):21–36

Yegin S, Kopec A, Kitts DD, Zawistowski J (2020) Dietary fiber: a functional food ingredient with physiological benefits. In: Dietary sugar, salt and fat in human health. Academic, London, pp 531–555

Zagoskina NV, Zubova MY, Nechaeva TL, Kazantseva VV, Goncharuk EA, Katanskaya VM et al (2023) Polyphenols in plants: structure, biosynthesis, abiotic stress regulation, and practical applications. Int J Mol Sci 24(18):13874

Zeb A (2020) Concept, mechanism, and applications of phenolic antioxidants in foods. J Food Biochem 44(9):e13394

Zoumpopoulou G, Kazou M, Alexandraki V, Angelopoulou A, Papadimitriou K, Pot B, Tsakalidou E (2018) Probiotics and prebiotics: an overview on recent trends. In: Probiotics and prebiotics in animal health and food safety, pp 1–34, Springer Cham, Switzerland

Chapter 4
Advances and Trends in Plant-Based Food Bioactives

Narimane Lammari, Marwa Boumaza, Mehdi Louaer, Hadjira Rabti, Ouahida Louaer, Abdelhamid Elaissari, and Abdeslam Hassen Meniai

Abstract Recently, a tremendous concern for the use of herbal medicine as a form of complementary or replacement therapy has been clearly recognized. In addition to uses for medicinal purposes, the application of plant-based bioactive product like plant oils and extracts is extending to food industry where they are replacing synthetic chemical products. However, the consideration of these products is often facing the problem of the complex chemical nature, the high chemical instability and the short residence time on the site of application. The application of nanotechnology for bioactive compounds encapsulation resolves such restrictions. According to the literature, several nanocarriers have been designed, among which polymeric nanoparticles, liposomes, solid-lipid nanoparticles and nanostructured lipid carriers received much attention. Therfore, this chapter aimed to provide an updated investigation of novelty detection research papers regarding the application of these nanocarriers for bioactive compounds encapsulation in the food industry. A priori the beneficial effects and challenges associated with the rational and appropriate use of plant-based products are described. Then, the current research findings reported in the last decade regarding the different techniques of preparation for nanocarriers are considered. Besides, the application of these colloidal systems in the food industry was highlighted.

Keywords Bioactive compound · Nanotechnology · Polymeric nanoparticles · Liposomes · Solid-lipid nanoparticles · Nanostructured-lipid carriers

N. Lammari (✉) · M. Boumaza · M. Louaer · H. Rabti · O. Louaer · A. H. Meniai
Environmental Process Engineering Laboratory, University Constantine 3, Salah Boubnider, Constantine, Algeria
e-mail: narimane.lammari@univ-constantine3.dz; mehdi.louaer@univ-constantine3.dz; hadjira.rabti@univ-constantine3.dz; ouahida.louaer@univ-constantine3.dz; abdeslam.meniai@univ-constantine3.dz

A. Elaissari
University Claude Bernard Lyon-1, CNRS, ISA-UMR 5280, Villeurbanne, France
e-mail: abdelhamid.elaissari@univlyon1.fr

S. Amir Ashraf, M. Adnan (eds.), *Food Bioactives and Nutraceuticals*,
https://doi.org/10.1007/978-981-97-9723-3_4

4.1 Introduction

In recent years, lifestyle changes have led to a prominent awareness among consumers of the close link between food consuming habits and possible dangerous induced consequences like diabetes, cancer and cardiovascular diseases (Moscatelli et al. 2023). For this reason, people are seeking out foods that offer health benefits beyond their fundamental nutritional value. Plant sourced foods are consumed worldwide and represent inherent substitutes for animal-based meat whose consumption is always linked to several metabolic disorders (Munialo and Vriesekoop 2023). Recently, researchers have identified bioactive compounds as specific components present in small quantities in foods which are recognized as having health-promoting effects beyond their basic nutritional value (Teodoro 2019). Bioactive compounds may be organic or inorganic molecules that appear to have valuable physiological, behavioral, and immunological properties (Hamzalıoğlu and Gökmen 2016). Many researchers have investigated their health benefits like antioxidant (Dhalaria et al. 2020), antiinflammatory (Zhao et al. 2023), anticarcinogenic (Chandrasekara and Shahidi 2018), and antimicrobial effects (Mickymaray 2019). To date, several bioactive compounds have been identified (Rashmi and Negi 2020). They change extensively in chemical structure and role and are categorized accordingly. Some examples of bioactive compounds are polyphenols, flavonoids, carotenoids, choline, carnitine, phytosterols, glucosinolates, phytoestrogens, vitamins and minerals (Albuquerque et al. 2020). Historically conventional extraction methods, such as maceration, decoction, digestion and soxhlet, have been applied for the extraction of these valuable resources (Azmir et al. 2013). Curent and emerging extraction approaches, including supercritical fluid-, green solvent-, enzyme-assisted- and ultrasonic- extractions have been developped to deal with challenges like the substantial amounts of solvents, the energy-intensive processes, the prolonged extraction time and the production of inherent hazardous waste, usually encountered when applying conventional extraction methods (Azmir et al. 2013; Wen et al. 2020). The performance and purpose of bioactive compounds are generally influenced by their instability to environmental conditions such as temperature, oxygen, light, pH, low gastrointestinal absorption and interaction with other food ingredients(Pai et al. 2022).

Overcoming these limitations is a critical issue for the widespread application of these natural ressources. Nanoencapsulation of bioactives products within naturally assembling structures and polymers has been in the interesting trends (Pateiro et al. 2021). Indeed, nanoncapsulation aims to enhance quality and safety of the valuable food constituents. It consists on incorporating bioactive compounds inside the nanostructure for protection as well as to promote their functionality (Rashid et al. 2022). Several nanocarriers such as polymeric nanoparticles, nanoliposomes, solid lipid nanoparticles (SLN) and nanostructered lipid carriers (NLC) have shown an ability to efficiently encapsulate the bioactive compounds (Lammari et al. 2021; Paolino et al. 2021).

In the present chapter an overview of the plant-based food bioactives compounds has been provided along with their potential health-promoting properties. The nanoencapsulation of bioactive products in several types of nanocarriers has also been addressed besides discussing their applications in food industry.

4.2 Health Beneficial Contibution of Plant-Based Bioactives

Medicinal plants have been used for healing since the dawn of time without need for systematic standards and analysis due to their efficacy and safety (Akhtar et al. 2023). Therefore, plant derived- bioactive molecules are more likely to succeed in toxicological analysis comparatively to the synthetic molecules (Mickymaray 2019). This explain the cumulative attention on the consumption of bioactive compounds in functional foods that are beneficial for human health (Jha et al. 2023). Bioactive compounds have the ability to alter more than one metabolic pathway or process and interact with one or more components of living tissue thereby providing a wide range of potential properties (Banwo et al. 2021). Most of these valuable sources are considered as secondary metabolites, i.e. they are not required to the growth and development of plants (e.g. protein synthesis, energy production, and others) but are necessary for the plant to survive in its environment by providing protection against UV radiation (flavonols), defense against microbes and herbivores (glucosinolates, polyphenols), or attraction for pollination (anthocyanins), etc. (Renard 2018). Primary metabolites such as carbohydrates, fatty acids and protein are also considered as bioactive compounds found in various plants and have been approved to affect health positively (Dixit et al. 2023). Various families of bioactive components have been identified and approved for their improving health benefits and promoting well-being health and are presented in Table 4.1. With the advances in science and analytical methods, more bioactive components are being continuously discovered and identified (Dixit et al. 2023).

Phenolic acids and flavonoids, amongst others, are the main classes of polyphenols, mainly found in vegetables, fruits, coffee, tea and cereals, among others. It has been proposed that phenolic acids show equivalent or even higher antioxidant potential than traditional antioxidants such as tocopherols and ascorbic acid in inhibiting oxidation and discoloration in meat products, oils, and lipid dispersions (Zhou and Elias 2013). Other biological potentials are attributed to phenolic acids, such as anticancer (Mumtaz et al. 2021), antidiabetic (Savych et al. 2021), as well as contributing to the management of cardiovascular diseases (El-Hussainy et al. 2016), inflammation (Paciello et al. 2020) and neurodegenerative disorders (Chang et al. 2019). Flavonoids are the most widespread and the largest plant polyphenolics class derived from plant- based food. They encompass six major subclasses which include (i) anthocyanidins (e.g. malvidin), (ii) flavanones (e.g. naringenin), (iii) flavones (apigenin and luteolin), (iv) isoflavones (e.g. glycitein), (v) flavan-3-ols (e.g. epicatechin), and (vi) flavonols (e.g. kaempferol). Favonoids have been reported to exhibit several biological benefits, such as free-radical scavenging ability (Zain

Table 4.1 List of bioactive compounds, sources and some important biological activities

Bioactive compound	Plant source	Biological activities	Mechanism	References
Phenolic acids				
Gallic acid	Tea leaves, grapes, berries and fruits	Antifungal	Antifungal activity comparable in potency to fluconazole.	Li et al. (2017)
Caffeic acid Chlorogenic acid	Kiwi, blueberry, cherry, apple, artichoke, potato and coffee	Anticancer	Chlorogenic and caffeic acids could inhibit the reduction of *Lachnospiraceae incertae sedis* and *Prevotella 9* and the increase of *Rumincoccaceae UCG-004* in hepatocellular carcinoma rats.	Zhang et al. (2017)
Flavonoid				
Quercetin		Antiviral	Quercetin inhibits influenza infection with a broad spectrum of strains with half maximal inhibitory concentration of 2.738 ± 1.931, 6.225 ± 0.467 and 7.756 ± 1.097 µg/ml, respectively.	Wu et al. (2016)
Pectolinarigenin Tangeretin	Hydrolyzed aqueous methanol extracts of *Kickxia aegyptiaca* L., *Citrus reticulata* Blanco, and *Anastatica hierochuntica* L.	Antiviral	Pectolinarigenin and angeretin showed the highest SARS-CoV-2 inhibition.	Al-Karmalawy et al. (2021)
Apigenin Luteolin Flavonoid C-glycoside	Oil palm (*Elaeis guineensis* Jacq.) leaves	Antioxidant	Significant correlation appears between the flavonoid compounds and antioxidant activity based on the Pearson correlation analysis.	Zain et al. (2020)
Carotenoids				
Lutein	–	Anti-inflammatory	Pre-treatment with lutein dose-dependently down-regulated IL-6, IL-1β and TNF mRNA expression, and decreased LPS-induced secretion of IL-6, IL-1β and TNF.	Rosanna et al. (2017)

Lycopene	Ethanol extract of red guava (*Psidium guajava* L.)	Anti-cancer	Lycopene induced cytostatic and cytotoxic effects on breast cancer cells by causing an apoptotic-like pathway.	Dos Santos et al. (2018)
Anthocyanins				
Cyanidin-3-glucoside, Cyanidin-3-rutinoside, Delphinidin-3-rutinoside Pelargonidin-3-rutinoside	Tamarillo (*Solanum betaceum* Cav.)	Antioxidant	Higher antioxidant activity in peels than in pulps were found when assessed by Ferric Reducing Ability of Plasma (FRAP) and Cupric Ion-Reducing Antioxidant Capacity (CUPRAC).	Diep et al. (2020)
Betalains	Amaranth (*Amaranthus tricolor* L.)	Antimicrobial	Results revealed a remarkable antimicrobial activity against four foodborne pathogens (*Staphylococcus aureus, Escherichia coli, Listeria monocytogenes* and *Salmonella* Typhimurium)	Hu et al. (2020)
	Beetroot and apple juice	Lowering blood pressure	Consuming a single dose of 500 g of beetroot and apple juice was able to lower blood pressure by 4–5 mmHg at 6-h.	Coles and Clifton (2012)
	Red beet (Beta vulgaris var. rubra)	Immunomodulatory effect	Betalain was able to enhance antioxidant property and specific and nonspecific immunity of mice.	Xiaowei et al. (2019)
Carbohydrates	Flax (*Linum usitatissimum* L.) seeds Salba-chia (*Salvia hispanica* L.) seeds	Reduce postprandial glycemia and satiety	Salba-chia and flax seeds supplementation may affect satiety and reduce postprandial glycemia by converting glucose into a slow-release carbohydrate.	Vuksan et al. (2017)
	Syzygium cumini seeds	Antilipidemic	Besides antioxidant, anti-inflammatory and antifibrotic activities, *Syzygium cumini* seed powder may ameliorate dyslipidemia and insulin resistance by increasing lipid metabolism in liver of high carbohydrate high fat diet rats.	Ulla et al. (2017)

(continued)

Table 4.1 (continued)

Bioactive compound	Plant source	Biological activities	Mechanism	References
Fatty acids				
	Olive oil	Reduced risk of cardiovascular disease	Olive oil appears to decrease DNA oxidation and plasma inflammatory biomarkers like	Sanchez-Rodriguez et al. (2019)
	Linseed oil Purslane oil Luffa oil Pumpkin oil	Antimicrobial Anticancer	The tested vegetable oils showed promising cytotoxic properties against the tested cancer cell lines, especially HeLa (cervical carcinoma) and NCI-H460 (non-small cell lung cancer), high antibacterial activities against *Enterobacter cloacae* and *Escherichia coli* with MIC and MBC values similar to the used positive controls (E211 and E224). Furthermore, they exhibited significant antifungal activities (*Aspergillus versicolor*, *Aspergillus niger*, and *Penicillium verrucosum* var. *cyclopium*).	Petropoulos et al. (2021)
Phytoestrogens				
Ferutinin	Ferula species	Anticancer Antioxidant	A significant reduction in cell viability and induction of apoptotic death in human Michigan Cancer Foundation-7 (MCF-7) breast cancer cell line was noticed. In addition, ferutinin reduced the lipid peroxidation of the liver tissue of mice and up-regulated the antioxidant enzymes (superoxide dismutase and catalase) in the mice liver and kidney cells.	Garmroudi et al. (2021)
Genistein Daidzein Equol	Soybean foods	Antioxidant	The tested phytoestrogens are effective free radical scavengers and exhibited high antioxidant potential *in vitro*.	Kładna et al. (2016)

Promensil	Red clover (*Trifolium pratense*) extract	Treatment of menopausal hot flushes	The results illustrated a statistical and clinically relevant decrease in hot flush frequency in treated women group compared to placebo.	Myers and Vigar (2017)
Phytosterols				
Sitosterol Cycloartenyl ferulate 24-methylenecycloartanyl ferulate Campesteryl ferulate	Rice bran Rice bran oil	Neuroprotective	The tested phytosterols were able to protect the dopaminergic neurons by activating the DAF-16/FOXO pathway and inhibiting of the apoptotic protein CED-3 overexpression.	Zhang et al. (2021)
Glucosinates				
Sinigrin Glucocochlearin Glucotropaeolin 4-methoxyglucobrassicin	*Lepidium latifolium* L. extract	Antimicrobial Anticancer	The results revealed potent antimicrobial effects against all tested pathogenic and food spoilage bacteria and fungi and high cytotoxic activities glioblastoma LN229 and bladder cancer UM-UC-3 cell lines.	Blažević et al. (2019)
Vitamins				
Vitamin B5 (pantothenic acid)	Peanuts, liver, and yolks	Stabilize carbohydrate, protein, lipid and nucleic acid metabolism	Vitamin B5 is the main precursor of acetyl-coenzyme A which is necessary for the cerebral synthesis of myelin and acetylcholine.	Ismail et al. (2020)

et al. 2020), anticancer (Nasrabadi and Zali 2019) and antiviral (Al-Karmalawy et al. 2021). Carotenoids, anthocyanins and betalains are natural pigments commonly used in the food industry. Carotenoids are lipid-soluble pigments, basically contained in orange- and yellow-colored fruits and vegetables like apricots, mangoes, carrots, pumpkins, and tomatoes (Samtiya et al. 2021). They are found also in cereal grains such as corn, wheat, barley and sorghum (Banwo et al. 2021). According to the structural changes in their polyene skeleton, carotenoids comprise carotenoids, lycopene and xanthophylls (Zabot et al. 2022). It has been demonstrated that lutein, zeaxanthin and β-cryptoxanthin have the ability to decrease cardiovascular diseases, down-regulate age-related muscular degeneration, protect against skin problems related to sunburn and strengthen the immune system (Benke and Benke 2014; Di et al. 2016). Anthocyanins are water-soluble pigments, mainly found in fruits and vegetables of blue, red and purple colors such as berries, grapes, currants and some tropical fruits (Khoo et al. 2017). Several biological properties are attributed to these pigments like anti inflammatory, anti diabetic (Nikbakht et al. 2021) and antioxidant (Diep et al. 2020). Betalains are nitrogenous water-soluble pigments, mainly present in the leaves, stems, seeds, flowers, fruits and roots of several plant families such as Cactaceae, Amaranthaceae, Portulacaceae, Basellaceae and Nyctaginaceae to which they confer their red-yellow pigmentation (Carreón-Hidalgo et al. 2022; Calva-Estrada et al. 2022). Several properties, including anticancer, antioxidant, antimicrobial, and antilipidemic activity have been attributed to these pigments (Gengatharan et al. 2016; Hu et al. 2020).

Carbohydrates represent the most important components of the human diet, along with fats and proteins. Based on their chemical structure, they are divided into four groups: monosaccharides, disaccharides, oligosaccharides and polysaccharides (Dixit et al. 2023). The foods that are rich of these compounds are wheat, pasta, potatoes, whole grains, nuts, beans, fruits and vegetables. The consumption of carbohydrates helps to decrease the incidence of, and mortality from, type 2 diabetes, cardiovascular diseases and some cancers (Seal et al. 2021). Fatty acids are lipid biomolecules that comprise saturated-, monounsaturated- and polyunsaturated-fatty acids. They are further classified, basing on the carbon number, into medium- and long-chains fatty acids (Dixit et al. 2023). Omega-3 fatty acid, known as healthy fat, is a polyunsaturated fatty acid made up of docosahexaenoic acid, eicosapentaenoic acid and alpha-linolenic acid. Higher consumption of omega-3 fatty acids is widely recommended as it is associated with lower level of triglycerides in the human body. Seafood, chia seeds, dairy products, flax and walnuts seeds are good sources of omega-3 fatty acids (Saini et al. 2021).

Phytoestrogens are another type of bioactive molecules, mainly present in several human foodstuffs including, vegetables (sprouts, beans, spinach, etc.), fruits (pear, plum, grape, etc.), wine, and tea. Their chemical structure is extremely similar to that of the mammalian estrogen and estradiol thus, they can bind to estrogen receptors, widely expressed in the uterus, breast, pituitary gland and hormone responsive tumors, and induce an estrogenic/antiestrogenic response in target tissues (Bacciottini et al. 2007). Other non receptor mediated beneficial properties are attributed to phytoestrogens including antioxidant and antiproliferative/antiangiogenic potentials (Garmroudi et al. 2021).

Phytosterols are bioactive compounds naturally found in vegetable oils (corn, soybean, sunflower and olive), oleaginous fruits, cereals (wheat germ and wheat bran). More than 250 types of phytosterols have already been identified; among which, beta-sitosterol, campesterol and stigmasterol are the sterols more commonly found in the plant-based diet (Cabral and Klein 2017). Due to their chemical structure strikingly similar to that of cholesterol, phytosterols are mainly used in treating hypercholesterolemia and preventing cardiovascular disorders (Poli et al. 2021). Glucosinolates are found in Brassica vegetables including broccoli, cabbage, cauliflower and various root vegetables like radish and turnip. Several studies have noticed the direct relationship between the consumption of these vegetables and the prevention of colon and rectal cancers (Melim et al. 2022). Choline is another valuable bioactive molecule involved in the synthesis of the neurotransmitter acetylcholine, the structure and the metabolism of phospholipids (phosphatidylcholine, sphingomyelin and lecithin). In addition, it has been revealed that choline has the ability to prevent the abnormal accumulation of fatty acids in the liver (hepatic steatosis) (do Nascimento et al. 2022). Plant foods that are rich in choline comprise those made from soybeans (tofu and soymilk), cooked dried beans, cruciferous vegetables, peanuts and peanut butter (Wallace et al. 2018). Vitamins are prominent cluster nutrients required, in limited amounts, for regular cell function, growth, and development. They are classified into two groups: water-soluble vitamins that are not stored in the body like vitamin C and B, and fat-soluble vitamins which are stored in the liver, muscles and fatty tissues like vitamin A, D, E, and K (Dixit et al. 2023). Minerals play a vital role in the regulation and growth of several physiological processes such as oxygen transport, muscle contraction, acid-base balance and maintenance of osmotic pressure (Dixit et al. 2023).

4.3 Role of Nanotechnology in Enhancing Plant-Based Bioactives Properties

The interest in the consumption of plant-based bioactives compounds has grown exponentially due to their prominent health benefits. Currently, these valuable compounds in several sectors are bound to escalate, thus unlocking new possibilities like food packaging field. For instance, numerous studies have demonstrated the efficiency of bioactive products over chemical preservatives in avoiding the growth of infectious agents and delaying food spoilage (Nogueira et al. 2020). However, during storage and handling, the inherent properties of bioactive components and functional foods including temperature, oxygen, pH, light, humidity as well as the gastrointestinal tract conditions like the acidity, enzymatic activities, and the presence of other nutrients, appreciably affects their biological performances and intended benefits (Santos et al. 2019). Parthasarathi et al. reported that the low water-solubility of vitamin E restricts its absorption in the gastrointestinal tract and its total bioavailability (Parthasarathi and Anandharamakrishnan 2016). In addition, it has been reported that the oxidation of vegetable oils rich in unsaturated fatty acids is one of the most undesirable reactions that happens during the heating of food products (Madhujith and Sivakanthan 2018). The derivatives formed upon oils-oxydation may be ineffective, dangerous or even toxic. For instance, several studies have noticed that the oxidation of safrole, pinene and limonene lead to the formation of carcinogenic metabolites (Bogusz and Al-Tufail 2008; Sarigiannis et al. 2011; Sköld et al. 2006). Besides plant oils, carotenoids are known for their sensibility when exposed to oxygen, light and temperature (Gul et al. 2015). Anthocyanins and betalains are also susceptible to be degraded during processing and storage due to light, pH, oxygen, heat and interaction with other nutrients, decreasing their bioavailability and biological performance (Enaru et al. 2021; Castro-Enríquez et al. 2020). Encapsulation techniques has been introduced to offer possible solutions to improve water solubility, bioavailability and stability of bioactive components (Delshadi et al. 2020). The function of nanoencapsulation is to enclose the bioactive materials inside a nanocarrier for protection as well as to enhance their functionality by retaining bioavailability and facilitating the controlled release of these valuable compounds while being delivered into functional food items (Akhavan et al. 2018). Nanocarriers are basicaly defined as elements with a particle size in nanoscale range (10^{-9} m), which have pericular physicochemical and functional properties compared to their macroscale counterparts, due to their large surface-to-volume ratios (Pateiro et al. 2021). Different techniques for nanoencapsulation of valuable food constituents in appropriate delivery systems are reported (Pateiro et al. 2021; Paolino et al. 2021). In this chapter, special attention has been devoted to polymeric and lipid based nanocarriers.

4.3.1 Polymeric Nanoparticles

Polymeric nanoparticles (NPs) have received tremendous concern over recent years due to their promissing properties resulting from their small size ranging from 1 to 1000 nm. They comprise both nanospheres and nanocapsules, which differ with respect to their structure (Zielińska et al. 2020). Nanospheres, recongnized as matrix system, are based on a continuous polymeric network in which the bioactive compound can be confined inside or adsorbed onto their surface. On the other hand, nanocapsules are the reservoir system in which the bioactive compound is retained in an oily core surrounded by a polymeric membrane (Lammari et al. 2021) (Fig. 4.1). Generally, the polymeric shell consists of natural or synthetic polymers. Biodegradable polymers like chitosan, poly(lactide-co-glycolide), poly (ε-caprolactone) and polylactidwere are widely used for this purpose (Froiio et al. 2020). Several techniques have been used to design NPs, while the spray drying, the nanoprecipitation and the ionic gelation methods are extensively used. Figure 4.1 shows the principle of each technique.

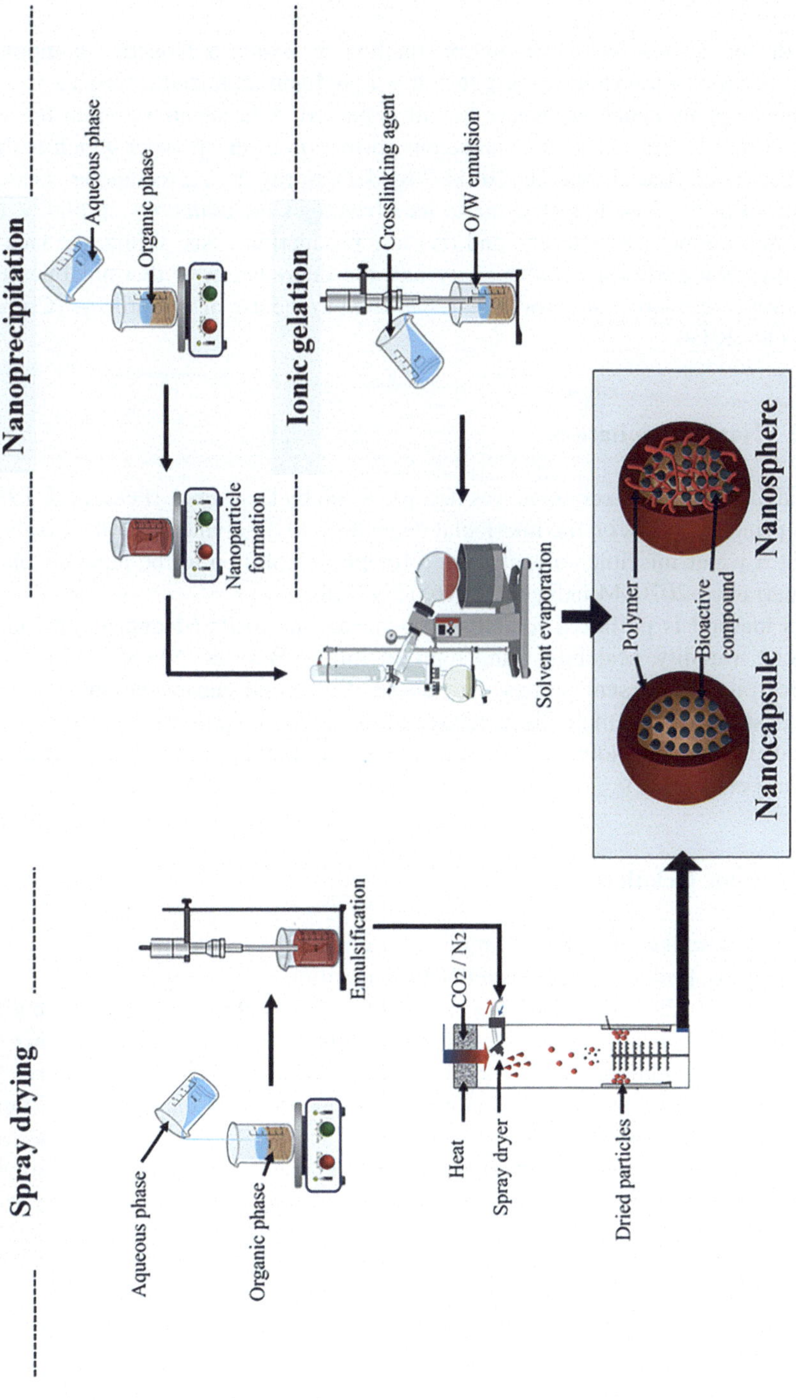

Fig. 4.1 Schematic illustration of some methods used for polymeric nanoparticles preparation

4.3.1.1 Spray Drying

Spray-drying is a solvent evaporation method in which a bioactive compound infeed-solution, -emulsion or -suspension was fed and atomized through a nozzle into fine drops in a chamber where hot air is passed. A fast evaporation of the solvent occurred leading to the instantaneous formation of dry powder which is then easily separated (Lammari et al. 2021) (Fig. 4.1). Spray drying technique is extensively used in the food industry due to its advantages including: (i) simplicity, (ii) ease of scaling up, (iii) rapidity, and (iv) low production costs. While, the lack of uniformity, the particles polydispersity and the degradation of thermolabile substances represent the main drawbacks associated to the use of this process (Corrêa-Filho et al. 2019).

4.3.1.2 Nanoprecipitation

The nanoprecipitation technique was first proposed by Fessi et al. (Fessi et al. 1989) and its principle relies on the interfacial deposition of a polymer after the displacement of a water miscible solvent from a lipophilic solution to the aqueous phase (Lammari et al. 2020; Miladi et al. 2016) (Fig. 4.1).

This method is preferred for NPs preparation due to its advantages including simplicity, rapidity, scalability, and reproducibility. Besides, NPs developped by nanoprecipitation present narrow size distribution, high encapsulation efficiency and high stability. The main disadvantage of the nanoprecipitation method consists on the use of organic solvents where residues may remain, even after purification (Lammari et al. 2021).

4.3.1.3 Ionic Gelation

Calvo et al. described the ionic gelation or coacervation process, which is becoming one of the most preferred methods for NPs preparation as it avoids the use of organic solvents or high shear forces (Calvo et al. 1997). This technique relies on the ionic crosslinking which happens between oppositely charged groups like for instance the positively charged amino groups of chitosan and negatively charged groups of tripolyphosphate, leading to the formation of nanoparticles in the nano-sized range (Nagavarma et al. 2012) (Fig. 4.1).

Like the nanoprecipitation method, the ionic gelation is simple, fast and reproducible. However, larger particle size and moderate encapsulation efficiency were often assessed.

4.3.2 *Liposomes*

Liposomes, also referred to as lipid vesicles, are intricate microscopic structures with spherical shape constructed from one or more layers of lipids, usually a combination of natural and synthetic phospholipids such as phosphatidylcholline, phosphatidylserine, phosphatidylethanolamine, phosphatidylinositol and phosphatidylglycerol (Nsairat et al. 2022). These lipid bilayers create closed structures that envelop a small volume of aqueous solution within their confines (Fig. 4.2) (Laouini et al. 2012). With respect to their size and lamellarity, liposomes comprise: (i) small unilamellar vesicles (SUV) with a particle size ranging from 20 to 100 nm, (ii) large unilamellar vesicles (LUV) with a size more than 100 nm and (iii) multilamellar vesicles (MLV) having a size greater than 0.5 μm (El-Asbahani et al. 2015).

Liposomes were first designed for drug delivery purposes by Bangham et al. (1965). Since then, various methods have been developed for liposome preparation, each designed to cater specific qualities and applications.

4.3.2.1 Thin Film Hydration

Thin film hydration is the first process which was initially used for liposomes developpement (Bangham et al. 1965). It implies the dissolution of primary phospholipid components in an organic solvent to establish a uniform mixture. The organic solvent is subsequently evaporated delicately under reduced pressure, leading to the formation of a thin film layer. Following this, the thin film is hydrated above the transition temperature of the chosen lipid using an aqueous buffer solution (Fig. 4.2) (Šturm and Poklar 2021). Concerning the integration of bioactive substances into liposomes, lipophilic components can be dissolved with the phospholipid mixture before thin film formation, whereas hydrophilic substances can be introduced into the hydration medium and then naturally incorporated into the liposomes during the hydration process (Šturm and Poklar 2021; Lombardo and Kiselev 2022).

The main drawback of this method in encapsulating bioactive products is the challenges to scaling-up and the generation of heterogeneous vesicles of large sizes requiring additional process such as extrusion and sonication in order to homogenize and reduce the vesicle size (Šturm and Poklar 2021). Besides, the use of organic solvents may be hazardeous. In fact, it has been proved that organic solvents residues remaining in the final product may contribute to short- and long-term health effects and reduce stability of the lipid vesicles (Kumar and Subhashini 2015).

4.3.2.2 Ethanol Injection

The solvent injection method, reported for the first time in 1973 (Batzri and Korn 1973), entails the injection of an organic solution, composed of lipids dissolved in an organic solvent (ethanol), into an aqueous solution resulting in liposomes

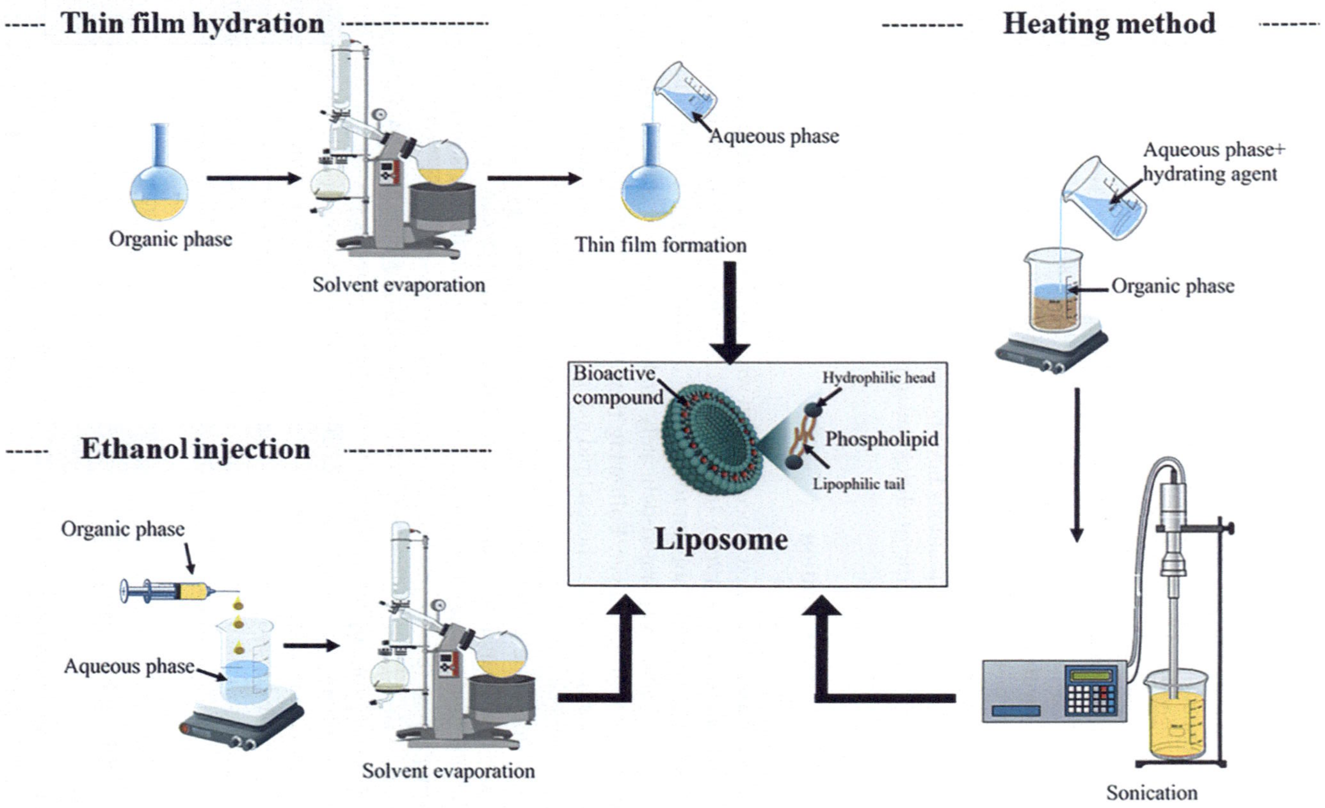

Fig. 4.2 Schematic illustration of some methods used for liposomes preparation

dispersion (Laouini et al. 2012). The solution is stirred during this step to ensure the precipitation of dissolved phospholipids and the subsequent development of bilayer phospholipid fragments. The organic solvent can be removed either by rotary evaporation or by centrifugation through a silica gel column (Jaafar-Maalej et al. 2011). After that, the solution is left to undergo hydration with continuous stirring (Fig. 4.2).

Some downsides are depicted to this method including the poor encapsulation efficiency of hydrophilic components, the continuous exposure of sensitive bioactive products like plant oils to organic solvents and high temperature which may affect both the stability and safety of the liposomes and the relatively limited solubility of lipids in ethanol (Šturm and Poklar 2021; Nkanga et al. 2019). The ethanol injection process is otherwise found to be scalable and quite useful for preparation of narrow size distribution of submicron liposomes (under 100 nm) without need to extrusion or sonication when compared to thin film hydration (Stano et al. 2004).

4.3.2.3 Heating Method

In order to avoid the use of toxic organic solvent, Mozafari et al. proposed a novel approach called heating method or Mozafari method (Mozafari and Mortazavi 2005). In this technique, a lipid mixture is hydrated in the presence of hydrating agent (glycerin or propylene glycol), which plays a crucial role in enhancing the stability of liposomes by preventing the coagulation and sedimentation of vesicles during the formation process. Subsequently, the combined components are heated up to 120 °C (Fig. 4.2) (Trucillo et al. 2020). This elevated temperature serves a dual purpose by not only fostering the formation of liposomes but also establishing an environment conducive to the effective integration of bioactive substances into the vesicles. In this process, bioactive substances can be loaded into the liposomes through several routes: (a) initially with liposomal ingredients and glycerol; (b) when the operating temperature exceeds the lipid transition temperature; (c) at room temperature after vesicle formation. The heating technique is simple, fast, scalable, non-toxic and increasingly trending technique for liposome preparation with high encapsulation efficiency (Nkanga et al. 2019). However, it is not suitable for the encapsulation of thermosensitive and bioactive components (Lammari et al. 2021).

4.3.3 *Solid Lipid Nanoparticles and Nanostructured Lipid Carriers*

In the early 1990s, Solid lipid nanoparticles (SLNs) were introduced as substitute carrier system to liposomes and NPs as they provided both lipid carriers' biological compatibility and stability of the solid matrix, preventing the drawbacks associated to those drug delivery systems (Sarangi and Padhi 2016). Essentially, SLNs were built upon a carefully selected solid lipid matrix chosen for its biocompatibility and biodegradability (Satapathy et al. 2021). This solid lipid matrix established a stable and protective environment, suitable for encapsulating either hydrophobic or hydrophilic components. Fatty acids (e.g. palmitic acid), triacylglycerols (e.g. trimyristin), waxes (e.g. cetyl palmitate) and acyl glycerols (e.g. glycerol monostearate) were commonly used materials to build up the lipid core (Akanda et al. 2023). The SLNs can be stabilized by using a range of surfactants such as phospholipids, poloxamer, tween 80, polyvinyl alcohol or combinations thereof (Lammari et al. 2021). In the aim to improve the loading capacity of SLNs, a second generation of lipid nanoparticles, known as nanostructured lipid carriers (NLCs), has been developped in 1999 (Mukherjee et al. 2009). In the composition of a NLC, a blend of liquid and solid lipids is used, unlike SLNs which exclusively contain solid lipids. This blend leads to the formation of a matrix with additional space for the loading of bioactive compounds, the enhanced stability for the encapsulated compound, as well as minimizing the probability of expulsion from the lipid phase (Pateiro et al. 2021; Patel 2013). Different methods are employed to produce SLNs and NLCs encapsulating plant-based products, while, high pressure homogenization and ultrasonication were widely used due to their simplicity and facility of scaling up (Fig. 4.3).

4.3.3.1 High Pressure Homogenization

The high pressure homogenization is a very popular and relatively easy technique. There are two approaches to SLNs preparation using the this process, namely, hot and cold processes homogenization (Akanda et al. 2023). In hot homogenization, active compounds and molten lipids are mixed at temperatures slightly higher than the solid lipid melting point. Under constant stirring, a preheated aqueous phase containing surfactants is added, forming a hot pre-emulsion which undergoes an homogenization at 500–1500 bars for 3–5 cycles creating a hot nanoemulsion which is cooled at room temperature allowing the formation of SLNs by crystallization of the lipid (Subroto et al. 2023). For cold homogenization, the process is practically the same, but in this case, the mixture of lipids and bioactive compounds is rapidly cooled using liquid nitrogen or dry ice creating solid lipid microparticles that are dispersed in a cold aqueous phase containing surfactants. The homogenization of the resultant pre-suspension produces SLNs (Akanda et al. 2023). In overall, high pressure homogenization technique is widely used for SLNs and NLCs production due to its advantages, including no need for organic solvents, the ease of scale-up and low production times (Khairnar et al. 2022). However, it encounters some issues related to the need of high energy input and the degradation of thermosensitive bioactive products (Akanda et al. 2023).

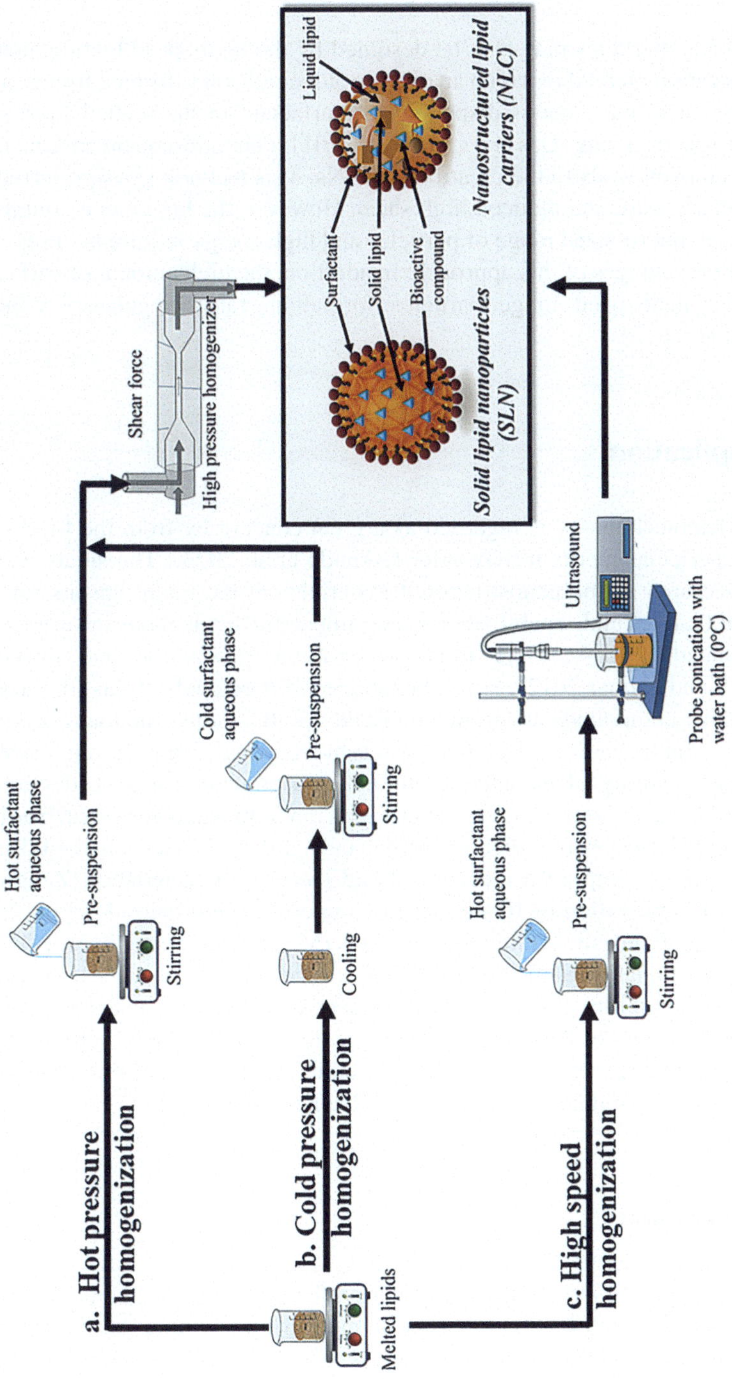

Fig. 4.3 Schematic illustration of some methods used for solid lipid nanoparticles and nanostructured lipid carriers preparation

4.3.3.2 High Speed Homogenization

Solid lipid nanoparticles may also be designed by the high-speed homogenization or ultrasonication method in which an oil-in-water emulsion is formed following the introduction of a hot aqueous dispersion of surfactant to the melted lipid phase under high-speed stirring (Das and Chaudhury 2011). The subsequent cooling of the obtained emulsion leads to the creation of SLNs. This technology is preferred due to its simplicity, ease and reduced high shear. However, the high metal contamination potential, micro-sized range of particles and high energy requirement represent the main disadvantages of this approach. In addition, the high amount of surfactant, required for nano-sized range particles formation, raises regulatory concerns (Akanda et al. 2023).

4.4 Application

Currently, nanotechnology is regarded as a great concern by from the food sector due to the pericular merits it may offer (Akanda et al. 2023). The small size and large surface area of the nanostructured materials are the main reasons for their exceptional features and capabilities for their impactful applications in various food industry-related processes like food preservation, production and packaging (Chaturvedi and Kumar 2012). Several examples of food applications for encapsulated bioactive compounds are shown in Table 4.2 illustrating the major improvements that nanotechnology has recently achieved, including: (i) an improved shelf-life and stability of bioactive compounds, a protection against degradation during processing and storage upon contact with environmental agents such as light, temperature, pH, oxygen or radicals (da Silva et al. 2019; Abbasi et al. 2014; Bashiri et al. 2020); (ii) an improved mechanical and thermal characteristics (Zeng et al. 2024); (iii) an integration of functional raw materials (Ghorbanzade et al. 2017); (iv) a masking of undesirable flavors and tastes (Marsanasco et al. 2015); (v) a protection of functional constituents from low gastric pH (Yi et al. 2014); (vi) an increase in the solubility of nutraceuticals (Pan et al. 2013); (vii) an enhancement in the biological activity of the encapsulated ingredients (antioxidants, antimicrobials, etc.) (Pinilla and Brandelli 2016; Mahmoud et al. 2016) and (vii) providing better quality and aspect for the food systems (Alikhani-Koupaei 2015).

4.5 Conclusion

In recent years, consumers have shifted their preferences towards plant-derived bioactive compounds, both for the growing demand for everything that is of natural origin and for the beneficial effects that these valuable products exhibit on human health. However, the poor water solubility and the instability under environmental

Table 4.2 Examples of bioactive components-encapsulated nanocarriers and their potential applications in the food sector

Bioactive compound	Nanocarrier	Physicochemical properties	Outcomes	References
Linseed oil	Chia mucilage-NPs	Size = 356 ± 2.8 nm z.pot = −22.7 ± 3.9 mV EE% = 50%	The enhanced thermal stability (up to 300 °C) and the increased solubility of linseed oil after encapsultion allows the enrichment of orange juice.	da Silva et al. (2019)
Vitamin D	Whey protein isolate-NPs	Size = 110–235 nm	Whey protein isolate-NPs delayed the degradation of vitamin D in presence of oxygen to be used as enriching agent in beverages.	Abbasi et al. (2014)
Cinnamon essential oil	NLCs	Size = 100–120 nm EE% more than 82.1%	NLCs were able to improve the protection and stability of cinnamon essential oil incorporated in milk and fruit juices at different pH conditions and pasteurization treatment.	Bashiri et al. (2020)
Quercetin	Carboxymethyl lotus root starch-NPs	/	The developped NPs effectively increased the thermal stability, mechanical strength, antioxidant and anti-fungal effects of quercetin NP/starch film against *Botrytis cinerea*, thus extending grape shelf life.	Zeng et al. (2024)

(continued)

Table 4.2 (continued)

Bioactive compound	Nanocarrier	Physicochemical properties	Outcomes	References
Fish oil	Liposome	Size = 300–500 nm EE% = 92.2 ± 0.2%	The obtained liposomes were capable to fortify yogurt without affecting the sensory characteristics of the final product.	Ghorbanzade et al. (2017)
Vitamin C + tocopherol	Liposome	Size = 13.7 ± 0.1um z.pot = +17.8 ± 0.6mv	The incorporation of the liposomal formulation conserved the bioactivity of orange juice even after pasteurization.	Marsanasco et al. (2015)
β-carotene	β-Lactoglobulin–dextran- NPs	Size = 60–70 nm	The developped nanoparticles were more stable to aggregation under gastric pH environment with good release and permeability characteristics.	Yi et al. (2014)
Curcumin	Sodium caseinate NPs	Size = 168.7 ± 10.2 nm EE% = 83.1%	Higher antioxidant activity was noticed for curcumin encapsulated in casein NPs due to the improved dispersibility.	Pan et al. (2013)
Catechin	Liposome	/	Enhanced antioxidant and antibacterial potentials of catechin on Chinese dried pork, after storage at room temperature for 25 days was noticed when compared to the un-encapsulated catechin.	Wu et al. (2018)

(continued)

Table 4.2 (continued)

Bioactive compound	Nanocarrier	Physicochemical properties	Outcomes	References
Nisin + garlic extract	Liposome	Size = 179 nm Z.pot = – 27.7 mV EE% = 82% (nisin); 90% (garlic extract)	Potent antimicrobial effect of nanoliposome-encapsulated nisin and garlic extract against *Listeria monocytogenes, Salmonella enteritidis, Escherichia coli* and *Staphylococcus aureus* in whole milk.	Pinilla and Brandelli (2016)
Orange peels extract	Maltodextrin-arabic gum-NPs	EE% = 63.8–97.6%	Orange peels extract nanoencapsulated form was considered as potential antioxidant ingredients even after 60 days storage of cake at 4 °C.	Mahmoud et al. (2016)
Rosemary essential oil	Liposome	/	Loaded-liposomes were more efficients in improving the quality of fresh-cut bananas and extending their shelf life than the free oil.	Alikhani-Koupaei (2015)

Z.pot, zeta potential; *EE%*, encapsulation efficiency

conditions have limited their applications and urged scientists to focus their attempts on developing advanced approaches to extend the shelf life of these bioactive molecules. Currently, the nanoencapsulation offers an enormous outlook to solve such issues. In this chapter, we reviewed bioactive compounds according to their types, biological properties, limitations, all together. The encapsulation of these valuable materials in various types of nanostructured carriers including polymeric nanoparticles, liposomes, solid lipid nanoparticles and nanostructured lipid carriers along with their preparation methods and applications in food industry were well discussed. The selection of the appropriate nanocarrier and the encapsulation process depends on several parameters, such as the nature and the sensitivity of bioactive compounds to be encapsulated and the costs of the technique. Each nanocarrier may provide different properties to the bioactive component in the final product in terms of size, structure, distribution, and bioavailability. Although impressive progress that has been noted in this concept, there are still some points not fully clarified like *in vivo* studies and industrial application in the aim of developping high quality food with desired nutritive and sensory features.

References

Abbasi A, Emam-Djomeh Z, Mousavi MAE, Davoodi D (2014) Stability of vitamin D3 encapsulated in nanoparticles of whey protein isolate. Food Chem 143:379–383

Akanda M, Mithu MSH, Douroumis D (2023) Solid lipid nanoparticles: an effective lipid-based technology for cancer treatment. J Drug Deliv Sci Technol 86:104709

Akhavan S, Assadpour E, Katouzian I, Jafari SM (2018) Lipid nano scale cargos for the protection and delivery of food bioactive ingredients and nutraceuticals. Trends Food Sci Technol 74:132–146

Akhtar M, Rafiullah M, Hossain M (2023) Antidiabetic activity of Cichorium intybus L water extract against streptozotocin-induced diabetic rats. J Umm Al-Qura Univ Appl Sci 9:565–571

Albuquerque TG, Nunes MA, Bessada SM, Costa HS, Oliveira MBP (2020) Biologically active and health promoting food components of nuts, oilseeds, fruits, vegetables, cereals, and legumes. In: Pico Y (ed) Chemical analysis of food, Second edn. Academic, London, pp 609–656

Alikhani-Koupaei M (2015) Liposomal and edible coating as control release delivery systems for essential oils: comparison of application on storage life of fresh-cut banana. Qual Assur Saf Crop Foods 7(2):175–185

Al-Karmalawy A, Farid M, Mostafa A, Ragheb A, Mahmoud H, Shehata M et al (2021) Naturally available flavonoid aglycones as potential antiviral drug candidates against SARS-CoV 2. Molecules 26:6559

Azmir J, Zaidul IS, Rahman M, Sharif K, Mohamed A, Sahena F et al (2013) Techniques for extraction of bioactive compounds from plant materials: a review. J Food Eng 117(4):426–436

Bacciottini L, Falchetti A, Pampaloni B, Bartolini E, Carossino A, Brandi M (2007) Phytoestrogens: food or drug? Clin Cases Miner Bone Metab 4(2):123–130

Bangham AD, Standish MM, Watkins JC (1965) Diffusion of univalent ions across the lamellae of swollen phospholipids. J Mol Biol 13(1):238–252

Banwo K, Olojede A, Adesulu-Dahunsi A, Verma D, Thakur M, Tripathy S et al (2021) Functional importance of bioactive compounds of foods with Potential health benefits: a review on recent trends. Food Biosci 43:101320

Bashiri S, Ghanbarzadeh B, Ayaseh A, Dehghannya J, Ehsani A, Adun P (2020) Essential oil-loaded nanostructured lipid carriers: the effects of liquid lipid type on the physicochemical properties in beverage models. Food Biosci 35:100526

Batzri S, Korn ED (1973) Single bilayer liposomes prepared without sonication. Biochim Biophys Acta Biomembr 298(4):1015–1019

Benke K, Benke K (2014) Experimental design issue for assessment of carotenoids lutein and zeaxanthin in age-related eye disease study 2 formulation for age-related macular degeneration. JAMA Ophthalmol 132:904

Blažević I, Đulović A, Maravić A, Čikeš Čulić V, Montaut S, Rollin P (2019) Antimicrobial and cytotoxic activities of Lepidium latifolium L. hydrodistillate, extract and its major sulfur volatile allyl isothiocyanate. Chem Biodivers 16(4):e1800661

Bogusz MJ, Al-Tufail M (2008) Chapter 18: Toxicological aspects of herbal remedies. In: Handbook of analytical separations, Elsevier, Oxford, vol 6, pp 589–610

Cabral C, Klein M (2017) Phytosterols in the treatment of hypercholesterolemia and prevention of cardiovascular diseases. Arq Bras Cardiol 109(5):475–482

Calva-Estrada S, Jiménez-Fernández M, Lugo-Cervantes E (2022) Betalains and their applications in food: the current state of processing, stability and future opportunities in the industry. Food Chem 4:100089

Calvo P, Remunan Lopez C, Vila-Jato JL, Alonso MJ (1997) Novel hydrophilic chitosan—polyethylene oxide nanoparticles as protein carriers. J Appl Polym Sci 63:125–132

Carreón-Hidalgo JP, Franco-Vásquez DC, Gómez-Linton DR, Pérez-Flores LJ (2022) Betalain plant sources, biosynthesis, extraction, stability enhancement methods, bioactivity, and applications. Food Res Int 151:110821

Castro-Enríquez DD, Montaño-Leyva B, Del Toro-Sánchez CL, Juaréz-Onofre JE, Carvajal-Millan E, Burruel-Ibarra SE et al (2020) Stabilization of betalains by encapsulation—a review. J Food Sci Technol 57(5):1587–1600

Chandrasekara A, Shahidi F (2018) Herbal beverages: bioactive compounds and their role in disease risk reduction - a review. J Tradit Complement Med 8(4):451–458

Chang W, Huang D, Lo Y, Tee Q, Kuo P, Wu J (2019) Protective effect of caffeic acid against Alzheimer's disease pathogenesis via modulating cerebral insulin signaling, beta-amyloid accumulation, and synaptic plasticity in hyperinsulinemic rats. J Agric Food Chem 67(27):7684–7693

Chaturvedi S, Kumar V (2012) Production techniques of lipid nanoparticles: a review. Res J Pharm Biol Chem Sci 3(2):525–541

Coles L, Clifton P (2012) Effect of beetroot juice on lowering blood pressure in free-living, disease-free adults: a randomized, placebo-controlled trial. Nutr J 11:106

Corrêa-Filho LC, Moldao-Martins M, Alves VD (2019) Advances in the application of microcapsules as carriers of functional compounds for food products. Appl Sci 9:571

da Silva Stefani F, de Campo C, Paese K, Stanisçuaski Guterres S, Haas Costa TM, Hickmann Flôres S (2019) Nanoencapsulation of linseed oil with chia mucilage as structuring material: characterization, stability and enrichment of orange juice. Food Res Int 120:872–879

Das S, Chaudhury A (2011) Recent advances in lipid nanoparticle formulations with solid matrix for oral drug delivery. AAPS PharmSciTech 12(1):62–76

Delshadi R, Bahrami A, Tafti AG, Barba FJ, Williams LL (2020) Micro and nano-encapsulation of vegetable and essential oils to develop functional food products with improved nutritional profiles. Trends Food Sci Technol 104:72–83

Dhalaria R, Verma R, Kumar D, Puri S, Tapwal A, Kumar V et al (2020) Bioactive compounds of edible fruits with their anti-aging properties: a comprehensive review to prolong human life. Antioxidants (Basel) 9:1123

Di PN, Di TP, Pandolfi A (2016) Carotenoids in cardiovascular disease prevention. JSM Atheroscler 1:1002

Diep T, Pook C, Yoo M (2020) Phenolic and anthocyanin compounds and antioxidant activity of Tamarillo (Solanum betaceum Cav.). Antioxidants (Basel) 9:169

Dixit V, Kamal WJ, Chole PB, Dayal D, Chaubey KK, Pal AK et al (2023) Functional foods: exploring the health benefits of bioactive compounds from plant and animal sources. J Food Qual 2023:5546753

do Nascimento R, Souza CM, Bastos T, Kaelle GC, de Oliveira S, Félix A (2022) Effects of an herbal source of choline on diet digestibility and palatability, blood lipid profile, liver morphology, and cardiac function in dogs. Animals 12:2658

Dos Santos RC, Ombredane AS, Souza JMT, Vasconcelos AG, Plácido A (2018) Lycopene-rich extract from red guava (Psidium guajava L.) displays cytotoxic effect against human breast adenocarcinoma cell line MCF-7 via an apoptotic-like pathway. Food Res Int 105:184–196

El-Asbahani A, Miladi K, Badri W, Sala M, Addi EHA, Casabianca H et al (2015) Essential oils: from extraction to encapsulation. Int J Pharm 483:220–243

El-Hussainy EHM, Hussein AM, Abdel-Aziz A, El-Mehasseb I (2016) Effects of aluminum oxide (Al2O3) nanoparticles on ECG, myocardial inflammatory cytokines, redox state, and connexin 43 and lipid profile in rats: possible cardioprotective effect of gallic acid. Can J Physiol Pharmacol 94:868–878

Enaru B, Drețcanu G, Pop TD, Stănilă A, Diaconeasa Z (2021) Anthocyanins: factors affecting their stability and degradation. Antioxidants (Basel) 10(12):1967

Fessi H, Puisieux F, Devissaguet JP, Ammoury N, Benita S (1989) Nanocapsule formation by interfacial polymer deposition following solvent displacement. Int J Pharm 55(1):R1–R4

Froiio F, Lammari N, Tarhini M, Alomari M, Louaer W, Meniai AH et al (2020) Chapter 16: Polymer-based nanocontainers for drug delivery. In: Smart nanocontainers micro and nano technologies, Elsevier, Oxford, pp 271–285

Garmroudi SNR, Karimi E, Oskoueian E, Homayouni-Tabrizi M, Iranshahi M (2021) Ferutinin: a phytoestrogen from ferula and its anticancer, antioxidant, and toxicity properties. J Biochem Mol Toxicol 35(4):e22713

Gengatharan A, Dykes G, Choo W (2016) Stability of betacyanin from red pitahaya (Hylocereus polyrhizus) and its potential application as a natural colourant in milk. Int J Food Sci Technol 51(2):427–434

Ghorbanzade T, Jafari SM, Akhavan S, Hadavi R (2017) Nano-encapsulation of fish oil in nano-liposomes and its application in fortification of yogurt. Food Chem 216:146–152

Gul K, Tak A, Singh AK, Singh P, Yousuf B, Wani AA (2015) Chemistry, encapsulation, and health benefits of β-carotene - a review. Yildiz F, editor. Cogent Food Agric 1(1):1018696

Hamzalıoğlu A, Gökmen V (2016) Chapter 18: Interaction between bioactive carbonyl compounds and asparagine and impact on icrylamide. In: Acrylamide in food. Academic, London, pp 355–376

Hu H, Yao X, Qin Y, Yong H, Liu J (2020) Development of multifunctional food packaging by incorporating betalains from vegetable amaranth (Amaranthus tricolor L.) into quaternary ammonium chitosan/fish gelatin blend films. Int J Biol Macromol 159:675–684

Ismail N, Kureishy N, Church SJ, Scholefield M, Unwin RD, Xu J et al (2020) Vitamin B5 (d-pantothenic acid) localizes in myelinated structures of the rat brain: potential role for cerebral vitamin B5 stores in local myelin homeostasis. Biochem Biophys Res Commun 522(1):220–225

Jaafar-Maalej C, Charcosset C, Fessi H (2011) A new method for liposome preparation using a membrane contactor. J Liposome Res 21(3):213–220

Jha V, Kadam P, Jain T, Bhargava A, Marick A, Saiya B et al (2023) Investigation of physicochemical properties and evaluation of the biological potential of essential oil extracted from Artemisia pallens. J Umm Al-Qura Univ Appl Sci. 9:494–507

Khairnar SV, Pagare P, Thakre A, Nambiar AR, Junnuthula V, Abraham MC et al (2022) Review on the scale-up methods for the preparation of solid lipid nanoparticles. Pharmaceutics 14(9):1886

Khoo H, Azlan A, Tang S, Lim S (2017) Anthocyanidins and anthocyanins: colored pigments as food, pharmaceutical ingredients, and the potential health benefits. Food Nutr Res 61(1):1361779

Kładna A, Berczyński P, Kruk I, Piechowska T, Aboul-Enein HY (2016) Studies on the antioxidant properties of some phytoestrogens. Luminescence 31(6):1201–1206

Kumar S, Subhashini D (2015) Health hazards of organic solvents. Res Rev J Chem 5(2):90–95

Lammari N, Louaer O, Meniai AH, Elaissari A (2020) Encapsulation of essential oils via nanoprecipitation process: overview, progress, challenges and prospects. Pharmaceutics 12:431

Lammari N, Louaer O, Hassen A, Fessi H, Elaissari A (2021) Plant oils: from chemical composition to encapsulated form use. Int J Pharm 601:120538

Laouini A, Jaafer-Maalej C, Limayem-Blouza I, Sfar S, Charcosset C, Fessi H (2012) Preparation, characterization and applications of Lliposomes: state of the art. J Colloid Sci Biotechnol 1(2):147–168

Li Z-J, Liu M, Dawuti G, Dou Q, Ma Y, Liu H-G (2017) Antifungal activity of gallic acid in vitro and in vivo. Phytother Res 31(7):1039–1045

Lombardo D, Kiselev MA (2022) Methods of liposomes preparation: formation and control factors of versatile nanocarriers for biomedical and nanomedicine application. Pharmaceutics 14(3):543

Madhujith T, Sivakanthan S (2018) Oxidative stability of edible plant oils. In: Mérillon J, Ramawat K (eds) Bioactive molecules in food, Reference series in phytochemistry. Springer, Cham, pp 1–23

Mahmoud KF, Ibrahim MA, Mervat E-D, Shaaban HA, Kamil MM, Hegazy NA (2016) Nano-encapsulation efficiency of lemon and orange peels extracts on cake shelf life. Am J Food Technol 11(3):63–75

Marsanasco M, Piotrkowski B, Calabró V, del Valle Alonso S, Chiaramoni NS (2015) Bioactive constituents in liposomes incorporated in orange juice as new functional food: thermal stability, rheological and organoleptic properties. J Food Sci Technol 52(12):7828–7838

Melim C, Lauro M, Pires I, Oliveira P, Cabral C (2022) The role of glucosinolates from cruciferous vegetables (Brassicaceae) in gastrointestinal cancers: from prevention to therapeutics. Pharmaceutics 14:190

Mickymaray S (2019) Efficacy and mechanism of traditional medicinal plants and bioactive compounds against clinically important pathogens. Antibiotics (Basel) 8(4):257

Miladi K, Sfar S, Fessi H, Elaissari A (2016) Nanoprecipitation process: From particle preparation to in vivo applications. In: Polymer nanoparticles for nanomedicines, Springer Cham, Switzerland, pp 17–53

Moscatelli F, De Maria A, Marinaccio LA, Monda V, Messina A, Monacis D et al (2023) Assessment of lifestyle, eating habits and the effect of nutritional education among undergraduate students in southern Italy. Nutrients 13:2894

Mozafari MR, Mortazavi SM (2005) Chapter 6: A new technique for the preparation of non-toxic liposomes and nanoliposomes: the heating method. In: Nanoliposomes: from fundamentals to recent developments. Trafford Publishing Ltd, Oxford, UK, pp 91–98

Mukherjee S, Ray S, Thakur R (2009) Solid lipid nanoparticles: a modern formulation approach in drug delivery system. Indian J Pharm Sci 71:349–358

Mumtaz M, Kausar F, Hassan M (2021) Anticancer activities of phenolic compounds from Moringa oleifera leaves: in vitro and in silico mechanistic study. Beni-Suef Univ J Basic Appl Sci 10:12

Munialo C, Vriesekoop F (2023) Plant-based foods as meat and fat substitutes. Food Sci Nutr 11(9):4898–4911

Myers S, Vigar V (2017) Effects of a standardised extract of Trifolium pratense (Promensil) at a dosage of 80mg in the treatment of menopausal hot flushes: a systematic review and meta-analysis. Phytomedicine 24:141–147

Nagavarma BVN, Yadav HKS, Ayaz A, Vasudha LS, Shivakumar HG (2012) Different techniques for preparation of polymeric nanoparticles-a review. Asian J Pharm Clin Res 5(3):16–23

Nasrabadi PN, Zali MR (2019) A detailed image of rutin underlying intracellular signaling pathways in human SW480 colorectal cancer cells based on miRNAs—lncRNAs—mRNAs—TFs interactions. J Cell Physiol 234(9):15570–15580

Nikbakht E, Singh I, Vider J (2021) Potential of anthocyanin as an anti-inflammatory agent: a human clinical trial on type 2 diabetic, diabetic at-risk and healthy adults. Inflamm Res 70:275–284

Nkanga CI, Bapolisi AM, Okafor NI, Krause RWM (2019) General perception of liposomes: formation, manufacturing and applications. In: Liposomes - advances and perspectives, IntechOpen, London pp 1–24

Nogueira GF, de Oliveira RA, Velasco JI, Fakhouri FM (2020) Methods of incorporating plant-derived bioactive compounds into films made with agro-based polymers for application as food packaging: a brief review. Polymers 12(11):2518

Nsairat H, Khater D, Sayed U, Odeh F, Al Bawab A, Alshaer W (2022) Liposomes: structure, composition, types, and clinical applications. Heliyon 8(5):e09394

Paciello F, Di PA, Rolesi R, Troiani D, Paludetti G, Grassi C et al (2020) Anti-oxidant and anti-inflammatory effects of caffeic acid: in vivo evidences in a model of noise-induced hearing loss. Food Chem Toxicol 143:111555

Pai S, Hebbar A, Selvaraj SA (2022) Critical look at challenges and future scopes of bioactive compounds and their incorporations in the food, energy, and pharmaceutical sector. Environ Sci Pollut Res 29:35518–35541

Pan K, Zhong Q, Baek SJ (2013) Enhanced dispersibility and bioactivity of curcumin by encapsulation in casein nanocapsules. J Agric Food Chem 61(25):6036–6043

Paolino D, Mancuso A, Cristiano M, Froiio F, Lammari N, Celia C et al (2021) Nanonutraceuticals: the new frontier of supplementary food. Nano 11(3):792

Parthasarathi S, Anandharamakrishnan C (2016) Enhancement of oral bioavailability of vitamin E by spray-freeze drying of whey protein microcapsules. Food Bioprod Process 100:469–476

Pateiro M, Gómez B, Munekata P, Barba F, Putnik P, Kovačević D et al (2021) Nanoencapsulation of promising bioactive compounds to improve their absorption, stability, functionality and the appearance of the final food products. Molecules 26(6):1547

Patel D (2013) Development & screening approach for lipid nanoparticle: a review. Int J Innov Pharm Sci 2(5):27–32

Petropoulos SA, Fernandes Â, Calhelha RC, Rouphael Y, Petrović J, Soković M et al (2021) Antimicrobial properties, cytotoxic effects, and fatty acids composition of vegetable oils from purslane, linseed, luffa, and pumpkin seeds. Appl Sci 11(12):5738

Pinilla CMB, Brandelli A (2016) Antimicrobial activity of nanoliposomes co-encapsulating nisin and garlic extract against Gram-positive and Gram-negative bacteria in milk. Innov Food Sci Emerg Technol 36:287–293

Poli A, Marangoni F, Corsini A, Manzato E, Marrocco W, Martini D et al (2021) Phytosterols, cholesterol control, and cardiovascular disease. Nutrients 13:2810

Rashid R, Wani SM, Manzoor S, Masoodi F, Altaf A (2022) Nanoencapsulation of pomegranate peel extract using maltodextrin and whey protein isolate. Characterisation, release behaviour and antioxidant potential during simulated invitro digestion. Food Biosci 50:102135

Rashmi H, Negi P (2020) Health benefits of bioactive compounds from vegetables. In: Swamy M (ed) Plant-derived bioactives. Springer, Singapore

Renard CMGC (2018) Extraction of bioactives from fruit and vegetables: state of the art and perspectives. LWT - Food Sci Technol 93:390–395

Rosanna CW, Leanderson P, Lundberg AK, Jonasson L (2017) Lutein exerts anti-inflammatory effects in patients with coronary artery disease. Atherosclerosis 262:87–93

Saini R, Prasad P, Sreedhar R, Naidu KA, Shang X, Keum Y (2021) Omega−3 polyunsaturated fatty acids (PUFAs): emerging plant and microbial sources, oxidative stability, bioavailability, and health benefits—a review. Antioxidants (Basel) 10:1627

Samtiya M, Aluko R, Dhewa T, Moreno-Rojas J (2021) Potential health benefits of plant food-derived bioactive components: an overview. Food Secur 10(4):839

Sanchez-Rodriguez E, Biel-Glesson S, Fernandez-Navarro J, Calleja M, Espejo-Calvo J, Gil-Extremera B et al (2019) Effects of virgin olive oils differing in their bioactive compound contents on biomarkers of oxidative stress and inflammation in healthy adults: a randomized double-blind controlled trial. Nutrients 11(3):561

Santos DI, Saraiva JMA, Vicente AA, Moldão-Martins M (2019) Methods for determining bioavailability and bioaccessibility of bioactive compounds and nutrients. In: Innovative thermal and non-thermal processing, bioaccessibility and bioavailability of nutrients and bioactive compounds. Elsevier, Oxford, pp 23–54

Sarangi MK, Padhi S (2016) Solid lipid nanoparticles- a review. J Crit Rev 3(3):5–12

Sarigiannis DA, Karakitsios SP, Gotti A, Liakos IL, Katsoyiannis A (2011) Exposure to major volatile organic compounds and carbonyls in European indoor environments and associated health risk. Environ Int 37(4):743–765

Satapathy MK, Yen T-L, Jan J-S, Tang R-D, Wang J-Y, Taliyan R et al (2021) Solid lipid nanoparticles (SLNs): an advanced drug delivery system targeting brain through BBB. Pharmaceutics 13(8):1183

Savych A, Marchyshyn S, Kyryliv M, Bekus I (2021) Cinnamic acid and its derivatives in the herbal mixtures and their antidiabetic activity. Farmacia 69(3):595–601

Seal CJ, Courtin CM, Venema K, de Vries J (2021) Health benefits of whole grain: effects on dietary carbohydrate quality, the gut microbiome, and consequences of processing. Compr Rev Food Sci Food Saf 20(3):2742–2768

Sköld M, Karlberg A-T, Matura M, Börje A (2006) The fragrance chemical β-caryophyllene—air oxidation and skin sensitization. Food Chem Toxicol 44(4):538–545

Stano P, Bufali S, Pisano C, Bucci F, Barbarino M, Santaniello M et al (2004) Novel camptothecin analogue (gimatecan)-containing liposomes prepared by the ethanol injection method. J Liposome Res 14(1–2):87–109
Šturm L, Poklar UN (2021) Basic methods for preparation of liposomes and studying their interactions with different compounds, with the emphasis on polyphenols. Int J Mol Sci 22(12):6547
Subroto E, Andoyo R, Indiarto R (2023) Solid lipid nanoparticles: review of the current research on encapsulation and delivery systems for active and antioxidant compounds. Antioxidants 12(3):633
Teodoro A (2019) Bioactive compounds of food: their role in the prevention and treatment of diseases. Oxidative Med Cell Longev 2019:3765986
Trucillo P, Campardelli R, Reverchon E (2020) Liposomes: from Bangham to supercritical fluids. PRO 8(9):1022
Ulla A, Alam MA, Sikder B, Sumi FA, Rahman MM, Habib ZF et al (2017) Supplementation of Syzygium cumini seed powder prevented obesity, glucose intolerance, hyperlipidemia and oxidative stress in high carbohydrate high fat diet induced obese rats. BMC Complement Altern Med 17(1):289
Vuksan V, Choleva L, Jovanovski E, Jenkins AL, Au-Yeung F, Dias AG et al (2017) Comparison of flax (Linum usitatissimum) and Salba-chia (Salvia hispanica L.) seeds on postprandial glycemia and satiety in healthy individuals: a randomized, controlled, crossover study. Eur J Clin Nutr 71(2):234–238
Wallace T, Blusztajn J, Caudill M, Klatt K, Natker E, Zeisel S et al (2018) Choline: the underconsumed and underappreciated essential nutrient. Nutr Today 53:240–253
Wen L, Zhang Z, Sun D-W, Sivagnanam SP, Tiwari BK (2020) Combination of emerging technologies for the extraction of bioactive compounds. Crit Rev Food Sci Nutr 60(11):1826–1841
Wu W, Li R, Li X, He J, Jiang S, Liu S et al (2016) Quercetin as an antiviral agent inhibits influenza A virus (IAV) entry. Viruses 8:6
Wu J, Guan R, Cao G, Liu Z, Wang Z, Shen H et al (2018) Antioxidant and antimicrobial effects of catechin liposomes on chinese dried pork. J Food Prot 81(5):827–834
Xiaowei B, Haixia H, Guangming D, Chenye W, Xuan Z, Wei R et al (2019) Antioxidant activities and immunomodulatory effects in mice of betalain in vivo. Food Sci 40(17):196–201
Yi J, Lam TI, Yokoyama W, Cheng LW, Zhong F (2014) Controlled release of β-carotene in β-lactoglobulin–dextran-conjugated nanoparticles' in vitro digestion and transport with Caco-2 monolayers. J Agric Food Chem 62(35):8900–8907
Zabot G, Rodrigues FS, Ody LP, Tres MV, Herrera E, Palacin H et al (2022) Encapsulation of bioactive compounds for food and agricultural applications. Polymers (Basel) 14:4194
Zain MC, Jakariah N, Yeoh J, Lee S, Shaari K (2020) Ultrasound-assisted extraction of polyphenolic contents and acid hydrolysis of favonoid glycosides from oil palm (Elaeis guineensis Jacq.) leaf: optimization and correlation with free radical scavenging activity. PRO 8:1540
Zeng Y-F, Chen Y-Y, Deng Y-Y, Zheng C, Hong C-Z, Li Q-M et al (2024) Preparation and characterization of lotus root starch based bioactive edible film containing quercetin-encapsulated nanoparticle and its effect on grape preservation. Carbohydr Polym 323:121389
Zhang Z, Wang D, Qiao S (2017) Metabolic and microbial signatures in rat hepatocellular carcinoma treated with caffeic acid and chlorogenic acid. Sci Rep 7:4508
Zhang J, Zhang L, Wu Z, Zhang P, Liu R, Chang M et al (2021) The dopaminergic neuroprotective effects of different phytosterols identified in rice bran and rice bran oil. Food Funct 12(21):10538–10549
Zhao Q, Zhu L, Wang S, Gao Y, Jin F (2023) Molecular mechanism of the anti-inflammatory effects of plant essential oils: a systematic review. J Ethnopharmacol 301:115829
Zhou L, Elias RJ (2013) Chapter 9: Understanding antioxidant and prooxidant mechanisms of phenolics in food lipids. In: Logan A, Nienaber U, Pan X (eds) Lipid oxidation. AOCS Press, USA, pp 297–321
Zielińska A, Carreiró F, Oliveira AM, Neves A, Pires B, Venkatesh DN et al (2020) Polymeric nanoparticles: production, characterization, toxicology and ecotoxicology. Molecules 25(16):3731

Chapter 5
Advances and Trends in Animal-Based Food Bioactives

Md Faruque Ahmad, Abdulrahman A. Alsayegh, Awais Ahmed, Syed Amir Ashraf, Abdul Haque, Abdullah Otayf, Muzaynah Al Khairat, Mohammed F. Bajahzer, Ehab Y. Elbendary, Nahla Kambal, and Fakhrudin Ali Ahmad

Abstract Biologically active substances that come from animals and have positive impacts on human health are known as animal-based bioactives. The potential health advantages of bioactive compounds derived from animals have received considerable attention. These compounds include collagen, iron, conjugated linoleic acid, vitamin B12, omega-3 fatty acids, carnosine, creatine, etc. These bioactive

M. F. Ahmad (✉)
Department of Clinical Nutrition, College of Nursing and Health Sciences, Jazan University, Jazan, Saudi Arabia

Department of Clinical Nutrition, College of Applied Medical Science, Jazan University, Jazan, Saudi Arabia
e-mail: mfahmad@jazanu.edu.sa

A. A. Alsayegh · A. Otayf · M. F. Bajahzer · E. Y. Elbendary · N. Kambal
Department of Clinical Nutrition, College of Nursing and Health Sciences, Jazan University, Jazan, Saudi Arabia
e-mail: aalsayegh@jazanu.edu.sa; ayateef@jazanu.edu.sa; mfbajahzer@jazanu.edu.sa; Eelbendary@jazanu.edu.sa; nhmohammed@jazanu.edu.sa

A. Ahmed
Department of Management, Shri JJT University, Vidyanagari, Churela, Rajasthan, India

S. A. Ashraf
Department of Clinical Nutrition, College of Applied Medial Sciences, University of Hail, Hail, Saudi Arabia

A. Haque
Department of Post-Harvest Engineering and Technology, Faculty of Agricultural Sciences, Aligarh Muslim University, Aligarh, India

M. Al Khairat
Department of Nutrition Science, College of Nursing and Allied Health Sciences, Howard University, Washington, DC, USA
e-mail: muzaynah.alkhairat@bison.howard.edu

F. A. Ahmad
Department of Basic and Applied Science, School of Engineering and Science, G.D Goenka University, Gurugram, Haryana, India

S. Amir Ashraf, M. Adnan (eds.), *Food Bioactives and Nutraceuticals*,
https://doi.org/10.1007/978-981-97-9723-3_5

substances exhibit potential health benefits such as immunomodulatory, anti-inflammatory, antibacterial, antioxidant, and anti-cancer properties. Advancements in this sector recently have illuminated the wide array of health advantages. Bioactive compounds derived from animals have increased scientific comprehension and created novel possibilities for their applications as functional foods, nutraceuticals, bioactive peptides to support gut health, and the development of drugs. To fully comprehend their modes of action, recommended dosages, and any side effects, more research is necessary. Developments in animal-based bioactives have demonstrated promise for use in nutrition, nutraceuticals, medicine, and functional foods. The objective of advancements and new trends is to develop innovative products from animal-based bioactive compounds that enhance overall health and wellness.

Keywords Animal based bioactive compounds · Omega-3 fatty acids · Vitamin B12 · Iron · Collagen · Nutraceutical and functional foods · Recent development

5.1 Introduction

Bioactive substances originating from animals that exist naturally and have particular physiological effects on humans are known as animal-based bioactive compounds (Tolve et al. 2021; Abeyrathne et al. 2022). These substances are well-known to have possible health advantages and have recently been the focus of numerous studies. They are present in fish, dairy products, meat, and other animal tissues, among other animal sources (Lozano and Trujillo 2012; Ashraf et al. 2020a). Animal-derived bioactive substances are frequently abundant in vital nutrients such fatty acids, proteins, minerals, and vitamins. These substances have special qualities that have the potential to improve human health. These qualities include immunomodulatory, anti-inflammatory, antibacterial, antioxidant, and anti-cancer effects (Chalamaiah et al. 2018; Lordan et al. 2017; de Alcântara Rodrigues et al. 2020; Yiğit 2020). Omega-3 fatty acids, collagen, iron, vitamin B12, conjugated linoleic acid, creatine, carnosine, and chondroitin sulfate are notable examples of bioactive substances derived from animals (Kulczyński et al. 2019; Awuchi et al. 2022; Fu et al. 2019). It is noteworthy that whereas bioactive substances derived from animals exhibit potential benefits for human health, additional investigation is still required to comprehensively comprehend their modes of action, ideal dosages, and possible adverse reactions (Lobine et al. 2022).

Advancements in the study of bioactive chemicals derived from animals have shown their potential uses in medicine and nutrition, among other domains. Researchers and scientists have been investigating the wide variety of bioactive substances obtained from animals and their possible advantages for human health (Mu et al. 2019; Ghosh et al. 2022). The creation of nutraceuticals and functional foods is also drawing attention to bioactive chemicals derived from animals and plants (Ahmad et al. 2011, 2013). These substances can be added to functional

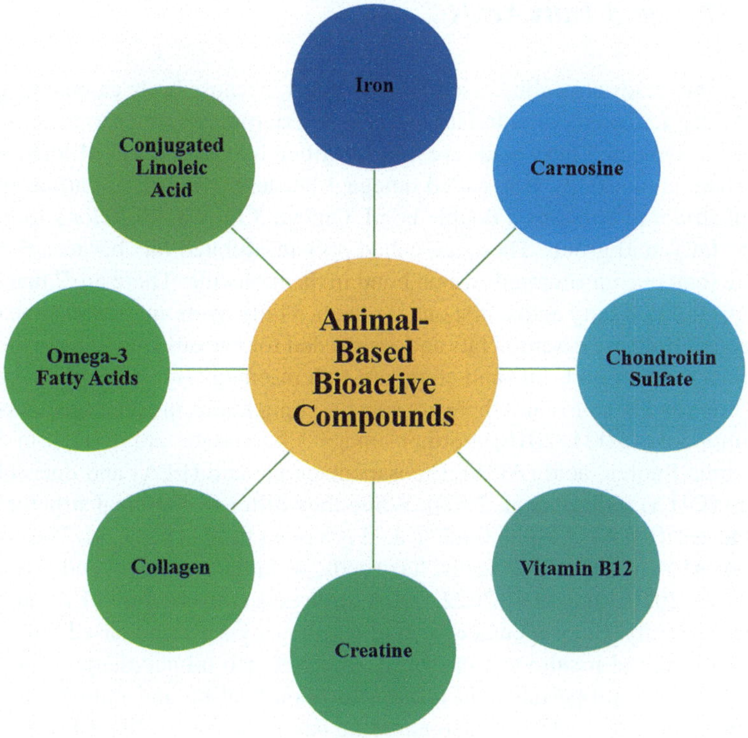

Fig. 5.1 Different animal-based bioactives those play potential role in health promotion

beverages, fortified foods, and dietary supplements to offer additional health advantages over simple nourishment (Joana Gil-Chávez et al. 2013). Furthermore, new developments in the study of animal-based bioactives include the use of marine-derived bioactives, peptides to support gut health, and applications in drug development (Šimat et al. 2020; Ashraf et al. 2020b). The goal of this study is to create novel products that promote general health and wellbeing. Different animal based bioactives can be seen in Fig. 5.1.

5.2 Animal-Based Bioactives

Bioactive substances originating from animals or animal products are known as animal-based bioactives. These bioactives may be beneficial to health and have a variety of physiological impacts. Here are a few examples of bioactives derived from animals:

5.2.1 Omega-3 Fatty Acids

Omega 3 fatty acids are called essential fatty acids, because body cannot synthesize omega 3 fatty acid on its own and it must rely on external resources to get this essential element. Omega 3 fatty acids are first identified by George and Mildred Burr in 1929 (Abate et al. 2022). It is called omega 3 because every third carbon from its chemical structure contains a double bond. Omega 3 fatty acids belong to polyunsaturated fatty acid group. These are called polyunsaturated fats because they contain more than one unsaturated carbon bond in the molecule. There are 2 main types of polyunsaturated fatty acids, first one is omega 3 fatty acids and second one omega 6 fatty acids. Both are essential fats and are needed for various functions in the body. Some diets contain both fats and some are rich in omega 6 fatty acids rather than omega 3 fats and because of it there are some complications in some special population (Simopoulos 2004, 2016). Further omega 3 fatty acids are divided in 3 main types. Alpha linoleic acid (ALA), Eicosapentaenoic acid (EPA) and docosahexaenoic acid (DHA) (Oliver et al. 2020). While they differ in chemical structure, their sources also differ. ALA alpha linoleic acid is a plant based fatty acid. The common source for ALA includes chia seeds, flax seeds, walnuts, canola oil and flaxseed oil (Oliver et al. 2020; Yuan et al. 2022). ALA often gets converted into EPA and DHA in limited quantity. Eicosapentaenoic acid is primarily a marine based fatty acid, it is commonly found it salmons, mackerel, sardines, fish oil supplements etc. it has anti-inflammatory properties, and it has effects on healing and immune response. Docosahexaenoic acid DHA is also a marine based fatty acid, like EPA it is found abundantly in fish oils, fatty fish etc. It is also present in high amounts in some of algae. DHA is important for cell membranes; it has special benefits for brain and eyes. For pregnant ladies and infants, it is the essential element that helps in brain development and cognition (Echeverría et al. 2017; Basak et al. 2020).

They are fat soluble compounds that have low melting point than the saturated fats. Because of this fat-soluble nature, it can cross blood brain barrier and help in brain development. DHA and EPA are liquid at room temperature; however, ALA typically has a melting point around -11 degree Celsius or 12 degrees Fahrenheit. Omega 3 fatty acids can easily get oxidized (Huang and Sathivel 2008), oxidation is a process in which when certain compounds meet with oxygen gets degraded and form free radicals and fatty acids. It loses its capability. So, it is necessary to keep these compounds in airtight opaque containers in a cool and dark place.

Health benefits of omega 3 fatty acids include cardio protection. As it stabilizes the cell membrane, because of this, triglyceride in the blood get reduced because of Omega 3 fatty acids (Adkins and Kelley 2010). Mechanism of action for reducing triglycerides is not fully understood. Cancer is a substantial global health concern, and its prevalence continues to be a challenge worldwide (Ahmad 2020, 2019; Wahab et al. 2021). Omega 3 fatty acids are known for some of their anticancer properties. Omega 3 fatty acids have anti-inflammatory properties and have ability to influence the cellular signaling and gene expression, because of this it has been considered as an important adjunct for cancer treatment and prevention (Wendel and Heller 2009; Laye et al. 2018). Anti-inflammatory nature also helps in reducing the pain and inflammation in arthritis and autoimmune disorders like rheumatoid arthritis. Omega 3 fatty acids are important for brain development of fetus and breast-fed infants (Lundqvist-Persson et al. 2010). Therefore, it is always good to add some supplements containing omega 3 fatty acids to mothers during pregnancy and breast feeding. Omega 3 fatty acids are commercially prepared from microalgae (Oliver et al. 2020) and then sold in the market as supplements (Covington 2004). Shelf life is shorter, and it becomes rancid if not properly stored in airtight, opaque container in the cool and dark place.

After COVID-19 pandemic, there is increased awareness about biologically active compounds that enhance the immunity (Ahmad et al. 2021a, b). People have started using omega 3 fatty acids as an essential element. There is specific understanding and interest of the potential effect of omega 3 fatty acids as immunomodulation and in mental health (Lange 2020). People are more aware for cardiac risk factors and increased triglyceride in the blood, so especially for elderly people the use of omega 3 fatty acids are increasing. Similarly, those having cardiac issues they usually shift to vegan diet, and they tend to align more on taking supplements (Mori 2014). The clinical anti-inflammatory benefit of omega 3 fatty acids is also attracting athletes, as they want their body to heal fast and feel little pain, all other pain medications have their side effects, omega 3 fatty acid being an essential nutrient, they tend towards it more than pain killer medicines, there is continuous research going on, in future there will be increase in the use of omega 3 fatty acids in multiple folds (Lawrence 2010; Lembke et al. 2014). Various characteristics of Omega 3 fatty acids can be seen in Table 5.1.

5.2.2 Iron

A mineral called iron is necessary for maintaining human health. It is involved in many physiological processes in the body, including as the synthesis of different enzymes and proteins, the transfer of oxygen, and the creation of energy (Carpenter and Mahoney 1992; Skalnaya and Skalny 2018). Fish, poultry, and red meat are among the foods highest in heme iron content. Although meals derived from animals are a wonderful source of iron, it's important to remember that a balanced diet can also contain sources of iron from plants. It greatly increases the human body iron reserves and is the more readily absorbed (Ahmad et al. 2022; Hambræus 1999). Conversely, non-heme iron can be found in diets high in both plant and animal sources. Legumes including beans and lentils, whole grains, fortified cereals, spinach, and nuts are plant-based sources of iron, while animal-based sources include red meat, poultry, and seafood (Wieringa et al. 2022; Khesin 2019).

Iron is an essential mineral, iron is needed for many body processes, such as the creation of red blood cells and the movement of oxygen throughout the body (Ems et al. 2017; Gupta 2014). Hemoglobin, a protein found in red blood cells, depends on iron for its proper structure and function, which is the main way that iron

Table 5.1 Summary of some characteristics and important facts of Omega 3 fatty acids

Characteristics	Facts	References
Discovered in	1929	Caramia (2008)
Discovered by	George and Mildred Burr	Caramia (2008)
Main sources	Fish and seafood like salmon, tuna, sardines, herring and mackerel. Seeds and nuts like flaxseeds, chia seeds and walnuts. Oils like soya bean oil, canola oil and flax seed oil. Some fortified foods like soy milk/ beverages, infant formulas, some yoghurts etc.	Covington (2004)
Industrially prepared from	Microalgae	Oliver et al. (2020)
Physical properties	Low melting point Easily gets oxidized. Fat soluble	
Site of action	Cell membranes of all tissues.	Surette (2008)
Mechanism of action	Reduces triglyceride and inflammation by cellular signaling and gene expression.	Surette (2008)
Main functions of omega 3 fatty acids	Cardio protective. Essential for brain development of embryo and infant during breast feeding. Anti-cancer. Improves cognitive function. Prevents age related macular degeneration leading to blindness in elderly. Helps in reducing inflammation in arthritis.	Covington (2004)
Half life	EPA-37 hours DHA- 48 hours	Krupa et al. (2020)
Elimination through	Eliminates through digestive system.	Tamargo and Tamargo (2016)
Daily recommended dose	EPA+DHA- not more than 5 gm/day.	Tamargo and Tamargo (2016)
Deficiency may lead to	Rough and scaly skin, red, swollen and itchy skin rash.	Covington (2004)
Overdose may cause	Bleeding, abdominal discomfort, low BP, suppressed immune function, impaired wound healing etc.	Huang and Sathivel (2008)
Adverse reactions	Bad taste, foul breathing, nausea, abdominal discomfort, diarrhea and (smelly sweat) bromhidrosis etc.	Covington (2004), Kannan et al. (2021), Yadav et al. (2023)
Interaction with other drugs	Causes bleeding when given with warfarin and other anticoagulants.	Covington (2004), Gross et al. (2017)

promotes oxygen delivery (Jensen 2009). The lungs hemoglobin is in charge of attaching to oxygen and transferring it to all of the body's tissues. Each molecule of hemoglobin is made up of four subunits of protein, each with an iron atom at its core (Perutz 1978; Farid et al. 2019). This iron atom can bond to oxygen molecules, resulting in an oxygenated hemoglobin complex. These oxygen molecules then attach to the iron atoms in hemoglobin, resulting in oxyhemoglobin. This oxygenated hemoglobin travels throughout the bloodstream to different tissues and organs (Gupta 2014; Bonaventura et al. 2013).

Many different enzymes in the body depend on iron to function properly. Iron is necessary for many enzymes to function properly and to be active because it is one of their cofactors (Lanz and Booker 2015). Drugs, toxic substances, and other endogenous substances are all metabolized by cytochrome P450 enzymes in the liver (Furge and Guengerich 2006). They are essential to the breakdown of chemicals like fatty acids, steroids, and toxins found in the environment, as well as the detoxification process. For cytochrome P450 enzymes to function, iron is a necessary cofactor (Stål et al. 1996; Song et al. 2023). Iron is necessary for nitric oxide synthase function, which enables nitric oxide production (Cooper 1999). The enzyme nitric oxide synthase is in charge of generating nitric oxide (NO), a signaling molecule that is involved in a number of physiological functions, such as neurotransmission and blood vessel dilatation (Förstermann and Sessa 2012; Ignarro 2002). In the same way, the enzymes catalase and peroxidase aid in the reduction of hydrogen peroxide, a reactive compound generated during a number of different metabolic activities (Cooper 1999; Nagababu et al. 2003). Iron is used by the enzymes catalase and peroxidase to help hydrogen peroxide break down into oxygen and water, avoiding the build-up of this potentially dangerous chemical. It highlights how important it is to keep your iron levels in control in order to maintain good enzyme activity and good overall health (Nagababu et al. 2003; Gaetani et al. 1989).

Decreases in hemoglobin levels or red blood cell counts result in anemia, a disorder that lowers the blood's ability to carry oxygen molecules. Iron deficiency anemia is a particular kind of anemia that is mostly dependent on iron (Ramakrishnan and Semba 2008; Jamnok et al. 2020). Nevertheless, iron is required for the immune system to operate properly. It aids in the generation of immunological cells, which support the body's defense against infections and external invaders (Nairz et al. 2014; Thakur et al. 2019). Moreover, for the best possible development of the brain and function, optimal intakes of iron are essential (Lozoff and Georgieff 2006). Neurotransmitter synthesis, which influences cognitive functions like memory, attention, and learning, is influenced by iron (Agarwal 2001; Pino et al. 2017). It's crucial to remember that the body can only excrete so much excess iron, so keeping things in balance is crucial. Iron overload and deficiency can both have adverse impacts on health. Eating a well-balanced diet rich in iron-rich foods is recommended to guarantee an adequate intake of iron. Heme iron is found in animal-based sources, and the body absorbs it more easily (Beutler et al. 2003; Muñoz et al. 2010; Puntarulo 2005). While dietary sources are generally the preferred way to obtain iron, in some cases, iron supplementation may be necessary under the guidance of a healthcare professional.

5.2.3 *Collagen*

Collagen is a protein found in abundance in the bodies of humans as well as animals. It is an essential component of connective tissues like skin, ligaments, tendons, bones, and cartilage. Collagen supports and strengthens many bodily parts (Nimni and Harkness 2018; Sherman et al. 2015). Collagen supplements are usually obtained from animals, such as cattle, fish, or chicken. They come in a variety of forms, including pills, and powders. Collagen is classified as an animal bioactive product because of its biological activity and numerous applications in disciplines like as medicine, cosmetics, and nutrition (Avila Rodríguez et al. 2018; Cherim et al. 2019).

Given its potential advantages for improving a number of facets of human health, collagen is frequently considered a health-promoting substance. One of the main ingredients in cartilage, which supports and cushions joints, is collagen. The primary protein in the bone matrix, collagen, acts as a scaffold to allow minerals like calcium and phosphorus to deposit (Arseni et al. 2018; Xu et al. 2021). It provides tensile flexibility as well as strength to the bones, enabling them to bear mechanical force without shattering. The process of bone repair requires collagen. Collagen provides a foundation for the growth of new bone tissue after a bone fracture (Viguet-Carrin et al. 2006; Gupta and Zioupos 2008). It facilitates the migration and multiplication of bone-repairing cells, allowing the damaged bone to regenerate and restructure (Singh et al. 2023; Salvatore et al. 2021). Increased bone mineral density, improved bone quality, and a decreased risk of fractures have all been proposed benefits of collagen supplementation as a potential strategy for managing or preventing osteoporosis (Garnero 2015; Viguet-Carrin et al. 2006; Henriksen and Karsdal 2024).

It has been proposed that collagen peptides encourage the synthesis of new collagen in the joints, which may lessen inflammation and increase joint mobility. Muscle repair and regeneration are aided by collagen (Shahrajabian and Sun 2024; Zhu et al. 2023). Collagen supplements are thought to boost overall athletic performance, lessen joint pain brought on by exercise, and encourage muscle recovery (Khatri et al. 2021). Collagen is also in charge of keeping the skin hydrated, firm, and elastic. Natural collagen production declines with age, causing wrinkles, sagging skin, and dryness of the skin. Supplements containing collagen are frequently promoted as a means of increasing collagen levels and promoting skin health, which may lessen the appearance of aging and increase skin suppleness (Khatri et al. 2021; Sharma et al. 2019; Sherman et al. 2015).

The structural protein collagen gives the hair shaft its flexibility, resilience, and strength. Collagen supplements are hypothesized to encourage thicker, healthier hair development and enhance the quality of existing hair (Hwang et al. 2023;

Lai-Cheong and Mcgrath 2013). Amino acids found in collagen peptides aid in the synthesis of keratin, the main protein found in hair strands (De Oliveira Martinez et al. 2020). Furthermore, collagen is a component of nails as well, supporting the integrity and strength of the structure. Sufficient amounts of collagen may help shield nails from breaking, splitting, and brittleness. it is also believing that by increasing nail thickness, stimulating nail development, and lessening nail brittleness, collagen supplements help enhance nail health (Hexsel et al. 2017; Yagoda and Gans 2014).

5.2.4 Conjugated Linoleic Acid (CLA)

Conjugated linoleic acid belongs to omega 6 fatty acid group. Conjugate linoleic acid is a group of octadecadienoic acid isomers which is extracted from linoleic acid. It contains 18 carbons with double bond at 9th and 12th carbon. There are many isomers for conjugate linoleic acid depending on its double bond, the most common are cis-9 trans 11 CLA and trans-10, cis-12 CLA. Common sources of CLA, include Beef, lamb, grass fed animal's dairy and meat (Chin et al. 1992). Cow milk is considered to have most of CLA (Collomb et al. 2006). According to some studies, it is commercially made by linoleic acid safflower oil (Kennedy et al. 2010). It is an odorless, colorless liquid at room temperature and lipid soluble (Uehara et al. 2008). Its site of action includes, adipose tissues, muscle tissues, liver, pancreas, immune system, and bones (Kennedy et al. 2010; Moon 2014). Details characteristics compiled in Table 5.2.

Main benefits of CLA include, being anticancer, an adjunct to weight loss medicines, increase insulin sensitivity. It reduces LDL cholesterol and helps in maintaining HDL cholesterol. LDL and VLDL are considered as bad cholesterol, that causes atherosclerotic plaque in the arteries like coronary, cerebral and peripheral, leading into deadly diseases like myocardial infarction, stroke and peripheral arterial disease (Kennedy et al. 2010; Dhar Dubey et al. 2019). CLA helps in reducing bad blood cholesterol and acts as a cardioprotective (den Hartigh 2019; Park 2009). It is anti-inflammatory and anti-obesity and athletes use it as it promotes healthy healing (Kim et al. 2016). It acts as antioxidant, because after various metabolic processes there are some free radicals develop in the body that causes damage to the tissues, when CLA meets these radicals, it neutralizes free radicals and they become harmless. CLA also has some anti-cancer property. It supports immune system and helps in maintain Bone mass density. According to some researchers the dose of CLA is 1.8–7 gm etc. (Benjamin, Sailas, et al. 2015). it is eliminated from urine and feces. It interacts with Lipid lowering medications, and it acts synergistically with statins, increases insulin resistance and causes in increased retinol levels when given together (Benjamin, Sailas, et al. 2015).

Selling and manufacturing CLA has been very profitable, and it has been expected to increase further. Because of its nature and health benefits many athletes use it as anti-inflammatory (Baghi et al. 2016). It has been used for reducing the fat

Table 5.2 Summary of some characteristics and important facts of conjugated linoleic acid

Characteristics	Facts	References
Discovered in	1987	Kennedy et al. (2010)
Discovered by	Michael W. Pariza and colleagues	Ha et al. (1987)
Main sources	Beef Lamb Dairy products Goat meat and dairy Sheep meat and dairy	Chin et al. (1992)
Industrially prepared from	Made form linoleic acid of safflower oils under alkaline conditions	Kennedy et al. (2010)
Physical properties	Liquid at room temperature, colorless and odorless, lipid soluble	Aydin (2005), Uehara et al. (2008)
Site of action	Adipose tissue, muscle tissues, liver, pancreas, immune system and bones	Kennedy et al. (2010)
Mechanism of action	Not fully understood	Kennedy et al. (2010), Pariza et al. (2000)
Main functions of CLA	Anti-cancer Helps in weight management Increases insulin sensitivity Reduces atherosclerosis/helps in reducing LDL cholesterol Antioxidant/anti-inflammatory Supports immune system by exerting some antimicrobial activity Helps in maintaining bone mass density Anti-obesity Reduces atherosclerosis	den Hartigh (2019), Kennedy et al. (2010), Collomb et al. (2006)
Half life	Depend on its different isomers	Lock et al. (2008)
Elimination through	Urine and feces	Lock et al. (2008)
Daily recommended dose	For weight management: 1.8–7 gm For exercise performance: 3–6 gm	Benjamin et al. (2015), Lehnen et al. (2015)
Overdose may cause	Stomach issues, extremely high doses can cause insulin resistance. Increase in liver fat may occur	RISerus et al. (2002)
Adverse reactions	Stomach upset, insulin resistance, liver fat development, allergic reactions and changes in blood sugar level	Vyas et al. (2012)
Interaction with other drugs	Synergistic activity with statin, increases insulin resistance and can increase retinol levels.	Benjamin et al. (2015)

since long. After any major illness and for health benefits, doctors start CLA as a supplement to health immediately after symptoms of disease starts subsiding. It is expected that in the future, CLA will have more attraction and benefit many people, especially after pandemic era, in which people are being sedentary and there is very limited mobility (Baghi et al. 2016; Liu et al. 2023).

5.2.5 *Creatine*

Creatine is an organic compound that is present in modest quantities in meals such as meat and fish. The human body also produces it in trace amounts, mostly in the liver, kidneys, and pancreas (Kreider et al. 2022; Wuertz and Reiser 2023; Wyss and Kaddurah-Daouk 2000). In energy metabolism, creatine is especially important, especially during brief, high-intensity exercises. The body changes creatine whether from food or supplements—into phosphocreatine. During vigorous physical activity, phosphocreatine acts as an easily accessible energy source that the muscles may use quickly (Rawson et al. 2002; Wax et al. 2021).

In order to improve the body's production of ATP, creatine is essential. The main energy unit of cells, adenosine triphosphate (ATP), is needed for many cellular functions, including the contraction of muscles during exercise (Guzun et al. 2011; Bogdanis et al. 2007). Creatine is absorbed into the bloodstream and absorbed by muscle cells when it is consumed or supplemented. Inside the muscle cells, creatine interacts with a phosphate molecule to generate phosphocreatine through an enzyme called creatine kinase (Brosnan and Brosnan 2016; Guzun et al. 2011; Wallimann et al. 2011). The muscles' reserves of ATP are rapidly used up during high-intensity exercise or any activity requiring a sudden surge in energy. This is a result of the quick breakdown of ATP, which releases energy for muscular contractions (Maclaren and Morton 2011; Atakan et al. 2021). However, the supply of ATP is scarce, and it must be replaced to maintain muscle function. People who use creatine supplements can raise the levels of phosphocreatine and creatine in their muscles, which expands the pool of phosphate groups that are easily available for ATP the resynthesis. This elevated phosphocreatine availability can improve the ability to engage in high-intensity, brief workouts and expedite ATP regeneration during recovery intervals (Cooper et al. 2012; Casey and Greenhaff 2000).

Creatine supplementation has been linked to improved muscle mass and strength improvements, notably among resistance training athletes. It may increase protein synthesis, decrease muscle protein degradation, and increase the water content of muscle cells, resulting in larger muscles (Aguiar et al. 2013; Gotshalk et al. 2002). Creatine has also been linked to improved muscle recovery following exercise (Baird et al. 2012; Jiaming and Rahimi 2021). Creatine has been linked to increased protein synthesis, the process by which the body produces new muscle proteins. Creatine may help with muscle repair and growth by increasing protein synthesis (Cooke et al. 2009; Farshidfar et al. 2017). Furthermore, there is rising study into creatine potential neuroprotective properties. Creatine supplementation has been linked to improved brain health and reduced cognitive loss associated with aging (Benton and Donohoe 2011; Ostojic et al. 2021). It's worth noting that, while there is research supporting creatine possible cognitive benefits, the effects may differ between individuals. Individual responses, dosage, length of supplementation, and baseline cognitive ability can all have an impact on the outcomes (Moriarty et al. 2023).

5.2.6 Vitamin B12

It is one of the vitamins that are needed for RBC production. It contains cobalt; hence it is called as cyanocobalamin (Qudsia and Riaz 2021). The history of identification of Vitamin B12 is very interesting and 2 times noble prize has been given because of working on Vitamin B12. It started when Minot and Murphy started to find out treatment of pernicious anemia, they were the first one who identified that the people who were given special diet containing liver, were having very good improvement on the pernicious anemia, after them Castle was the one who identified a gastric component involved, which he named as intrinsic factor. Finally, the vitamin was isolated and found by Dorothy Hodgkin, which eventually earned her a Nobel prize in 1964 (Minot and Murphy 1926, 1948).

The main sources of Vitamin B12 include Beef liver, chicken, turkey, fishes like salmon, yoghurt, cheese, and egg yolk. The main sources are basically from animal origin, so deficiency of vitamin B12 is very common in people who are vegan and do not use vitamin B12 fortified supplements (Watanabe and Bito 2018; Watanabe 2007). It is also found to be deficient in those cases who got chronic gastritis and their parietal cells of stomach has been damaged, or they got stomach sling surgery to reduce obesity (Vasavada and Sanghavi 2020). It is commercially prepared by bacterial fermentation and then added to the fortified food supplements. Physical properties of vitamin B12 include its bright red color, it is sparingly soluble in water and in crystalline form. Because of being in crystal form it has been used in developing the X-ray films. It stimulates reticulocyte that produce the red blood cells, hence it is one of the essential components for development of RBCs. It is essential for DNA synthesis, that is another reason for its use in pernicious and megaloblastic anemia. Vitamin B12 is found to be an ignored cause of neuropathy in older adults and patients who have malabsorption (Badar 2022; Allen 2012; Chan et al. 2016). It is very important in management of complications of diabetes and diabetic neuropathy as well (Didangelos et al. 2021). According to sources its half-life is about 6 days. When it is given orally, it binds with intrinsic factor, the glycoprotein presents in the gastric lining. Then it is taken to liver and after metabolizing it becomes available for protein binding in the blood. If given as injection, it is readily available for protein binding in the blood. It is excreted in urine and feces. Daily recommended dose for vitamin B12 is age and situation dependent. The requirement increases with pregnant ladies and older adults (Reyes et al. 1923; Vasavada and Sanghavi 2020). Allergic reaction has been reported with vitamin B12, it can cause itching, erythema and urticaria. Caution should be observed specifically for those people who have allergies with cobalt containing compounds. Drugs like omeprazole, metformin, colchicine and neomycin decreases absorption of vitamin B12. Hence those patients who are using these medications since long should always be provided vitamin B12 supplement (Elgharably et al. 2023; Brescoll and Daveluy 2015; Guéant et al. 2022) (Table 5.3).

People have increased awareness about vitamin B12, and its deficiency as compared to past, because of it vitamin B12 use has been increasing. People who are

Table 5.3 Summary of some characteristics and important facts of vitamin B12

Characteristics	Facts	References
Discovered in	1948	Minot and Murphy (1926)
Discovered by	Dorothy Hodgkin	Scott and Molloy (2012), Okuda (1999)
Main sources	Beef liver, chicken and turkey, some fishes like salmon etc., yoghurt and cheese, egg yolk,	Watanabe (2007), Watanabe and Bito (2018)
Industrially prepared from	Fermentation of bacteria	Scott and Molloy (2012)
Physical properties	Bright red in color due to cobalt content, it can form crystal thus it can be useful in X ray film development. Light, heat and acidic conditions degrade the vitamin B12, it is not 100% water soluble.	Reyes et al. (1923), Vasavada and Sanghavi (2020)
Site of action	Stomach, nerves and stimulates reticulocytes.	Reyes et al. (1923)
Mechanism of action	Methionine synthesis and Methylmalonyl-CoA.	Reyes et al. (1923)
Main functions of vitamin B12	Formation of RBCs. Protecting the neurological function, production of energy, mood regulation etc.	Scott and Molloy (2012)
Half life	Approximately 6 days.	Reyes et al. (1923), Vasavada and Sanghavi (2020)
Elimination through	Urine and feces.	Vasavada and Sanghavi (2020), Reyes et al. (1923)
Daily recommended dose	Infants- 0.4 mcg, adults 2.6 mcg	Reyes et al. (1923), Vasavada and Sanghavi (2020)
Deficiency may lead to	Pernicious anemia, megaloblastic anemia, demyelination and neuropathy.	Scott and Molloy (2012)
Overdose may cause	No overdose	Vasavada and Sanghavi (2020), Reyes et al. (1923)
Adverse reactions	Allergic reaction like itching, erythema & wheals.	Reyes et al. (1923), Vasavada and Sanghavi (2020)
Interaction with other drugs	Omeprazole, colchicine, metformin, chloramphenicol, and aminoglycosides like neomycin decrease the absorption of vitamin B12.	Vasavada and Sanghavi (2020)

vegan and they understand the need of vitamin B12, they use vitamin B12 supplements. People having problems like chronic gastritis, diabetes, they are using vitamin B12 on daily basis. Effectiveness of vitamin B12 in behavioral diseases is also established, so people are using it for prevention of psychological disorders. Vitamin B12 has been used to fortify foods for children, to save them from megaloblastic and pernicious anemia (Rizzo et al. 2016; Světnička et al. 2022; Sipponen et al. 2003).

5.2.7 *Carnosine*

Carnosine is primarily found in animal tissues, particularly in skeletal muscle, heart, and brain. The body produces it by combining the amino acids histidine and beta-alanine, and the enzyme carnosine synthase controls this process (Kohen et al. 1988; Solana-Manrique et al. 2022). Superoxide and hydroxyl radicals are two examples of reactive oxygen species (ROS) that carnosine is able to communicate with and scavenge (Cripps et al. 2017; Ma et al. 2020). This scavenging action aids in preventing oxidative damage to lipids, proteins, and DNA that is caused by reactive oxygen species (ROS) (Prokopieva et al. 2016; Hipkiss 2009)... Certain metal ions, like copper and iron, can be bound or chelated by carnosine. Iron and copper, two transition metals, can take part in processes that produce extremely reactive free radicals. Carnosine can inhibit these metals' involvement in oxidative reactions and lower the generation of free radicals by chelating them (Mozdzan et al. 2005; Abate et al. 2022; Chmielewska et al. 2020). Carnosine's antioxidant action is further influenced by its capacity to prevent the production of advanced glycation end products (AGEs). AGEs have the ability to cause inflammation and oxidative damage. Carnosine helps lessen the production of free radicals and the ensuing oxidative damage by blocking the synthesis of AGEs (Houjeghani et al. 2018; Song et al. 2021; Caruso et al. 2019).

It has been discovered that carnosine affects the brain's glutamate and dopamine neurotransmitter systems. One factor that may lead to neurodegenerative illnesses is imbalances in neurotransmitter levels (Banerjee and Poddar 2020; Solana-Manrique et al. 2022). Carnosine may aid in protecting neurons from the excitotoxicity and other neurochemical imbalances by controlling neurotransmitter activity. It has been demonstrated that carnosine has anti-inflammatory qualities, influencing immunological responses and lowering the synthesis of pro-inflammatory chemicals. Carnosine helps shield neurons from inflammatory damage by reducing inflammation. Although carnosine has demonstrated possible neuroprotective effects in a number of studies, more investigation is required to completely comprehend its modes of action and its particular uses in a variety of neurological disorders.

5.3 Recent Advancements and Trends in Animal-Based Compounds

Recent advancements in the research of animal-based bioactive compounds have focused on several key areas:

5.3.1 Marine-Derived Bioactives

Research on the bioactive substances present in marine animals like fish, algae, and shellfish has become increasingly popular (Hamed et al. 2015). These substances include polysaccharides, peptides, omega-3 fatty acids, and other bioactive molecules that have demonstrated promise in decreasing inflammatory disorders, enhancing general wellbeing, and promoting cardiovascular health. Additionally, studies are looking into how marine bioactives might help treat diseases like diabetes, obesity, and neurodegenerative disorders (Elbandy 2022; Lordan et al. 2011; Alves et al. 2020).

The use of bioactives obtained from marine sources in cosmetics and cosmeceutical product development is growing (Martins et al. 2014). Astaxanthin, collagen peptides, and other marine antioxidants are among the compounds being used because of their ability to decrease wrinkles, increase skin suppleness, and shield the skin from UV ray damage (Soni et al. 2023). Furthermore, the development of biopharmaceuticals and medication discovery have demonstrated the encouraging potential of marine-derived bioactives. Marine peptides and compounds generated from sponges are examples of chemicals with distinct biological activity. These activities include antibacterial, anticancer, and anti-inflammatory qualities (Soni et al. 2023; Mohiuddin 2019). Various advances and trends in animal based food has been depicted in Fig. 5.2.

5.3.2 Functional Foods and Nutraceuticals

Recent research has focused on the production of nutraceuticals and functional foods generated from animal-based bioactives. These products have been designed to contain bioactive substances with unique physiological effects in an effort to offer health benefits beyond basic nutrition (Vlaicu et al. 2023; Meng et al. 2022). For instance, the potential antihypertensive, antioxidant, and antibacterial effects of bioactive peptides originating from animal sources are being investigated. These are but just some of the nutraceuticals and functional foods made from animal bioactive substances (Bhat et al. 2015; Albenzio et al. 2017).

Nutraceuticals and functional foods are currently in a state of constant evolution to satisfy shifting customer preferences and needs. The aging population and growing interest in mental health have led to an increasing focus on nutraceuticals and functional foods that enhance brain function and cognitive wellness (Wichansawakun et al. 2022; Kaur et al. 2020). The potential of components like vitamins, antioxidants, and omega-3 fatty acids to improve memory, focus, and general cognitive function is being researched. The development of nutraceuticals and functional meals aimed at improving sports nutrition and performance is still in demand (Stavrinou et al. 2020; Power et al. 2022). Fitness enthusiasts and athletes are searching for items that promote hydration, energy, endurance, and muscle

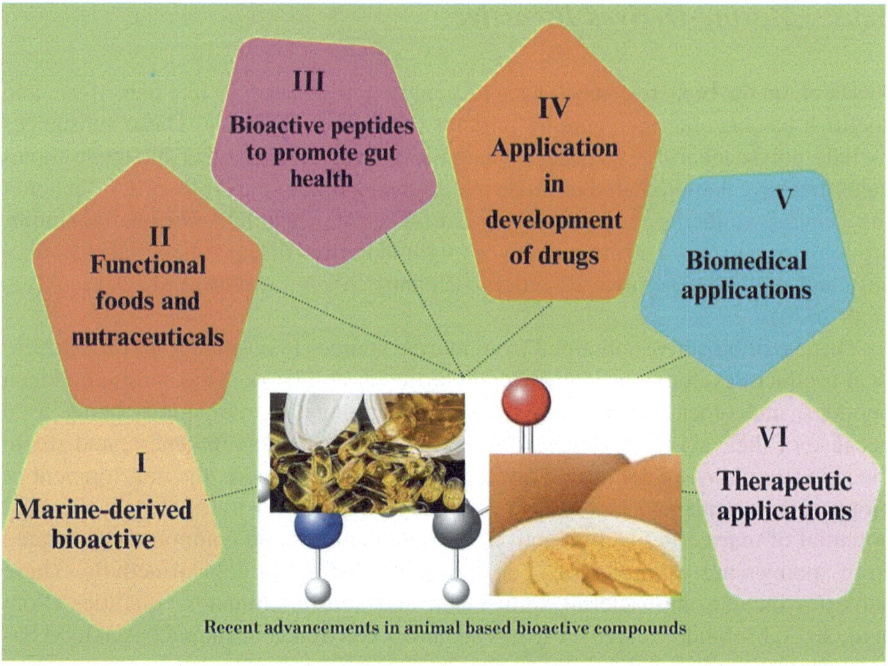

Fig. 5.2 Recent advancement and trends in various applications of animal-derived bioactive compoundsAbbreviations: I: bioactive substances present in marine animals such as polysaccharides, peptides, omega-3 fatty acids; II: Effort to offer health benefits beyond basic nutrition such as bioactive peptides, vitamins, antioxidants, to promote health; III: Bioactive peptides to promote gut health such as casein-derived peptides, such as lactoferrin have antibacterial and immunomodulatory activities, maintaining gut microbiota balance; IV: Application in development of drugs such as bioactive peptides and proteins used as therapeutic carriers; V: Animal-based bioactive compounds have found applications in numerous biomedical areas such as collagen and elastin, derived from animal tissues, are applied in tissue engineering and regenerative medicine, promote cell growth and tissue repair; VI: Animal-based bioactive compounds have revealed potential therapeutic claims for instance certain peptides from animal sources have exhibited cytotoxic properties on cancer cells or the ability to inhibit tumor growth. These compounds can be discovered for their potential in cancer management and drug development

recovery. Functional drinks are becoming more and more popular. Examples include herbal teas, fortified waters, and functional juices (Nazir et al. 2019; Ghoshal and Kansal 2019). These beverages are supplemented with vitamins, minerals, antioxidants, and other bioactive substances to provide targeted health advantages (Ghoshal and Kansal 2019; Gupta et al. 2023). Functional beverages offer convenience and are considered as a functional ingredient into daily activities. These developments reflect changing consumer tastes and an increasing understanding of the link between diet, nutrition, and overall health (Gupta et al. 2023). The functional foods and nutraceuticals sector is constantly evolving to fulfill these demands while providing targeted health advantages (Sun-Waterhouse 2011).

5.3.3 Bioactive Peptides to Promote Gut Health

There is considerable interest in animal-derived bioactive peptides for intestinal health. Peptides derived from casein, collagen, and fish have been investigated for possible prebiotic and probiotic benefits, gut microbiota modification, and gut barrier function enhancement (Mireku-Gyimah 2021). Milk proteins like casein and whey include bioactive peptides that have been linked to a variety of gastrointestinal health benefits (Fernández-Tomé and Hernández-Ledesma 2020; Marcone et al. 2017). Casein-derived peptides, such as lactoferrin and lactoferricin, have antibacterial and immunomodulatory activities, maintaining gut microbiota balance and lowering the risk of gastrointestinal infections (Nagpal et al. 2011; Sienkiewicz et al. 2022). Furthermore, fish proteins offer bioactive peptides with antioxidant, anti-inflammatory, and antibacterial properties. These peptides may lessen inflammation, shield the gut from oxidative stress, and encourage the development of healthy gut flora. Peptides made from fish proteins, such as those found in cod, salmon, and tuna, are among examples (da Rocha et al. 2018; Khora 2013). In addition, gelatin is a protein derived from animal collagen, usually from pig or cow sources. Peptides produced from gelatin have been investigated for possible benefits to gut health. These peptides may lessen intestinal permeability, enhance the development of good gut bacteria, and enhance the function of the gut barrier (Scaldaferri et al. 2014; Gómez-Guillén et al. 2011). It is essential to remember that additional study is required to completely comprehend the modes of action and ideal dosages of these bioactive peptides for the development of gut health.

5.3.4 Application in Development of Drugs

To improve the effectiveness and targeted administration of drugs, bioactive substances derived from animals are being added to drug delivery systems. Bioactive peptides and proteins, for example, are being employed as therapeutic carriers to provide regulated release and enhanced bioavailability. The goal of these developments is to maximize the delivery of bioactive substances for a range of biological uses (Rein et al. 2013; Greek 2013; Van Meer et al. 2015).

A bioactive substance called heparin is extracted from the mucosal tissues of animals. It is frequently taken as an anticoagulant drug to stop blood clots from forming. Heparin is frequently used in operations and functions by preventing the blood's clotting factors from activating (Oduah et al. 2016). Furthermore, a wide range of bioactive substances with possible medical uses can be found in snake venom. Peptides found in snake venom, for instance, have been researched for their antibacterial, analgesic, and anticancer qualities. Certain peptides found in snake venom, such as those found in the venom of the Brazilian pit viper, have been made into medications that are used to treat heart failure and hypertension (Bocian and

Hus 2020; Sarray et al. 2013; Frangieh et al. 2021). Animal-derived antimicrobial peptides are also being studied as possible substitutes for conventional antibiotics because of their distinct modes of action and perhaps decreased susceptibility to the development of resistance. It is noteworthy that the utilization of bioactive substances derived from animals for therapeutic purposes frequently necessitates meticulous extraction and purification procedures to guarantee safety and effectiveness (De Simeis and Serra 2021; Li et al. 2023).

5.4 Challenges in Recent Advancements in Animal-Based Bioactive Compounds

Extraction or use of animal tissues or byproducts is frequently required for the application of bioactive substances derived from animals. This can lead to ethical quandaries, particularly when animals get injured or put in uncomfortable situations. A major challenge is finding a balance between the necessity for scientific progress and ethical considerations. It is imperative to guarantee the sustainability of bioactive substances derived from animals (Archibald 2018; Dothel et al. 2013). Overuse of animal resources can endanger biodiversity and cause ecological imbalances. In order to reduce adverse effects on the environment, researchers must take into account the long-term viability of sourcing these substances (Vabre et al. 2017). While standardization and quality control are also important issues. The composition, potency, and efficacy of bioactive chemicals derived from animals can vary greatly, and they constitute a diverse group. Because these substances are complicated, it can be difficult to establish standardization standards and quality control systems for them (Pathak et al. 2024). Furthermore, it is an essential task to ensuring efficacy and safety in pharmaceutical and biological applications. It is a difficult job to identify and describe novel bioactive chemicals obtained from animals. Nuclear magnetic resonance spectroscopy and mass spectrometry techniques are needed to identify the individual substances. Furthermore, deciphering these constituents, mechanisms of action and its therapeutic uses is a difficult task that requires extensive research (Renou et al. 2004; Dunn et al. 2011).

Certain people may experience allergic responses or negative effects from bioactive substances derived from animals. The safety of these substances must be guaranteed by evaluating and reducing the risk of allergic reactions, especially when they are utilized in food or medicinal goods. Regulations govern the use of bioactive substances derived from animals in a variety of uses, including medications and nutritional supplements. It can be difficult and time-consuming to meet regulatory standards for safety, efficacy, and quality control; substantial research and documentation are needed (Abd Rahim et al. 2023; Gallo et al. 2020; Muthukumar et al. 2020).

5.5 Conclusion

Significant progress has been made in the study of bioactive compounds derived from animals in recent years, which has shown promise for a range of dietary and therapeutic uses. Numerous important substances derived from animals, such as iron, collagen, vitamin B12, conjugated linoleic acid, creatine, and omega-3 fatty acids, show promise as health-promoting substances. These animal-based compounds have been studied for their potential roles in various diseases such as anti-cancer, immunomodulatory, anti-inflammatory, antibacterial, antioxidant, and against cardiovascular diseases, neurological disorders as well as in wound healing. While there has been progress in the extraction and application of animal-based bioactive substances, ethical considerations and sustainable sourcing methods must be followed. All things considered, new directions in the study of bioactive substances derived from animals have created opportunities for the creation of drugs, possible therapeutic interventions, and the investigation of natural remedies for a range of medical ailments. To fully utilize these substances while maintaining their safety and effectiveness, scientists, medical professionals, and industry stakeholders must continue their research and work together.

References

Abate C, Aiello D, Cordaro M, Giuffrè O, Napoli A, Foti C (2022) Binding ability of L-carnosine towards Cu2+, Mn2+ and Zn2+ in aqueous solution. J Mol Liq 368:120772

Abd Rahim MH, Hazrin-Chong NH, Harith HH, Wan WAAQI, Sukor R (2023) Roles of fermented plant-, dairy-and meat-based foods in the modulation of allergic responses. Food Sci Human Wellness 12:691–701

Abeyrathne EDNS, Nam K, Huang X, Ahn DU (2022) Plant-and animal-based antioxidants' structure, efficacy, mechanisms, and applications: a review. Antioxidants (Basel) 11:1025

Adkins Y, Kelley DS (2010) Mechanisms underlying the cardioprotective effects of omega-3 polyunsaturated fatty acids. J Nutr Biochem 21:781–792

Agarwal KN (2001) Iron and the brain: neurotransmitter receptors and magnetic resonance spectroscopy. Br J Nutr 85:S147–S150

Aguiar AF, Januário RSB, Junior RP, Gerage AM, Pina FLC, Do Nascimento MA, Padovani CR, Cyrino ES (2013) Long-term creatine supplementation improves muscular performance during resistance training in older women. Eur J Appl Physiol 113:987–996

Ahmad MF (2019) Ganoderma lucidum: a macro fungus with phytochemicals and their pharmacological properties. In: Plant and human health, Volume 2: Phytochemistry and molecular aspects, Cham, Switzerland, pp 491–515

Ahmad MF (2020) Ganoderma lucidum: a rational pharmacological approach to surmount cancer. J Ethnopharmacol 260:113047

Ahmad MF, Ashraf SA, Ahmad FA, Ansari JA, Siddiquee MRA (2011) Nutraceutical market and its regulation. Am J Food Technol 6:342–347

Ahmad MF, Ashraf SA, Azad ZAA, Panda BP (2013) Production and development of nutraceuticals using bacillus subtilis ncim 2708 under solid state fermentation by response surface methodology. Eur Sci J 1:773–778

Ahmad MF, Ali M, Alsayegh ARA, Ahmad S, Alam N, Wahab S, Ali MS, Athar MT (2021a) A current novel perspective approach for coronavirus disease-2019 pandemic outbreak. J Adv Pharm Technol Res 12:311

Ahmad MF, Wahab S, Ahmad FA, Alam MI, Ather H, Siddiqua A, Ashraf SA, Shaphe MA, Khan MI, Beg RA (2021b) A novel perspective approach to explore pros and cons of face mask in prevention the spread of SARS-CoV-2 and other pathogens. Saudi Pharm J 29:121–133

Ahmad S, Ain HBU, Tufail T, Maqsood M, Bibi S, Ahmad B, Ahmad S, Nasir M, Mushtaq Z, Khan RS (2022) Evaluating the effect of animal-based iron sources on iron deficiency anemia: effect of animal-based iron sources on iron deficiency anemia HOW TO CITE. Pak BioMed J 5:29–33

Albenzio M, Santillo A, Caroprese M, Della Malva A, Marino R (2017) Bioactive peptides in animal food products. Food Secur 6:35

Allen LH (2012) Vitamin b-12. Adv Nutr 3:54–55

Alves A, Sousa E, Kijjoa A, Pinto M (2020) Marine-derived compounds with potential use as cosmeceuticals and nutricosmetics. Molecules 25:2536

Archibald K (2018) Animal research is an ethical issue for humans as well as for animals. J Anim Ethics 8:1–11

Arseni L, Lombardi A, Orioli D (2018) From structure to phenotype: impact of collagen alterations on human health. Int J Mol Sci 19:1407

Ashraf SA, Adnan M, Patel M, Siddiqui AJ, Sachidanandan M, Snoussi M, Hadi S (2020a) Fish-based bioactives as potent nutraceuticals: exploring the therapeutic perspective of sustainable food from the sea. Mar Drugs 18:265

Ashraf SA, Elkhalifa AEO, Siddiqui AJ, Patel M, Awadelkareem AM, Snoussi M, Ashraf MS, Adnan M, Hadi S (2020b) Cordycepin for health and wellbeing: a potent bioactive metabolite of an entomopathogenic medicinal fungus Cordyceps with its nutraceutical and therapeutic potential. Molecules 25:2735

Atakan MM, Li Y, Koşar ŞN, Turnagöl HH, Yan X (2021) Evidence-based effects of high-intensity interval training on exercise capacity and health: a review with historical perspective. Int J Environ Res Public Health 18:7201

Avila Rodríguez MI, Rodríguez Barroso LG, Sánchez ML (2018) Collagen: a review on its sources and potential cosmetic applications. J Cosmet Dermatol 17:20–26

Awuchi CG, Chukwu CN, Iyiola AO, Noreen S, Morya S, Adeleye AO, Twinomuhwezi H, Leicht K, Mitaki NB, Okpala COR (2022) Bioactive compounds and therapeutics from fish: revisiting their suitability in functional foods to enhance human wellbeing. Biomed Res Int 2022:3661866

Aydin R (2005) Conjugated linoleic acid: chemical structure, sources and biological properties. Turk J Vet Anim Sci 29:189–195

Badar A (2022) Neuropsychiatric disorders associated with vitamin B12 deficiency: an autobiographical case report. Cureus 14:e21476

Baghi AN, Mazani M, Nemati A, Amani M, Alamolhoda S, Mogadam RA (2016) Anti-inflammatory effects of conjugated linoleic acid on young athletic males. JPMA J Pak Med Assoc 66:280–284

Baird MF, Graham SM, Baker JS, Bickerstaff GF (2012) Creatine-kinase-and exercise-related muscle damage implications for muscle performance and recovery. J Nutr Metab 2012:960363

Banerjee S, Poddar MK (2020) Carnosine research in relation to aging brain and neurodegeneration: a blessing for geriatrics and their neuronal disorders. Arch Gerontol Geriatr 91:104239

Basak S, Mallick R, Duttaroy AK (2020) Maternal docosahexaenoic acid status during pregnancy and its impact on infant neurodevelopment. Nutrients 12:3615

Benjamin S, Prakasan P, Sreedharan S, Wright A-DG, Spener F (2015) Pros and cons of CLA consumption: an insight from clinical evidences. Nutr Metab (Lond) 12:1–21

Benton D, Donohoe R (2011) The influence of creatine supplementation on the cognitive functioning of vegetarians and omnivores. Br J Nutr 105:1100–1105

Beutler E, Hoffbrand AV, Cook JD (2003) Iron deficiency and overload. ASH Educ Program Book 2003:40–61

Bhat Z, Kumar S, Bhat HF (2015) Bioactive peptides of animal origin: a review. J Food Sci Technol 52:5377–5392
Bocian A, Hus KK (2020) Antibacterial properties of snake venom components. Chem Pap 74:407–419
Bogdanis GC, Papaspyrou A, Maridaki M (2007) Muscle metabolism and fatigue during sprint exercise: effects of creatine supplementation. Serb J Sports Sci 1:37–57
Bonaventura C, Henkens R, Alayash AI, Banerjee S, Crumbliss AL (2013) Molecular controls of the oxygenation and redox reactions of hemoglobin. Antioxid Redox Signal 18:2298–2313
Brescoll J, Daveluy S (2015) A review of vitamin B12 in dermatology. Am J Clin Dermatol 16:27–33
Brosnan ME, Brosnan JT (2016) The role of dietary creatine. Amino Acids 48:1785–1791
Caramia G (2008) The essential fatty acids omega-6 and omega-3: from their discovery to their use in therapy. Minerva Pediatr 60:219–233
Carpenter CE, Mahoney AW (1992) Contributions of heme and nonheme iron to human nutrition. Crit Rev Food Sci Nutr 31:333–367
Caruso G, Fresta CG, Musso N, Giambirtone M, Grasso M, Spampinato SF, Merlo S, Drago F, Lazzarino G, Sortino MA (2019) Carnosine prevents Aβ-induced oxidative stress and inflammation in microglial cells: a key role of TGF-β1. Cells 8:64
Casey A, Greenhaff PL (2000) Does dietary creatine supplementation play a role in skeletal muscle metabolism and performance? Am J Clin Nutr 72:607S–617S
Chalamaiah M, Yu W, Wu J (2018) Immunomodulatory and anticancer protein hydrolysates (peptides) from food proteins: a review. Food Chem 245:205–222
Chan CQH, Low LL, Lee KH (2016) Oral vitamin B12 replacement for the treatment of pernicious anemia. Front Med 3:38
Cherim M, Mustafa A, Cadar E, Lupaşcu N, Paris S, Sirbu R (2019) Collagen sources and areas of use. Eur J Med Nat Sci 2:8–13
Chin SF, Liu W, Storkson JM, Ha YL, Pariza MW (1992) Dietary sources of conjugated dienoic isomers of linoleic acid, a newly recognized class of anticarcinogens. J Food Compos Anal 5:185–197
Chmielewska K, Dzierzbicka K, Inkielewicz-Stępniak I, Przybyłowska M (2020) Therapeutic potential of carnosine and its derivatives in the treatment of human diseases. Chem Res Toxicol 33:1561–1578
Collomb M, Schmid A, Sieber R, Wechsler D, Ryhänen E-L (2006) Conjugated linoleic acids in milk fat: variation and physiological effects. Int Dairy J 16:1347–1361
Cooke MB, Rybalka E, Williams AD, Cribb PJ, Hayes A (2009) Creatine supplementation enhances muscle force recovery after eccentrically-induced muscle damage in healthy individuals. J Int Soc Sports Nutr 6:13
Cooper CE (1999) Nitric oxide and iron proteins. Biochim Biophys Acta 1411:290–309
Cooper R, Naclerio F, Allgrove J, Jimenez A (2012) Creatine supplementation with specific view to exercise/sports performance: an update. J Int Soc Sports Nutr 9:33
Covington MB (2004) Omega-3 fatty acids. Am Fam Physician 70:133–140
Cripps MJ, Hanna K, Lavilla C Jr, Sayers SR, Caton PW, Sims C, De Girolamo L, Sale C, Turner MD (2017) Carnosine scavenging of glucolipotoxic free radicals enhances insulin secretion and glucose uptake. Sci Rep 7:13313
da Rocha M, Alemán A, Baccan GC, López-Caballero ME, Gómez-Guillén C, Montero P, Prentice C (2018) Anti-inflammatory, antioxidant, and antimicrobial effects of underutilized fish protein hydrolysate. J Aquat Food Prod Technol 27:592–608
de Alcântara Rodrigues I, Ferrari RG, Panzenhagen PHN, Mano SB, Conte-Junior CA (2020) Antimicrobial resistance genes in bacteria from animal-based foods. Adv Appl Microbiol 112:143–183
De Oliveira Martinez JP, Cai G, Nachtschatt M, Navone L, Zhang Z, Robins K, Speight R (2020) Challenges and opportunities in identifying and characterising keratinases for value-added peptide production. Catalysts 10:184

De Simeis D, Serra S (2021) Actinomycetes: a never-ending source of bioactive compounds—An overview on antibiotics production. Antibiotics 10:483
den Hartigh LJ (2019) Conjugated linoleic acid effects on cancer, obesity, and atherosclerosis: a review of pre-clinical and human trials with current perspectives. Nutrients 11:370
Dhar Dubey KK, Sharma G, Kumar A (2019) Conjugated linolenic acids: implication in cancer. J Agric Food Chem 67:6091–6101
Didangelos T, Karlafti E, Kotzakioulafi E, Margariti E, Giannoulaki P, Batanis G, Tesfaye S, Kantartzis K (2021) Vitamin B12 supplementation in diabetic neuropathy: a 1-year, randomized, double-blind, placebo-controlled trial. Nutrients 13:395
Dothel G, Vasina V, Barbara G, De Ponti F (2013) Animal models of chemically induced intestinal inflammation: predictivity and ethical issues. Pharmacol Ther 139:71–86
Dunn WB, Broadhurst DI, Atherton HJ, Goodacre R, Griffin JL (2011) Systems level studies of mammalian metabolomes: the roles of mass spectrometry and nuclear magnetic resonance spectroscopy. Chem Soc Rev 40:387–426
Echeverría F, Valenzuela R, Hernandez-Rodas MC, Valenzuela A (2017) Docosahexaenoic acid (DHA), a fundamental fatty acid for the brain: new dietary sources. Prostaglandins Leukot Essent Fat Acids 124:1–10
Elbandy M (2022) Anti-inflammatory effects of marine bioactive compounds and their potential as functional food ingredients in the prevention and treatment of neuroinflammatory disorders. Molecules 28:2
Elgharably N, Al Abadie M, Al Abadie M, Ball PA, Morrissey H (2023) Vitamin B group levels and supplementations in dermatology. Dermatol Reports 15(1):9511
Ems T, St Lucia K, Huecker MR (2017) Biochemistry, iron absorption. StatPearls Publishing, Treasure Island, FL
Farid Y, Bowman NS, Lecat P (2019) Biochemistry, hemoglobin synthesis
Farshidfar F, Pinder MA, Myrie SB (2017) Creatine supplementation and skeletal muscle metabolism for building muscle mass-review of the potential mechanisms of action. Curr Protein Pept Sci 18:1273–1287
Fernández-Tomé S, Hernández-Ledesma B (2020) Gastrointestinal digestion of food proteins under the effects of released bioactive peptides on digestive health. Mol Nutr Food Res 64:2000401
Förstermann U, Sessa WC (2012) Nitric oxide synthases: regulation and function. Eur Heart J 33:829–837
Frangieh J, Rima M, Fajloun Z, Henrion D, Sabatier J-M, Legros C, Mattei C (2021) Snake venom components: tools and cures to target cardiovascular diseases. Molecules 26:2223
Fu Y, Therkildsen M, Aluko RE, Lametsch R (2019) Exploration of collagen recovered from animal by-products as a precursor of bioactive peptides: successes and challenges. Crit Rev Food Sci Nutr 59:2011–2027
Furge LL, Guengerich FP (2006) Cytochrome P450 enzymes in drug metabolism and chemical toxicology: an introduction. Biochem Mol Biol Educ 34:66–74
Gaetani GF, Galiano S, Canepa L, Ferraris AM, Kirkman HN (1989) Catalase and glutathione peroxidase are equally active in detoxification of hydrogen peroxide in human erythrocytes. Blood 73(1):334–339
Gallo M, Ferrara L, Calogero A, Montesano D, Naviglio D (2020) Relationships between food and diseases: what to know to ensure food safety. Food Res Int 137:109414
Garnero P (2015) The role of collagen organization on the properties of bone. Calcif Tissue Int 97:229–240
Ghosh S, Sarkar T, Pati S, Kari ZA, Edinur HA, Chakraborty R (2022) Novel bioactive compounds from marine sources as a tool for functional food development. Front Mar Sci 9:832957
Ghoshal G, Kansal SK (2019) The emerging trends in functional and medicinal beverage research and its health implication. In: Functional and medicinal beverages. Academic, Amsterdam, The Netherlands, pp 41–71

Gómez-Guillén M, Giménez B, López-Caballero MA, Montero M (2011) Functional and bioactive properties of collagen and gelatin from alternative sources: a review. Food Hydrocoll 25:1813–1827
Gotshalk LA, Volek JS, Staron RS, Denegar CR, Hagerman FC, Kraemer WJ (2002) Creatine supplementation improves muscular performance in older men. Med Sci Sports Exerc 34:537–543
Greek R (2013) Animal models in drug development. In: New insights into toxicity and drug testing. In Tech, Croatia/London, pp 123–152
Gross BW, Gillio M, Rinehart CD, Lynch CA, Rogers FB (2017) Omega-3 fatty acid supplementation and warfarin: a lethal combination in traumatic brain injury. J Trauma Nurs | JTN 24:15–18
Guéant J-L, Guéant-Rodriguez R-M, Alpers DH (2022) Vitamin B12 absorption and malabsorption. Vitam Horm 119:241–274
Gupta C (2014) Role of iron (Fe) in body. IOSR J Appl Chem 7:38–46
Gupta H, Zioupos P (2008) Fracture of bone tissue: the 'hows' and the 'whys'. Med Eng Phys 30:1209–1226
Gupta A, Sanwal N, Bareen MA, Barua S, Sharma N, Olatunji OJ, Nirmal NP, Sahu JK (2023) Trends in functional beverages: functional ingredients, processing technologies, stability, health benefits, and consumer perspective. Food Res Int 170:113046
Guzun R, Timohhina N, Tepp K, Gonzalez-Granillo M, Shevchuk I, Chekulayev V, Kuznetsov AV, Kaambre T, Saks V (2011) Systems bioenergetics of creatine kinase networks: physiological roles of creatine and phosphocreatine in regulation of cardiac cell function. Amino Acids 40:1333–1348
Ha Y, Grimm N, Pariza M (1987) Anticarcinogens from fried ground beef: heat-altered derivatives of linoleic acid. Carcinogenesis 8:1881–1887
Hambræus L (1999) Animal-and plant-food-based diets and iron status: benefits and costs. Proc Nutr Soc 58:235–242
Hamed I, Özogul F, Özogul Y, Regenstein JM (2015) Marine bioactive compounds and their health benefits: a review. Compr Rev Food Sci Food Saf 14:446–465
Henriksen K, Karsdal M (2024) Type I collagen. In: Biochemistry of collagens, laminins and elastin. Academic Press: Cambridge, MA
Hexsel D, Zague V, Schunck M, Siega C, Camozzato FO, Oesser S (2017) Oral supplementation with specific bioactive collagen peptides improves nail growth and reduces symptoms of brittle nails. J Cosmet Dermatol 16:520–526
Hipkiss AR (2009) Carnosine and its possible roles in nutrition and health. Adv Food Nutr Res 57:87–154
Houjeghani S, Kheirouri S, Faraji E, Jafarabadi MA (2018) L-Carnosine supplementation attenuated fasting glucose, triglycerides, advanced glycation end products, and tumor necrosis factor–α levels in patients with type 2 diabetes: a double-blind placebo-controlled randomized clinical trial. Nutr Res 49:96–106
Huang J, Sathivel S (2008) Thermal and rheological properties and the effects of temperature on the viscosity and oxidation rate of unpurified salmon oil. J Food Eng 89:105–111
Hwang SB, Park HJ, Lee B-H (2023) Collagen hydrolysate from the scales of Mozambique tilapia (Oreochromis mossambicus) improve hair and skin health by alleviating oxidative stress and inflammation and promoting hair growth and extracellular matrix factors. Mar Drugs 21:475
Ignarro L (2002) Nitric oxide as a unique signaling molecule in the. J Physiol Pharmacol 53:503–514
Jamnok J, Sanchaisuriya K, Sanchaisuriya P, Fucharoen G, Fucharoen S, Ahmed F (2020) Factors associated with anaemia and iron deficiency among women of reproductive age in Northeast Thailand: a cross-sectional study. BMC Public Health 20:1–8
Jensen FB (2009) The dual roles of red blood cells in tissue oxygen delivery: oxygen carriers and regulators of local blood flow. J Exp Biol 212:3387–3393
Jiaming Y, Rahimi MH (2021) Creatine supplementation effect on recovery following exercise-induced muscle damage: a systematic review and meta-analysis of randomized controlled trials. J Food Biochem 45:e13916

Joana Gil-Chávez G, Villa JA, Fernando Ayala-Zavala J, Basilio Heredia J, Sepulveda D, Yahia EM, González-Aguilar GA (2013) Technologies for extraction and production of bioactive compounds to be used as nutraceuticals and food ingredients: an overview. Compr Rev Food Sci Food Saf 12:5–23

Kannan N, Rao A, Nair A (2021) Microbial production of omega-3 fatty acids: an overview. J Appl Microbiol 131:2114–2130

Kaur H, Agarwal S, Agarwal M, Agarwal V, Singh M (2020) Therapeutic and preventive role of functional foods in process of neurodegeneration. IJPSR 11:2882–2891

Kennedy A, Martinez K, Schmidt S, Mandrup S, Lapoint K, Mcintosh M (2010) Antiobesity mechanisms of action of conjugated linoleic acid. J Nutr Biochem 21:171–179

Khatri M, Naughton RJ, Clifford T, Harper LD, Corr L (2021) The effects of collagen peptide supplementation on body composition, collagen synthesis, and recovery from joint injury and exercise: a systematic review. Amino Acids 53:1493–1506

Khesin A (2019) Living well with hemochromatosis: a healthy diet for reducing iron intake, managing symptoms, and feeling great. Simon and Schuster. Berkeley, CA

Khora SS (2013) Marine fish-derived bioactive peptides and proteins for human therapeutics. Int J Pharm Pharm Sci 5:31–37

Kim JH, Kim Y, Kim YJ, Park Y (2016) Conjugated linoleic acid: potential health benefits as a functional food ingredient. Annu Rev Food Sci Technol 7:221–244

Kohen R, Yamamoto Y, Cundy KC, Ames BN (1988) Antioxidant activity of carnosine, homocarnosine, and anserine present in muscle and brain. Proc Natl Acad Sci 85:3175–3179

Kreider RB, Jäger R, Purpura M (2022) Bioavailability, efficacy, safety, and regulatory status of creatine and related compounds: a critical review. Nutrients 14:1035

Krupa K, Fritz K, Parmar M (2020) Omega-3 fatty acids

Kulczyński B, Sidor A, Gramza-Michałowska A (2019) Characteristics of selected antioxidative and bioactive compounds in meat and animal origin products. Antioxidants (Basel) 8:335

Lai-Cheong JE, Mcgrath JA (2013) Structure and function of skin, hair and nails. Medicine 41:317–320

Lange KW (2020) Omega-3 fatty acids and mental health. Global Health Journal 4:18–30

Lanz ND, Booker SJ (2015) Auxiliary iron–sulfur cofactors in radical SAM enzymes. Biochim Biophys Acta 1853:1316–1334

Lawrence GD (2010) The fats of life: essential fatty acids in health and disease. Rutgers University Press, New Brunswick

Laye S, Nadjar A, Joffre C, Bazinet RP (2018) Anti-inflammatory effects of omega-3 fatty acids in the brain: physiological mechanisms and relevance to pharmacology. Pharmacol Rev 70:12–38

Lehnen TE, Da Silva MR, Camacho A, Marcadenti A, Lehnen AM (2015) A review on effects of conjugated linoleic fatty acid (CLA) upon body composition and energetic metabolism. J Int Soc Sports Nutr 12:36

Lembke P, Capodice J, Hebert K, Swenson T (2014) Influence of omega-3 (n3) index on performance and wellbeing in young adults after heavy eccentric exercise. J Sports Sci Med 13:151

Li S, Hofstra N, Van De Schans MG, Yang J, Li Y, Zhang Q, Ma L, Strokal M, Kroeze C, Chen X (2023) Riverine antibiotics from animal production and wastewater. Environ Sci Technol Lett 10:1059–1067

Liu X, Du Z, Wang J, Wang Q, Zheng Y, Niu L, Hao C, Xue D, Zhang Y (2023) Association between trans fatty acids and COVID-19: a multivariate Mendelian randomization study. J Med Virol 95:e28455

Lobine D, Rengasamy KR, Mahomoodally MF (2022) Functional foods and bioactive ingredients harnessed from the ocean: current status and future perspectives. Crit Rev Food Sci Nutr 62:5794–5823

Lock A, Rovai M, Gipson T, De Veth M, Bauman D (2008) A conjugated linoleic acid supplement containing trans-10, cis-12 conjugated linoleic acid reduces milk fat synthesis in lactating goats. J Dairy Sci 91:3291–3299

Lordan S, Ross RP, Stanton C (2011) Marine bioactives as functional food ingredients: potential to reduce the incidence of chronic diseases. Mar Drugs 9:1056–1100
Lordan R, Tsoupras A, Zabetakis I (2017) Phospholipids of animal and marine origin: structure, function, and anti-inflammatory properties. Molecules 22:1964
Lozano MC, Trujillo M (2012) Chemical residues in animal food products: an issue of public health. In: Public health–methodology, environmental and systems issues. Rijeka, Croatia, pp 163–175
Lozoff B, Georgieff MK (2006) Iron deficiency and brain development. Semin Pediatr Neurol 13(3):158–165
Lundqvist-Persson C, Lau G, Nordin P, Strandvik B, Sabel K-G (2010) Early behaviour and development in breast-fed premature infants are influenced by omega-6 and omega-3 fatty acid status. Early Hum Dev 86:407–412
Ma H, Zhao J, Meng H, Hu D, Zhou Y, Zhang X, Wang C, Li J, Yuan J, Wei Y (2020) Carnosine-modified fullerene as a highly enhanced ROS scavenger for mitigating acute oxidative stress. ACS Appl Mater Interfaces 12:16104–16113
Maclaren D, Morton J (2011) Biochemistry for sport and exercise metabolism. Wiley, Hoboken, NJ
Marcone S, Belton O, Fitzgerald DJ (2017) Milk-derived bioactive peptides and their health promoting effects: a potential role in atherosclerosis. Br J Clin Pharmacol 83:152–162
Martins A, Vieira H, Gaspar H, Santos S (2014) Marketed marine natural products in the pharmaceutical and cosmeceutical industries: tips for success. Mar Drugs 12:1066–1101
Meng W, Mu T, Marco G-V (2022) Seaweeds and microalgal biomass: the future of food and nutraceuticals. In: Future foods. MA, USA
Minot GR, Murphy WP (1926) Treatment of pernicious anemia by a special diet. JAMA J Am Med Assoc 87:470–476
Minot GR, Murphy WP (1948) Treatment of pernicious anemia by a special diet. Blood 3:8–21
Mireku-Gyimah NA (2021) Bioactivities and health-promoting properties of bioactive peptides. In: Bioactive peptides. CRC Press, Boca Raton, FL
Mohiuddin AK (2019) Skin aging & modern age anti-aging strategies. Int J Clin Dermatol Res 7:209–240
Moon H-S (2014) Biological effects of conjugated linoleic acid on obesity-related cancers. Chem Biol Interact 224:189–195
Mori TA (2014) Omega-3 fatty acids and cardiovascular disease: epidemiology and effects on cardiometabolic risk factors. Food Funct 5:2004–2019
Moriarty T, Bourbeau K, Dorman K, Runyon L, Glaser N, Brandt J, Hoodjer M, Forbes SC, Candow DG (2023) Dose–response of creatine supplementation on cognitive function in healthy young adults. Brain Sci 13:1276
Mozdzan M, Szemraj J, Rysz J, Nowak D (2005) Antioxidant properties of carnosine re-evaluated with oxidizing systems involving iron and copper ions. Basic Clin Pharmacol Toxicol 96:352–360
Mu N, Mehar JG, Mudliar SN, Shekh AY (2019) Recent advances in microalgal bioactives for food, feed, and healthcare products: commercial potential, market space, and sustainability. Compr Rev Food Sci Food Saf 18:1882–1897
Muñoz M, García-Erce JA, Remacha ÁF (2010) Disorders of iron metabolism. Part II: iron deficiency and iron overload. J Clin Pathol
Muthukumar J, Selvasekaran P, Lokanadham M, Chidambaram R (2020) Food and food products associated with food allergy and food intolerance–an overview. Food Res Int 138:109780
Nagababu E, Chrest FJ, Rifkind JM (2003) Hydrogen-peroxide-induced heme degradation in red blood cells: the protective roles of catalase and glutathione peroxidase. Biochim Biophys Acta 1620:211–217
Nagpal R, Behare P, Rana R, Kumar A, Kumar M, Arora S, Morotta F, Jain S, Yadav H (2011) Bioactive peptides derived from milk proteins and their health beneficial potentials: an update. Food Funct 2:18–27

Nairz M, Haschka D, Demetz E, Weiss G (2014) Iron at the interface of immunity and infection. Front Pharmacol 5:152

Nazir M, Arif S, Khan RS, Nazir W, Khalid N, Maqsood S (2019) Opportunities and challenges for functional and medicinal beverages: current and future trends. Trends Food Sci Technol 88:513–526

Nimni ME, Harkness RD (2018) Molecular structure and functions of collagen. In: Collagen. CRC Press. Boca Raton, FL

Oduah EI, Linhardt RJ, Sharfstein ST (2016) Heparin: past, present, and future. Pharmaceuticals 9:38

Okuda K (1999) Discovery of vitamin B12 in the liver and its absorption factor in the stomach: a historical review. J Gastroenterol Hepatol 14:301–308

Oliver L, Dietrich T, Marañón I, Villarán MC, Barrio RJ (2020) Producing omega-3 polyunsaturated fatty acids: a review of sustainable sources and future trends for the EPA and DHA market. Resources 9:148

Ostojic SM, Korovljev D, Stajer V (2021) Dietary creatine and cognitive function in US adults aged 60 years and over. Aging Clin Exp Res 33:3269–3274

Pariza MW, Park Y, Cook M (2000) Mechanisms of action of conjugated linoleic acid: evidence and speculation (44457). Proc Soc Exp Biol Med 223:8–13

Park Y (2009) Conjugated linoleic acid (CLA): good or bad trans fat? J Food Compos Anal 22:S4–S12

Pathak M, Mahunta PN, Sagar NA, Seow E-K, Agrawal RK (2024) Animal products standardization, safety, and traceability. In: Engineering applications in livestock production. Elsevier. United states

Perutz MF (1978) Hemoglobin structure and respiratory transport. Sci Am 239:92–125

Pino JM, da Luz MH, Antunes HK, Giampá SQDC, Martins VR, Lee KS (2017) Iron-restricted diet affects brain ferritin levels, dopamine metabolism and cellular prion protein in a region-specific manner. Front Mol Neurosci 10:145

Power R, Nolan JM, Prado-Cabrero A, Roche W, Coen R, Power T, Mulcahy R (2022) Omega-3 fatty acid, carotenoid and vitamin E supplementation improves working memory in older adults: a randomised clinical trial. Clin Nutr 41:405–414

Prokopieva V, Yarygina E, Bokhan N, Ivanova S (2016) Use of carnosine for oxidative stress reduction in different pathologies. In: Oxidative medicine and cellular longevity, vol 2016, p 2939087

Puntarulo S (2005) Iron, oxidative stress and human health. Mol Asp Med 26:299–312

Qudsia F, Riaz S (2021) Therapeutic effect of folate and cobalamin in diabetics. In: Type 2 diabetes-from pathophysiology to cyber systems. IntechOpen, London

Ramakrishnan U, Semba RD (2008) Iron deficiency and anemia. In: Nutrition and health in developing countries. Totowa (NJ), pp 479–505

Rawson E, Clarkson P, Price T, Miles M (2002) Differential response of muscle phosphocreatine to creatine supplementation in young and old subjects. Acta Physiol Scand 174:57–65

Rein MJ, Renouf M, Cruz-Hernandez C, Actis-Goretta L, Thakkar SK, Da Silva Pinto M (2013) Bioavailability of bioactive food compounds: a challenging journey to bioefficacy. Br J Clin Pharmacol 75:588–602

Renou J-P, Bielicki G, Deponge C, Gachon P, Micol D, Ritz P (2004) Characterization of animal products according to geographic origin and feeding diet using nuclear magnetic resonance and isotope ratio mass spectrometry. Part II: beef meat. Food Chem 86:251–256

Reyes V, Martínez O, Hernández G (1923) National center for biotechnology information. Plant Breeding. Universidad Autónoma Agraria Antonio Narro, Calzada Antonio Narro

Riserus U, Arner P, Brismar K, Vessby B (2002) Treatment with dietary trans 10 cis 12 conjugated linoleic acid causes isomer-specific insulin resistance in obese men with the metabolic syndrome. Diabetes Care 25:1516–1521

Rizzo G, Laganà AS, Rapisarda AMC, La Ferrera GMG, Buscema M, Rossetti P, Nigro A, Muscia V, Valenti G, Sapia F (2016) Vitamin B12 among vegetarians: status, assessment and supplementation. Nutrients 8:767

Salvatore L, Gallo N, Natali ML, Terzi A, Sannino A, Madaghiele M (2021) Mimicking the hierarchical organization of natural collagen: toward the development of ideal scaffolding material for tissue regeneration. Front Bioeng Biotechnol 9:644595
Sarray S, Luis J, Ayeb M, Marrakchi N (2013) Snake venom peptides: promising molecules with anti-tumor effects. InTech, London
Scaldaferri F, Lopetuso LR, Petito V, Cufino V, Bilotta M, Arena V, Stigliano E, Maulucci G, Papi M, Emiliana CM (2014) Gelatin tannate ameliorates acute colitis in mice by reinforcing mucus layer and modulating gut microbiota composition: emerging role for 'gut barrier protectors' in IBD? United European Gastroenterol J 2:113–122
Scott JM, Molloy AM (2012) The discovery of vitamin B12. Ann Nutr Metab 61:239–245
Shahrajabian MH, Sun W (2024) Mechanism of action of collagen and epidermal growth factor: a review on theory and research methods. Mini Rev Med Chem 24:453–477
Sharma S, Dwivedi S, Chandra S, Srivastava A, Vijay P (2019) Collagen: a brief analysis. Oral Maxillofac Pathol J 10:11–17
Sherman VR, Yang W, Meyers MA (2015) The materials science of collagen. J Mech Behav Biomed Mater 52:22–50
Sienkiewicz M, Jaśkiewicz A, Tarasiuk A, Fichna J (2022) Lactoferrin: an overview of its main functions, immunomodulatory and antimicrobial role, and clinical significance. Crit Rev Food Sci Nutr 62:6016–6033
Šimat V, Elabed N, Kulawik P, Ceylan Z, Jamroz E, Yazgan H, Čagalj M, Regenstein JM, Özogul F (2020) Recent advances in marine-based nutraceuticals and their health benefits. Mar Drugs 18:627
Simopoulos AP (2004) Omega-6/omega-3 essential fatty acid ratio and chronic diseases. Food Rev Intl 20:77–90
Simopoulos AP (2016) Evolutionary aspects of the dietary omega-6/omega-3 fatty acid ratio: medical implications. In: Evolutionary thinking in medicine: from research to policy and practice. Cham, pp 119–134
Singh D, Rai V, Agrawal DK (2023) Regulation of collagen I and collagen III in tissue injury and regeneration. Cardiol Cardiovasc Med 7:5
Sipponen P, Laxen F, Huotari K, Härkönen M (2003) Prevalence of low vitamin B12 and high homocysteine in serum in an elderly male population: association with atrophic gastritis and Helicobacter pylori infection. Scand J Gastroenterol 38:1209–1216
Skalnaya MG, Skalny AV (2018) Essential trace elements in human health: a physician's view, vol 224. Publishing House of Tomsk State University, Tomsk, pp 1–222
Solana-Manrique C, Sanz FJ, Martínez-Carrión G, Paricio N (2022) Antioxidant and neuroprotective effects of carnosine: therapeutic implications in neurodegenerative diseases. Antioxidants (Basel) 11:848
Song Q, Liu J, Dong L, Wang X, Zhang X (2021) Novel advances in inhibiting advanced glycation end product formation using natural compounds. Biomed Pharmacother 140:111750
Song Y-S, Zaitoun IS, Wang S, Darjatmoko SR, Sorenson CM, Sheibani N (2023) Cytochrome P450 1B1 expression regulates intracellular iron levels and oxidative stress in the retinal endothelium. Int J Mol Sci 24:2420
Soni N, Dhandhukia P, Thakker JN (2023) Carotenoids from marine bacteria: a natural antioxidant and UV protectant. Int J Adv Health Sci (IJHS) 10:49–62
Stål P, Johansson I, Ingelman-Sundberg M, Hagen K, Hultcrantz R (1996) Hepatotoxicity induced by iron overload and alcohol: studies on the role of chelatable iron, cytochrome P450 2E1 and lipid peroxidation. J Hepatol 25:538–546
Stavrinou PS, Andreou E, Aphamis G, Pantzaris M, Ioannou M, Patrikios IS, Giannaki CD (2020) The effects of a 6-month high dose omega-3 and omega-6 polyunsaturated fatty acids and antioxidant vitamins supplementation on cognitive function and functional capacity in older adults with mild cognitive impairment. Nutrients 12:325
Sun-Waterhouse D (2011) The development of fruit-based functional foods targeting the health and wellness market: a review. Int J Food Sci Technol 46:899–920

Surette ME (2008) The science behind dietary omega-3 fatty acids. CMAJ 178:177–180
Světnička M, Sigal A, Selinger E, Heniková M, El-Lababidi E, Gojda J (2022) Cross-sectional study of the prevalence of cobalamin deficiency and vitamin B12 supplementation habits among vegetarian and vegan children in the Czech Republic. Nutrients 14:535
Tamargo J, Tamargo M (2016) Pharmacokinetics and safety profile of omega-3 polyunsaturated fatty acids. In: Omega-3 fatty acids: keys to nutritional health. Cham, pp 541–584
Thakur A, Joshi V, Thakur N (2019) Immunology and its relation with food components: an overview. Int J Food Ferment Technol 9:1–16
Tolve R, Tchuenbou-Magaia F, Di Cairano M, Caruso MC, Scarpa T, Galgano F (2021) Encapsulation of bioactive compounds for the formulation of functional animal feeds: the biofortification of derivate foods. Anim Feed Sci Technol 279:115036
Uehara H, Suganuma T, Negishi S, Uda Y, Furukawa Y, Ueno S, Sato K (2008) Physical properties of two isomers of conjugated linoleic acid. J Am Oil Chem Soc 85:29–36
Vabre P, Gatimel N, Moreau J, Gayrard V, Picard-Hagen N, Parinaud J, Leandri RD (2017) Environmental pollutants, a possible etiology for premature ovarian insufficiency: a narrative review of animal and human data. Environ Health 16:1–18
Van Meer PJ, Graham ML, Schuurman H-J (2015) The safety, efficacy and regulatory triangle in drug development: impact for animal models and the use of animals. Eur J Pharmacol 759:3–13
Vasavada A, Sanghavi D (2020) Cyanocobalamin
Viguet-Carrin S, Garnero P, Delmas P (2006) The role of collagen in bone strength. Osteoporos Int 17:319–336
Vlaicu PA, Untea AE, Varzaru I, Saracila M, Oancea AG (2023) Designing nutrition for health—incorporating dietary by-products into poultry feeds to create functional foods with insights into health benefits, risks, bioactive compounds, food component functionality and safety regulations. Food Secur 12:4001
Vyas D, Kadegowda AKG, Erdman RA (2012) Dietary conjugated linoleic acid and hepatic steatosis: species-specific effects on liver and adipose lipid metabolism and gene expression. J Nutr Metab 2012:932928
Wahab S, Alshahrani MY, Ahmad MF, Abbas H (2021) Current trends and future perspectives of nanomedicine for the management of colon cancer. Eur J Pharmacol 910:174464
Wallimann T, Tokarska-Schlattner M, Schlattner U (2011) The creatine kinase system and pleiotropic effects of creatine. Amino Acids 40:1271–1296
Watanabe F (2007) Vitamin B12 sources and bioavailability. Exp Biol Med 232:1266–1274
Watanabe F, Bito T (2018) Vitamin B12 sources and microbial interaction. Exp Biol Med 243:148–158
Wax B, Kerksick CM, Jagim AR, Mayo JJ, Lyons BC, Kreider RB (2021) Creatine for exercise and sports performance, with recovery considerations for healthy populations. Nutrients 13:1915
Wendel M, Heller AR (2009) Anticancer actions of omega-3 fatty acids-current state and future perspectives. Anticancer Agents Med Chem 9:457–470
Wichansawakun S, Chupisanyarote K, Wongpipathpong W, Kaur G, Buttar HS (2022) Antioxidant diets and functional foods attenuate dementia and cognition in elderly subjects. In: Functional foods and nutraceuticals in metabolic and non-communicable diseases. United States, pp 533–549
Wieringa FT, Dijkhuizen MA, Berger J (2022) Sources of iron: diet, supplemental, and environmental. In: Nutritional anemia. Springer, Cham
Wuertz S, Reiser S (2023) Creatine: a valuable supplement in aquafeeds? Rev Aquac 15:292–304
Wyss M, Kaddurah-Daouk R (2000) Creatine and creatinine metabolism. Physiol Rev 80:1107–1213
Xu N, Peng X-L, Li H-R, Liu J-X, Cheng J-S-Y, Qi X-Y, Ye S-J, Gong H-L, Zhao X-H, Yu J (2021) Marine-derived collagen as biomaterials for human health. Front Nutr 8:702108
Yadav H, Mahalvar A, Pradhan M, Yadav K, Sahu KK, Yadav R (2023) Exploring the potential of phytochemicals and nanomaterial: a boon to antimicrobial treatment. Med Drug Discov 17:100151

Yagoda M, Gans E (2014) A nutritional supplement formulated with peptides, lipids, collagen and hyaluronic acid optimizes key aspects of physical appearance in nails, hair and skin. J Nutr Food Sci 5

Yiğit AA (2020) Animal and plant-based milk and their antioxidant properties. Vet J Mehmet Akif Ersoy Univ 4:113–122

Yuan Q, Xie F, Huang W, Hu M, Yan Q, Chen Z, Zheng Y, Liu L (2022) The review of alpha-linolenic acid: sources, metabolism, and pharmacology. Phytother Res 36:164–188

Zhu C, Karvar M, Koh DJ, Sklyar K, Endo Y, Quint J, Samandari M, Tamayol A, Sinha I (2023) Acellular collagen–glycosaminoglycan matrix promotes functional recovery in a rat model of volumetric muscle loss. Regen Med 18:623–633

Chapter 6
Advances and Trends in Food Bioactives of Microbial Origin

Mahdieh Iranmanesh, Mitesh Patel, Malvi Surti, and Naheed Mojgani

Abstract Bioactive compounds include both essential and non-essential compounds, which have health effect and can be gained by food chain as natural constituents. These compounds are produced by a diverse range of microorganisms including eukaryotes, archaea, and bacteria and have found potential applications in pharmaceutical, agrochemicals, nutraceuticals and food industries. Some of the biotherapeutic compounds of microbial origin include antimicrobial, antioxidants, immunosuppressive, anticancer, and anti-inflammatory compounds. Among the bioactive compounds of microbial origin, the compounds produced by bacteria including both Gram-positive and Gram-negative are most widely studied. A number of bacteria yielding compounds, such as peptides, polyphenols, polyunsaturated fatty acids, polysaccharides, amino acids, vitamins, etc. have the potential to bestow health benefits when consumed through a food source. Among the Gram-positive bacteria, lactic acid bacteria (LAB) well known for their health benefits are highly favored in food industry owing to their capability to produce a wide range of bioactive compounds during fermentation including bacteriocins, exoploysaccahrides, lactic acid etc. With the recent advancement in the field of biotechnology, a number of research studies has investigated production of a vast array of novel bioactive compounds having immense potential for application in various sectors most importantly the food sector. In recent years, vast number of research has focused on exploring new ways to cultivate and use microbes to produce these valuable com-

M. Iranmanesh
Department of Food Science and Technology, Science and Research Branch, Islamic Azad University, Tehran, Iran

M. Patel · M. Surti
Research and Development Cell (RDC), Parul Institute of Applied Sciences, Parul University, Waghodia, Vadodara, Gujarat, India

Department of Biotechnology, Parul Institute of Applied Sciences, Parul University, Waghodia, Vadodara, Gujarat, India

N. Mojgani (✉)
Biotechnology Department, Razi Vaccine and Serum Research Institute, Agricultural Research, Education and Extension Organization, Karaj, Iran

S. Amir Ashraf, M. Adnan (eds.), *Food Bioactives and Nutraceuticals*,
https://doi.org/10.1007/978-981-97-9723-3_6

pounds for food applications. Presently, Artificial Intelligence (AI) and Machine learning (ML) has allowed prediction and identification of new bioactive compounds and has aided with the scale-up production of bioactive compounds via optimized bio-process strategy. In this comprehensive review we cover potential aspects of microbial-derived bioactive compounds emphasizing especially on the role of lactic acid bacteria as a functional food component.

Keywords Bioactive compounds · Metabolites · Probiotics · Antimicrobial agents · Antioxidants · Fermentation · Functional foods

6.1 Introduction

Over the course of time, the function of food has evolved from merely providing flavor and nourishment to a significant contributor to human health through enhanced functionality. Diet is essential for overall health, acting as a primary defense against a variety of diseases. With the increasing awareness of the direct link between food and health, the demand for functional foods has increased significantly. Beyond its nutritional values, the health benefits of a food product come from the presence of some naturally occurring bioactive ingredients, mainly known as bioactive compounds (Sorrenti et al. 2023). Bioactive compounds are the food constituents that can have a positive impact on the whole body or target specific tissues or cells. However, unlike nutrients, we do not classify these compounds as essential, and there are currently no recommended daily intake values for them. These compounds, when present in food, can serve various purposes, including: (i) enhancing the quality of traditional food by improving its nutritional, sensory, and technological attributes; (ii) creating functional foods offering physiological benefits in terms of essential nutritional elements; (iii) developing nutraceuticals, which are isolated components derived from food or agro-industrial wastes having health benefits; and (iv) developing films for use in smart active and bioactive food packaging solutions (de Oliveira Filho et al. 2021; Reque and Brandelli 2021). In the last couple of years, the application of bioactive compounds in foods has significantly increased, which can be attributed to their beneficial health effects, such as their role in enhancing and protecting the immune system, exhibiting anti-inflammatory properties, mitigating oxidative damage to cells, and the prevalence of chronic noncommunicable diseases (Silva et al. 2019; Alongi and Anese 2021).

Bioactive compounds are often derived in small amounts from plant and animal sources. Among various food sources, these compounds occur naturally in vegetables, fruits, nuts, oils, and whole grains. Some of these bioactive compounds include polyphenols, triterpenes, phytosterols, terpenoids, polysaccahrides, capsaicinoids, carotenoids, lycopene, lignin, tocopherols, flavonoids, alkaloids, tanins, indoles, saponins, glucosinolates, etc. (Makkar et al. 2021). Additionally, these compounds could be obtained from agro-industrial wastes like mango seeds, grape skins, apples, avocados, eggplant peels, etc. (Shirahigue and Ceccato-Antonini 2020). Apart from these sources, microbes such as fungi, yeasts, and bacteria are another major source of these bioactive molecules (Câmara et al. 2020). In the pursuit of low-molecular-weight natural bioactive compounds, microorganisms of three primary categories have been extensively studied, namely eukaryotes, archaea, and bacteria. Several fields of medicine, food, agrochemicals, and industries have benefited from these compounds because of their proven benefits. Some of the significant advantages of these compounds include their antioxidant, inhibitory, and anti-inflammatory properties, suppression or activation of receptor functions, induction and inhibition of gene expression, etc. (Correia et al. 2012). Furthermore, a few microbial bioactive metabolites, such as avermectin, nanomycin, rokitamycin, tilmicosin, etc., were shown to possess chemotherapeutic effects, showing their potential as a safe and effective biotherapeutic agent (Liaqat et al. 2021; Giurini et al. 2024). Here in this chapter, we summarize the importance of functional foods and some of the major microbial bioactive compounds, highlighting their inhibitory, antioxidant, anti-inflammatory, and immune-modulatory properties. In our review, we highlight past and present studies describing their therapeutic role, with a particular focus on their anti-cancer, anti-hypertensive, and anti-diabetic properties.

6.2 Functional Foods and Nutraceuticals

Functional foods are foods that have physiological benefits in addition to nutritional value. Functional ingredients in foods can be derived from plants, such as antioxidants, vitamins, fibers, minerals, omega-3-fatty acids, or from bacteria, such as probiotics (Palanivelu et al. 2022). A probiotic LAB's beneficial health effects have been linked to the presence of the microbial metabolites produced during fermentation (Abdul Hakim et al. 2023). Functional foods are those that contain beneficial bacteria (probiotics) in order to enhance the health benefits of that particular food. Despite the lack of a universally accepted definition, functional foods are generally understood as food products that deliver various health benefits when included in one's diet. Some of the main types of functional foods include probiotics, prebiotics, symbiotics (mixtures of probiotics and prebiotics), and nutraceuticals (Damián et al. 2022). These compounds are also called nutraceuticals with a nutrient effect (Biesalski et al. 2009). "Nutraceutical" is a term coined in 1989 by Stephen DeFelice (1992). It is defined "as food or parts of food that provide medical or health benefits, including the prevention and treatment of disease" (Dillard and German 2000; Ashraf et al. 2020). Nutraceuticals can be obtained from food, cereals, beverages, and herbal products. There is increasing interest in identifying and integrating bioactive components into food products to meet medically established standards and support population health.

6.2.1 Microbial Bioactive Compounds

A diverse range of microorganisms, including archeae, bacteria, molds, and yeast, play a role in the bioconversion and biotransformation of various food constituents. Microbial food sources are gaining popularity owing to their sustainable production, potential health benefits, and increasing health consciousness among consumers (Fig. 6.1). As a result, functional foods, which offer health benefits beyond basic nutrition, have become increasingly popular, leading to increased research into food-derived bioactive compounds.

Metabolism is a process that includes catabolism and anabolism. It is required for life-sustaining chemical transformations in any organism, as metabolic intermediates, which are small molecules termed metabolites. The significance of these microorganisms lies in their ability to produce a large number of metabolites with proven health benefits (Pessione and Cirrincione 2016). Metabolites are divided into two groups: microbial primary metabolites like ethanol, lactic acid, and amino acids, which are essential for their growth, multiplication, reproduction, and development, while secondary microbial metabolites such as pigments, growth hormones, antibiotics, and mycotoxins are beneficial for other living organisms for their improved metabolic processes (Gobbetti et al. 2010). These bioactive metabolites are recognized as significant sources of life-saving drugs (Demain 2014). In humans and animals, both endogenous and exogenous microbes have health impacts. This is owing to their capacity to produce a range of bioactive components with health benefits. Among the endogenous symbiotic microorganisms, also

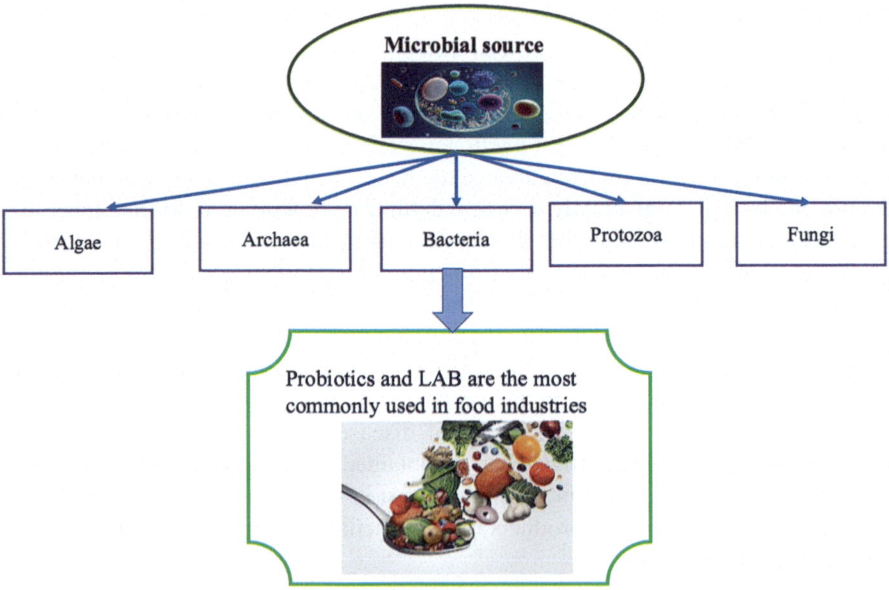

Fig. 6.1 Microbial source of bioactive compounds

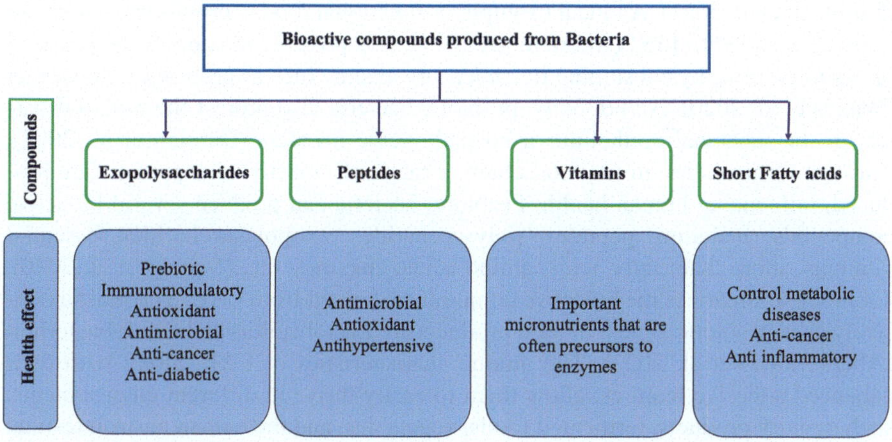

Fig. 6.2 Various bioactive compounds produced from bacteria and their health effect

referred to as "microbiota," has the potential to modulate gene expressions, induce preferential food consumption, affect pH levels, redox potentials, and the balance between pro-inflammatory and anti-inflammatory cytokines (Belkaid and Hand 2014). All of these ultimately could have an impact on brain functions, metabolism, immune functions, and several homeostatic processes. While the exogenous microbiota achieved through bacteria and yeasts derived from food sources can also impact our nutritional parameters, metabolism, oxidative status, immune response, appetite, and behavior (Arora et al. 2013). As described earlier, microbial derived bioactive metabolites represent a diverse array of metabolites that exert beneficial effects on the environment, including humans, plants, and animals. The most fundamental microbial metabolites, including exopolysaccharides, fatty acids, vitamins, amino acids, peptides, etc., have a wide range of applications in the medical sector (Fig. 6.2).

6.3 Bioactive Compounds Produced by Probiotic Bacteria

Fermented products commonly contain bacteria derived from food, such as Lactic Acid Bacteria (LAB). These bacteria, also referred to as probiotics, are well known for their safety and health-promoting properties. As functional food ingredients, these beneficial bacteria are widely categorized and are well known for delivering health benefits (Begum et al. 2017; Palanivelu et al. 2022). The term probiotic was first hypothesized by the Russian Nobel laureate E´ lie Metchnikoff in 1908, who laid the concept that the manipulation of the intestinal microbiome composition with certain beneficial bacteria could improve health and prolong life (Vasiljevic and Shah 2008). Specifically, FAO/WHO defines probiotics as "live microorganisms that confer health benefits on their hosts when administered adequately."

(Salminen et al. 2021). An ideal example of functional food is probiotics, which are artificially added to food (Anonymous 2017). It is possible to support the function of our microbiota by consuming live microorganisms, such as probiotics, in our diet (Wieërs et al. 2020). Additionally, probiotic bacteria also protect the mucosal barrier of the intestinal epithelium from pathogenic invaders (Damián et al. 2022). Through fermentation in the food chain, a range of bioactive components are produced, influencing human health. Probiotic bacteria can produce several bioactive compounds, including peptides, polysaccharides, exopolysaccharides, essential vitamins, short-chain fatty acids, amino acids, enzymes, etc. (Mathur et al. 2020). Table 6.1 summarizes the bioactive compounds derived from LAB. The most extensively used probiotic genera in food products are Lactobacillus and Bifidobacterium (Abdel Tawab et al. 2023). The unique characteristics of LAB, along with their enhanced adhesive features, allow them to easily thrive in different environments, such as dairy products, fermented foods, vegetables, and the human gastrointestinal tract (Bintsis 2018). In the food and dairy industry, LABs are highly useful since they produce organic acids and other metabolites necessary for flavor production, inhibit spoilage bacteria, and serve other purposes during fermentation. (Ayivi et al. 2020). In the below sections, we describe some of the essential properties and health benefits of bioactive compounds produced by probiotic LAB species, especially in food sources.

6.3.1 Antibacterial Activity

A number of secondary metabolites produced by probiotic bacteria show a wide spectrum of inhibitory actions against pathogens and harmful bacteria, such as hydrogen peroxide, acids, bacteriocin-like inhibitory substances (BLIS), and bacteriocins (O′ Toole and Cooney 2008). Bacteriocins are among these inhibitory metabolites that have been extensively studied. Bacteriocins are small, helical, heat-stable, and ribosomal synthesized peptides produced by certain bacteria (Indira et al. 2019). As antimicrobial compounds, these compounds form pore-forming structures, perturb membranes, or inhibit cellular division (Lopetuso et al. 2019). Diop et al. (2007) identified bacteriocin from lactic acid bacteria (LAB) isolated from traditional fermented food from Senegal. The identified bacteriocin from *Lactococcus lactis subsp. lactis* and *Enterococcus faecium* showed activity against *Listeria monocytogenes* and *Bacillus coagulans,* while only bacteriocin from *Lactococcus lactis* has activity against *Bacillus cereus*. Another study evaluated the antimicrobial activity of LAB isolated from home-made fermented vegetables against *Escherichia coli, Staphylococcus aureus, Yersinia enterocolitica,* and *Bacillus cereus* (Kazemipoor et al. 2012). In addition, the antimicrobial activity of LAB against fungi and yeast was evaluated by Cizeikiene et al. (2013). They showed that organic acids and bacteriocins like inhibitory substances (BLIS) produced by *Lactobacillus sakei, Pediococcus acidilactici,* and *Pediococcus pentosaceus* exhibit inhibitory activity against *Fusarium culmorum, Penicillium chrysogenum,*

Table 6.1 Bioactive compounds produced by LAB during food fermentation

Bioactive compounds	LAB species	Food Product	References
Bacteriocin	*Lactobacillus plantarum*	Traditional fermented foods such as "tempeh", "tempoyak" and "tapai"	Kormin et al. (2001)
	Lactobacillus	Carrot fermentation	Joshi et al. (2006)
	Weissella cibaria	Fermented meat and fish	Pringsulaka et al. (2012)
	Lactococcus lactis 19.3, *Lactobacillus plantarum* 26.1, *Enterococcus durans* 41.2,	Dairy products	Grosu-Tudor et al. (2014)
	P. pentosaceus	Fermented sausage of Thailand	Swetwiwathana and Visessanguan (2015)
Antioxidant activity	*Leuconostoc mesenteroides* ssp. *cremoris*, *Lactobacillus jensenii* and *Lactobacillus acidophilus*	Dairy starter cultures	Virtanen et al. (2007)
	Bifidobacterium bifidum, Bifidobacterium lactis	Korean fermented food products	Kang et al. (2021)
	Lactobacillus pentosus; L. plantarum and Pediococcus pentosaceus	Traditional fermented tea	Unban et al. (2021)
	Pediococus pentosaceus	Fermented meat products	Lepecka et al. (2023)
	Lactobacillus plantarum NCU001563 and *Streptococcus thermophilus*	Soymilk	Madjirebaye et al. (2022)
	Lactobacillus plantarum	Fermented Tibetan yak milk	Aili et al. (2023)
Anti-inflammatory activity	*Streptococcus salivarius*	Soymilk	Liao et al. (2010)
	L. plantarum S-SU1 and L. lactis S-SU2	Fermented fish	Kuda et al. (2014)
	Lactobacillus paracasei	Fermented milk	Zagato et al. (2014)
	Lactobacillus helveticus	Koumiss	Rong et al. (2015)
	L. paraplantarum	Fermented foods	Devi et al. (2018)
	L. plantarum KU15149	Homemade diced-radish kimchi	Han et al. (2020)
	Lactiplantibacillus plantarum	Traditional fermented meat	Kostelac et al. (2023)

(continued)

Table 6.1 (continued)

Bioactive compounds	LAB species	Food Product	References
Exopolysaccharides	*L. plantarum 70810*	Fermented soymilk	Li et al. (2014)
	Lactobacillus delbrueckii ssp. bulgaricus OLL1073R-1	Yogurt	Makino et al. (2016)
	Lactobacillus fermentum, Lactobacillus plantarum	Malaysian fermented foods	Khalil et al. (2018a)
	L.plantarum YO175, L.plantarum OF101, P. pentosaceus OF31, W. confusa OF126 and W. confusa WS90	Cereal-based fermented foods	Adesulu-Dahunsi et al. (2018b)
	L. plantarum, L. fermentum, L. crispatus, L. reuteri, and L. pentosus	Tempoyak	Khalil et al. (2018b)
Amino acids	*Lactobacillus casei*	Fermented sour-milk drink	Kakimov et al. (2017)
Vitamins	*Lactobacillus casei*	Fermented banana-milk drink	Drywien et al. (2015)
Short fatty acids	*Lactobacillus acidophilus and L. casei*	Dahi	Yadav et al. (2007)
	Bifidobacterium animalis subsp. lactis HN019 in co-culture with Streptococcus thermophilus TA040 and Lactobacillus delbrueckii subsp. bulgaricus LB340	Fermented milks	Florence et al. (2012)
	Pediococcus pentosaceus	Fermented milk	Balakrishnan and Agrawal (2014).
	Lactobacillus plantarum IIA-2C12 and L. acidophilus IIA-2B4.	Fermented sausages.	Arief et al. (2016)
	Bifidobacterium bifidum, Bifidobacterium lactis	Korean fermented food products	Kang et al. (2021)

Aspergillus fumigatus, Aspergillus versicolor, Penicillium expansum, Aspergillus niger, Debaryomyces hansenii, and *Candida parapsilosis*. They also used the mentioned LAB strains as biopreservatives in bread to lower ropiness. The identified LAB as *Lactobacillus animalis, Lactobacillus rhamnosus, Lactobacillus fermentum,* and *Lactobacillus reuteri* showed activity against all the pathogens. The antimicrobial activity of *L. lactis and L. paracasei* isolated from buttermilk against *Salmonella enteritidis* was related to hydrogen peroxide production (Iranmanesh et al. 2014).

(i) **Antioxidant activities**

Antioxidants could be supplied through food sources or from bacteria present in food. One of the major bacterial groups applied to dairy and non-dairy food products includes LAB species. These beneficial bacterial species possess significant antioxidant activity against reactive oxygen species (ROS). ROS can damage body cells and cause a lot of human diseases, such as cancer, atherosclerosis, and diabetes (Beckman and Ames 1998). ROS is unstable and rapidly reacts with other groups or substances in the body, damaging cells or resulting in disease. Although the body has its own defense system against ROS, it is not enough to prevent damage; therefore, more antioxidants are required to prevent these damages and restore vital functions of our body.

A study examined the antioxidant activity of 25 LAB strains used for milk fermentation by Virtanen et al. (2007). Results indicated that among isolated *Leuconostoc mesenteroides* ssp. *cremoris* strains, *Lactobacillus jensenii* and *Lactobacillus acidophilus* exhibited the highest antioxidant activity. It has been shown that patients with type 2 diabetes have decreased levels of superoxide dismutase (SOD), glutathione peroxidase (GPx), and catalase (CAT), which scavenge reactive oxygen species (Rahbani-Nobar et al. 1999). In a study conducted by Ejtahed et al. (2012), antioxidant status was examined in patients with type 2 diabetes. According to these researchers, consumption of probiotic yogurt containing *L. acidophilus* and *B. lactis* (300 g/d for 6 weeks) improved fasting blood glucose and antioxidant status, while serum malondialdehyde (MDA) concentration decreased.

A study of fermented cereals such as adlay, chestnut, lotus seed, and walnut by *B. subtilis* and *L. plantarum* demonstrated antioxidant activity in these cereals, according to Wang et al. (2014). After fermentation, they found that total flavonoids and phenolics increased, indicating antioxidant activity. It was found that phenolic content was higher in cereals fermented by *Lactobacillus*, while flavonoids were higher in samples fermented by *B. subtilis*. Additionally, DPPH radicals were inhibited by Bacillus fermented cereal, and walnuts provided the highest effective DPPH radical scavenging. In goji juice fermented by *B. velezensis*, *B. licheniformis*, and *L. reuteri* mixed with *L. rhamnosus* and *L. plantarum,* the mixture showed significant enhancement in antioxidant capacity (Liu et al. 2019). There was a significant increase in total phenolics and flavonoids in this study. In another study, the antioxidant activity of 23 LAB strains isolated from raw fermented meat products was assayed by Lepecka et al. (2023). According to their findings, *P. pentosaceus* and *L. plantarum* had the highest DPPH radical and ABTS radical scavenging, respectively.

(ii) **Anti-inflammatory activity**

Inflammation is a biological response to harmful stimuli in the human body and is related to the activity of some enzymes such as cyclooxygenase, lipoxygenase, tyrosine kinase, phospholipase A2, and protein kinase C (Ciumărnean et al. 2020). Inflammation can also cause other diseases such as diabetes, asthma, cardiovascular diseases, and cancer. Microorganisms reduce inflammation by producing compounds like omega-3 fatty acids, cyclic peptides, antimicrobial peptides, oligosaccharides, and polysaccharides (Jenab et al. 2020).

The anti-inflammatory effect of LAB in yogurt fermented by St. thermophilus CRL807 and *L. delbrueckii subsp. bulgaricus* CRL86 in vivo was analyzed by Del Carmen et al. (2016). They found that mice receiving fermented yogurt had lower inflammation than those not given fermented milk. In addition, the group of mice that received fermented yogurt had a lesser number of infiltrating cells in their large intestines. In addition, they had fewer thickenings of the muscle layer than the inflamed control groups. In a study by Kim et al. (2023), it was indicated that L. brevis isolated from fermented foods in South Korea produced γ-aminobutyric acid (GABA). γ-Aminobutyric acid (GABA) is a non-protein amino acid that can improve the production of the anti-inflammatory mediator TGF-β1 and reduce the production of inflammatory mediators, such as interleukin-1 beta (IL-1β), tumor necrosis factor-alpha (TNF-α), interferon-gamma (IFN-γ), and interleukin-12 (IL-12), in streptozotocin-treated mice, thus exhibiting anti-inflammatory effects (Soltani et al. 2011). GABA can inhibit Nitric Oxide (NO) and inducible NO synthase (iNOS) production, as well as p-nuclear factor kappa B (NF-kB) activity, showing anti-inflammatory activity (Kim et al. 2023). Similar results showed that L. plantarum isolated from traditional fermented meat products could reduce the DPPH scavenging activity, and also TNF-α production was decreased (Kostelac et al. 2023).

6.3.2 *Exopolysaccharides (EPS)*

A number of LAB species produce exopolysaccharides that act as prebiotics, favoring the growth of probiotic microorganisms (Silva et al. 2019). Prebiotics are defined as "a nondigestible food ingredient that beneficially affects the host through selectively stimulating the growth and/or activity of some bacteria in the colon and therefore improves host health" (Gibson and Roberfroid 1995). Later, the International Association of Probiotics and Prebiotics (ISAPP) modified the definition and stated that "a dietary prebiotic is a selectively fermented ingredient that results in specific changes in the composition and/or activity of the gastrointestinal microbiota, thus conferring benefit(s) upon host health" (ISAPP 2008). Recent studies have explored the health effects of prebiotics, which are the specific substrates utilized by the microbial population residing in the gut and the probiotic microorganisms, consequently supporting the growth of beneficial bacteria that may provide various health advantages (Gibson et al. 2017). Some of the extensively studied prebiotics associated with health benefits include inulin, fructo-oligosaccharides (FOS), oligosaccharides, galacto oligosaccharides (GOS), and lactulose (Bamigbade et al. 2022). Prebiotics are primarily sourced from naturally occurring foods such as breads, cereals, onions, garlic, artichokes, etc. (Varzakas et al. 2018). Several LAB species are essential sources of naturally occurring prebiotics.

The term 'exopolysaccharide' is a mixture of two separate words, '*exo*' means 'external' and '*polysaccharide*' means 'a carbohydrate consisting of multiple sugar monomers. As biopolymers, EPS are natural polymers found in several living systems, including plants, animals, and microbes. Exopolysaccharides (EPS) are produced by bacteria and can be used as natural bio thickeners in the food industry. In addition, EPS has beneficial effects on health, such as modulating the gut microbiome (Jurášková et al. 2022). The main EPS produced by LAB that acts as a prebiotic is dextran (Kothari et al. 2015). Levan is another exopolysaccharide (a soluble dietary fiber) with prebiotic properties (Santos-Moriano et al. 2015). The prebiotic characteristic of EPS produced by 37 LAB strains was studied by Tsuda and Miyamoto (2010). They showed that *P. acidilactici, L. casei* L-49, *L. plantarum, and L. delbrueckii subsp. lactis* exhibited prebiotic activities. Additionally, low prebiotic activities were found with inulin in tested LAB. Miyamoto et al. (2023), indicated EPS produced by *L. mesenteroides* is an indigestible α-glucan which can change gut microbial composition and improve glucose metabolism and energy homeostasis. EPS health benefits are summarized further in the below sections.

(i) **Immunomodulatory properties**

The immune system defends the host against foreign particles, and can be acquired through innate immunity. The first line of defense is innate immunity and comprises skin, phagocytic cells, mucosal barriers, and lymphoid cells, whereas the acquired immune system includes antigen-presenting cells, B cells, and T cells (Rajoka et al. 2020). EPSs modulate the immune system by altering immune system activity in different ways, like cytokines (Singh and Saini 2017). Therefore, the behavior of the immune system can be determined by monitoring cytokine production (levels of NF-α, IL-6, and IL-1β). In a study, it was shown that the EPS produced by *L. paracasei subsp. paracasei* and *L. plantarum* induced -6, TNF-α, and IL-1β production by 101EP and 102EP in Raw 264.7 in a dose-dependent manner (Liu et al. 2011). Furthermore, these EPS were found to promote macrophage growth and induce pro-inflammatory responses in a murine macrophage cell line, Raw 264.7. RAW264.7 cells are extensively used in immunological research of polysaccharide via increasing proliferation, phagocytic ability, NO production, and cytokine secretion. In fact, RAW264.7 cells' viability indicates cytotoxicity and immune activation. The EPS of *B. licheniformis* and *L. mesenteroides*, isolated from Korean fermented kimchi, also enhanced secretion of the anti-inflammatory cytokine IL-10 and prevented secretion of the pro-inflammatory cytokine IL-6 in lipopolysaccharide-stimulated RAW264.7 macrophages by 40.7% and 32.1%, respectively (Kook et al. 2019). Additionally, EPS isolated from *L. helveticus* fermented milk showed immunostimulatory activity (You et al. 2020). They showed that EPS can increase proliferation of RAW264.7 macrophages and enhance phagocytosis, acid phosphatase activity, nitric oxide production, and cytokine production in macrophages.

(ii) **Antioxidant properties**

EPS can reduce malondialdehyde (MDA) production and increase the activities of superoxide dismutase (SOD) and glutathione peroxidase (GSH-Px) (Huang et al. 2020a). The polysaccharides can show antioxidant activity in various ways, such as regulation of antioxidant enzyme activity through the Nrf2/ARE pathway, removing free radicals, and antagonizing NO (Mu et al. 2021). The EPS isolated from *B. animalis* exhibited activity for anti-lipid peroxidation by inhibiting MDA production, as well as stronger scavenging capacity of hydroxyl radicals and superoxide radicals than ascorbic acid (Xu et al. 2011). They also showed that EPS activity is dose-dependent, and at the concentrations of 0.10 and 0.2 mg/mL, the scavenging activity was 40.54 ± 1.40% and 68.11 ± 1.40%, respectively. The mechanism of scavenging superoxide anion may be related to the separation energy of the O-H bond. The higher the number of electron-withdrawn groups attached to polysaccharides, the weaker the O-H bond dissociation energy. Similarly, EPS isolated from *L. plantarum* prevented MDA formation and increased SOD activities and total antioxidant capacities (T-AOC) in a dose-dependent manner (Zhang et al. 2013). In addition, EPS from *Weissella cibaria* isolated from gari (a Nigerian traditional fermented cassava mash) showed antioxidant capacities in a dose-dependent manner. While the EPS and ascorbic acid increased, the scavenging activity also increased (Adesulu-Dahunsi et al. 2018a). They also found that at a concentration of 4 mg/mL, the scavenging activity of the ascorbic acid was 83%, whereas the EPS activity reached up to 88%. All studies indicated that EPS antioxidant activity was higher than the known antioxidant ascorbic acid.

(iii) **Antimicrobial properties**

The antimicrobial activity of EPS against both Gram-positive and Gram-negative bacteria has been studied recently. It seems this antimicrobial activity is due to disrupting the bacterial cell envelope, especially the peptidoglycan layer, by EPS (Abdalla et al. 2021). In addition, EPS might enable the accumulation of secondary metabolites in the growth medium that could adversely affect both Gram-positive and Gram-negative pathogens. In some manner, functional groups in EPS structure interact with bacterial cells, resulting in their antimicrobial activity (Zhou et al. 2019a). The EPS isolated from had activity against seven species of food-spoilage and infection bacteria, including *E. coli, S.* Typhimurium, *P. aeruginosa, V. parahaemolyticus, S. aureus, B. subtilis*, and *B. cereus* (Wu et al. 2010). The results showed that 80 ppm EPS could inhibit all tested bacteria, and the appropriate dose for inhibition was 20 ppm. The EPS produced by *L. kefiranofaciens* isolated from kefir against *L. monocytogenes* and *S. Enteritidis* was evaluated by Jeong et al. (2017). They showed that both pathogens were inhibited by 1% and 2.5% EPS-supplemented nutrient broth; in addition, EPS exhibits bactericidal effects on pathogenic bacteria. The antimicrobial activity of Lactiplantibacillus, Lactococcus, Lacticaseibacillus, and Lactobacillus species isolated from bovine milk against Gram-negative and Gram-positive bacteria was evaluated by Yildiz and Karatas (2018). In their study they found that EPS from these bacteria had antibacterial actions against *B. cereus, S. typhi, E. coli, S. aureus*, and *P. aeruginosa*, while EPS from *Lactiplantibacillus plantarum* exhibited the strongest antibacterial activity. The antimicrobial activity may be related to the sulfate group present in EPS and provides negative charge that helps exert stronger interaction against the cell wall of Gram-positive bacteria that are positively charged (Nehal et al. 2019).

(iv) **Anti-cancer properties**

EPS from Lactobacillus species have shown anti-cancer activities. Cancer is an abnormal growth of tissue caused by irretrievable damage to DNA caused by mutations in certain proto-oncogenes and tumor suppressor genes in cells that exhibit uncontrolled division, leading to a progressive increase in the number of dividing cells (Abd El Ghany et al. 2015). According to the Global Cancer Observatory, cancer is the second leading cause of death in the world and led to about 9.6 million deaths in 2018 (Bray et al. 2018). The inhibitory effects of EPS from *L. helveticus* on three cancer cell lines and on human colonic epithelial HCoEpiC cells indicated that higher concentration (600 μg/mL) and treatment time resulted in higher inhibition in HepG-2 and BGC-82340 about 98 ± 0.62% and 38.75 ± 0.74% respectively (Li et al. 2015). Whereas, at the lowest concentration (50 μg/mL) and incubation period (24 h), the inhibition of c-EPS against HepG-2 and BGC-823 cells were 9.03 ± 0.81% and 8.85 ± 0.78%, respectively. The EPS of *L. pantheris* can induce NO, TNF-α, IL-6 in Ana-1 cells and peritoneal macrophage cells (Sheng et al. 2022). In addition, EPS inhibited the proliferation of human colon cancer (HCT-116), ovarian cancer (A-2780), and gastric cancer (BCG-803) cell lines. The inhibitory activity depends on concentration and time. At a dose of 500 μg/mL, the HCT-116, A-2780, and BCG-803 cells were inhibited at rates of 45.68, 71.55, and 54.50%, respectively. Overall, the anticancer activity of EPS is affected by factors such as chemical composition, MW, chain conformation, and even charge characteristics (Li et al. 2015).

(v) **Anti-diabetic properties**

Diabetes mellitus (DM) is a disorder that causes chronically increased blood glucose levels and is divided into two types: type 1 DM (T1DM) and type 2 DM (T2DM). The occurrence of T2DM is greater than that of T1DM (Consultation 1999). The number of people with DM worldwide reached 422 million in 2014 (Gao 2017). The main incidence of T2DM is disorders of glucose metabolism. Hydrolysis of starch by α-amylase produced glucose, which causes increased blood glucose, and eventually damaged organs (Xie et al. 2008). Additionally, insulin resistance in liver and skeletal muscle cells is mostly related to lacking glycogenesis and dysregulation of insulin signaling (Kazeem and Davies 2016). So, inhibitors of α-amylase and insulin-releasing agents are quite important in T2DM management. In a study, it was reported that the EPS of *L. plantarum* could lower cholesterol levels (42.24%) and α-amylase inhibition (40%) (Dilna et al. 2015). They also found that the inhibition of α-amylase was dose dependent, and at concentrations of 100 and 800 mg/mL, the inhibition was 5% and 40%, respectively. The antidiabetic activity of EPS from *Strep. thermophilus* and *L. bulgaricus* isolated from Labaneh yogurt (naturally fermented without industrial starter cultures) was evaluated by Al-Nabulsi et al. (2022). They found that EPS of *S. thermophilus* and *L. bulgaricus* inhibited α-amylase activity at 47.8% and 53.8%, respectively, at 1 mg/mL, which increased slightly to 60% and 65.7%, respectively, at 5 mg/mL. Furthermore, the EPS of *S. thermophilus* and *L. bulgaricus* had inhibition activities against α-glucosidase of 11.9% and 9.2%, respectively, at 1 mg/mL, which increased to 18.8% and 19.7%, respectively, at 5 mg/mL. They suggested that inhibition may be related to binding the EPS to the active sites of diabetic enzymes. This may result in hypoglycemic effect and limiting their progression. Huang et al. (2020b) also showed that EPS of *L. plantarum* isolated from pickled cabbage inhibited the activity of α-amylase, resulted in decreasing the blood glucose level in T2DM. In addition, the effect of different concentrations of EPS on insulin-resistant HepG2 cells indicated that the supernatant glucose concentration in insulin-resistant HepG2 cells reduced when the EPS concentration decreased from 10–2 to 10–8 M. Although the glucose concentration in HepG2 cells supernatants increased when EPS concentration was more decreased up to 10–10 M. Therefore, only an appropriate concentration of EPS could alleviate cell insulin resistance.

6.3.3 Amino Acids and Peptides

Microorganisms are able to synthesize a large number of nitrogen-bearing molecules like amino acids, amino acid derivatives, and oligopeptides. LAB proteolytic system comprises three fundamental components: cell envelope-bound or cell wall-bound proteinase (CEP), peptide transporters, and intracellular peptidases. CEP is considered the most critical element of the LAB proteolytic system. It plays a pivotal role in protein breakdown, cleaving them into peptides that range from 4 to 30 amino acid residues. Six distinct types of CEPs identified within LAB include (i) PrtP, first characterized in *L. paracasei subsp. paracasei*, *L. lactis subsp. cremoris*, and *L. lactis subsp. lactis;* (ii) PrtB present in *L. delbrueckii subsp. bulgaricus*; (iii) PrtH, found in *L. helveticus;* (iv) PrtS, identified in *S. thermophiles*; (v) PrtR located in *L. rhamnosus;* and (vi) PrtL, found in *L. delbrueckii subsp. lactis* (Rizwan et al. 2023). Among the genus Lactobacillus, *L. helveticus* is the most proteolytic species, possessing the capability to produce diverse bioactive peptides owing to the presence of four different CEPs, each with unique specificities (Raveschot et al. 2018). PrtP genes have been found located on a plasmid in *L. lactis* or on the chromosome of bacteria like *L. bulgaricus, L. johnsonii, L. casei, L. rhamnosus, L. acidophilus*, and *S. thermophilus* (Liu et al. 2010). Proteinases PrtP expression is affected by various factors, like the specific LAB strain and environmental conditions.

The cleavage activity of CEP generates peptides that are released into the extracellular environment, from where they are subsequently taken up by the cells, where they further undergo degradation into amino acids (Savijoki et al. 2006). The production of bioactive peptides is the result of hydrolytic reactions by the proteins involved in this process. The enzymes that produce bioactive peptides are cell wall-bound proteinases (PrtP family) and can be extracellular (Laan and Konings 1989). These peptides can be produced from food proteins and have various health effects such as anti-hypertensive, anti-thrombotic, cholesterol-lowering, metal-chelating, antimicrobial, antioxidant, and immune-modulating (Raveschot et al. 2018).

(i) **Antimicrobial activity**

The peptides with antimicrobial activity are commonly referred to as bacteriocins (mentioned earlier) produced from LAB and divided into three groups: class I—the antibiotics; class II—the heat-stable unmodified bacteriocins; and class III—the larger heat-labile bacteriocins (Nes and Holo 2000).

According to Sutyak et al. (2008), bacteriocin from *B. amyloliquefaciens* isolated from the yogurt-flavored cultured beverage showed antimicrobial activity against *L. monocytogenes*, *G. vaginalis,* and *S. agalactiae.* They indicated that bacteriocin was stable at high temperatures and retained after an hour at 100 °C as well as across the pH range of 2–10. The unknown protein was identical to bacteriocin subtilosin, produced by *B. subtilis.* The *B. cereus* isolated from Ricotta could produce an antimicrobial peptide in the presence of casein, antimicrobially active against several species of *Bacillus* (*B. cereus* ATCC 11778 (BC45) and *B. thuringiensis* USDA HD22) and *L. monocytogenes* (Ouertani et al. 2018). They identified the proteolytic enzymes implicated in the cleavage of the β-casein protein to synthesize antimicrobial peptides and also found two predicted enzymes with caseinolytic activity that can cleave after Lys. and Phe. Simons et al. (2020) found that two strains named *L. sakei* and *L. lactis* isolated from food produced bacteriocin against *L. monocytogenes*. They also found that the bacteriocin of *L. lactis* was NisZ (a natural variant of Nisin A, which has been approved by the World Health Organization as a food preservative). Moreover, the total produced Nisin Z content was 2.32 mg/L, and when the bacterial growth temperature was increased from 30 to 37 °C, the activity against *L. monocytogenes* of the purified Nisin Z was enhanced from 55% up to 65%, with the same yield (around 2 mg/L).

(ii) **Antioxidant activity**

The antioxidant activity of fermented cereals with selected LAB was evaluated by Coda et al. (2012). They used some strains such as *L. alimentarius*, *L. brevis*, *L. sanfranciscensis*, and *L. hilgardii* for cereal fermentation. The results indicated that higher antioxidant activity was found in whole wheat, spelt, rye, and Kamut sourdoughs. Moreover, the identified antioxidant peptides showed DPPH radical-scavenging activity and also inhibited linoleic acid oxidation. The antioxidant activity of peptides from *L. acidophilus* 20552 ATCC (T2) or *L. helveticus* CH 5 (T3), which were used as yogurt starters, was analyzed by Taha et al. (2017). The results showed that both strains could hydrolyze all caseins, while whey proteins could not hydrolyze. Additionally, *L. helveticus* has the ability to hydrolyze L-LG compared to other strains. Both strains could produce low-molecular-weight bioactive peptides from β-casein and fewer came from K- and αs1 casein during fermentation with antioxidant activity. They also reported that antioxidant activity was the highest in milk fermented with *L. acidophilus* (63.99%), and the lowest was seen in samples fermented with *S. thermophilus* and *L. delbrueckii* ssp. *bulgaricus* (39.43%). Another study showed that the use of *L. plantarum* in tomatoes can produce peptides with antioxidant activity (Mechmeche et al. 2017). They suggested that a high content of total and soluble proteins resulted in a higher antioxidant activity. Furthermore, they found a relationship between antioxidant activity and the amino acid composition of tomatoes. The major amino acids of tomatoes were glutamic acid, aspartic acid, arginine, lysine, leucine, and phenylalanine, which are known as antioxidant amino acids (Sarmadi and Ismail 2010).

(iii) **Anti-hypertensive activity**

Compounds that inhibit the angiotensin-converting enzyme (ACE) are used medically to treat human hypertension. The effect of antihypertensive peptides from 5 LAB, including *L. casei, L. acidophilus, S. thermophilus, L. bulgaricus*, and B. longum, for fermentation of soymilk and then added protease, prozyme 6, was evaluated (Tsai et al. 2006). The ACE-inhibitory activity in samples only fermented by LAB was decreased from 9.28 to 2.89 mg powder/mL, whereas 10 h of fermentation with Prozyme 6 facilitated resulted in a lowered IC 50 of ACE from 9.28 to 0.75 mg powder/mL, and decreased to 0.66 mg powder/mL as fermentation time was up to 30 h. Additionally, the molecular weights of ACE-inhibitory peptides ranged from 1300 to 65 Da with four fractions. Also, SHR systolic blood pressure reduced to 19.0 mmHg after 8 weeks of oral administration in a spontaneously hypertensive rat. The production of ACE-inhibiting peptides during milk fermentation by *L. acidophilus, L. casei, L. delbrueckii ssp. bulgaricus, L. helveticus, L. paracasei, L. lactis, L. mesenteroides*, and *P. acidilactici* was tested in a study (Elkhtab et al. 2017). The results indicated that ACE-inhibitory activities were higher after 72 h of incubation than shorter incubation times due to increased proteolysis to a certain level. The highest activity was found in milk fermented by *L. casei*. The isolated peptides of this strain contained 7–12 amino acid residues, with molecular weights of 780.3–1,325.83 Da. In another study, the ability of ACE activity of 34 LAB in fermented whey proteins was compared (Daliri et al. 2018). Among the selected strains, *P. acidilactici* peptides showed the strongest ACE inhibitory activity of 84.7 ± 0.67% (IC50 = 19.78 ± 1.73 μg/mL). They also found that most peptides were generated from the C-terminus of β-CN than the N-terminal due to the presence of several phosphoserine residues in that region.

6.3.4 Vitamins

All living organisms require vitamins as micronutrients. Their function is to act as precursors of intracellular coenzymes, which regulate vital biochemical reactions in the cell. Since humans cannot synthesize most vitamins, they must be obtained from other sources, such as the gut microbiota and diet. Though the adequate requirement for vitamins supplied by diet exists, the risk of low micronutrient intakes in humans was still reported. The B group or B-complex vitamins include thiamine (B1), riboflavin (B2), niacin (B3), pyridoxine (B6), pantothenic acid (B5), biotin (B7 or H), folate (B11–B9 or M), and cobalamin (B12). Humans suffer from deficiencies of these vitamins rather often because they are water-soluble and heat-resistant, so they are easily removed or destroyed during cooking or food processing. *L. lactis* and *S. thermophilus* can synthesize folate; therefore, fermented dairy products, such as yogurt, contain more of this vitamin than non-fermented dairy products (LeBlanc et al. 2007). According to a study, pure and mixed cultures of *S. thermophilus* ST5 and *L. helveticus* R0052 or *B. longum* R0175 did not significantly alter the amount of B1 or B6 in fermented soy beverages (Champagne et al. 2010). The count of vitamin B-group in fermented cashew apple juice with 5 strains, including *L. acidophilus, L. casei, L. plantarum, L. mesenteroides*, and *B. longum*, was analyzed by Kaprasob et al. (2018). The amount of B1 in all fermented samples decreased, although the count of B1 in *L. acidophilus* fermented samples was high (5.44 mg/L) after 48 h, but its value was less than its control. The count of B2 did not differ significantly and was the highest in samples fermented by *L. casei* (137.15 mg/L). The highest B3 was detected in samples fermented with *L. casei* and *L. acidophilus* for 48 h (50.22 and 49.53 mg/L, respectively). Though the level of later vitamin was increased in all fermented samples due to the potential of LAB to synthesize vitamin B3 (Champagne et al. 2010). The vitamin B6 of *L. casei* and *B. longum* fermented samples increased, but the vitamin B6 of *L. acidophilus* and *L. mesenteroides* reduced during 48 h of fermentation. The highest content of B12 was found in samples fermented with *L. plantarum* for 24 h (41.55 mg/L), then by L. casei at 48 h (36.08 mg/L), and *L. acidophilus* at 24 h (35.71 mg/L). The level of B12 in fruits and vegetables is low and only synthesized by bacteria (Shakirzyanova et al. 2002). The presence of B12 depends on the metabolic pathway, which can convert glycerol into propanediol. In another study, the production of vitamins by 5 strains of Lactobacillus (KGL2, KGL3A, KGL4, RNS4, and WTS4) isolated from rice-based traditional fermented foods of Garo Hills, Meghalaya, India was evaluated (Hati et al. 2019). The highest production of B2 (0.72 μg/mL) was found in the KGL3A strain after 36 h. while, RNS4 reported maximum B2 production (0.55 μg/mL) after 24 h, then decreased after 36 h and 48 h, respectively. Additionally, WTS4 produced 0.092 μg/mL folate after 24 h and reduction after 36 h, whereas KGL2, KGL3A, KGL4, and RNS4 showed folate production in the range from 0.05 to 0.082 μg/mL after 12 h, then reduced between 24 and 36 h. The results of B12 production indicated that KGL2 has the highest 0.058 μg/mL production, followed by KGL3A (0.046 μg/mL) after 36 h of incubation.

6.3.5 Fatty Acids (FAs)

Fermentation has been indicated as a tool to enhance the nutritional value of foods and beverages, such as short-chain fatty acids (SCFAs), which have emerged as some of the most studied compounds in the last decade due to their beneficial effect on human health. SCFAs are small organic monocarboxylic acids with different chain lengths, ranging from two to six carbon atoms. They are prevalently produced by gut microbiota as end-products of dietary polysaccharides fermentation, including fiber and resistant starch. The most representative SCFAs are acetic acid (C2), propionic acid (C3), and butyric acid (C4). The SCFAs may be efficiently involved in glycemic control and also play a relevant role in the management of metabolic diseases, including obesity and diabetes. This is where food intake control is crucial. Moreover, propionate reduces liponeogenesis and cholesterol synthesis (Kimura et al. 2011). The production of SCFAs by *L. rhamnosus* strain ASCC 1520 in fermented soymilk added to the diet of mice was evaluated (Dai et al. 2019). The results indicated that propionic acid significantly increased after 4-week administration of *L. rhamnosus*-fermented soymilk ($P < 0.01$). Since soymilk contains considerable amounts of oligosaccharides, such as raffinose and stachyose, which are not digested by humans as they lack β-galactosidase, therefore, soymilk fermented by *L. rhamnosus* may improve the SCFA *in vivo*. The effect of yogurt fermented in various LABs including *S. thermophilus* and *L. bulgaricus* (control, C); (2) *S. thermophilus*, *L. bulgaricus*, and *B. bifidum* (C-Bb); (3) *S. thermophilus*, *L. bulgaricus*, and *L. acidophilus* (C-La); and (4) *S. thermophilus*, *L. bulgaricus*, and *L. gasseri* (C-Lg) on SCFA was analyzed (Chang et al. 2021). Results showed that acetate levels were significantly higher in C-Bb, C-La, and C-Lg yogurts than in C yogurt. Whereas the levels of propionate, butyrate, and lactate were not significantly different between various fermented yogurts. Furthermore, the expression of *MUC2* (a major mucin protein in the mucus layer) and *CDX2* (regulates the expression of *MUC2)* was significantly higher in HT29-MTX cells treated with C-La ($P < 0.005$) as well as C-Bb and C-Lg ($P < 0.05$) yogurts than in those treated with C yogurt. Mucin, a substance produced by the mucus layer on the intestinal epithelium, enhances gut immunity by protecting the epithelial barrier from pathogens. The HT29-MTX cells are human colon epithelium-derived goblet cells that produce mucin and are widely used to observe mucin expression.

6.4 Large Scale Production of Microbial Bioactive Compounds

Natural bioactive compounds encompass a wide range of structures and functions. They offer a valuable array of molecules suitable for the development of nutraceuticals, functional foods, and food additives. At the same time, a multitude of challenges, such as low production titers, difficulties in product isolation, and structural identification in the microbial production of these bioactive compounds, have hindered their large-scale production (Sorrenti et al. 2023).

A number of different strategies have been applied for the production of bioactive peptides by Lactobacillus species or their CEPs. These include functionalization of food products via fermentation, extraction and purification of peptides from fermentation broth, and the use of purified or partially purified CEPs from lactobacillus species. The strategy employed for bioactive compound production depends on the targeted application (Raveschot et al. 2018).

One of the essential techniques used for bioactive compound production from a microbial source includes fermentation. Fermentation is one of the oldest approaches used for product transformation into value-added products through microorganisms. Solid-state fermentation, submerged fermentation, and liquid fermentation are the most commonly used fermentation methods. Fermentation choice depends on the product to be obtained (Subramaniyam and Vimala 2012). Fermentation processes such as submerged fermentation (SmF) and solid-state fermentation (SSF) are used to produce bioactive compounds of industrial interest from a variety of substrates, including waste products. SmF, or liquid fermentation process, includes cultivating the microorganisms in a liquid medium or put them in a water source with nutrients, while SSF is based on the growth of the microorganism and product formation on solid particles in the absence (or near absence) of water (Pandey 2003). Several research studies have shown SSF to be superior to the SmF method, as this method gives higher yields and productivity along with better product characteristics. Moreover, in SSF, low-cost agricultural and agro-industrial residues are used as substrates; therefore, capital and operating costs are lower than SmF (Hölker and Lenz 2005). The main problem of this system is the scaling-up of the process, largely due to heat transfer and culture homogeneity problems (Di Luccio et al. 2004; Mitchell et al. 2000).

In order to develop a successful bioprocess for SSF conditions, some significant factors must be considered. These factors include the selection of a suitable microorganism strain and the solid support to be used. The selection of microorganisms in the SSF process depends on the desired end product (Martins et al. 2011). Through the aid of genetic engineering tools such as the CRISPR-Cas system, particularly CRISPR/Cas9, modifications could be incorporated into the genomes of various microbes. These modifications facilitate strain improvement, leading to enhanced productivity of bioactive compounds and resilience in large-scale fermenters (Wu et al. 2020). These approaches have been used for developing commercially viable strains that produce glutamic acid, lysine, threonine, phenylalanine, etc., with glutamic acid and lysine being the highest annual producers by *E. coli* and *Corneybacterium glutamicum* (Wang et al. 2016; Heider and Wendisch 2015).

To improve bioactive compound production yield, a suitable choice of substrate or mixtures of substrates with appropriate nutrients is crucial (Singh nee' Nigam 2009). Low-cost support materials for microorganism growth are favorable, and these materials can be obtained from residual natural materials produced in agricultural and agro-industrial activities. These residue compounds can also be used as a source of carbon, which could reduce fermentation costs.

Some factors can affect the SSF system, such as pretreatment and particle-size of substrates, medium ingredients, supplementation of growth medium, sterilization of SSF medium, moisture content, water activity (AW), inoculum density, temperature, pH, agitation, and aeration (Singh nee' Nigam and Pandey 2009). Among these factors, moisture content and AW play a significant role in SSF.

Finally, during the SSF process, the selection of a suitable downstream process for the obtained product plays an important role. SSF components can be extracted with solvents (aqueous or other solvent mixtures). In extraction, the type of solvent and its concentration, the ratio of solvent to the solid, and pH are important factors that affect the final product (Martins et al. 2011).

6.5 Control Factors in Scaling-Up of Process

Microorganisms are highly sensitive to their environment, and increasing production levels may interfere with their growth and metabolic functions. This may result in reduced bioactive compound yields. Additionally, the types of substrates employed, food treatment methods, and duration of fermentation are critical elements that impact the fermentation process. In addition, bioactive compounds are very sensitive to environmental factors such as temperature and moisture. As expected, control of such factors becomes difficult during large-scale production (Singh nee' Nigam 2009).

Incubation temperature control in any bioreactor is difficult. In solid substrate fermentation systems, heat transfer occurs via convection and conduction. Solid fermenting biomass uses air as its convective medium, and so the rate of heat exchange is much smaller than in liquid submerged cultivations. Increasing the temperature affects the rate of microbial growth and moisture loss in the process, which results in yield production. While extreme heat production may result in the progressive drying of the solid substrate with time, requiring 10 or more days for secondary metabolites to be synthesised optimally. This is a serious problem, which may affect the yield of products (Singh nee' Nigam 2009). Another factor crucial to solid substrates is moisture. However, control of moisture is difficult due to the combined effects of heat production and moisture release due to microbial respiration (Singh nee' Nigam 2009). With advancements in fermentation techniques and process optimization, some of the challenges are addressed and bioprocess conditions optimized, yet overcoming remaining challenges is still a critical focus in microbial biotechnology (Sadh et al. 2018).

6.6 Encapsulation of Bioactive Compounds

Encapsulation is a technique that allows for the temporary confinement of an active substance within a material that forms a shell. A number of polymers are used for the encapsulation of bioactive compounds, including chitosan, gums (xanthan gum, gum Arabic, etc.), maltodextrin, pectin, starch, whey protein, cellulose, sodium alginate, cellulose galactomannan, etc. (Mirmazloum et al. 2021; Sodeinde et al. 2021). Microcapsules are an effective method of incorporating bioactive compounds into food and beverages, facilitating the creation of innovative functional foods.

Microencapsulation of microbial bioactive compounds involves coating these sensitive ingredients with protective materials to enhance their stability and bioavailability of bioactive compounds, facilitating the controlled delivery and release of various functionally active substances (Jyothi et al. 2010). This technique is particularly useful for preserving probiotics and prebiotics, ensuring they remain active and effective during storage and consumption. Various techniques exist for microcapsule production, including physical methods like freeze-drying, spray drying, solvent evaporation, and precipitation using supercritical fluids. Additionally, physical-chemical methods encompass internal and external ionic gelation, complex coacervation, and liposomes. Furthermore, microencapsulation can also be achieved through chemical methods, such as molecular inclusion complexation and interfacial polymerization (Tyagi et al. 2021). The choice of microencapsulation techniques depends on multiple factors, such as the type of bioactive compound to be encapsulated, their stability and sensitivity, the encapsulating agent, and the related costs of the method (Mahdavi et al. 2014). Most importantly, each method could yield distinct outcomes in the final product, affecting the properties of the bioactive compound in terms of shape, structure, size distribution, and bioavailability (Zabot et al. 2022).

Co-encapsulation represents an additional method that integrates various functional compounds, enabling a single food matrix to exhibit a blend of advantageous and synergistic effects (Chen et al. 2013). During a study conducted by Zaeim et al. (2019), probiotics and prebiotics were successfully encapsulated together. Similarly, Bao et al. (2020) used the co-encapsulation technique to combine the advantageous properties α-tocopherol and resveratrol within a single matrix. Additionally, probiotics and proanthocyanidin rich cinnamon extracts were co-encapsulated to protect them from the harsh effects of the gastrointestinal tract (Holkem et al. 2020). Co-encapsulation is still a relatively unexplored field that necessitates further research to explore its application in the development of functional foods.

Food products can benefit greatly from nanoencapsulation technology (Murthy et al. 2018). This technology has overcome and surpassed some of the limitations associated with bioactive compounds, such as their instability, solubility, bioaccessibility, absorption, and bioavailability. Pateiro et al. (2021) reported that nanoencapsulation is an effective method for safeguarding bioactive substances and is conducive to the development of specifically designed foods, enabling controlled delivery and enhanced bioaccessibility of functional components to target tissues.

6.7 Application of Artificial Intelligence (AI) and Machine Learning (ML)

A healthy diet is one of the most effective ways to avoid a number of diseases. While foods enriched with bioactive compounds could enhance nutritional and health benefits. However, screening of food bioactive compounds includes extraction, isolation, and purification. This is laborious, time-consuming, and expensive, though it may not guarantee the existence of the required bioactive compounds. AI and ML have been recommended and applied recently to accelerate bioactive compound identification and extraction.

AI and ML have been successfully applied to the fields of medicine, environment, food, and other fields. In the food sector, different computational methods have been used for food classification, food calorie estimation, quality detection, and food contamination (Zhou et al. 2019b). While considering the importance of functional foods and microbial bioactive compounds, AI and ML algorithms could be used for: (i) exploring functional components in food; (ii) predicting the bioactivity of microbial compounds based on their chemical structure and properties; (iii) identifying promising microbial strains yielding specific desired bioactive compounds; and (iv) optimizing fermentation processes and predicting optimal conditions for maximum microbial growth and yield of bioactive compounds.

In a research study, feedforward neural networks (FNN), the most popular type of artificial neural network (ANN), were used to determine the bioactivity of fish protein-derived peptides. The results of the study indicated that ANN can be used to predict the biological properties of these peptides (Li et al. 2006). During a study, Ekpenyong et al. (2021) used an AI model to identify optimal fermentation conditions to maximize desired compound production. Response surface methodology (RSM) and ANN methods were compared in terms of their predictive modeling capabilities to optimize fermentation conditions for producing a surface/bioactive glycolipopeptide. The condition factors such as temperature, pH, agitation, and duration were compared by the ANN method to predict and maximize product yield using the matrix of a central composite rotatable design (CCRD). Their studies showed that for optimizing factor levels for maximum glycolipopeptide production, bioprocess conditions such as 32 °C, pH 7.6, agitation speed of 130 rpm, and a fermentation time of 66 h at a combined desirability function of 0.872 were required. Kobayashi et al. (2023) used AI to understand the growth of LAB, *Limosilactobacillus fermentum*. They designed a deep neural network (DNN) using amino acids (AAs) as explanatory variables and LAB growth as the objective variable. The most accurate DNN model was obtained with high accuracy and low mean square errors and predicted that Asp would affect LAB growth. Bayesian optimization (BO) using this model recommended an optimal growth media comprising maximum amounts of Asn, Asp, Lys, Thr, and Tyr and minimum amounts of Gln, Pro, and Ser. Furthermore, this proposed media was empirically validated to promote LAB growth. While Jawan and his colleagues (2021) employed RSM and ANN to enhance bacteriocin production from the *L. lactis* strain. They found that the ANN (R2 = 0.98) methodology provided better estimation points and data fitting than RSM (R2 = 0.79). Furthermore, BLIS production in optimal medium (717.13 ± 0.76 AU/mL) was about 1.40-fold higher than that obtained in non-optimized (520.56 ± 3.37 AU/mL) medium. In recent years, AI and ML have been able to unravel the potential of microbial bioactive compounds intended primarily for food use, as knowledge progresses. Computationally generated data could lead to the discovery of bioactive compounds and microbial strains with high yields.

6.8 Safety Aspects of Bioactive Compounds

With emerging advancements in biotechnology, a vast array of bioactive compounds have been discovered that are finding their way into our food. However, with these advancements, the safety of these compounds for human, animal, and environmental health becomes paramount importance. This urges rigorous safety evaluations. Bioactive compounds obtained during production or added to food may also fall within this "novel food" category. Although many of these bioactive compounds have been used for a long time and are categorized as safe, some concerns might exist regarding their safety and quality. According to reports, although a number of products derived from LAB offer advantages to consumers, conversely, certain compounds, such as biogenic amines (including tyramine and histamine), can negatively impact human health, leading to issues such as allergies, hypertensive crises, and headaches.

Owing to the complex microbial metabolic processes involved in the production of these compounds, a diverse array of chemical and biological activities are expected. Risk assessment of these constituents is related to the food matrix, which may affect how the compounds work in the body. In order to ensure the safety of novel constituents, a multifaceted approach is highly needed to assess the possible potential risks, including their toxicity, allergenicity, and genotoxicity. Nevertheless, because most toxicological investigations are conducted *in vitro* and in animals, the level of proof supporting bioactive intake safety must be increased. To date, just a few investigations into the potential toxicological impact on humans have been undertaken.

6.9 Limitations and Challenges

Although bioactive compounds have found wide applications due to their beneficial and potential health advantages, despite these benefits, there are several factors that hinder their wide use. These compounds are usually unstable in light, humidity, oxygen, and temperature. It is vital that these compounds withstand acidic stomach conditions and resist enzymes throughout the gastrointestinal tract (Wen et al. 2017).

Factors such as molecular instability, solubility, bio-accessibility, absorption, and bioavailability for a number of these bioactive compounds have limited their wide technological applications (Rashwan et al. 2022; Tarone et al. 2020). Another major factor hindering their application is their instability during food processing, storage, and digestion. Moreover, bioactive compounds could be unstable, and their efficacy might be diminished or lost entirely after isolation. To address these challenges, innovative technologies could be developed to mitigate or prevent the loss of functionality of these compounds. Enhancing the understanding of these factors is crucial for improving bioactive compounds technological applications (Nicolescu et al. 2023; Zhou et al. 2022).

6.10 Future Prospective

In the last couple of years, tremendous interest has emerged owing to the significant health benefits associated with microbial bioactive compounds. Furthermore, consumer demand for healthy foods drives the ongoing development of innovative food products and production methods. These products are environmentally friendly, cost-effective, sustainable, and more efficient than their chemical counterparts. Additionally, a thorough characterization of LAB pertinent to food utilizing transcriptomic, proteomic, and metabolomics studies can offer substantial, evidence-based validation of their advantageous effects, thereby serving as a promising approach for the development of more effective functional foods. While detailed investigations into the influence of fermented foods on human physiology represent a compelling field for future research. Among the various emerging synthesis methods, genetically engineered microbes show considerable promise as a viable alternative to traditional chemical processes.

Nanotechnology presents a promising avenue for enhancing the stability of bioactive compounds, thereby expanding their industrial applications (Taouzinet et al. 2023). The integration of these compounds into nanocomposites and the creation of nanocarriers can increase their effectiveness, protect the active substances, improve solubility, and facilitate more precise targeting within the body (Giaconia et al. 2020). As with any nanoencapsulation process, it is crucial to identify the appropriate system for encapsulating the bioactive compound. This involves careful selection of encapsulating materials and techniques to enhance compound preservation. Additionally, evaluating the characterization methods of nanostructures is essential to ensure the retention of the bioactivities of the incorporated compounds (Medeiros et al. 2020). Achieving an optimal combination of biomaterials, methods, and bioactive compounds for food applications necessitates extensive research, experimentation, and innovation.

6.11 Conclusions

Recent advances in the microbial production of bioactive compounds in food focus on the biosynthesis and enhancement of safe, efficacious, and stable bioactive compounds harboring specific desirable health benefits. To date, numerous studies have shown significant progress in understanding diverse microorganisms capable of producing these compounds. These compounds are increasingly utilized in functional foods and supplements for their health benefits. While ongoing research with technological advancement has allowed scientists and researchers to explore future trends that include employing biotechnological approaches and optimizing fermentation processes. This will improve the yield and efficacy of these bioactive compounds.

References

Abd El Ghany K, Hamouda R, Abd Elhafez E, Mahrous H, Salem-Bekhit M, Hamza HA (2015) A potential role of lactobacillus acidophilus LA1 and its exopolysaccharides on cancer cells in male albino mice. Biotechnol Biotechnol Equip 29(5):977–983

Abdalla AK, Ayyash MM, Olaimat AN, Osaili TM, Al-Nabulsi AA, Shah NP, Holley R (2021) Exopolysaccharides as antimicrobial agents: mechanism and spectrum of activity. Front Microbiol 12:664395

Abdel Tawab FI, Abd Elkadr MH, Sultan AM, Hamed EO, El-Zayat AS, Ahmed MN (2023) Probiotic potentials of lactic acid bacteria isolated from Egyptian fermented food. Sci Rep 13(1):16601

Abdul Hakim BN, Xuan NJ, Oslan SNH (2023) A comprehensive review of bioactive compounds from lactic acid bacteria: potential functions as functional food in dietetics and the food industry. Food Secur 12(15):2850

Adesulu-Dahunsi AT, Jeyaram K, Sanni AI (2018a) Probiotic and technological properties of exopolysaccharide producing lactic acid bacteria isolated from cereal-based nigerian fermented food products. Food Control 92:225–231

Adesulu-Dahunsi AT, Sanni AI, Jeyaram K (2018b) Production, characterization and in vitro antioxidant activities of exopolysaccharide from Weissella cibaria GA44. LWT 87:432–442

Aili L, Liu C, Han X, Zheng J, Zhang G, Qi X, Peng D, Liu L (2023) Tibetan Plateau yak milk: a comprehensive review of nutritional values, health benefits, and processing technology. Food Chem X 20. https://doi.org/10.1016/j.fochx.2023.100919

Al-Nabulsi AA, Jaradat ZW, Al Qudsi FR, Elsalem L, Osaili TM, Olaimat AN et al (2022) Characterization and bioactive properties of exopolysaccharides produced by Streptococcus thermophilus and lactobacillus bulgaricus isolated from labaneh. LWT 167:113817

Alongi M, Anese M (2021) Re-thinking functional food development through a holistic approach. J Funct Foods 81:104466

Anonymous (2017) Probiotics market by products (functional foods, dietary supplements, specialty nutrients, animal feed) applications and ingredients – Global Trends & Forecasts to 2017

Arief II, Afiyah DN, Wulandari Z, Budiman C (2016) Physicochemical properties, fatty acid profiles, and sensory characteristics of fermented beef sausage by probiotics Lactobacillus plantarum IIA-2C12 or Lactobacillus acidophilus IIA-2B4. J Food Sci 81(11):M2761–M2769

Arora T, Singh S, Sharma RK (2013) Probiotics: interaction with gut microbiome and antiobesity potential. Nutrition 29(4):591–596

Ashraf SA, Adnan M, Patel M, Siddiqui AJ, Sachidanandan M, Snoussi M, Hadi S (2020) Fish-based bioactives as potent nutraceuticals: exploring the therapeutic perspective of sustainable food from the sea. Mar Drugs 18(5):265

Ayivi RD, Gyawali R, Krastanov A, Aljaloud SO, Worku M, Tahergorabi R et al (2020) Lactic acid bacteria: food safety and human health applications. Dairy 1(3):202–232

Balakrishnan G, Agrawal R (2014) Antioxidant activity and fatty acid profile of fermented milk prepared by Pediococcus pentosaceus. J Food Sci Technol 51:4138–4142

Bamigbade GB, Subhash AJ, Kamal-Eldin A, Nyström L, Ayyash M (2022) An updated review on prebiotics: insights on potentials of food seeds waste as source of potential prebiotics. Molecules 27(18):5947

Bao H, Ni Y, Dong H, Liang L (2020) α-Tocopherol and resveratrol in emulsion-filled whey protein gels: co-encapsulation and in vitro digestion. Int Dairy J 104:104649

Beckman KB, Ames BN (1998) The free radical theory of aging matures. Physiol Rev 78(2):547–581

Begum PS, Madhavi G, Rajagopal S, Viswanath B, Razak MA, Venkataratnamma V (2017) Probiotics as functional foods: potential effects on human health and its impact on neurological diseases. Int J Nutr Pharmacol Neurol Dis 7(2):23–33

Belkaid Y, Hand TW (2014) Role of the microbiota in immunity and inflammation. Cell 157(1):121–141

Biesalski HK, Dragsted LO, Elmadfa I, Grossklaus R, Müller M, Schrenk D et al (2009) Bioactive compounds: definition and assessment of activity. Nutrition 25(11–12):1202–1205
Bintsis T (2018) Lactic acid bacteria as starter cultures: an update in their metabolism and genetics. AIMS Microbiol 4(4):665
Bray F, Ferlay J, Soerjomataram I, Siegel RL, Torre LA, Jemal A (2018) Global cancer statistics 2018: GLOBOCAN estimates of incidence and mortality worldwide for 36 cancers in 185 countries. CA Cancer J Clin 68(6):394–424
Câmara JS, Albuquerque BR, Aguiar J, Corrêa R, Gonçalves JL, Granato D, Pereira J, Barros L, Ferreira I (2020) Food bioactive compounds and emerging techniques for their extraction: polyphenols as a case study. Foods 10(1):37. https://doi.org/10.3390/foods10010037
Champagne CP, Tompkins TA, Buckley ND, Green-Johnson JM (2010) Effect of fermentation by pure and mixed cultures of Streptococcus thermophilus and Lactobacillus helveticus on isoflavone and B-vitamin content of a fermented soy beverage. Food Microbiol 27(7):968–972
Chang YH, Jeong CH, Cheng WN, Choi Y, Shin DM, Lee S, Han SG (2021) Quality characteristics of yogurts fermented with short-chain fatty acid-producing probiotics and their effects on mucin production and probiotic adhesion onto human colon epithelial cells. J Dairy Sci 104(7):7415–7425
Chen Q, McGillivray D, Wen J, Zhong F, Quek SY (2013) Co-encapsulation of fish oil with phytosterol esters and limonene by milk proteins. J Food Eng 117(4):505–512
Ciumărnean L, Milaciu MV, Runcan O, Vesa ŞC, Răchişan AL, Negrean V et al (2020) The effects of flavonoids in cardiovascular diseases. Molecules 25(18):4320
Cizeikiene D, Juodeikiene G, Paskevicius A, Bartkiene E (2013) Antimicrobial activity of lactic acid bacteria against pathogenic and spoilage microorganism isolated from food and their control in wheat bread. Food Control 31(2):539–545
Coda R, Rizzello CG, Pinto D, Gobbetti M (2012) Selected lactic acid bacteria synthesize antioxidant peptides during sourdough fermentation of cereal flours. Appl Environ Microbiol 78(4):1087–1096
Consultation, W. H. O. (1999) Definition, diagnosis and classification of diabetes mellitus and its complications
Correia RT, Borges KC, Medeiros MF, Genovese MI (2012) Bioactive compounds and phenolic-linked functionality of powdered tropical fruit residues. Food Sci Technol Int 18(6):539–547
Dai S, Pan M, El-Nezami HS, Wan JM, Wang MF, Habimana O et al (2019) Effects of lactic acid bacteria-fermented soymilk on isoflavone metabolites and short-chain fatty acids excretion and their modulating effects on gut microbiota. J Food Sci 84(7):1854–1863
Daliri EBM, Lee BH, Park BJ, Kim SH, Oh DH (2018) Antihypertensive peptides from whey proteins fermented by lactic acid bacteria. Food Sci Biotechnol 27:1781–1789
Damián MR, Cortes-Perez NG, Quintana ET, Ortiz-Moreno A, Garfias Noguez C, Cruceño-Casarrubias CE et al (2022) Functional foods, nutraceuticals and probiotics: a focus on human health. Microorganisms 10(5):1065
de Oliveira Filho JG, Braga ARC, de Oliveira BR, Gomes FP, Moreira VL, Pereira VAC, Egea MB (2021) The potential of anthocyanins in smart, active, and bioactive eco-friendly polymer-based films: a review. Food Res Int 142:110202
DeFelice SL (1992) The nutraceutical initiative: a recommendation for U.S. Economic and regulatory reforms. Genet Eng News 12:13–15
Del Carmen S, De Moreno De Leblanc A, LeBlanc JG (2016) Development of a potential probiotic yoghurt using selected anti-inflammatory lactic acid bacteria for prevention of colitis and carcinogenesis in mice. J Appl Microbiol 121(3):821–830
Demain AL (2014) Importance of microbial natural products and the need to revitalize their discovery. J Ind Microbiol Biotechnol 41(2):185–201
Devi SM, Kurrey NK, Halami PM (2018) In vitro anti-inflammatory activity among probiotic lactobacillus species isolated from fermented foods. J Funct Foods 47:19–27

Di Luccio M, Capra F, Ribeiro NP, Vargas GD, Freire DM, de Oliveira D (2004) Effect of temperature, moisture, and carbon supplementation on lipase production by solid-state fermentation of soy cake by Penicillium simplicissimum. Appl Biochem Biotechnol 113:173–180

Dilna SV, Surya H, Aswathy RG, Varsha KK, Sakthikumar DN, Pandey A, Nampoothiri KM (2015) Characterization of an exopolysaccharide with potential health-benefit properties from a probiotic Lactobacillus plantarum RJF4. LWT-Food Sci Technol 64(2):1179–1186

Dillard CJ, German JB (2000) Phytochemicals: nutraceuticals and human health. J Sci Food Agric 80:1744–1756. https://doi.org/10.1002/1097-0010(20000915)80:12<1744::AID-JSFA725>3.0.CO;2-W

Diop MB, Dubois Dauphin R, Tine E, Ngom A, Destain J, Thonart P (2007) Bacteriocin producers from traditional food products. Biotechnol Agron Soc Environ 11(4)

Drywien M, Frackiewicz J, Górnicka M, Gadek J, Jalosinska M (2015) Effect of probiotic and storage time of thiamine and riboflavin content in the milk drinks fermented by Lactobacillus casei KNE-1. Rocz Panstw Zakl Hig 66(4):373–377

Ejtahed HS, Mohtadi-Nia J, Homayouni-Rad A, Niafar M, Asghari-Jafarabadi M, Mofid V (2012) Probiotic yogurt improves antioxidant status in type 2 diabetic patients. Nutrition 28(5):539–543

Ekpenyong M, Asitok A, Antai S, Ekpo B, Antigha R, Ogarekpe N (2021) Statistical and artificial neural network approaches to modeling and optimization of fermentation conditions for production of a surface/bioactive glyco-lipo-peptide. Int J Pept Res Ther 27:475–495

Elkhtab E, El-Alfy M, Shenana M, Mohamed A, Yousef AE (2017) New potentially antihypertensive peptides liberated in milk during fermentation with selected lactic acid bacteria and kombucha cultures. J Dairy Sci 100(12):9508–9520

Florence ACR, Béal C, Silva RC, Bogsan CS, Pilleggi ALO, Gioielli LA, Oliveira MN (2012) Fatty acid profile, trans-octadecenoic, α-linolenic and conjugated linoleic acid contents differing in certified organic and conventional probiotic fermented milks. Food Chem 135(4):2207–2214

Gao X (2017) Current situation and thinking of prevention and treatment of diabetes mellitus. J Sowthwest Med Univ 40(1):2–4

Giaconia MA, dos Passos Ramos S, Pereira CF, Lemes AC, De Rosso VV, Braga ARC (2020) Overcoming restrictions of bioactive compounds biological effects in food using nanometer-sized structures. Food Hydrocoll 107:105939

Gibson GR, Roberfroid MB (1995) Dietary modulation of the human colonic microbiota: introducing the concept of prebiotics. J Nutr 125(6):1401–1412

Gibson GR, Hutkins R, Sanders ME, Prescott SL, Reimer RA, Salminen SJ et al (2017) Expert consensus document: the International Scientific Association for Probiotics and Prebiotics (ISAPP) consensus statement on the definition and scope of prebiotics. Nat Rev Gastroenterol Hepatol 14(8):491–502

Giurini EF, Godla A, Gupta KH (2024) Redefining bioactive small molecules from microbial metabolites as revolutionary anticancer agents. Cancer Gene Ther 31(2):187–206

Gobbetti M, Di Cagno R, de Angelis M (2010) Functional microorganisms for functional food quality. Crit Rev Food Sci Nutr 50(8):716–727

Grosu-Tudor SS, Stancu MM, Pelinescu D, Zamfir M (2014) Characterization of some bacteriocins produced by lactic acid bacteria isolated from fermented foods. World J Microbiol Biotechnol 30:2459–2469

Han KJ, Lee JE, Lee NK, Paik HD (2020) Antioxidant and anti-inflammatory effect of probiotic Lactobacillus plantarum KU15149 derived from Korean homemade diced-radish kimchi. J Microbiol Biotechnol 30(4):591

Hati S, Patel M, Mishra BK, Das S (2019) Short-chain fatty acid and vitamin production potentials of Lactobacillus isolated from fermented foods of Khasi Tribes, Meghalaya, India. Ann Microbiol 69:1191–1199

Heider SA, Wendisch VF (2015) Engineering microbial cell factories: metabolic engineering of Corynebacterium glutamicum with a focus on non-natural products. Biotechnol J 10(8):1170–1184

Holkem AT, Favaro-Trindade CS, Lacroix M (2020) Study of anticancer properties of proanthocyanidin-rich cinnamon extract in combination with Bifidobacterium animalis subsp. lactis BLC1 and resistance of these free and co-encapsulated materials under in vitro simulated gastrointestinal conditions. Food Res Int 134:109274

Hölker U, Lenz J (2005) Solid-state fermentation—are there any biotechnological advantages? Curr Opin Microbiol 8(3):301–306

Huang H, Chen F, Long R, Huang G (2020a) The antioxidant activities in vivo of bitter gourd polysaccharide. Int J Biol Macromol 145:141–144

Huang Z, Lin F, Zhu X, Zhang C, Jiang M, Lu Z (2020b) An exopolysaccharide from lactobacillus plantarum H31 in pickled cabbage inhibits pancreas α-amylase and regulating metabolic markers in HepG2 cells by AMPK/PI3K/Akt pathway. Int J Biol Macromol 143:775–784

Indira M, Venkateswarulu TC, Abraham Peele K, Nazneen Bobby M, Krupanidhi S (2019) Bioactive molecules of probiotic bacteria and their mechanism of action: a review. 3 Biotech 9(8):306

International Association of Probiotics and Prebiotics (ISAPP) (2008) 6th Meeting of the International Scientific Association of Probiotics and Prebiotics, London

Iranmanesh M, Ezzatpanah H, Mojgani N (2014) Antibacterial activity and cholesterol assimilation of lactic acid bacteria isolated from traditional Iranian dairy products. LWT-Food Sci Technol 58(2):355–359

Jawan R, Abbasiliasi S, Tan JS, Kapri MR, Mustafa S, Halim M, Ariff AB (2021) Evaluation of the estimation capability of response surface methodology and artificial neural network for the optimization of Bacteriocin-like inhibitory substances production by Lactococcus lactis Gh1. Microorganisms 9(3):579

Jenab A, Roghanian R, Emtiazi G (2020) Bacterial natural compounds with anti-inflammatory and immunomodulatory properties (mini review). Drug Des Devel Ther 14:3787–3801

Jeong D, Kim DH, Kang IB, Kim H, Song KY, Kim HS, Seo KH (2017) Characterization and antibacterial activity of a novel exopolysaccharide produced by lactobacillus kefiranofaciens DN1 isolated from kefir. Food Control 78:436–442

Joshi VK, Sharma S, Rana NS (2006) Production, purification, stability and efficacy of bacteriocin from isolates of natural lactic acid fermentation of vegetables. Food Technol Biotechnol 44(3):435–439

Jurášková D, Ribeiro SC, Silva CC (2022) Exopolysaccharides produced by lactic acid bacteria: from biosynthesis to health-promoting properties. Food Secur 11(2):156

Jyothi NVN, Prasanna PM, Sakarkar SN, Prabha KS, Ramaiah PS, Srawan GY (2010) Microencapsulation techniques, factors influencing encapsulation efficiency. J Microencapsul 27(3):187–197

Kakimov A, Kakimova Z, Mirasheva G, Bepeyeva A, Toleubekova S, Jumazhanova M et al (2017) Amino acid composition of sour-milk drink with encapsulated probiotics. Annu Res Rev Biol 18(1):1–7

Kang CH, Kim JS, Park HM, Kim S, Paek NS (2021) Antioxidant activity and short-chain fatty acid production of lactic acid bacteria isolated from Korean individuals and fermented foods. 3 Biotech 11(5):217

Kaprasob R, Kerdchoechuen O, Laohakunjit N, Somboonpanyakul P (2018) B vitamins and prebiotic fructooligosaccharides of cashew apple fermented with probiotic strains Lactobacillus spp., Leuconostoc mesenteroides and Bifidobacterium longum. Process Biochem 70:9–19

Kazeem MI, Davies TC (2016) Anti-diabetic functional foods as sources of insulin secreting, insulin sensitizing and insulin mimetic agents. J Funct Foods 20:122–138

Kazemipoor M, Radzi CWJWM, Begum K, Yaze I (2012) Screening of antibacterial activity of lactic acid bacteria isolated from fermented vegetables against food borne pathogens. arXiv preprint arXiv:1206.6366

Khalil ES, Manap MY, Mustafa S, Amid M, Alhelli AM, Aljoubori A (2018a) Probiotic characteristics of exopolysaccharides-producing lactobacillus isolated from some traditional Malaysian fermented foods. CyTA-J Food 16(1):287–298

Khalil ES, Abd Manap MY, Mustafa S, Alhelli AM, Shokryazdan P (2018b) Probiotic properties of exopolysaccharide-producing Lactobacillus strains isolated from tempoyak. Molecules 23(2):398

Kim YL, Nguyen TH, Kim JS, Park JY, Kang CH (2023) Isolation of γ-aminobutyric acid (GABA)-producing lactic acid bacteria with anti-inflammatory effects from fermented foods in Korea. Fermentation 9(7):612

Kimura I, Inoue D, Maeda T, Hara T, Ichimura A, Miyauchi S et al (2011) Short-chain fatty acids and ketones directly regulate sympathetic nervous system via G protein-coupled receptor 41 (GPR41). Proc Natl Acad Sci 108(19):8030–8035

Kobayashi Y, Chiou TY, Konishi M (2023) Artificial intelligence-assisted analysis reveals amino acid effects and interactions on Limosilactobacillus fermentum growth. Biosci Biotechnol Biochem 87(9):1068–1076

Kormin S, Rusul G, Radu S, Ling FH (2001) Bacteriocinproducing lactic Acid bacteria isolated from traditional fermented food. Malays J Med Sci 8(1):63–68. PMCID: PMC3433967

Kook SY, Lee Y, Jeong EC, Kim S (2019) Immunomodulatory effects of exopolysaccharides produced by Bacillus licheniformis and Leuconostoc mesenteroides isolated from Korean kimchi. J Funct Foods 54:211–219

Kostelac D, Gerić M, Gajski G, Frece J (2023) Probiotic bacteria isolated from fermented meat displays high antioxidant and anti-inflammatory potential. Mutagenesis 38(1):58–63

Kothari D, Tingirikari JMR, Goyal A (2015) In vitro analysis of dextran from Leuconostoc mesenteroides NRRL B-1426 for functional food application. Bioact Carbohydr Diet Fibre 6(2):55–61

Kuda T, Kawahara M, Nemoto M, Takahashi H, Kimura B (2014) In vitro antioxidant and anti-inflammation properties of lactic acid bacteria isolated from fish intestines and fermented fish from the Sanriku Satoumi region in Japan. Food Res Int 64:248–255

Laan H, Konings WN (1989) Mechanism of proteinase release from Lactococcus lactis subsp. cremoris Wg2. Appl Environ Microbiol 55:3101–3106

LeBlanc JG, de Giori GS, Smid EJ, Hugenholtz J, Sesma F (2007b) Folate production by lactic acid bacteria and other food-grade microorganisms. In: Watson RR, Preedy V, eds. Bioactive foods as dietary interventions for liver and gastrointestinal disease (1st ed). Academis Press, Cambridge, MA, pp 329–339

Lepecka A, Szymański P, Okoń A, Zielińska D (2023) Antioxidant activity of environmental lactic acid bacteria strains isolated from organic raw fermented meat products. LWT 174:114440

Li L, Wang J, Zhao M, Cui C, Jiang Y (2006) Artificial neural network for production of antioxidant peptides derived from bighead carp muscles with alcalase. Food Technol Biotechnol 44(3):441–448

Li C, Li W, Chen X, Feng M, Rui X, Jiang M, Dong M (2014) Microbiological, physicochemical and rheological properties of fermented soymilk produced with exopolysaccharide (EPS) producing lactic acid bacteria strains. LWT-Food Sci Technol 57(2):477–485

Li W, Xia X, Tang W, Ji J, Rui X, Chen X et al (2015) Structural characterization and anticancer activity of cell-bound exopolysaccharide from Lactobacillus helveticus MB2-1. J Agric Food Chem 63(13):3454–3463

Liao CL, Huang HY, Sheen LY, Chou CC (2010) Anti-inflammatory activity of soymilk and fermented soymilk prepared with lactic acid bacterium and bifidobacterium. J Food Drug Anal 18(3):5

Liaqat H, Kim KJ, Park SY, Jung SK, Park SH, Lim S, Kim JY (2021) Antioxidant effect of wheat germ extracts and their antilipidemic effect in palmitic acid-induced steatosis in HepG2 and 3T3-L1 cells. Food Secur 10(5):1061

Liu M, Bayjanov JR, Renckens B, Nauta A, Siezen RJ (2010) The proteolytic system of lactic acid bacteria revisited: a genomic comparison. BMC Genomics 11(1):36

Liu CF, Tseng KC, Chiang SS, Lee BH, Hsu WH, Pan TM (2011) Immunomodulatory and antioxidant potential of Lactobacillus exopolysaccharides. J Sci Food Agric 91(12):2284–2291

Liu Y, Cheng H, Liu H, Ma R, Ma J, Fang H (2019) Fermentation by multiple bacterial strains improves the production of bioactive compounds and antioxidant activity of goji juice. Molecules 24(19):3519

Lopetuso LR, Giorgio ME, Saviano A, Scaldaferri F, Gasbarrini A, Cammarota G (2019) Bacteriocins and bacteriophages: therapeutic weapons for gastrointestinal diseases? Int J Mol Sci 20(1):183

Madjirebaye P, Xiao M, Mahamat B, Xiong S, Mueed A, Wei B et al (2022) In vitro characteristics of lactic acid bacteria probiotics performance and antioxidant effect of fermented soymilk. Food Biosci 49:101952

Mahdavi SA, Jafari SM, Ghorbani M, Assadpoor E (2014) Spray-drying microencapsulation of anthocyanins by natural biopolymers: a review. Dry Technol 32(5):509–518

Makino S, Sato A, Goto A, Nakamura M, Ogawa M, Chiba Y et al (2016) Enhanced natural killer cell activation by exopolysaccharides derived from yogurt fermented with Lactobacillus delbrueckii ssp. bulgaricus OLL1073R-1. J Dairy Sci 99(2):915–923

Makkar R, Behl T, Kumar A, Nijhawan P, Arora S (2021) Emerging therapeutic effects of herbal plants in rheumatoid arthritis. Endocr Metab Immune Disord Drug Targets 21(4):617–625

Martins S, Mussatto SI, Martínez-Avila G, Montañez-Saenz J, Aguilar CN, Teixeira JA (2011) Bioactive phenolic compounds: production and extraction by solid-state fermentation. A review. Biotechnol Adv 29(3):365–373

Mathur H, Beresford TP, Cotter PD (2020) Health benefits of lactic acid bacteria (LAB) fermentates. Nutrients 12(6):1679

Mechmeche M, Kachouri F, Ksontini H, Hamdi M (2017) Production of bioactive peptides from tomato seed isolate by Lactobacillus plantarum fermentation and enhancement of antioxidant activity. Food Biotechnol 31(2):94–113

Medeiros I, de Oliveira GLR, de Queiroz JLC, de Carvalho Gomes C, de Carvalho FMC, de Souza Lima MCJ et al (2020) Safety and bioactive potential of nanoparticles containing Cantaloupe melon (Cucumis melo L.) carotenoids in an experimental model of chronic inflammation. Biotechnol Rep (Amst) 28:e00567

Mirmazloum I, Ladányi M, Omran M, Papp V, Ronkainen VP, Pónya Z et al (2021) Co-encapsulation of probiotic Lactobacillus acidophilus and Reishi medicinal mushroom (Ganoderma lingzhi) extract in moist calcium alginate beads. Int J Biol Macromol 192:461–470

Mitchell DA, Krieger N, Stuart DM, Pandey A (2000) New developments in solid-state fermentation: II. Rational approaches to the design, operation and scale-up of bioreactors. Process Biochem 35(10):1211–1225

Miyamoto J, Shimizu H, Hisa K, Matsuzaki C, Inuki S, Ando Y et al (2023) Host metabolic benefits of prebiotic exopolysaccharides produced by Leuconostoc mesenteroides. Gut Microbes 15(1):2161271

Mu S, Yang W, Huang G (2021) Antioxidant activities and mechanisms of polysaccharides. Chem Biol Drug Des 97(3):628–632

Murthy KC, Monika P, Jayaprakasha GK, Patil BS (2018) Nanoencapsulation: an advanced nanotechnological approach to enhance the biological efficacy of curcumin. In: Advances in plant phenolics: from chemistry to human health. American Chemical Society, Washington, DC, pp 383–405

Nehal F, Sahnoun M, Smaoui S, Jaouadi B, Bejar S, Mohammed S (2019) Characterization, high production and antimicrobial activity of exopolysaccharides from Lactococcus lactis F-mou. Microb Pathog 132:10–19

Nes IF, Holo H (2000) Class II antimicrobial peptides from lactic acid bacteria. Pept Sci 55(1):50–61

Nicolescu A, Babotă M, Barros L, Rocchetti G, Lucini L, Tanase C et al (2023) Bioaccessibility and bioactive potential of different phytochemical classes from nutraceuticals and functional foods. Front Nutr 10:1184535

O′ Toole PW, Cooney JC (2008) Probiotic bacteria influence the composition and function of the intestinal microbiota. Interdiscip Perspect Infect Dis 2008(1):175285

Ouertani A, Chaabouni I, Mosbah A, Long J, Barakat M, Mansuelle P et al (2018) Two new secreted proteases generate a casein-derived antimicrobial peptide in Bacillus cereus food born isolate leading to bacterial competition in milk. Front Microbiol 9:1148

Palanivelu J, Thanigaivel S, Vickram S, Dey N, Mihaylova D, Desseva I (2022) Probiotics in functional foods: survival assessment and approaches for improved viability. Appl Sci 12(1):455

Pandey A (2003) Solid-state fermentation. Biochem Eng J 13(2–3):81–84

Pateiro M, Gómez B, Munekata PE, Barba FJ, Putnik P, Kovačević DB, Lorenzo JM (2021) Nanoencapsulation of promising bioactive compounds to improve their absorption, stability, functionality and the appearance of the final food products. Molecules 26(6):1547

Pessione E, Cirrincione S (2016) Bioactive molecules released in food by lactic acid bacteria: encrypted peptides and biogenic amines. Front Microbiol 7:876

Pringsulaka O, Thongngam N, Suwannasai N, Atthakor W, Pothivejkul K, Rangsiruji A (2012) Partial characterisation of bacteriocins produced by lactic acid bacteria isolated from Thai fermented meat and fish products. Food Control 23(2):547–551

Rahbani-Nobar ME, Rahimi-Pour A, Rahbani-Nobar M, Adi-Beig F, Mirhashemi SM (1999) Total antioxidant capacity, superoxide dismutase and glutathione peroxidase in diabetic patients. Med J Islamic Acad Sci 12(4):109–114

Rajoka MSR, Wu Y, Mehwish HM, Bansal M, Zhao L (2020) Lactobacillus exopolysaccharides: new perspectives on engineering strategies, physiochemical functions, and immunomodulatory effects on host health. Trends Food Sci Technol 103:36–48

Rashwan AK, Younes HA, Karim N, Taha EM, Mozafari MR, Tawfeuk HZ, Chen W (2022) Health properties of bioactive food compounds-loaded micro and nano-encapsulation systems: a review. SVU-Int J Agric Sci 4(2):178–203

Raveschot C, Cudennec B, Coutte F, Flahaut C, Fremont M, Drider D, Dhulster P (2018) Production of bioactive peptides by lactobacillus species: from gene to application. Front Microbiol 9. https://doi.org/10.3389/fmicb.2018.02354

Reque PM, Brandelli A (2021) Encapsulation of probiotics and nutraceuticals: applications in functional food industry. Trends Food Sci Technol 114:1–10

Rizwan D, Masoodi FA, Wani SM et al (2023) Bioactive peptides from fermented foods and their relevance in COVID-19 mitigation. Food Prod Process Nutr 5:53. https://doi.org/10.1186/s43014-023-00165-w

Rong J, Zheng H, Liu M, Hu X, Wang T, Zhang X et al (2015) Probiotic and anti-inflammatory attributes of an isolate lactobacillus helveticus NS8 from Mongolian fermented koumiss. BMC Microbiol 15:1–11

Sadh PK, Kumar S, Chawla P, Duhan JS (2018) Fermentation: a boon for production of bioactive compounds by processing of food industries wastes (by-products). Molecules 23(10):2560

Salminen S, Collado MC, Endo A, Hill C, Lebeer S, Quigley EM et al (2021) The International Scientific Association of Probiotics and Prebiotics (ISAPP) consensus statement on the definition and scope of postbiotics. Nat Rev Gastroenterol Hepatol 18(9):649–667

Santos-Moriano P, Fernandez-Arrojo L, Poveda A, Jimenez-Barbero J, Ballesteros AO, Plou FJ (2015) Levan versus fructooligosaccharide synthesis using the levansucrase from Zymomonas mobilis: effect of reaction conditions. J Mol Catal B Enzym 119:18–25

Sarmadi BH, Ismail A (2010) Antioxidative peptides from food proteins: a review. Peptides 3:49–56

Savijoki K, Ingmer H, Varmanen P (2006) Proteolytic systems of lactic acid bacteria. Appl Microbiol Biotechnol 71(4):394–406. https://doi.org/10.1007/s00253-006-0427-1

Shakirzyanova MP, Ruzieva DM, Abdul'myanova LI, Seidametova EA, Gulyamova TG (2002) The ability of some strains of propionic acid bacteria isolated in Uzbekistan to synthesize vitamin B12. Microbiology (00262617) 71(4)

Sheng S, Fu Y, Pan N, Zhang H, Xiu L, Liang Y et al (2022) Novel exopolysaccharide derived from probiotic *Lactobacillus pantheris* TCP102 strain with immune-enhancing and anticancer activities. Front Microbiol 13:1015270

Shirahigue LD, Ceccato-Antonini SR (2020) Agro-industrial wastes as sources of bioactive compounds for food and fermentation industries. Ciência rural 50:e20190857

Silva LA, Lopes Neto JHP, Cardarelli HR (2019) Exopolysaccharides produced by Lactobacillus plantarum: technological properties, biological activity, and potential application in the food industry. Ann Microbiol 69:321–328

Simons A, Alhanout K, Duval RE (2020) Bacteriocins, antimicrobial peptides from bacterial origin: overview of their biology and their impact against multidrug-resistant bacteria. Microorganisms 8(5):639

Singh nee' Nigam P (2009) Production of bioactive secondary metabolites. In: Singh nee' Nigam P, Pandey A (eds). Biotechnology for agro-industrial residues utilisation. Springer, Dordrecht, 466p

Singh nee' Nigam P, Pandey A (2009) Solid-state fermentation technology for bioconversion of biomass and agricultural residues. In: Biotechnology for agro-industrial residues utilization: utilization of agro-residues, Springer pp 197–221. https://doi.org/10.1007/978-1-4020-9942-7

Singh P, Saini P (2017) Food and health potentials of exopolysaccharides derived from Lactobacilli. Microbiol Res J Int 22:1–14

Sodeinde KO, Ojo AM, Olusanya SO, Ayanda OS, Adeoye AO, Dada TM, Lawal OS (2021) Cellulose isolated from Delonixregia pods: characterisation and application in the encapsulation of vitamin a. Ind Crop Prod 160:113138

Soltani N, Qiu H, Aleksic M, Glinka Y, Zhao F, Liu R et al (2011) GABA exerts protective and regenerative effects on islet beta cells and reverses diabetes. Proc Natl Acad Sci 108(28):11692–11697

Sorrenti V, Burò I, Consoli V, Vanella L (2023) Recent advances in health benefits of bioactive compounds from food wastes and by-products: biochemical aspects. Int J Mol Sci 24(3):2019

Subramaniyam R, Vimala R (2012) Solid state and submerged fermentation for the production of bioactive substances: a comparative study. Int J Sci Nat 3(3):480–486

Sutyak KE, Wirawan RE, Aroutcheva AA, Chikindas ML (2008) Isolation of the Bacillus subtilis antimicrobial peptide subtilosin from the dairy product-derived bacillus amyloliquefaciens. J Appl Microbiol 104(4):1067–1074

Swetwiwathana A, Visessanguan W (2015) Potential of bacteriocin-producing lactic acid bacteria for safety improvements of traditional Thai fermented meat and human health. Meat Sci 109:101–105

Taha S, El Abd M, De Gobba C, Abdel-Hamid M, Khalil E, Hassan D (2017) Antioxidant and antibacterial activities of bioactive peptides in buffalo's yoghurt fermented with different starter cultures. Food Sci Biotechnol 26:1325–1332

Taouzinet L, Djaoudene O, Fatmi S, Bouiche C, Amrane-Abider M, Bougherra H et al (2023) Trends of nanoencapsulation strategy for natural compounds in the food industry. PRO 11(5):1459

Tarone AG, Cazarin CBB, Junior MRM (2020) Anthocyanins: new techniques and challenges in microencapsulation. Food Res Int 133:109092

Tsai JS, Lin YS, Pan BS, Chen TJ (2006) Antihypertensive peptides and γ-aminobutyric acid from prozyme 6 facilitated lactic acid bacteria fermentation of soymilk. Process Biochem 41(6):1282–1288

Tsuda H, Miyamoto T (2010) Production of exopolysaccharide by lactobacillus plantarum and the prebiotic activity of the exopolysaccharide. Food Sci Technol Res 16(1):87–92

Tyagi T, Garlapati PK, Yadav P, Naika M, Mallya A, Kandangath Raghavan A (2021) Development of nano-encapsulated green tea catechins: studies on optimization, characterization, release dynamics, and in-vitro toxicity. J Food Biochem 45(11):e13951

Unban K, Chaichana W, Baipong S, Abdullahi AD, Kanpiengjai A, Shetty K, Khanongnuch C (2021) Probiotic and antioxidant properties of lactic acid bacteria isolated from indigenous fermented tea leaves (Miang) of North Thailand and promising application in synbiotic formulation. Fermentation 7(3):19

Varzakas T, Kandylis P, Dimitrellou D, Salamoura C, Zakynthinos G, Proestos C (2018) Innovative and fortified food: probiotics, prebiotics, GMOs, and superfood. In: Preparation and processing of religious and cultural foods. Woodhead Publishing, Duxford, pp 67–129

Vasiljevic T, Shah NP (2008) Probiotics—from Metchnikoff to bioactives. Int Dairy J 18(7):714–728
Virtanen T, Pihlanto A, Akkanen S, Korhonen H (2007) Development of antioxidant activity in milk whey during fermentation with lactic acid bacteria. J Appl Microbiol 102(1):106–115
Wang CY, Wu SJ, Shyu YT (2014) Antioxidant properties of certain cereals as affected by food-grade bacteria fermentation. J Biosci Bioeng 117(4):449–456
Wang Y, Li Q, Zheng P, Guo Y, Wang L, Zhang T et al (2016) Evolving the L-lysine high-producing strain of Escherichia coli using a newly developed high-throughput screening method. J Ind Microbiol Biotechnol 43(9):1227–1235
Wen P, Zong MH, Linhardt RJ, Feng K, Wu H (2017) Electrospinning: a novel nano-encapsulation approach for bioactive compounds. Trends Food Sci Technol 70:56–68
Wieërs G, Belkhir L, Enaud R, Leclercq S, Philippart de Foy JM, Dequenne I et al (2020) How probiotics affect the microbiota. Front Cell Infect Microbiol 9:454
Wu MH, Pan TM, Wu YJ, Chang SJ, Chang MS, Hu CY (2010) Exopolysaccharide activities from probiotic bifidobacterium: immunomodulatory effects (on J774A. 1 macrophages) and antimicrobial properties. Int J Food Microbiol 144(1):104–110
Wu X, Zha J, Koffas MA (2020) Microbial production of bioactive chemicals for human health. Curr Opin Food Sci 32:9–16
Xie W, Zhang Y, Wang N, Zhou H, Du L, Ma X et al (2008) Novel effects of macrostemonoside A, a compound from Allium macrostemon Bung, on hyperglycemia, hyperlipidemia, and visceral obesity in high-fat diet-fed C57BL/6 mice. Eur J Pharmacol 599(1–3):159–165
Xu R, Shen Q, Ding X, Gao W, Li P (2011) Chemical characterization and antioxidant activity of an exopolysaccharide fraction isolated from Bifidobacterium animalis RH. Eur Food Res Technol 232:231–240
Yadav H, Jain S, Sinha PR (2007) Production of free fatty acids and conjugated linoleic acid in probiotic dahi containing Lactobacillus acidophilus and Lactobacillus casei during fermentation and storage. Int Dairy J 17(8):1006–1010
Yildiz H, Karatas N (2018) Microbial exopolysaccharides: resources and bioactive properties. Process Biochem 72:41–46
You X, Li Z, Ma K, Zhang C, Chen X, Wang G et al (2020) Structural characterization and immunomodulatory activity of an exopolysaccharide produced by Lactobacillus helveticus LZ-R-5. Carbohydr Polym 235:115977
Zabot GL, Schaefer Rodrigues F, Polano Ody L, Vinícius Tres M, Herrera E, Palacin H et al (2022) Encapsulation of bioactive compounds for food and agricultural applications. Polymers 14(19):4194
Zaeim D, Sarabi-Jamab M, Ghorani B, Kadkhodaee R (2019) Double layer co-encapsulation of probiotics and prebiotics by electro-hydrodynamic atomization. LWT 110:102–109
Zagato E, Mileti E, Massimiliano L, Fasano F, Budelli A, Penna G, Rescigno M (2014) Lactobacillus paracasei CBA L74 metabolic products and fermented milk for infant formula have antiinflammatory activity on dendritic cells in vitro and protective effects against colitis and an enteric pathogen in vivo. PLoS One 9(2):e87615
Zhang L, Liu C, Li D, Zhao Y, Zhang X, Zeng X et al (2013) Antioxidant activity of an exopolysaccharide isolated from Lactobacillus plantarum C88. Int J Biol Macromol 54:270–275
Zhou Y, Cui Y, Qu X (2019a) Exopolysaccharides of lactic acid bacteria: structure, bioactivity and associations: a review. Carbohydr Polym 207:317–332. https://doi.org/10.1016/j.carbpol.2018.11.093
Zhou L, Zhang C, Liu F, Qiu Z, He Y (2019b) Application of deep learning in food: a review. Compr Rev Food Sci Food Saf 18(6):1793–1811
Zhou L, Li K, Duan X, Hill D, Barrow C, Dunshea F et al (2022) Bioactive compounds in microalgae and their potential health benefits. Food Biosci 49:101932

Chapter 7
Functional Aspects of Dietary Fibers in Nutraceuticals

Abdul Haque, Tariq Khan, Irfan Khan, Md Faruque Ahmad, and Syed Amir Ashraf

Abstract Dietary fibers primarily consist of carbohydrates and are derived from plant-based sources. Two distinct categories exist depending on their solubility in water, namely soluble and insoluble dietary fibers. These substances are indigestible and/or absorbable by the human body. They play a crucial function in the maintenance of optimal health. Recent research indicates that high-fiber diets, particularly those containing soluble dietary fibers, can effectively manage weight and protect against obesity, ulcerative colitis, diabetes, cardiovascular diseases, hypertension, gastrointestinal diseases, diverticulosis, hemorrhoids, and colon cancer. Although subject to ongoing debate, a proposition suggests that fiber may have potential benefits for those diagnosed with diabetes mellitus, namely in managing their blood glucose levels following a meal and enhancing carbohydrate metabolism. This chapter explores the functional aspects of dietary fiber in nutraceuticals, emphasizing its role in promoting human health and well-being. The chapter covers the various types of dietary fiber, their sources, and their potential applications in formulating functional food products. The chapter also delves into the physiological effects of dietary fiber consumption, its impact on gut microbiota, its potential role in preventing chronic diseases, and the harmful effects of consuming excessive dietary fiber.

A. Haque (✉) · T. Khan
Department of Post-harvest Engineering and Technology, Faculty of Agricultural Sciences, Aligarh Muslim University, Aligarh, Uttar Pradesh, India

I. Khan
Department of Bioengineering, Integral University, Lucknow, Uttar Pradesh, India

M. F. Ahmad
Department of Clinical Nutrition, College of Nursing and Health Sciences, Jazan University, Jazan, Saudi Arabia
e-mail: mfahmad@jazanu.edu.sa

S. A. Ashraf
Department of Clinical Nutrition, College of Applied Medial Sciences, University of Hail, Hail, Saudi Arabia

S. Amir Ashraf, M. Adnan (eds.), *Food Bioactives and Nutraceuticals*,
https://doi.org/10.1007/978-981-97-9723-3_7

Keywords Dietary fibers · Health · Cardiovascular diseases · Gastrointestinal diseases · Functional food products · Nutraceuticals

7.1 Introduction

Previous definition for dietary fiber solely took into account the indigestible components of plant cell walls (Stribling and Ibrahim 2023). Dietary fibers are comprised of polysaccharides resistant to digestion in the small intestine, with some undergoing digestion in the colon. These fibers consist of cellulose, hemicelluloses, and lignin. Whole grains and bran are primary sources of dietary fibers. Another definition characterizes dietary fibers as components of plant foods that resist digestion by human digestive enzymes (Zhang et al. 2024). Modern definitions of fiber now encompass all plant parts you can eat and similar extracted or synthetic carbohydrates that bypass digestion in the small intestine and ferment in the large intestine (Stribling and Ibrahim 2023). The expansions have led to the inclusion of indigestible oligosaccharides in many classifications. The European Union, United Kingdom, United States, and several other nations have willingly accepted its incorporation, and it was made an elective component of the Codex Alimentarius Commission's (CAC) definition in 2009 (Stribling and Ibrahim 2023). Dietary fiber with a low molecular weight (LMW) includes oligosaccharides, a type of short-chain carbohydrates. One type of carbohydrate that is commonly known as low molecular weight (LMW) fiber is oligosaccharides (Thompson and Brick 2016). These carbohydrates occur naturally in grains made from wheat, pasta, bread, pulses, and legumes; they serve as energy stores for plants. The designation of oligosaccharides as dietary fiber is a matter of great contention because their main association is with their microbial effects, not their impact on bowel function. Oligosaccharide and polysaccharide forms of inulin are both considered low molecular weight (LMW) dietary fibers (Thompson and Brick 2016; Westenbrink et al. 2013). Because it is a naturally occurring, indigestible carbohydrate polymer, the Codex definition automatically classifies it as dietary fiber. Consuming dietary fiber provides a multitude of health advantages, diminishing the likelihood of getting ailments like stroke, coronary heart disease, diabetes, hypertension, obesity, and specific gastrointestinal issues (Haque et al. 2023). Furthermore, consuming more dietary fibers has been linked to various benefits including better blood glucose levels, improved immune function, balanced serum lipid concentrations, enhanced diabetes management, reduced blood pressure, promotion of regular bowel movements, aid in weight loss (Haque et al. 2023), and the development of nutraceutical food products by the food industry's research and development division, such as sustained-release glucose and high-fiber biscuits (Bouazizi et al. 2020). A typical 'Western' diet is characterized by high levels of saturated fat, refined sugar, and salt, and a low intake of fiber-rich plant-based foods (Statovci et al. 2017), resulting in inadequate consumption of dietary fiber. This is a worldwide issue, as just 9% of persons in the United Kingdom are able to get the recommended daily consumption of dietary fiber, which is 30 g (Mayor 2019). As a result, attempts have been undertaken to increase the amount of dietary fiber in regularly consumed food products by adding extracted or synthesized fiber, such as oligosaccharides and inulin (Stribling and Ibrahim 2023). The health impacts of extracted/synthetic fiber compared to intrinsic fiber are indeterminate, and product labeling fails to distinguish between intrinsic and added fiber, potentially leading to consumer perplexity.

Significant quantities of extracted or synthesized oligosaccharides and inulin can be added to food items. These additives serve multiple purposes, including altering the fiber content, acting as sweeteners, replacing fats, modifying texture, and retaining moisture. These applications have noteworthy ramifications for individuals with functional bowel disorders (FBDs), which impact more than 33% of the population and are associated with a declined quality of life as well as large healthcare expenses (Sperber et al. 2021). Individuals with FBDs may experience symptoms such as bloating, stomach pain, and excessive flatulence when they consume low molecular weight (LMW) dietary fiber (Stribling and Ibrahim 2023). Oligosaccharides and inulin, which are the primary constituents eliminated in the low FODMAP (fermentable oligosaccharides, disaccharides, monosaccharides, and polyols) diet, a widely employed dietary treatment for alleviating symptoms in individuals with functional bowel disorders, give rise to apprehensions. There is ongoing debate in the field of human nutrition about the physiological effects of added/functional fibers compared to naturally occurring fibers. Several research studies within the realm of food science have investigated the incorporation of refined fiber powders derived from various sources into food items (Li and Komarek 2017). The nutritional world recognizes the importance of dietary fiber (DF) in human health. However, there are ongoing controversies about whether added or functional fiber offers the same or similar biological or physiological benefits as its original or endogenous form. There are concerns about the possibility of gastrointestinal issues and reduced nutrient absorption caused by the presence of fiber from functional sources added to processed foods (Grundy et al. 2016). Nevertheless, the substantial clinical evidence endorsing the advantageous impacts of dietary fiber on human health has motivated food scientists and engineers to create innovative products that are enhanced or fortified with fiber. The main method used to accomplish this is by utilizing separated and purified fiber components, usually known as functional fiber (Adams et al. 2018). There is a clear and urgent requirement to establish connections and coordinate research endeavors across different fields, creating interdisciplinary teams to tackle and promote dietary fiber research.

The chapter provides a comprehensive exploration of the functional attributes of dietary fiber in nutraceuticals, with a keen focus on its contribution to enhancing human health and overall well-being. The content encompasses an in-depth examination of various types of dietary fiber, their diverse sources, and their potential applications in the development of functional food products. Additionally, the chapter delves into the physiological effects of dietary fiber consumption, examining its influence on gut microbiota, its potential role in preventing chronic diseases, and the adverse effects associated with excessive dietary fiber intake. The comprehensive coverage and thoughtful analysis make this chapter a valuable contribution to the understanding of dietary fiber's multifaceted roles in the realm of nutraceuticals.

7.2 Dietary Fiber Classification and Its Nutraceutical Sources

Dietary fiber, sometimes called roughage or bulk, is the non-digestible part of plant-based foods that moves through the digestive system without being broken down. Fiber is essential for maintaining digestive health, promoting satiety, regulating blood sugar levels, and preventing the onset of chronic illnesses. Fiber can be classified into different categories depending on its solubility, fermentability, and physiological impact.

7.2.1 Soluble Fiber

Soluble fiber, known for its capacity to dissolve in water and create a gel-like substance in the digestive system, has a crucial function in preserving digestive well-being. This gel functions as a hindrance, impeding the process of digestion and assimilation of nutrients such as carbohydrates, lipids, and cholesterol. Soluble fiber assists in the regulation of blood sugar levels, making it very advantageous for persons who are controlling diabetes or aiming to minimize post-meal blood sugar surges (Guan et al. 2021). Soluble fibers like β-glucans, mucilage, pectin, and gum, sourced from plant inner tissues, dissolve in water. In the colon, they create a gel broken down by gut bacteria into short chain fatty acids (SCFAs) and gases. Examples of foods rich in soluble fibers include oats, barley, fruits, peas, beans, legumes, and root vegetables (Yusuf et al. 2022). Soluble fibers are easily digestible fibers that can increase the volume of feces by stimulating the growth of good bacteria in the gut, resulting in the production of SCFAs. SCFAs are really important for keeping your colon healthy. They can start a chain reaction of good things happening in your body. These qualities can help regulate the consistency of feces by softening rigid stools in cases of constipation and solidifying loose or watery stools in diarrhea (Dai and Chau 2017).

Pectin is a group of polysaccharides made mainly of d-galacturonic acid, found in plant cell walls, acting as a kind of a binding agent between cells. It easily mixes with water and is mainly broken down by bacteria in the colon. Pectin's gel-forming properties enable it to impede stomach emptying and influence the rate at which food traverses the small intestine. Pectin's distinctive characteristic renders it a prevalent component in several food items, including jellies and jams (Williams et al. 2019; Samarasinghe et al. 2022). Citrus peels and apple pomace are effective commercial sources for acquiring pectin (Zannini et al. 2021).

β-glucans are a group of polysaccharides (complex carbohydrates) found in the cell walls of bacteria, fungi, yeasts, algae, lichens, and plants, particularly in cereal grains such as oats, barley, and mushrooms. Consuming these β-glucans brings various benefits, such as reducing blood serum cholesterol levels and controlling blood glucose levels, particularly advantageous for individuals with diabetes. The effectiveness of these health benefits is commonly associated with the quantity and size of solubilized β-glucans present in the gastrointestinal tract (Geng et al. 2022). Soluble dietary fiber encompasses hydrocolloids made from polysaccharides, which are difficult for human enzymes to digest. This group has different types of gums: gum arabic, seed gums like locust bean gum and guar gum, fermented gums like xanthan gum and gellan gum, and seaweed gums. Gums are special fibers made inside certain cells of plants, unlike usual plant cell wall parts. They consist of highly branched polysaccharides with the ability to bind water and form gels (Dhingra et al. 2012). Mucilages refer to water-soluble fibers present in the outside coating of plantain seeds. These fibers consist of unique mucilage cells that contribute to various properties of the seeds (Dybka-Stępień et al. 2021). The commonly used mucilages include psyllium husk, yellow mustard mucilage, and flaxseed gum. Psyllium husk is rich in soluble fibers, constituting about 70%, which is also referred to as psyllium gum. This natural substance acts as a gentle laxative, promoting better colon health and functionality. Yellow mustard mucilage serves as a reliable stabilizer for emulsions. On the other hand, flaxseed gum, known for its low viscosity, finds practical application in dairy products (Agrawal and Agrawal 2021). Several plant-derived oligosaccharides resist digestion in humans, categorized as low-molecular-weight soluble dietary fiber (Stribling and Ibrahim 2023). Inulin, found abundantly in plants like rye, onion, and chicory, serves as a reserve carbohydrate. It can dissolve in warm water, doesn't get digested in the small intestine, but bacteria in the large intestine quickly break it down into SCFAs. Food producers commonly use inulin as a replacement for sugar and fat in a range of food items such as desserts and yogurt ((PhD) TFT 2019).

7.2.2 *Insoluble Fibers*

Insoluble fibers, sourced from plant outside layers, possess unique traits. Unlike soluble fibers, they resist water dissolution and bacterial fermentation in the colon. These properties aid in stool bulk and facilitate regular bowel movements (Giorgetti et al. 2023). Insoluble fibers come from a variety of plant sources. Cellulose, found in potatoes, corn bran, fruit skins, many green vegetables, and certain fruits, is a major contributor. Hemicellulose is found in whole grains, while lignin is plentiful in nuts and seeds (Giorgetti et al. 2023). Insoluble fibers promote fecal bulk, facilitating the alleviation of constipation and promoting regular digestion. The increase in stool volume is caused by the water being held inside the fiber structure. Moreover, insoluble fibers are associated with enhanced peristalsis in the intestines, which might aid in the prevention and relief of constipation (Dai and Chau 2017). Cellulose, a polysaccharide and the predominant organic substance on earth, plays an essential role as a primary structural element in the plant cell walls (Cui et al. 2011). Cellulose is in many foods like fruits, veggies, and cereals because it is important for plant cell walls. It is a polysaccharide composed of a chain of thousand glucose molecules. Cellulose exhibits a high degree of insolubility due to its closely packed chains, rendering it challenging for human enzymes to enzymatically degrade during the process of digestion. Grains and fruits include around 25% of dietary fiber, while vegetables and nuts contain approximately 33%. This fiber is primarily composed of cellulose (Dai and Chau 2017). Cellulose, a highly regarded dietary fiber, has undergone significant research and is linked to numerous health advantages, including relieving constipation and assisting with weight control. Because it is found a lot in plant foods, it is the most common type of fiber in diets of animals like us. As a result, it has been extensively studied in the field of dietary fiber (Samarasinghe et al. 2022). Hemicelluloses constitute a varied group of polysaccharides present in plant cell walls. They consist of xylose, pentose sugars, and arabinose, displaying a wide range of sugar units in varying amounts and with different substituents (Huang et al. 2021). Hemicellulose typically fills the gaps between cellulose fibers in plant cell walls, and it can sometimes join with pectin to create the cell wall's structural framework. Cereals are recognized for their significant content of hemicelluloses (Arzami et al. 2022). Lignin is present in the majority of raw materials utilized in the food industry. The composition includes polysaccharides that establish covalent bonds with phenolic compounds. Lignin, along with cellulose and hemicelluloses, forms lignin-carbohydrate complexes (LCC) in plant tissues, which have a role in the formation of the plant cell wall. Lignin forms covalent connections, such as glycosidic, ether, and ester bonds, with polysaccharides by attaching to the monosaccharide residues of the polysaccharide using its monolignols (Tao et al. 2020). Lignin, as a non-soluble dietary fiber, aids in prevention of the formation of bile stones and lowers cholesterol levels (Samarasinghe et al. 2022). Foods like coconut shells, oat husks, rice husks, almond shells, and hazelnut shells are rich sources of lignin (Tao et al. 2020).

7.2.3 Fermentable Fiber

Fermentable fiber, which is a category of dietary fiber, is broken down by bacteria in the colon through fermentation. This process results in the generation of gases and SCFAs such as acetate, propionate, and butyrate. These SCFAs have many good effects on health (Nogal et al. 2021). One key advantage is their role in providing energy for colon cells, which helps to maintain the integrity of the intestinal lining and support optimal digestive function (Nogal et al. 2021). By virtue of their anti-inflammatory effects, SCFAs contribute to the decrease of inflammation not only in the digestive tract but also throughout the body. Also, these things are good for your gut because they help good bacteria grow and stop bad bacteria from growing too much, which keeps your gut healthy (Silva et al. 2020). Fermentable fiber include inulin, fructo-oligosaccharides (FOS), and resistant starch, which are found in various plant-based foods (Kumar et al. 2018). Legumes, whole grains, onions, garlic, and bananas are examples of dietary sources that contain a significant amount of fermentable fiber. By integrating these nutrients into their diet, people can bolster intestinal health, diminish inflammation, and foster general well-being (Khalid et al. 2022).

7.2.4 *Non-fermentable Fiber*

Non-fermentable fiber, another significant component of dietary fiber, differs from its fermentable counterpart in that it resists fermentation in the colon and passes through the digestive tract relatively intact. Although it does not undergo fermentation to produce SCFAs like fermentable fiber, non-fermentable fiber still serves essential functions in promoting digestive health (Puhlmann and de Vos 2022). Its primary role lies in providing bulk to the stool, which helps to regulate bowel movements and prevent constipation. By adding volume and weight to the stool, non-fermentable fiber facilitates its passage through the intestines, contributing to regularity in bowel movements (Dhingra et al. 2012). Non-fermentable fiber includes certain types of insoluble fiber, such as cellulose and lignin, which are abundant in various plant-based foods. Whole grains, nuts, seeds, and certain fruits and veggies are rich in non-fermentable fiber. Incorporating these fiber-rich foods into one's diet can help maintain digestive regularity and support overall gastrointestinal health (Dhingra et al. 2012).

7.3 Biological and Physiological Properties of Dietary Fiber

The biological and physiological properties of dietary fiber refer to the effects and characteristics of fiber in the body's biological and physiological processes. This includes how fiber interacts with the digestive system, its impact on nutrient absorption, its ability to promote bowel regularity, its influence on blood sugar levels, its role in lowering cholesterol, and its effects on satiety and weight management. These properties contribute to the overall health benefits associated with consuming dietary fiber. Dietary fiber, found abundantly in plant-based foods, possesses several key biological and physiological properties essential for optimal health (He et al. 2022). Various studies suggest that the physical and chemical features of dietary fiber, such as how much water it holds, its ability to swell in water, its capacity to adsorb glucose and cholesterol, and its visibility, are closely linked to how they work in our bodies. Hence, the relationship between these physicochemical and physiological function are discussed (He et al. 2022).

7.3.1 Bulk Formation and Bowel Regularity

The ability of dietary fiber to hold water is influenced by its structure. As the contact area and hydrophilic characteristics of dietary fiber rise, water retention within the hydrophilic portion or the fiber network structure improves, resulting in enhanced water holding capacity (WHC). The WHC of DF varies, ranging from 2.01 to 25.03 g/g. Among different types of foods, vegetables exhibit the highest WHC (7.10–18.90 g), followed by fruits (3.38–13.08 g) and grains (2.01–6.23 g) (He et al. 2022). Vegetable dietary fiber can hold a lot of water because its surface is uneven, loose, and has holes, just as Xie et al. (2017a) studied that Angelica keiskei's root DF has high WHC. According to Zheng et al. (Zheng and Li 2018), the dietary fiber content in coconut is measured at (13.08 ± 0.36 g), surpassing the levels found in many vegetables. The greater water retention ability of coconut is due to its more porous dietary fiber structure, which offers increased surface area for water retention compared to most vegetables. In the field of food research and industrial production, a variety of processing techniques are commonly utilized to modify the structure of dietary fiber. The objective is to improve specific physicochemical properties. The excellent water holding capacity (WHC) of dietary fiber (DF) is a significant factor contributing to its beneficial effects on human health. Likewise, another crucial physicochemical property of DF is its water swelling capacity (WSC). DF exhibits WSC due to the following explanations: (Stribling and Ibrahim 2023) water possesses surface tension strength, allowing it to be retained within the capillary structure of dietary fiber; (Zhang et al. 2024) hydrogen bonds and hydrophilic groups with dipoles attract and bind water molecules; (Thompson and Brick 2016) the extent of this interaction is directly associated with the contact area of DF. Dietary fiber has varying water WSC, from 0.95 to 23.90 ml/g. Vegetable DF ranges from 5.35 to 18.70 g/g, fruit DF from 3.20 to 10.17 g/g, and grain DF from 0.95 to 7.54 g/g. Coconut DF has the highest WSC, reaching (10.17 ± 0.40) g/g due to its exposure of more hydrogen bonds and dipole forms (Zheng and Li 2018). Xie et al. (2017b) found that the WSC of dietary fiber from purple potatoes is influenced by processing conditions. High pressure treatment enhances WSC (18.30 ml/g) due to changes in chemical and structural properties induced by this process. Overall, the WSC of DF contributes to its functional characteristics similarly to water holding capacity. Because of its high WHC and WSC, dietary fiber can absorb water and expand within the digestive tract. This property is attributed to its ability to hold significant amounts of water and swell when in contact with it. This swelling action increases the volume and softness of stool, adding bulk to its composition (Hervik and Svihus 2019). The bulkier stool stimulates the walls of the colon, triggering the natural contractions necessary for bowel movements. This process facilitates the passage of stool through the digestive tract. Fiber aids in the prevention of constipation by increasing stool bulk. Constipation, marked by infrequent bowel movements and difficulty passing stool, can cause discomfort, bloating, and straining during defecation. The increased bulk of stool due to fiber consumption promotes regularity in bowel movements by enhancing gastrointestinal motility (Ho et al. 2012).

7.3.2 *Viscosity and Nutrient Absorption*

Viscosity refers to the resistance of fluids to flow when subjected to an external force. Several elements impact it, such as the chemical composition, arrangement, particle size, surface area, and manufacturing process. In a study by Gallaher et al. (1999), rats were given muffins made from whole oatmeal or instant oatmeal. They checked the viscosity of the stuff inside the intestines afterward. They found that the stuff from whole oatmeal was thicker compared to the stuff from instant oatmeal. The variation is ascribed to the elevated concentration of viscous fiber, specifically β-glucan, found in whole oats. β-glucan possesses a greater number of hydrophilic groups, which augment the thickening of the water film within the intestinal contents. Ullah et al. (2018) observed that the viscosity of okara stabilized with thermal pre-treatment increased. The viscosity of dietary fiber has the potential to mitigate obesity and diabetes by retarding fat formation and gastric emptying, while also improving insulin sensitivity. Dietary fiber also plays an essential role in influencing nutrient absorption and related physiological responses due to its excellent nutrient absorption capacity. Soluble fiber, known for its ability to dissolve in water, undergoes hydration when it comes into contact with moisture in the digestive tract (Lattimer and Haub 2010). This process transforms the fiber into a viscous gel-like substance. The gel formation alters the physical properties of the digestive contents, creating a thicker and more cohesive environment. This change in viscosity affects the movement and interaction of nutrients within the gastrointestinal (GI) tract (Lattimer and Haub 2010). Within this gel matrix, the digestion of carbohydrates is hindered, particularly complex carbohydrates found in foods like grains, legumes, and certain fruits. The gel acts as a barrier, slowing down the access of digestive enzymes to carbohydrates. As a result, the breakdown of carbohydrates into simpler sugars, such as glucose, is delayed (Lovegrove et al. 2017). This delayed digestion results in a gradual release of sugars into the bloodstream during absorption, prolonging the duration over which nutrients are absorbed into the bloodstream. The gradual release of sugars into the bloodstream helps regulate and stabilize blood sugar levels after meals, resulting in a smoother and more controlled increase. By moderating the rate of glucose absorption, soluble fiber helps prevent sharp spikes and subsequent crashes in blood sugar levels. This control is very important for people with diabetes or those who might get insulin resistance (Lovegrove et al. 2017).

7.3.3 *Prebiotic Characteristics*

The prebiotic effects of certain fibers represent a fascinating aspect of their interaction with the gut microbiota, shaping gut health and influencing various physiological processes. Prebiotics refer to dietary components, like specific types of fiber, that are indigestible and selectively indorse the growth and function of useful bacteria in the gastrointestinal tract (Guarino et al. 2020). By providing a favorable environment for beneficial bacteria to thrive, prebiotics contribute to the maintenance of a diverse and balanced gut microbiota, which is necessary for optimal digestive health and overall well-being. Prebiotic fibers are fermented by helpful bacteria in the colon, such as *Bifidobacteria and Lactobacilli*, while being less accessible to harmful or pathogenic microorganisms. This selective fermentation process stimulates the growth and activity of useful bacteria, enhancing their predominance within the gut microbiota. During fermentation, beneficial bacteria metabolize prebiotic fibers to produce SCFAs. These SCFAs function as a basis of energy for cells in the colon, support the integrity of the gut barrier, and elicit diverse metabolic and immunomodulatory effects (Guarino et al. 2020).

7.3.4 Glucose Adsorption Capacity (GAC)

The GAC of DF is linked to its physical characteristics. Glucose tends to adhere to the network structure of DF, thereby reducing its contact with the human intestinal tract (Moghaddam et al. 2018). The GAC of dietary fiber in food typically ranges from 0.17 to 4.65 mmol/g, with research primarily concentrating on vegetable-derived dietary fiber. Huang et al. (2018) detected arise in the glucose adsorption capacity (GAC) of dietary fiber (DF) from garlic stems after ultrasonic treatment. This improvement is linked to the enhancement of the honeycomb network structure. Yu et al. (2018) found that insoluble dietary fiber (IDF) from carrots, extracted using a complex enzyme method, had a glucose adsorption capacity (GAC) of 2.43 mmol/g. They showed that under hydrostatic pressure, the GAC increased significantly to 2.63 mmol/g. This capacity is linked to the glucose metabolism function of dietary fiber, which absorbs sugar and generates SCFAs through fermentation, potentially aiding in diabetes prevention. In general, GAC is important for regulating sugar for metabolic health, and soluble fiber plays a significant role in modulating this process. Fiber from foods like oats, barley, beans, and certain fruits turns into a jelly-like stuff when it mixes with water in your stomach. This jelly slows down how fast your body digests and takes in carbohydrates, including glucose. The viscous substance produced by soluble fiber also slows down the emptying of your stomach, causing glucose to be released into your bloodstream more gradually after meals (Giuntini et al. 2022). This gradual release helps prevent rapid spikes in blood sugar levels. Soluble fiber has been shown to improve insulin sensitivity, allowing cells to more effectively respond to insulin. Insulin is the hormone tasked with moving glucose from the bloodstream into cells, where it is utilized for energy production (Giuntini et al. 2022).

7.3.5 *Cholesterol Adsorption Capacity (CAC)*

CAC refers to the quantity of cholesterol absorbed by 1 g of a sample. The CAC of dietary fiber (DF) involves both physical and chemical adsorption mechanisms. Physical adsorption is primarily influenced by factors such as particle size, specific surface area, porosity, and reaction temperature of DF (Mirmiran et al. 2016; Nandi et al. 2019). Chemical adsorption of cholesterol by DF is influenced by factors such as the charge and hydrophobic groups present in the DF. CAC varies widely, ranging from 0.03 to 37.10 mg/g. Most research on CAC has concentrated on vegetable DF, with reported values typically falling within the range of 12.94–37.10 mg/g. Bamboo shoot shell fiber can bind bile acids and cholesterol, reducing cholesterol levels by about 12.94 mg/g (Luo et al. 2017). In a study by Liu et al. (2016), it was discovered that soluble dietary fiber (SDF) found in soybean hulls exhibits notable viscosity, leading to enhanced cholesterol adsorption capacity (CAC) of 7.62 mg/g. This effect contributes to reducing blood cholesterol levels. Dietary fiber with significant CAC holds potential for preventing and treating obesity-related conditions. This is because dietary fiber serves as a key component in the effort to reduce LDL cholesterol levels by virtue of its ability to bind to cholesterol in the digestive tract, thereby facilitating its excretion from the body. This mechanism is fundamental in mitigating the risk of CVD, as elevated LDL cholesterol levels are strongly associated with a heightened risk of heart problems and related complications (Chang et al. 2017). By incorporating soluble fiber-rich foods into the diet, individuals can harness the cholesterol-lowering effects of fiber to maintain optimal lipid profiles and promote cardiovascular well-being (Soliman 2019). Embracing a dietary pattern rich in soluble fiber sources, such as oats, barley, legumes, and certain fruits, represents a proactive and effective strategy for managing cholesterol levels and decreasing the risk of heart problems (Soliman 2019).

7.4 Nutraceutical Functions of Dietary Fiber

Dietary fiber plays a pivotal role in promoting health and well-being through its nutraceutical functions. Figure 7.1 illustrates various roles of dietary fiber in promoting health and well-being through by nutraceutical functions. Firstly, soluble fibers have been linked to cholesterol reduction, contributing to cardiovascular health. Additionally, dietary fiber aids in maintaining optimal blood glucose levels, making it instrumental in diabetes management. Its ability to enhance digestive health by promoting regular bowel movements and preventing constipation underscores another crucial aspect of its nutraceutical function. Furthermore, dietary fiber acts as a prebiotic, fostering the growth of beneficial gut bacteria and supporting a balanced microbiome. This intricate interplay contributes to improved immune function and overall gut health. As a nutraceutical, dietary fiber also exhibits satiety-inducing properties, aiding in weight management and promoting a healthy body

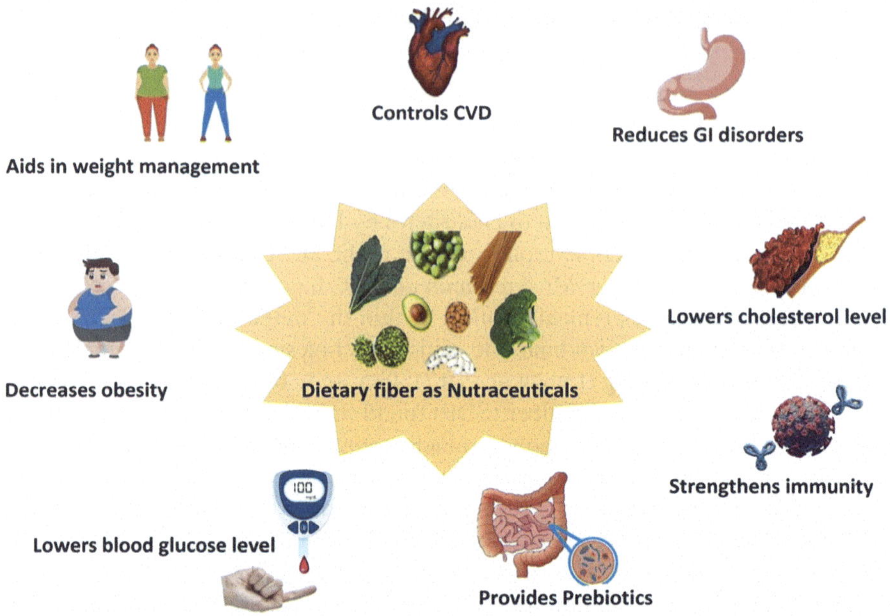

Fig. 7.1 Dietary fiber's role in promoting health by its nutraceutical functions

composition. In essence, the nutraceutical functions of dietary fiber encompass a spectrum of health benefits, making it an integral component of a wholesome and nutritionally rich diet.

Dietary fibers are polysaccharides that resist digestion in the small intestine, although some may undergo digestion in the colon. They consist of cellulose, hemicelluloses, and lignin. Primary sources of dietary fibers include whole grains and bran (Stribling and Ibrahim 2023). Dietary fibers are components found in plant-based foods that resist digestion by human digestive enzymes. Incorporating dietary fiber into one's daily nutrition offers numerous health profits, including declining the risk of several conditions such as stroke, coronary heart problems, diabetes, hypertension, obesity, and certain gastrointestinal problems (Anderson et al. 2009). Moreover, consuming more dietary fibers is associated with enhancements in blood sugar levels, immune system function, serum lipid levels, and diabetes management. It also reduces blood pressure, supports regular bowel movements, aids in weight management, and drives research and growth efforts in the food industry. For example, food industry research and development units create therapeutic food products such as sustained-release glucose and high-fiber biscuits (Ngaha et al. 2023). Increasing fiber intake can delay or mitigate the peak response to carbohydrate consumption. When the same amount of carbohydrates is consumed, additional dietary fibers may lower blood glucose levels by 10–20%. Given their beneficial impact on metabolic control in diabetes, cardiovascular health, and other risk factors, it is expected that higher dietary fiber consumption in daily meals could yield positive outcomes (Giuntini et al. 2022).

7.4.1 Preventive Role of Dietary Fiber in Cholesterol and Cardiovascular Diseases

Dietary fiber plays an important preventive role in cholesterol and cardiovascular diseases (CVD) by influencing various physiological processes within the body (McRae 2017). Understanding this connection involves delving into the mechanisms by which dietary fiber affects cholesterol levels and overall cardiovascular health. Dietary fiber, has been shown to decrease LDL cholesterol levels. It does so by binding to bile acids in the intestines, which are then excreted. To replace these bile acids, the liver uses cholesterol, thus lowering the circulating levels of LDL cholesterol. High-fiber foods tend to be more filling and satiating, leading to a reduced overall calorie intake (Soliman 2019). This property aids in weight management, as maintaining a healthy weight is essential for preventing CVD. Moreover, obesity is often associated with dyslipidaemia (abnormal lipid levels), insulin resistance, and inflammation, all of which raise the risk of CVD. Soluble fiber can make it take longer for your body to digest and use carbohydrate, which helps avoid quick rises in blood sugar after eating. Maintaining stable blood sugar levels is crucial for reducing the risk of diabetes, which is a significant contributor to CVD (Giuntini et al. 2022). Chronic inflammation is a significant contributor to the initiation and progression of atherosclerosis, which is the leading cause of most CVD. Consumption of fiber-rich diets, particularly those abundant in fruits, vegetables, and whole grains, has been associated with lower levels of inflammatory markers within the body (Giuntini et al. 2022). According to the FDA, diets rich in whole grains, fruits, and vegetables, while low in cholesterol and saturated fat, are associated with a decreased risk of CVD (Hussain et al. 2020). For a large portion of the global population, it's recommended to aim for a daily dietary fiber intake ranging from 25 to 35 g, with about 6 g being soluble dietary fiber (SDF). Numerous research studies suggest a connection between regular intake of dietary fiber and reduced risks of cardiovascular diseases. However, recent studies present compelling data indicating that with every additional 10 g of dietary fiber incorporated into a diet, the mortality rate from coronary heart disease declines by 17–35% (Lattimer and Haub 2010). The main things that can raise your risk of heart disease are having high cholesterol, high levels of fats in your blood, being overweight, having high blood pressure, and having type 2 diabetes (Streppel et al. 2008; Petrie et al. 2018). A study has discovered that adding dietary fiber from vegetable waste to meat items is a promising method to increase their nutritional content and may lower the risk of cholesterol and CVD. Fiber-enriched meat products can enhance a balanced and heart-healthy diet by utilizing the cholesterol-lowering and health-promoting qualities of soluble fibers, as well as the sensory and gastronomic advantages of different components (Haque et al. 2023).

7.4.2 Role of Dietary Fiber in Lowering Blood Glucose Level and Diabetes

Dietary fiber is essential for managing blood glucose levels and preventing or managing diabetes. Its mechanism involves slowing down the digestion and absorption of carbohydrates. As a result, when consuming fiber-rich foods, glucose is released into the bloodstream more steadily, thereby preventing rapid spikes in blood sugar levels. Dietary fiber has been shown to improve insulin sensitivity, allowing cells to better respond to insulin and therefore better regulate blood sugar levels. Foods with lots of fiber usually have a lower glycemic index (GI). That means they make your blood sugar go up more slowly and not as much as foods with a higher GI (Giuntini et al. 2022). The mechanism by which dietary fiber reduces blood glucose levels operates as follows: The network formed by dietary fiber acts as a physical obstacle, impeding the diffusion of glucose. The fibrous particles catch glucose in the fiber linkage. When the fiber structure loosens up and gets more porous, with more of its surface functional groups showing, it makes the connection with glucose stronger, which boosts GAC (He et al. 2022). Research indicates that those eating more dietary fiber, especially from cereals, may have a lower risk of type 2 diabetes. Adding β-glucan or psyllium to the daily diet of people with type 2 diabetes might slightly decrease fasting blood glucose levels and the percentage of glycosylated hemoglobin (McRae 2018).

7.4.3 Role of Dietary Fiber in Prevention of Constipation

Fiber adds bulk to the stool, making it easier for the intestines to move waste through the digestive tract. The extra bulk from fiber softens stool, making it easier to pass. Fiber also helps keep bowel movements regular. It adds substance to the stool, which stimulates the muscles of the intestines, encouraging them to contract and move waste along (Bellini et al. 2021). Constipation may predispose individuals to the formation of hemorrhoids and diverticulosis (small pouches in the colon wall). Sufficient fiber consumption can mitigate these conditions by fostering regularity in bowel movements and lessening straining during defecation (Ho et al. 2012). In scientific research, it is proposed that the gastrointestinal system serves as the body's largest barrier tissue. In a healthy state, the environment within the intestines is vital for facilitating the digestion of food and impacting the immune system within the mucosal lining through a mutually beneficial relationship with microorganisms. However, dysbiosis, characterized by an imbalance in gut microbiota, can disrupt the integrity of the intestinal barrier, increasing susceptibility to various illnesses. The occurrence of intestinal disorders is intricately linked to lifestyle and dietary habits. People who eat lots of fiber can help control the bacteria in their gut, avoid constipation, and keep their gut wall strong, which helps keep their stomach and intestines healthy (He et al. 2022).

7.4.4 Dietary Fiber in Weight Management and Obesity Control

Dietary fiber is essential for weight management and combating obesity. It helps increase satiety, stabilize blood sugar levels, and support digestive health, making it a fundamental component of any plan for weight loss or weight maintenance. The ingestion of dietary fiber in food leads to the formation of bulk, inducing a sensation of fullness in the stomach and prolonging the interval before further eating. Hormones secreted in the gastrointestinal tract, stimulated by DF, assist in regulating satiety and promoting gastrointestinal emptying (Costabile et al. 2018). Scientists from various regions globally have investigated the weight management advantages of dietary fiber (Puupponen-Pimiä et al. 2002). A meta-analysis of studies examining various types of fiber revealed a combined weight loss of 2.494 kg (Thompson et al. 2017). Another meta-analysis revealed that the use of chitosan supplements resulted in a total body weight loss of approximately 1.814 pounds (Mhurchu et al. 2005). A study says that eating fiber can help your stomach and intestines stay healthy and lower the chances of getting diseases related to obesity. Soluble, viscous fibers hinder cholesterol absorption and lower serum cholesterol levels. Consuming fiber increases satiety and reduces food intake in subsequent meals. Alterations in gut microbiota and metabolites generated from gut fermentation contribute to lowering the risk of obesity-related conditions. Dietary fibers impact gut microbiota and generate metabolites such as SCFAs, potentially explaining their role in preventing and managing obesity (Dayib et al. 2020).

7.4.5 Dietary Fiber in Prevention of Inflammatory Bowel Disease

Dietary fiber is notably influential in preventing inflammatory bowel disease (IBD), which encompasses a range of disorders marked by persistent inflammation in the digestive tract, including Crohn's disease and ulcerative colitis. Fiber-rich foods, provide essential nutrients and promote gastrointestinal health through several mechanisms (Di Rosa et al. 2022). Dietary fiber acts as a prebiotic, serving as fuel for helpful gut bacteria. These bacteria produce SCFAs during the fermentation of fiber, which help maintain a healthy gut environment by supporting the integrity of the intestinal barrier and regulating immune function. SCFAs also possess anti-inflammatory properties, which can help mitigate inflammation in the gut and reduce the risk of developing IBD (Di Rosa et al. 2022). In 2011, Benjamin and colleagues found that individuals with active Crohn's disease who were administered 15 g of fructo-oligosaccharides (FOS) per day exhibited reduced proportions of interleukin (IL-6)-positive lamina propria dendritic cells (DC), increased levels of IL-10 staining in the DC, and no alteration in IL-12p40 production (Benjamin et al. 2011). Although the study observed alterations in DC function, the researchers concluded that no therapeutic benefit was attained in patients with active Crohn's disease. Possible reasons for this lack of benefit may include the quantity of dietary fiber, the brief duration of the trial (4 weeks), and the use of FOS as the sole treatment. In another study, researchers examined the effects of germinated barley foodstuff (GBF) on serum TNF-α, IL-6, and IL-8 levels in UC patients who were in remission. During the two-month period, participants in the treatment group received 30 g of GBF daily through oral administration in addition to standard drug therapy. Throughout the study, TNF-α, IL-6, and IL-8 levels all decreased in the GBF group compared to the control group (Faghfoori et al. 2011). In 2014, the same group of researchers piloted a subsequent study to assess the impact of germinated barley foodstuff (GBF) supplementation on serum C-reactive protein (CRP) levels and clinical symptoms in UC patients. The results showed a significant reduction in serum CRP levels in the GBF group, suggesting that combining GBF with standard medications may effectively alleviate inflammation and clinical symptoms in UC patients (Faghfoori et al. 2014). In a recent crossover trial with 17 ulcerative colitis (UC) patients in remission or with minimal disease activity, patients were split into two groups. One group ate a diet low in fat but high in fiber, while the other group followed a typical American diet but with some improvements. The study found that the low-fat, high-fiber diet helped decrease signs of inflammation and gut bacteria imbalance (Fritsch et al. 2021).

7.4.6 Dietary Fiber Role in Gastrointestinal and Colonic Fermentation

Dietary fiber plays a key role in gastrointestinal and colonic fermentation, which is the process by which gut bacteria break down complex carbohydrates that cannot be digested by human enzymes. Dietary fiber acts as a source of nourishment for the vast population of microorganisms inhabiting the human intestinal tract, commonly referred to as the gut microbiota. These microbes have the enzymes necessary to break down various types of fiber that human digestive enzymes cannot process. Gut microbes ferment dietary fiber by metabolizing it into SCFAs. This fermentation process primarily occurs in the colon (Holscher 2017). In a study found that, dietary fiber consists of undigested carbohydrates that bypass digestion and absorption in the small intestine, reaching the large intestine where they undergo partial or complete fermentation by colonic microbiota. The fermentation process yields SCFAs, including acetic, propionic, and butyric acids, which have been associated with favorable alterations in microbiota composition. Choosing this diet may decrease the risk of developing conditions such as colon issues and health problems like obesity, diabetes, kidney disease, and inflammation in the body (Wang et al. 2019).

Researchers recently discovered that different kinds of fiber have distinct impacts on the function of the intestinal barrier. These effects are influenced by bacteria and SCFAs, according to the findings of the study. Soluble fibers, such as inulin, boosted the growth of *Actinobacteria*, *Proteobacteria*, *Blautia*, and *Atopobium*, whereas insoluble fibers, such as cellulose, enhanced the abundance of Bacteroidetes, *Euryarchaeota*, *Phascolarctobacterium*, and *Coprococcus*. When compared to insoluble fiber, the consumption of soluble fiber led to increased levels of short-chain fatty acids (SCFA), as well as increased microbial diversity and community richness. The results of these studies contributed to an improvement in the function of the intestinal barrier (Chen et al. 2019).

7.4.7 Dietary Fiber in Boosting Immunity

Dietary fibers play a significant role in human health, particularly in influencing immune function. Before undergoing microbial fermentation in the colon, dietary fibers affect the intestine by modulating intestinal barrier function and immune responses. Specific fibers like galactooligosaccharides (GOS), inulins, pectins, and β-galactomannan have been shown to strengthen intestinal epithelial barrier function and regulate immune cell cytokine secretion. Additionally, fibers can directly interact with pattern recognition receptors (PRRs) such as C-type lectin receptors (CLRs) and Toll-like receptors (TLRs), thereby influencing immune cell activation (Fu et al. 2022). After microbial fermentation, dietary fibers produce various immunological metabolites, with SCFAs being extensively studied. SCFAs profoundly affect the host immune system through diverse mechanisms. They interact with specific G protein-coupled receptors (GPCRs), inhibit histone deacetylase (HDAC) activity, and act as ligands for the aryl hydrocarbon receptor (AhR). SCFAs regulate numerous immune cell functions, including cytokine secretion by dendritic cells and lymphocytes, regulatory T cell differentiation, bone marrow haematopoiesis, effector T cell activity, maintenance of epithelial barrier integrity, IgA secretion, modulation of mast cell degranulation, and activation of innate lymphoid cells (ILCs) (Fu et al. 2022). The introduction of a diverse range of fruits and vegetables into an infant's diet throughout the first year of life has been linked to higher concentrations of fecal butyrate. Infants who have the highest concentrations of butyrate and propionate in their feces have a significantly lower likelihood of developing allergies and asthma as they age (Roduit et al. 2019). Consuming dietary fiber enhances immune function throughout pregnancy (Zerfu and Mekuria 2019). An adequately balanced diet that is abundant in fiber can promote general well-being, including the enhancement of immunological function. Dietary fiber can enhance the immune system indirectly by fostering a healthy gut flora, diminishing inflammation, and facilitating nutrient absorption. This, in turn, may aid the body in effectively responding to infections, such as COVID-19 (Calcuttawala 2022).

7.5 Dietary Fiber-Enriched Food as Nutraceuticals

Dietary fiber-enriched foods play a pivotal role in the realm of nutraceuticals, offering a multifaceted approach to promoting health and well-being. These foods, rich in soluble and insoluble fibers, serve as potent nutraceutical agents that contribute to various physiological benefits. Soluble fibers, such as β-glucans and pectins, are known to help regulate blood sugar levels and lower cholesterol, making them invaluable in managing conditions like diabetes and cardiovascular diseases. Insoluble fibers, which are plentiful in whole grains and vegetables, help keep your digestive system healthy by making sure you have regular bowel movements and don't get constipated. Additionally, fiber-enriched foods act as prebiotics, supporting the growth of beneficial gut bacteria, which is increasingly recognized for its influence on immune function and overall gut health. By intertwining nutrition and pharmaceutical principles, dietary fiber-enriched foods emerge as essential components of a balanced and health-conscious diet, showcasing their potential as nutraceuticals in the pursuit of holistic well-being. Dietary fiber-enriched foods encompass a wide array of nutraceutical options that contribute to overall health and disease prevention. One exemplary source is oat-based products, fortified with beta-glucans that have been linked to cholesterol reduction and cardiovascular health. Similarly, flaxseeds, rich in soluble and insoluble fiber, serve as powerful nutraceuticals due to their potential in improving digestive function and regulating blood sugar levels. Whole grains, such as quinoa and brown rice, not only provide essential nutrients but also deliver a substantial dose of dietary fiber, promoting satiety and aiding in weight management. Additionally, fruits like apples and berries, with their skin intact, offer a bounty of fiber, supporting digestive well-being and providing antioxidants. These dietary fiber-enriched options exemplify the fusion of nutrition and pharmaceutical benefits, showcasing the versatility of these foods as nutraceuticals in promoting a holistic approach to health. Table 7.1 shows few examples of incorporating nutraceutical dietary fiber derived from diverse plant-based sources in a spectrum of food commodities and their associated benefits. The following text discusses specific fiber supplements that are readily available as nutraceuticals and their primary functions in the body.

Table 7.1 Examples of incorporating nutraceutical dietary fiber derived from diverse plant-based sources in a spectrum of food commodities and their associated benefits

Food products	Nutraceutical DF sources	Fiber type	Benefits	References
Cakes	Pineapple pulp waste	Soluble and insoluble DF	Raised the nutritional content	Singh (2016)
Snacks	Mushroom powder	Soluble and insoluble DF	Enhanced the nutritional and functional characteristics	Keerthana et al. (2020)
Biscuit	Dragon fruit powder	Soluble and insoluble DF	Enhanced the nutritional and functional properties, providing 20% of the recommended daily intake of dietary fiber	Pawde et al. (2020)
Bread	Mixed psyllium fiber	Soluble DF	Provided nutritional value and antioxidants	Wojciechowicz-Budzisz et al. (2023)
Meat rolls	Psyllium husk	Soluble DF	Decreased cholesterol levels and increased dietary fiber content	PDF (n.d.)
Bread	Chia seed flour	Soluble DF	Improved the nutritional attributes	Verdú et al. (2017)
Fruit-based beverage	Basil seed	Insoluble DF	Enhanced the nutrients and improved the storage stability	Hajmohammadi et al. (2016)
Beef sausage	Flaxseed and tomato powders	Soluble and insoluble DF	Enhanced the essential fatty acids and protein and brought lycopene and fiber	Ghafouri-Oskuei et al. (2020)
Cookie	Flaxseed flour	Soluble and insoluble DF	Improved the overall nutritional value	Kaur et al. (2019)
Noodle	Fenugreek seed powder	Soluble DF	Enriched the nutritional value, particularly fiber and protein	Ahmad et al. (2022)

7.5.1 *Psyllium Husk*

Research indicates that the incorporation of 10 g/day of psyllium for a period of 4 weeks resulted in a lessening of LDL cholesterol ranging from 6% to 24% and total cholesterol decreasing by 2% to 20% (Agrawal et al. 2007). The most significant decreases were observed in individuals with elevated initial cholesterol levels. Research has shown that psyllium husk, a soluble fiber with a high viscosity, can regulate blood glucose levels and reduce LDL cholesterol. Constipation is efficiently prevented by its remarkable water-holding capacity, which helps soften firm stools. Hemorrhoids sufferers find relief by increasing their water consumption and consuming 7–14 g of psyllium husk daily (Garg 2017). Psyllium is the only fiber supplement that has strong clinical data supporting its use for treating chronic constipation, which justifies its recommendation (McRorie 2015). Psyllium husk was used in a nutraceutical chewing candy (CCN) formulation using fermented milk permeate (MP), which serves as a source of galactooligosaccharides (GOS) and viable lactic acid bacteria (LAB). We also added psyllium husk, which has helpful thickening properties, and apple by-products, which are full of healthy compounds called phenolics, to the mix (Zokaityte et al. 2021).

7.5.2 Chia Seeds

Salba-chia, scientifically referred to as *Salvia hispanica* L., is renowned for its advantageous impact on Type 2 diabetes, decrease of obesity, and management of total weight. Studies on Chia seeds reveal that a 100-g portion includes 34–40 g of dietary fiber, which fulfills the entire daily recommended intake for adults (Khalid et al. 2023). Chia seeds are packed with beneficial nutrients, like soluble fiber, protein, polyphenols, and polyunsaturated fatty acids like ω-3 and ω-6. Consequently, they are useful in treating diabetes, cancer, hypertension, obesity, and cardiovascular diseases (CVDs) (Khalid et al. 2023). According to the findings of a study that focused on muffins that were supplemented with chia seeds, there was a decrease in the amount of body weight growth, the accumulation of visceral fat, total cholesterol levels, blood glucose, triglycerides, and LDL cholesterol within the participants. A number of markers, including ferritin (FER), hematocrit (HCT), hemoglobin (Hb), platelet count, and HDL cholesterol, showed signs of recovery at the same time (Rabail et al. 2022).

7.5.3 Black Cumin Seeds

On the basis of the nutritional analysis that was compiled from a variety of sources, it was determined that the protein content ranged from 20% to 85%, the fat content was 38.20%, the fiber content ranged from 7% to 94%, and the total carbohydrate content was 31.94%. Glutamate, arginine, and aspartate are among the amino acids that have been identified as being of particular significance. Cysteine and methionine, on the other hand, are the main and minor amino acids, respectively. Additionally, black cumin seeds contain significant amounts of vital minerals including Fe, Cu, Zn, P, and Ca, in addition to vitamins like B_1, B_3, B_6, and B_9. These characteristics make black cumin seeds an excellent source of nutrition (Yimer et al. 2019). In addition, the phytochemical analysis of *N. sativa* revealed the existence of a great deal of phytoconstituents, the majority of which were alkaloids, saponins, sterols, and essential oil. There is still a lack of information regarding the chemical identification and biological validation of a significant number of these constituents. Linoleic acid (64.6% of the seeds) and palmitic acid (20.4% of the seeds) are the two most abundant fatty acids found in the seeds of N. sativa, which constitute around 26–34% fixed oil combined. Furthermore, the essential oil content of the seed oil ranges from 0.4% to 2.5% (Ghahramanloo et al. 2017). It is clear that thymoquinone is the most significant component of the essential oil, regardless of the other active ingredients that have been recorded up to this point. In terms of therapeutic effects, this bioactive molecule exhibits a wide range of potential applications (Shafodino et al. 2022).

According to the findings of a study that was carried out by Bensiameur-Touati and colleagues, the administration of 0.2 g per kilogram of Nigella sativa (NS) led to an increase in the severity of dyslipidemia in diabetic rats. A decrease in triglycerides (TGs), cholesterol, and low-density lipoprotein cholesterol (LDL-C) was observed in conjunction with arise in high-density lipoprotein cholesterol (HDL-C). This improvement was characterized by an increase in HDL-C (Bensiameur-Touati et al. 2017). In another study, the immune function of male Wistar rats was improved by the incorporation of *Nigella sativa* (black seeds) to their diet (Mahmoud et al. 2021). Within the context of a separate experiment, the incorporation of black seed (*Nigella sativa L.*) into baked goods, specifically cakes and muffins, was found to have significant potential as a functional resource for the management of obesity (Kamandloo et al. 2023).

7.5.4 Basil Seeds

The perennial aromatic herb *Ocimum basilicum* L. is a member of the Labiatae family and is commonly known as basil or sweet basil. Throughout history, this herb has served multiple purposes. You can use the leaves, either fresh or dried, to give cuisine a distinct scent and flavor. Also, basil is used in making cheese, liqueurs, vinegars, cocktails, teas, and a host of other things. Many different products in the food, medicine, and beauty sectors use essential oils that are derived from plants, such as flowers and leaves (Calderón Bravo et al. 2021). The seeds are a typical component in many desserts, including ice cream and drinks. For nutraceutical purposes, they are also added to baked goods, either whole or ground, to supply dietary fiber. Moreover, the seeds are utilized to improve fruit-based beverages for both aesthetic and practical reasons (Hajmohammadi et al. 2016). Due to their high fiber content, the seeds show great promise as a functional element. Scientists have studied basil seed mucilage extensively, and what they've found is that it has a wide range of useful qualities, including the ability to emulsify, froth, thicken, stabilize, increase viscosity, and gel (Naji-Tabasi and Razavi 2017). Even though studies show that basil seeds have lots of good nutrients and can bring big health benefits like helping with diabetes, fighting bacteria, protecting against damage in the body, and even possibly fighting cancer, they're not commonly eaten as food (Gajendiran et al. 2016). Numerous applications can be found for the mucilage that is derived from basil seeds. By interacting with protein, it operates as a water-binding agent in low-salt meat products (Lee and Chin 2020), functions as a fat substitute in sponge cakes with a remarkable 75% reduction in fat content (Song et al. 2017), and acts as a gelling and stabilizing agent in pudding (milk protein gel), ice cream, and low-fat yogurt (Kim et al. 2020; BahramParvar et al. 2012; Song and Kim 2019). All of these functions are accomplished through its interaction with milk protein. By enhancing rheological characteristics, reducing syneresis, and imparting high gel strength, this interaction has been proved to be very effective. Additionally, the mucilage is utilized as an ingredient in order to advance the physicochemical and sensory aspects of bread and a variety of other bakery goods (Israr et al. 2017).

7.5.5 *Flax Seeds*

Flax seed, renowned for its nutraceutical properties, is a small yet potent seed that offers significant health advantages. Flax seeds are rich in vital nutrients such as ω-3 fatty acids, fiber, and antioxidants, which have been acknowledged for their ability to improve general health. The ω-3 fatty acids present in flax seeds promote cardiovascular well-being, while the abundant fiber content aids in maintaining digestive function and managing weight. In addition to their nutritional benefits, flax seeds contain abundant amounts of lignans, which are powerful antioxidants recognized for their anti-inflammatory qualities. Flax seeds, due to their composition and research indicating possible benefits in regulating blood sugar and preventing cancer, are considered a versatile nutraceutical. In addition, their ability to adapt makes them a versatile ingredient that can be added to a variety of dishes, further confirming their reputation as a highly desired functional component. Flax seeds are gaining popularity in nutraceuticals because of the growing focus on personalized nutrition and overall health (Sharma et al. 2019; Rizvi et al. 2022). It is considered that the biological features of flax seed gum have a good impact on human health, which has led to its utilization in the management of hyperglycemia, the reduction of cholesterol levels, the treatment of diabetes, and the fight against obesity (Lorenc et al. 2022). For infants who have a hypersensitivity to milk, the process of developing hypoallergenic formulations requires the incorporation of a number of different components, including hydrolyzed protein, milk that has been fortified with calcium and vitamins, and other components. The incorporation of flaxseeds into the formulation presents an opportunity for the development of these value-added dairy products (Akande et al. 2010). Indian yogurt, called Dahi, was made better by adding microencapsulated flaxseed oil powder in three different amounts: 1%, 2%, and 3%. This might be a good way to give people more ω-3 fatty acids (Goyal et al. 2016). A wide variety of food products can be made using extrusion technology, including ready-to-eat meals, snacks, confectionary goods, and crisp bread. Extruded meals are highly popular among consumers because they are convenient, affordable, visually appealing, and have a unique texture (Anton et al. 2009). Consumers may be inclined to integrate extruded flaxseed into their diet in order to get the health benefits associated with it (Wu et al. 2007).

7.5.6 Fenugreek

Fenugreek seeds are well-known for their composition, which includes fiber, gum, numerous chemical elements, and volatile compounds. The composition of these seeds includes 45.4% dietary fiber, with 32% being insoluble and 13.3% soluble. The gum consists of galactose and mannose, both of which are associated with a hypoglycemic impact (Roberts 2011). An amount of 25 g of powdered fenugreek seeds was given as a dietary supplement for a duration of roughly 1 month. The findings demonstrated a noteworthy decrease in both pre-meal and post-meal blood glucose levels. Therefore, the study determined that a diet rich in high-fiber fenugreek is effective in managing diabetes (Gaddam et al. 2015). Fenugreek seeds are commonly used for various therapeutic benefits, including promoting wound healing, aiding digestion, addressing sinus and lung congestion, reducing inflammation and infections, providing relief, treating hair-related problems, and exhibiting aphrodisiac effects (Srinivasa and Naidu 2021). By incorporating fenugreek into bread, Losso et al. demonstrated that the incorporation of fenugreek into food can contribute to a reduction in the amount of sugar that is present in the blood (Losso et al. 2009). Fenugreek seeds, whether they were consumed in their raw, soaking, or germinated forms, were shown to result in a significant reduction in total lipids, serum total cholesterol, and LDL cholesterol, according to research conducted by Hussein and colleagues. The levels of triglycerides and HDL cholesterol in the serum, on the other hand, did not show any significant changes (Hussein et al. 2011). Increased levels of total proteins, fibers, iron, zinc, calcium, vitamin B2, carotene, vitamin E, and vitamin C were seen in individuals who consumed basal diets that included fenugreek leaves, seeds (both dry and germinated), and wheat flour supplemented with germinated fenugreek powder at amounts ranging from 5% to 10% (Wani and Kumar 2018). Excessive ingestion of fiber can have negative consequences on the body, such as bloating, gas, and constipation, despite its benefits. Additionally, it might cause diminished assimilation of vital minerals like iron and calcium, leading to shortages. Studies have found a correlation between increased consumption of dietary fiber and problems such as reduced bone density and impaired absorption of minerals in males (Lattimer and Haub 2010).

7.6 Potential Negative Aspects of High Fiber Consumption

An excessive consumption of dietary fiber may have negative effects, despite the fact that it lowers plasma cholesterol levels, glycemic index, and the risk of developing colon cancer. There is a theory that suggests that consuming a greater quantity of fiber (more than 30 g/day) could result in intestinal chelation, which would result in the elimination of important elements, particularly polyvalent cations such as iron, zinc, copper, magnesium, and calcium (Shah et al. 2009). In addition, an excessive consumption of dietary fiber may be a factor in the development of phytobezoars, which may lead to obstruction of the gastrointestinal tract. Furthermore, there is a possibility that the efficiency of energy and nutrient digestion will diminish, but there are certain points of view that consider this to be a desirable component, particularly in relation to therapeutic weight loss practices (Jalili et al. 2000). In recent clinical research, a diet that is high in pure soluble fiber has been found to be associated with an increased risk of developing hepatocellular carcinomas. However, it was noted that the synthesis of a considerable amount of butyrate, which was the result of the fermentation of inulin, was the single factor that led to the development of these liver tumors in laboratory mice that had intestinal dysbiosis. According to these findings, taking dietary supplements that include soluble fiber in addition to the amount that one consumes may have negative consequences (Bretin and Chassaing 2019).

7.7 Conclusion

In conclusion, this discussion exploring the functional aspects of dietary fiber in nutraceuticals has shed light on the profound impact of these compounds on human health. From their role in promoting digestive health and regulating blood sugar levels to their potential in preventing chronic diseases, the multifaceted benefits of dietary fiber have become increasingly evident. As the nutraceutical industry continues to evolve, understanding the functional properties of dietary fiber becomes crucial for formulating effective and holistic health-promoting products. The scientific insights gleaned from these articles contribute to the expanding body of knowledge in nutrition and underscore the importance of incorporating diverse dietary fibers into our diets for overall well-being. The convergence of research in this field catalyzes future innovations in nutraceutical development, fostering a more comprehensive approach to improving public health through dietary interventions.

References

Adams S, Sello C, Qin GX, Che D, Han R (2018) Does dietary fiber affect the levels of nutritional components after feed formulation? Fibers [Internet] 6(2):29. Available from: http://www.mdpi.com/2079-6439/6/2/29

Agrawal R, Agrawal R (2021) Psyllium: a source of dietary fiber. Diet Fibers

Agrawal AR, Tandon M, Sharma PL (2007) Effect of combining viscous fibre with lovastatin on serum lipids in normal human subjects. Int J Clin Pract [Internet] 61(11):1812–1818. Available from: https://onlinelibrary.wiley.com/doi/10.1111/j.1742-1241.2007.01512.x

Ahmad Z, Ilyas M, Ameer K, Khan MA, Waseem M, Shah FUH et al (2022) The influence of fenugreek seed powder addition on the nutritional, antioxidant, and sensorial properties of value-added noodles. J Food Qual 2022:1–10

Akande KE, Doma UD, Agu HO, Adamu HM (2010) Major antinutrients found in plant protein sources: their effect on nutrition. Pak J Nutr [Internet] 9(8):827–832. Available from: https://www.scialert.net/abstract/?doi=pjn.2010.827.832

Anderson JW, Baird P, Davis RH, Ferreri S, Knudtson M, Koraym A et al (2009) Health benefits of dietary fiber. Nutr Rev 67(4):188–205

Anton AA, Gary Fulcher R, Arntfield SD (2009) Physical and nutritional impact of fortification of corn starch-based extruded snacks with common bean (Phaseolus vulgaris L.) flour: effects of bean addition and extrusion cooking. Food Chem [Internet] 113(4):989–996. Available from: https://linkinghub.elsevier.com/retrieve/pii/S0308814608010261

Arzami AN, Ho TM, Mikkonen KS (2022) Valorization of cereal by-product hemicelluloses: fractionation and purity considerations. Food Res Int 151:110818

BahramParvar M, Razavi SMA, Mazaheri TM (2012) Optimising the ice cream formulation using basil seed gum (O cimum basilicum L.) as a novel stabiliser to deliver improved processing quality. Int J Food Sci Technol [Internet] 47(12):2655–2661. Available from: https://ifst.onlinelibrary.wiley.com/doi/10.1111/j.1365-2621.2012.03148.x

Bellini M, Tonarelli S, Barracca F, Rettura F, Pancetti A, Ceccarelli L et al (2021) Chronic constipation: is a nutritional approach reasonable? Nutr 13(10):3386

Benjamin JL, Hedin CRH, Koutsoumpas A, Ng SC, McCarthy NE, Hart AL et al (2011) Randomised, double-blind, placebo-controlled trial of fructo-oligosaccharides in active Crohn's disease. Gut 60(7):923–929

Bensiameur-Touati K, Kacimi G, Haffaf EM, Berdja S, Aouichat-Bouguerra S (2017) In vivo subacute toxicity and antidiabetic effect of aqueous extract of Nigella sativa. Evid Based Complement Alternat Med [Internet] 2017:1–13. Available from: https://www.hindawi.com/journals/ecam/2017/8427034/

Bouazizi S, Montevecchi G, Antonelli A, Hamdi M (2020) Effects of prickly pear (Opuntia ficus-indica L.) peel flour as an innovative ingredient in biscuits formulation. LWT [Internet] 124:109155. Available from: https://linkinghub.elsevier.com/retrieve/pii/S0023643820301432

Bretin A, Chassaing B (2019) Inuline: la belle et la bête. médecine/sciences [Internet] 35(2):100–102. Available from: https://www.medecinesciences.org/10.1051/medsci/2019014

Calcuttawala F (2022) Nutrition as a key to boost immunity against COVID-19. Clin Nutr Espen 49:17–23

Calderón Bravo H, Vera Céspedes N, Zura-Bravo L, Muñoz LA (2021) Basil seeds as a novel food, source of nutrients and functional ingredients with beneficial properties: a review. Foods [Internet] 10(7):1467. Available from: https://www.mdpi.com/2304-8158/10/7/1467

Chang S, Cui X, Guo M, Tian Y, Xu W, Huang K et al (2017) Insoluble dietary fiber from pear pomace can prevent high-fat diet-induced obesity in rats mainly by improving the structure of the gut microbiota. J Microbiol Biotechnol 27(4):856–867

Chen T, Chen D, Tian G, Zheng P, Mao X, Yu J et al (2019) Soluble fiber and insoluble fiber regulate colonic microbiota and barrier function in a piglet model. Biomed Res Int 2019:1–12

Costabile G, Griffo E, Cipriano P, Vetrani C, Vitale M, Mamone G et al (2018) Subjective satiety and plasma PYY concentration after wholemeal pasta. Appetite 125:172–181

Cui SW, Nie S, Roberts KT (2011) Functional properties of dietary fiber. Compr Biotechnol Second Ed 4:517–525

Dai FJ, Chau CF (2017) Classification and regulatory perspectives of dietary fiber. J Food Drug Anal 25(1):37–42

Dayib M, Larson J, Slavin J (2020) Dietary fibers reduce obesity-related disorders: mechanisms of action. Curr Opin Clin Nutr Metab Care 23(6):445–450

Dhingra D, Michael M, Rajput H, Patil RT (2012) Dietary fibre in foods: a review. J Food Sci Technol 49(3):255–266

Di Rosa C, Altomare A, Imperia E, Spiezia C, Khazrai YM, Guarino MPL (2022) The role of dietary fibers in the management of IBD symptoms. Nutrients 14(22):4775

Dybka-Stępień K, Otlewska A, Góźdź P, Piotrowska M (2021) The renaissance of plant mucilage in health promotion and industrial applications: a review. Nutrients 13(10):3354

Faghfoori Z, Navai L, Shakerhosseini R, Somi MH, Nikniaz Z, Norouzi MF (2011) Effects of an oral supplementation of germinated barley foodstuff on serum tumour necrosis factor-alpha, interleukin-6 and -8 in patients with ulcerative colitis. Ann Clin Biochem 48(Pt 3):233–237

Faghfoori Z, Shakerhosseini R, Navai L, Somi MH, Nikniaz Z, Abadi A (2014) Effects of an oral supplementation of germinated barley foodstuff on serum CRP level and clinical signs in patients with ulcerative colitis. Heal Promot Perspect 4(1):116–121

Fritsch J, Garces L, Quintero MA, Pignac-Kobinger J, Santander AM, Fernández I et al (2021) Low-fat, high-fiber diet reduces markers of inflammation and Dysbiosis and improves quality of life in patients with ulcerative colitis. Clin Gastroenterol Hepatol 19(6):1189–1199.e30

Fu J, Zheng Y, Gao Y, Xu W (2022) Dietary fiber intake and gut microbiota in human health. Microorganisms 10(12):2507

Gaddam A, Galla C, Thummisetti S, Marikanty RK, Palanisamy UD, Rao PV (2015) Role of fenugreek in the prevention of type 2 diabetes mellitus in prediabetes. J Diabetes Metab Disord [Internet] 14(1):74. Available from: https://link.springer.com/10.1186/s40200-015-0208-4

Gajendiran A, Thangaraman V, Thangamani S, Ravi D, Abraham J (2016) Antimicrobial, antioxidant and anticancer screening of *Ocimum Basilicum* seeds. Bull Pharm Res [Internet] 6(3):114–119. Available from: http://journal.appconnect.in/wp-content/uploads/2017/02/ReprintBPR137.pdf

Gallaher DD, Wood KJ, Gallaher CM, Marquart LF, Engstrom AM (1999) Intestinal contents supernatant viscosity of rats fed oat-based muffins and cereal products. Cereal Chem 76(1):21–24

Garg P (2017) Psyllium husk should be taken at higher dose with sufficient water to maximize its efficacy. J Acad Nutr Diet [Internet] 117(5):681. Available from: https://linkinghub.elsevier.com/retrieve/pii/S2212267217302253

Geng L, Li M, Zhang G, Ye L (2022) Barley: a potential cereal for producing healthy and functional foods. Food Qual Saf 6

Ghafouri-Oskuei H, Javadi A, Saeidi Asl MR, Azadmard-Damirchi S, Armin M (2020) Quality properties of sausage incorporated with flaxseed and tomato powders. Meat Sci 161:107957

Ghahramanloo K, Kamalidehghan B, Akbari Javar H, Teguh Widodo R, Majidzadeh K, Noordin MI (2017) Comparative analysis of essential oil composition of Iranian and Indian Nigella sativa L. extracted using supercritical fluid extraction and solvent extraction. Drug Des Devel Ther [Internet] 11:2221. Available from: https://www.dovepress.com/comparative-analysis-of-essential-oil-composition-of-iranian-and-india-peer-reviewed-article-DDDT–2226

Giorgetti GM, Eramo A, Clemente V, Ramos GP, Ramos O (2023) High fiber diet. Colon Diverticular Dis:201–208

Giuntini EB, Sardá FAH, de Menezes EW (2022) The effects of soluble dietary fibers on glycemic response: An overview and futures perspectives. Food Secur 11(23):3934

Goyal A, Sharma V, Sihag MK, Singh AK, Arora S, Sabikhi L (2016) Fortification of dahi (Indian yoghurt) with omega-3 fatty acids using microencapsulated flaxseed oil microcapsules. J Food Sci Technol [Internet] 53(5):2422–2433. Available from: http://link.springer.com/10.1007/s13197-016-2220-1

Grundy MML, Edwards CH, Mackie AR, Gidley MJ, Butterworth PJ, Ellis PR (2016) Re-evaluation of the mechanisms of dietary fibre and implications for macronutrient bioaccessibility, digestion and postprandial metabolism. Br J Nutr [Internet] 116(5):816–833. Available from: https://www.cambridge.org/core/product/identifier/S0007114516002610/type/journal_article

Guan ZW, Yu EZ, Feng Q (2021) Soluble dietary fiber, one of the most important nutrients for the gut microbiota. Molecules 26(22)

Guarino MPL, Altomare A, Emerenziani S, Di Rosa C, Ribolsi M, Balestrieri P et al (2020) Mechanisms of action of prebiotics and their effects on gastro-intestinal disorders in adults. Nutrients 12(4)

Hajmohammadi A, Pirouzifard M, Shahedi M, Alizadeh M (2016) Enrichment of a fruit-based beverage in dietary fiber using basil seed: effect of Carboxymethyl cellulose and Gum Tragacanth on stability. LWT [Internet] 74:84–91. Available from: https://linkinghub.elsevier.com/retrieve/pii/S0023643816304364

Haque A, Ahmad S, Azad ZRAA, Adnan M, Ashraf SA (2023) Incorporating dietary fiber from fruit and vegetable waste in meat products: a systematic approach for sustainable meat processing and improving the functional, nutritional and health attributes. PeerJ [Internet] 11:e14977. Available from: https://peerj.com/articles/14977

He Y, Wang B, Wen L, Wang F, Yu H, Chen D et al (2022) Effects of dietary fiber on human health. Food Sci Human Wellness 11(1):1–10

Hervik AK, Svihus B (2019) The role of fiber in energy balance. J Nutr Metab 2019:1–11

Ho KS, Tan CYM, Daud MAM, Seow-Choen F (2012) Stopping or reducing dietary fiber intake reduces constipation and its associated symptoms. World J Gastroenterol 18(33):4593–4596

Holscher HD (2017) Dietary fiber and prebiotics and the gastrointestinal microbiota. Gut Microbes 8(2):172–184

Huang L, Ding X, Zhao Y, Li Y, Ma H (2018) Modification of insoluble dietary fiber from garlic straw with ultrasonic treatment. J Food Process Preserv 42(1):e13399

Huang LZ, Ma MG, Ji XX, Choi SE, Si C (2021) Recent developments and applications of hemicellulose from wheat straw: a review. Front Bioeng Biotechnol 9:690773

Hussain S, Jõudu I, Bhat R (2020) Dietary fiber from underutilized plant resources—a positive approach for valorization of fruit and vegetable wastes. Sustain For 12(13):5401

Hussein AMS, Abd El-Azeem AS, Hegazy AM, Afifi AA, Ragab GH (2011) Physiochemical, sensory and nutritional properties of corn-fenugreek flour composite biscuits. Aust J Basic Appl Sci 5(4):84–95

Israr T, Rakha A, Rashid S, Shehzad A, Ahmed W, Sohail M (2017) Effect of basil seed gum on physico-chemical and rheological properties of bread. J Food Process Preserv [Internet]. 41(5):e13128. Available from: https://onlinelibrary.wiley.com/doi/10.1111/jfpp.13128

Jalili T, Wildman REC, Medeiros DM (2000) Nutraceutical roles of dietary fiber. J Nutraceuticals, Funct Med Foods [Internet] 2(4):19–34. Available from: http://www.tandfonline.com/doi/full/10.1300/J133v02n04_03

Kamandloo F, Miran M, Salami M (2023) Evaluation of black seed (Nigella sativa L.) cake and its protein in muffins as a valuable potential functional source for obesity control. Food Funct Food Sci Obes [Internet] 1(6):7. Available from: https://www.ffhdj.com/index.php/FoodScienceandObesity/article/view/1107

Kaur P, Sharma P, Kumar V, Panghal A, Kaur J, Gat Y (2019) Effect of addition of flaxseed flour on phytochemical, physicochemical, nutritional, and textural properties of cookies. J Saudi Soc Agric Sci 18(4):372–377

Keerthana K, Anukiruthika T, Moses JA, Anandharamakrishnan C (2020) Development of fiber-enriched 3D printed snacks from alternative foods: A study on button mushroom. J Food Eng 287:110116

Khalid W, Arshad MS, Jabeen A, Muhammad Anjum F, Qaisrani TB, Suleria HAR (2022) Fiber-enriched botanicals: a therapeutic tool against certain metabolic ailments. Food Sci Nutr 10(10):3203–3218

Khalid W, Arshad MS, Aziz A, Rahim MA, Qaisrani TB, Afzal F et al (2023) Chia seeds (Salvia hispanica L.): a therapeutic weapon in metabolic disorders. Food Sci Nutr [Internet] 11(1):3–16. Available from: https://onlinelibrary.wiley.com/doi/10.1002/fsn3.3035

Kim SY, Hyeonbin O, Lee P, Kim YS (2020) The quality characteristics, antioxidant activity, and sensory evaluation of reduced-fat yogurt and nonfat yogurt supplemented with basil seed gum as a fat substitute. J Dairy Sci [Internet] 103(2):1324–1336. Available from: https://linkinghub.elsevier.com/retrieve/pii/S0022030219310549

Kumar CG, Sripada S, Poornachandra Y (2018) Status and future prospects of Fructooligosaccharides as nutraceuticals. Role Mater Sci Food Bioeng:451–503

Lattimer JM, Haub MD (2010) Effects of dietary fiber and its components on metabolic health. Nutrients [Internet] 2(12):1266–1289. Available from: http://www.mdpi.com/2072-6643/2/12/1266

Lee CH, Chin KB (2020) Physical properties and structural changes of myofibrillar protein gels prepared with basil seed gum at different salt levels and application to sausages. Foods [Internet] 9(6):702. Available from: https://www.mdpi.com/2304-8158/9/6/702

Li YO, Komarek AR (2017) Dietary fibre basics: health, nutrition, analysis, and applications. Food Qual Saf [Internet] 1(1):47–59. Available from: https://academic.oup.com/fqs/article-lookup/doi/10.1093/fqs/fyx007

Liu C, Lin XL, Wan Z, Zou Y, Cheng FF, Yang XQ (2016) The physicochemical properties, in vitro binding capacities and in vivo hypocholesterolemic activity of soluble dietary fiber extracted from soy hulls. Food Funct 7(12):4830–4840

Lorenc F, Jarošová M, Bedrníček J, Smetana P, Bárta J (2022) Structural characterization and functional properties of flaxseed hydrocolloids and their application. Foods [Internet] 11(15):2304. Available from: https://www.mdpi.com/2304-8158/11/15/2304

Losso JN, Holliday DL, Finley JW, Martin RJ, Rood JC, Yu Y et al (2009) Fenugreek bread: a treatment for diabetes mellitus. J Med Food [Internet] 12(5):1046–1049. Available from: http://www.liebertpub.com/doi/10.1089/jmf.2008.0199

Lovegrove A, Edwards CH, De Noni I, Patel H, El SN, Grassby T et al (2017) Role of polysaccharides in food, digestion, and health. Crit Rev Food Sci Nutr 57(2):237–253

Luo X, Wang Q, Zheng B, Lin L, Chen B, Zheng Y et al (2017) Hydration properties and binding capacities of dietary fibers from bamboo shoot shell and its hypolipidemic effects in mice. Food Chem Toxicol 109:1003–1009

Mahmoud HS, Almallah AA, Gad EL-Hak HN, Aldayel TS, HMA A, Khaled HE (2021) The effect of dietary supplementation with Nigella sativa (black seeds) mediates immunological function in male Wistar rats. Sci Rep [Internet] 11(1):7542. Available from: https://www.nature.com/articles/s41598-021-86721-1

Mayor S (2019) Eating more fibre linked to reduced risk of non-communicable diseases and death, review finds. BMJ [Internet] 11590. Available from: https://www.bmj.com/lookup/doi/10.1136/bmj.l159

McRae MP (2017) Dietary fiber is beneficial for the prevention of cardiovascular disease: an umbrella review of meta-analyses. J Chiropr Med 16(4):289–299

McRae MP (2018) Dietary fiber intake and type 2 diabetes mellitus: an umbrella review of meta-analyses. J Chiropr Med 17(1):44–53

McRorie JW (2015) Evidence-based approach to fiber supplements and clinically meaningful health benefits, part 2. Nutr Today [Internet] 50(2):90–97. Available from: https://journals.lww.com/00017285-201503000-00009

Mhurchu CN, Dunshea-Mooij C, Bennett D, Rodgers A (2005) Effect of chitosan on weight loss in overweight and obese individuals: a systematic review of randomized controlled trials. Obes Rev 6(1):35–42

Mirmiran P, Bahadoran Z, Moghadam SK, Vakili AZ, Azizi F (2016) A prospective study of different types of dietary fiber and risk of cardiovascular disease: Tehran lipid and glucose study. Nutrients 8(11):686

Moghaddam SM, Brick MA, Echeverria D, Thompson HJ, Brick LA, Lee R et al (2018) Genetic architecture of dietary fiber and oligosaccharide content in a middle American panel of edible dry bean. Plant Genome 11(1):170074

Naji-Tabasi S, Razavi SMA (2017) Functional properties and applications of basil seed gum: an overview. Food Hydrocoll [Internet] 73:313–325. Available from: https://linkinghub.elsevier.com/retrieve/pii/S0268005X16309547

Nandi I, Sengupta A, Ghosh M (2019) Effects of dietary fibres extracted from defatted sesame husk, rice bran & flaxseed on hypercholesteromic rats. Bioact Carbohydrates Diet Fibre 17:100176

Ngaha WD, Tchabo W, Matsinkou RS, Nyame LK, Fombang EN (2023) Formulation of low glycaemic index biscuits suitable for diabetics from unripe banana, okra, and stevia leaves/jujube fruit. Food Prod Process Nutr 5(1):1–9

Nogal A, Valdes AM, Menni C (2021) The role of short-chain fatty acids in the interplay between gut microbiota and diet in cardio-metabolic health. Gut Microbes 13(1):1–24

Pawde S, Talib MI, Parate VR, Talib I (2020) Development of Fiber-Rich Biscuit by Incorporating Dragon Fruit Powder. Int J Fruit Sci 20:S1620–S1628

(PDF) Optimization and quality evaluation of dietary fiber rich chicken meat rolls incorporated with psyllium husk

Petrie JR, Guzik TJ, Touyz RM (2018) Diabetes, hypertension, and cardiovascular disease: clinical insights and vascular mechanisms. Can J Cardiol 34(5):575–584

(PhD) TFT (2019) Direct and indirect actions of inulin as prebiotic polysaccharide: a review. Cient Period Nutr

Puhlmann ML, de Vos WM (2022) Intrinsic dietary fibers and the gut microbiome: rediscovering the benefits of the plant cell matrix for human health. Front Immunol 13:954845

Puupponen-Pimiä R, Aura AM, Oksman-Caldentey KM, Myllärinen P, Saarela M, Mattila-Sandholm T et al (2002) Review. Trends Food Sci Technol 1(13):3–11

Rabail R, Sultan MT, Khalid AR, Sahar AT, Zia S, Kowalczewski PŁ et al (2022) Clinical, nutritional, and functional evaluation of chia seed-fortified muffins. Molecules [Internet] 27(18):5907. Available from: https://www.mdpi.com/1420-3049/27/18/5907

Rizvi Q-u-e-H, Shams R, Pandey VK, Dar AH, Tripathi A, Singh R (2022) A descriptive review on nutraceutical constituents, detoxification methods and potential health benefits of flaxseed. Appl Food Res [Internet] 2(2):100239. Available from: https://linkinghub.elsevier.com/retrieve/pii/S2772502222001949

Roberts KT (2011) The potential of fenugreek (Trigonella foenum-graecum) as a functional food and nutraceutical and its effects on Glycemia and Lipidemia. J Med Food [Internet] 14(12):1485–1489. Available from: http://www.liebertpub.com/doi/10.1089/jmf.2011.0002

Roduit C, Frei R, Ferstl R, Loeliger S, Westermann P, Rhyner C et al (2019) High levels of butyrate and propionate in early life are associated with protection against atopy. Allergy 74(4):799–809

Samarasinghe K, Dharmadasa CR, Waisundara VY, Samarasinghe K, Dharmadasa CR, Waisundara VY (2022) Introductory chapter: the basics of dietary fibers. Diet Fibers

Shafodino FS, Lusilao JM, Mwapagha LM (2022) Phytochemical characterization and antimicrobial activity of Nigella sativa seeds. Nevárez-Moorillón GV, editor. PLoS One [Internet] 17(8):e0272457. Available from: https://dx.plos.org/10.1371/journal.pone.0272457

Shah M, Chandalia M, Adams-Huet B, Brinkley LJ, Sakhaee K, Grundy SM et al (2009) Effect of a high-fiber diet compared with a moderate-fiber diet on calcium and other mineral balances in subjects with type 2 diabetes. Diabetes Care [Internet] 32(6):990–995. Available from: https://diabetesjournals.org/care/article/32/6/990/28697/Effect-of-a-High-Fiber-Diet-Compared-With-a

Sharma M, Dadhwal K, Gat Y, Kumar V, Panghal A, Prasad R et al (2019) A review on newer techniques in extraction of oleaginous flaxseed constituents. OCL [Internet] (26):14. Available from: https://www.ocl-journal.org/10.1051/ocl/2019006

Silva YP, Bernardi A, Frozza RL (2020) The role of short-chain fatty acids from gut microbiota in gut-brain communication. Front Endocrinol (Lausanne) 11:508738

Singh R (2016) Development of fiber enriched bakery products by incorporating fruit pulp waste powder and their acceptability evaluation. Int J Adv Res Biol Sci 3(6):222–226

Soliman GA (2019) Dietary fiber, atherosclerosis, and cardiovascular disease. Nutrients 11(5):1155

Song K, Kim Y (2019) Effect of mucilage extracted from Basil (Ocimum basilicum L.) seeds on physicochemical and rheological properties in low-fat milk protein gel. J Food Process Preserv [Internet] 43(11) Available from: https://onlinelibrary.wiley.com/doi/10.1111/jfpp.14191

Song KY, O H, Joung KY, Shin SY, Kim YS (2017) Effects of basil (Ocimum Basilicum L.) seed mucilage substituted for fat source in sponge cake: physicochemical, structural, and retrogradation properties. Ital J Food Sci 29(4):681–696

Sperber AD, Bangdiwala SI, Drossman DA, Ghoshal UC, Simren M, Tack J et al (2021) Worldwide prevalence and burden of functional gastrointestinal disorders, results of Rome foundation global study. Gastroenterology [Internet] 160(1):99–114.e3. Available from: https://linkinghub.elsevier.com/retrieve/pii/S001650852030487X

Srinivasa UM, Naidu MM (2021) Fenugreek (Trigonella foenum-graecum L.) seed: promising source of nutraceutical. In: Studies in natural products chemistry, pp 141–184. Available from: https://linkinghub.elsevier.com/retrieve/pii/B9780323910958000143

Statovci D, Aguilera M, MacSharry J, Melgar S (2017) The impact of Western diet and nutrients on the microbiota and immune response at mucosal interfaces. Front Immunol [Internet] 8. Available from: http://journal.frontiersin.org/article/10.3389/fimmu.2017.00838/full

Streppel MT, Ocké MC, Boshuizen HC, Kok FJ, Kromhout D (2008) Dietary fiber intake in relation to coronary heart disease and all-cause mortality over 40 y: the Zutphen study. Am J Clin Nutr 88(4):1119–1125

Stribling P, Ibrahim F (2023) Dietary fibre definition revisited - the case of low molecular weight carbohydrates. Clin Nutr ESPEN [Internet] 55:340–356. Available from: https://linkinghub.elsevier.com/retrieve/pii/S2405457723001079

Tao J, Li S, Ye F, Zhou Y, Lei L, Zhao G (2020) Lignin–An underutilized, renewable and valuable material for food industry. Crit Rev Food Sci Nutr 60(12):2011–2033

Thompson HJ, Brick MA (2016) Perspective: closing the dietary fiber gap: an ancient solution for a 21st century problem. Adv Nutr [Internet] 7(4):623–626. Available from: https://linkinghub.elsevier.com/retrieve/pii/S2161831322007499

Thompson SV, Hannon BA, An R, Holscher HD (2017) Effects of isolated soluble fiber supplementation on body weight, glycemia, and insulinemia in adults with overweight and obesity: a systematic review and meta-analysis of randomized controlled trials. Am J Clin Nutr 106(6):1514–1528

Ullah I, Yin T, Xiong S, Huang Q, Zia-ud-Din ZJ et al (2018) Effects of thermal pre-treatment on physicochemical properties of nano-sized okara (soybean residue) insoluble dietary fiber prepared by wet media milling. J Food Eng 237:18–26

Verdú S, Barat JM, Grau R (2017) Improving bread-making processing phases of fibre-rich formulas using chia (Salvia hispanica) seed flour. LWT 84:419–425

Wang M, Wichienchot S, He X, Fu X, Huang Q, Zhang B (2019) In vitro colonic fermentation of dietary fibers: fermentation rate, short-chain fatty acid production and changes in microbiota. Trends Food Sci Technol 88:1–9

Wani SA, Kumar P (2018) Fenugreek: a review on its nutraceutical properties and utilization in various food products. J Saudi Soc Agric Sci [Internet] 17(2):97–106. Available from: https://linkinghub.elsevier.com/retrieve/pii/S1658077X15301065

Westenbrink S, Brunt K, van der Kamp JW (2013) Dietary fibre: challenges in production and use of food composition data. Food Chem [Internet] 140(3):562–567. Available from: https://linkinghub.elsevier.com/retrieve/pii/S0308814612014318

Williams BA, Mikkelsen D, Flanagan BM, Gidley MJ (2019) "Dietary fibre": moving beyond the "soluble/insoluble" classification for monogastric nutrition, with an emphasis on humans and pigs. J Anim Sci Biotechnol 10(1):1–12

Wojciechowicz-Budzisz A, Pejcz E, Spychaj R, Harasym J (2023) Mixed psyllium fiber improves the quality, nutritional value, polyphenols and antioxidant activity of Rye bread. Food Secur 12(19):3534

Wu W, Huff HE, Hsieh F (2007) processing and properties of extruded flaxseed-corn puff. J Food Process Preserv [Internet] 31(2):211–226. Available from: https://onlinelibrary.wiley.com/doi/10.1111/j.1745-4549.2007.00105.x

Xie F, Wang Y, Wu J, Wang Z (2017a) Insoluble dietary fibers from Angelica keiskei by-product and their functional and morphological properties. Starch - Stärke 69(3–4):1600122

Xie F, Li M, Lan X, Zhang W, Gong S, Wu J et al (2017b) Modification of dietary fibers from purple-fleshed potatoes (Heimeiren) with high hydrostatic pressure and high pressure homogenization processing: a comparative study. Innov Food Sci Emerg Technol 42:157–164

Yimer EM, Tuem KB, Karim A, Ur-Rehman N, Anwar F, Nigella sativa L. (2019) (Black cumin): A promising natural remedy for wide range of illnesses. Evid Based Complement Alternat Med [Internet] 2019:1–16. Available from: https://www.hindawi.com/journals/ecam/2019/1528635/

Yu G, Bei J, Zhao J, Li Q, Cheng C (2018) Modification of carrot (Daucus carota Linn. var. Sativa Hoffm.) pomace insoluble dietary fiber with complex enzyme method, ultrafine comminution, and high hydrostatic pressure. Food Chem 257:333–340

Yusuf K, Saha S, Umar S (2022) Health benefits of dietary fiber for the management of inflammatory bowel disease. Biomedicines 10(6)

Zannini D, Dal Poggetto G, Malinconico M, Santagata G, Immirzi B (2021) Citrus pomace biomass as a source of pectin and lignocellulose fibers: from waste to upgraded biocomposites for mulching applications. Polymers 13(8):1280

Zerfu TA, Mekuria A (2019) Pregnant women have inadequate fiber intake while consuming fiber-rich diets in low-income rural setting: evidences from analysis of common "ready-to-eat" stable foods. Food Sci Nutr 7(10):3286–3292

Zhang S, Ma Q, Deng M, Jia X, Huang F, Dong L et al (2024) Composition, structural, physicochemical and functional properties of dietary fiber from different milling fractions of black rice bran. LWT [Internet]:115743. Available from: https://linkinghub.elsevier.com/retrieve/pii/S0023643824000227

Zheng Y, Li Y (2018) Physicochemical and functional properties of coconut (Cocos nucifera L) cake dietary fibres: effects of cellulase hydrolysis, acid treatment and particle size distribution. Food Chem 257:135–142

Zokaityte E, Siriakovaite K, Starkute V, Zavistanaviciute P, Lele V, Mozuriene E et al (2021) Characteristics of nutraceutical chewing candy formulations based on fermented milk permeate, psyllium husk, and apple by-products. Foods [Internet] 10(4):777. Available from: https://www.mdpi.com/2304-8158/10/4/777

Chapter 8
Functional Aspects of Polyunsaturated Fatty Acids as Nutraceuticals

Hilal Şimşek and Aslı Uçar

Abstract Over the last few decades, nutraceuticals have become a popular research topic as the focus of nutrition science has shifted from providing basic nutrition to promoting health and ensuring well-being. There is great interest in nutraceuticals for reasons such as an aging population and increasing global awareness of health promotion or disease prevention. Polyunsaturated fatty acids are one of the nutrients with remarkable nutraceutical potential. Many polyunsaturated fatty acid metabolites, especially essential fatty acids and eicosanoids synthesized from eicosapentaenoic acid or arachidonic acid play an active role in the nutraceutical properties attributed to polyunsaturated fatty acids. Polyunsaturated fatty acids have nutraceutical roles in diseases including cardiovascular diseases, type-2 diabetes, non-alcoholic fatty liver disease, neurodegeneration, and cancer, through several biochemical pathways, especially antioxidant and anti-inflammatory effects. In this chapter, the functional aspects of polyunsaturated fatty acids as nutraceuticals and the mechanisms on which they are based are summarized according to the current scientific literature.

Keywords Nutraceuticals · Polyunsaturated fatty acids · Functional foods · Dietary supplements

H. Şimşek
University of Ankara, Graduate School of Health Sciences, Department of Nutrition and Dietetics, Ankara, Turkey

Department of Nutrition and Dietetics, Faculty of Health Sciences, Niğde Ömer Halisdemir University, Niğde, Turkey
e-mail: hllsimsek@ankara.edu.tr

A. Uçar (✉)
University of Ankara, Faculty of Health Sciences, Department of Nutrition and Dietetics, Ankara, Turkey
e-mail: aucar@ankara.edu.tr

S. Amir Ashraf, M. Adnan (eds.), *Food Bioactives and Nutraceuticals*,
https://doi.org/10.1007/978-981-97-9723-3_8

8.1 Introduction

The scope of nutritional science has evolved over the past few decades, from providing essential nutrition to promoting health, preventing disease, and ensuring well-being. One of the prominent topics in this context is nutraceuticals (Bassaganya-Riera et al. 2020; Kingsley Nwosu and Ikechukwu 2020; Neufeld et al. 2023). The concept of nutraceutical, first defined by De Felice in 1989, covers food or nutrients that contain a hybrid of pharmaceutical and nutritional content. Since the term nutraceutical does not yet have acknowledged definitions of food supplements, herbal products, or functional/enriched foods, it is often considered together with food supplements. Although the boundaries of these concepts have not been clarified by regulatory authorities, food supplements focus on sustaining or compensating for nutrients, while nutraceuticals focus on preventing or alleviating the pathology of any health problem (Puri et al. 2022; Daliu et al. 2019). Nutraceuticals contain nutrients or extracts obtained from foods or natural sources such as herbs with prophylactic or therapeutic properties (Chopra et al. 2022; Elkhalifa et al. 2021). There has been increased interest in nutraceuticals, which are more than ordinary food but can also be described as a potential alternative to drug treatment. The growing interest in nutraceuticals over the past few decades is associated with an aging population, increased life expectancy, and attention to issues of health promotion or disease prevention (Kingsley Nwosu and Ikechukwu 2020; Daliu et al. 2019). In addition, the global growth of the nutraceutical market in recent years is attributed to reasons such as the impact of COVID-19, the importance given to health protection, and increased awareness (Puri et al. 2022; Chopra et al. 2022; Lordan 2021). Polyunsaturated fatty acids (PUFAs) are among the most researched nutraceuticals for their antioxidant, anti-inflammatory, anti-aggregatory, and immunomodulatory effects. As the main functional areas of nutraceutical researched PUFAs in the literature, the topics studied in this chapter are cardiovascular disease (CVD), Type 2 Diabetes (T2DM), Non-Alcoholic Fatty Liver Disease (NAFLD), neurodegeneration, and cancer. Particularly from a non-communicable disease perspective, PUFAs have significant nutraceutical potential. This chapter reviews the functional aspects and mechanisms of action of PUFAs as nutraceuticals, considering current evidence.

8.2 Polyunsaturated Fatty Acids

Polyunsaturated fatty acids (PUFA), which contain two or more double bonds in the hydrocarbon chain, are characterized as n-3 PUFA or n-6 PUFA, depending on the position of the first double bond relative to the methyl (or omega) end. The main n-3 PUFA in food sources are α-linolenic acid (ALA) (18:3 Δ9, 12, 15), docosahexaenoic acid (DHA) (22:6 Δ4, 7, 10, 13, 16, 19), eicosapentaenoic acid (EPA) (20:5 Δ5, 8, 11, 14, 17), and docosapentaenoic acid (DPA) (22:5 Δ7, 10, 13, 16, 19), while n-6 PUFA include linoleic acid (LA) (18:2 Δ9, 12) and arachidonic acid (AA)

(20,4 Δ5, 8, 11, 14). These two families of PUFAs have special importance for human health because they cover the essential fatty acids LA and ALA (Lichtenstein 2023; Abedi and Sahari 2014). These fatty acids cannot be synthesized in humans due to the absence of enzymes that provide desaturation beyond the Δ9 position, and therefore their dietary intake is considered essential. The essential fatty acids LA and ALA are used to synthesize the other long-chain PUFAs mentioned above in the human body when consumed through a diet (Patel et al. 2020).

The adequate intake (AI) for ALA is 1.1 g/day for adult (>19 age) women and 1.6 g/day for men, while for LA it is 12 g/day for women and 17 g/day for men aged 19–50 (Trumbo et al. 2002). The dietary recommendation is as follows: 0.5% of daily energy for ALA and 4% for LA. Additionally, the positive effect of ALA intake above the recommended value (0.6–1% of energy) has also been previously noted (Lenighan et al. 2019; Food and Agriculture Organization of the United Nations 2010). The main sources of essential fatty acids are oilseeds and vegetable oils. Vegetable oils such as corn, soybean, rice bran, or plant seeds are the main sources of LA, while the best sources of ALA are seed oils such as flaxseed, perilla, or canola oil. Animal-based foods stand out as dietary sources of other PUFAs synthesized from essential fatty acids. Meat, dairy products, or eggs are good sources of AA since animals can synthesize AA from LA. The best-known source of EPA and DHA is seafood, especially oily fish such as salmon or tuna. Also, cod liver oil, krill oil, or fish oil are good sources of EPA and DHA (Patel et al. 2020; Kapoor et al. 2021; Turpeinen and Merimaa 2011; Gibbs and Ian 2008). Since plant-based nutrition has become increasingly important in recent years for ecological, ethical, cultural, or economic reasons, plant-derived PUFAs are among the new research topics. Algal extracts, seaweed, or genetically modified plants are still ongoing research topics in this context (Rizzo et al. 2023). The contents of n-3 and n-6 fatty acids in selected plant and animal-based foods are presented in Table 8.1.

8.2.1 Metabolism of Polyunsaturated Fatty Acids

Common PUFAs and synthesis from essential fatty acids are summarized in Fig. 8.1. An important issue in this context is that the synthesis of EPA and DHA from ALA in humans is far from providing the daily requirement. As summarized in Fig. 8.1, the LA and ALA metabolic pathways are identical. Therefore, excessive consumption of one leads to competition for desaturation and elongation enzymes in the synthesis of EPA or arachidonic acid (AA). It is well known that excessive LA consumption suppresses EPA and DHA synthesis (Patel et al. 2020; Hussein et al. 2005; Goyens et al. 2005; Burdge and Wootton 2002; Burdge et al. 2002). Beyond the competitive common enzymatic reactions of PUFAs, the ratio of n-6 to n-3 PUFA is also of nutritional importance due to the synthesis of different eicosanoid series with opposite effects. Eicosanoids, described as autocrine/paracrine hormones, are synthesized from 20-carbon PUFAs (ARA and EPA) and play a key role

Table 8.1 Contents of n-3 and n-6 fatty acids in selected plant and animal-based foods (g/100 g)

	n-6		n-3		
Source	LA	ARA	ALA	EPA	DHA
Oilseeds					
Flaxseed	5.26	–	19.4	–	–
Pumpkin seeds	20.5	–	0.05	–	–
Nuts (raw)	4.4	–	0.07	–	–
Walnut (raw)	35.5	–	6.2	–	–
Oils					
Canola	17.8	–	7.45	–	–
Flaxseed (cold pressed)	14.2	–	53.4	–	–
Soybean	50.9	–	6.62	–	–
Corn	51.9	–	1.04	–	–
Safflower	13.6	–	0.15	0.001	–
Seafood					
Salmon (pink, raw)	0.06	0.01	0.045	0.18	0.33
Tuna (bluefin, raw)	0.05	0.04	–	0.13	0.89
Fish haddock (raw)	0.01	0.01	0.002	0.04	0.09
Cod liver	0.94	0.94	0.94	6.9	11
Other animal-based foods					
Kidney (cooked)	1.14	0.01	0.03	–	–
Chicken breast (cooked)	0.60	0.09	0.03	0.004	0.01
Egg, chicken (raw, whole)	1.64	0.003	0.04	–	–
Milk (whole fat)	0.08	0.004	0.01	0.001	–

Data from the USDA National Nutrient Database for Standard Reference open source (USDA 2024) and Turkish National Food Composition Database (TURKOMP) (TURKOMP 2023). Abbreviations: LA linoleic acid, AA arachidonic acid, ALA α-linolenic acid, EPA eicosapentaenoic acid, DHA docosahexaenoic acid

in the physiology of several health effects attributed to PUFAs. Eicosanoids, consisting of prostaglandins (PGs), thromboxanes (TXs), leukotrienes (LTs), and lipoxins (LXs), have game-changing functions in many diseases such as cancer, upper respiratory tract diseases, atherosclerosis, and arthritis, especially through various biochemical pathways such as inflammatory processes, vasoconstriction or vasodilation (Litwack 2022). (These biochemical mechanisms are explained in Sect. 8.3 below.) For the balance of eicosanoids synthesized from n-6 and n-3 PUFAs, it is recommended that the ratio of n-6/n-3 PUFAs in the diet is between 5:1 and 10:1 (Abedi and Sahari 2014; Tortosa-Caparrós et al. 2017). However, it is known that in industrialized countries dominated by the Western diet, this ratio can rise to 15–20:1 (Harwood 2019; Hall and Harwood 2017). It has been suggested that this phenomenon may be one of the reasons why non-communicable diseases (NCDs) such as diabetes, chronic inflammatory diseases, and cardiovascular diseases are more common in industrialized societies (Fedacko et al. 2012; Singh et al. 2022). There are concerns that higher dietary n-6 PUFA intake will increase the risk of CVD due to the potential pro-inflammatory and pro-thrombotic effects in parallel with high AA

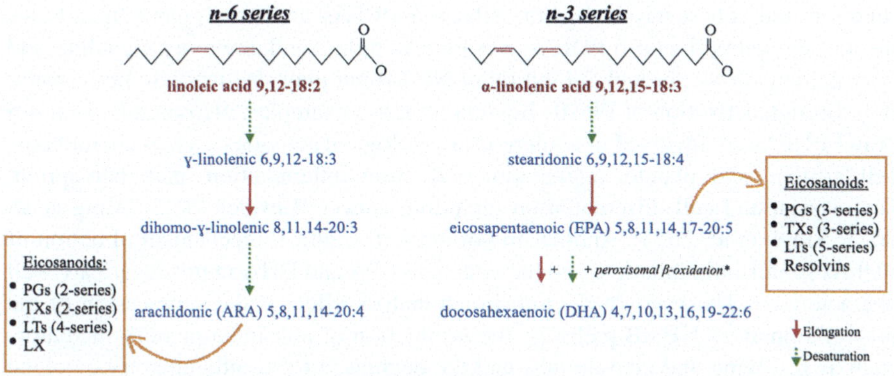

Fig. 8.1 Biosynthesis of n-6 and n-3 PUFA series from linoleic acid and α-linolenic acid respectively. *Intermediate steps and products in synthesizing DHA from EPA are not shown. Abbreviations: *PGs* prostaglandins, *TXs* thromboxanes, *LTs* leukotrienes, *LX* lipoxin

intake and the eicosanoids synthesized from AA (Ramsden et al. 2016). On the other hand, according to stable isotope studies, it has been suggested that the conversion of AA from LA is limited in humans and that increasing dietary LA intake does not have significant effects on plasma and adipose tissue AA levels (Hussein et al. 2005; Bjermo et al. 2012). Moreover, in a meta-analysis in this context, it was previously reported that there was no relationship between LA intake and the concentration of inflammatory markers (Johnson and Fritsche 2012). The results of research in this context are explained in more detail under Sect. 8.4.

8.3 Mechanisms of Action as Nutraceuticals

The biochemical mechanisms underlying the health effects of PUFAs are associated with different series of eicosanoids. While 2 series of PGs and TXs are metabolized from AA, 3 series of them are produced from EPA (Fig. 8.1). These different series of eicosanoids of n-6 and n-3 origin have significantly opposite physiological effects. For example, the 2-series of PGs have a pro-arrhythmic effect, while the 3-series has an anti-arrhythmic effect. Similarly, for TXs, the 2-series functions as a platelet activator or vasoconstrictor, while the 3-series functions as a platelet inhibitor or vasodilator (Litwack 2022; Schmitz and Ecker 2008). LTB4 (derivative from n-6) increases vascular permeability, is a potent chemotactic agent for leukocytes, and accelerates reactive oxygen species (ROS) production. Additionally, TNFα can induce the production of inflammatory cytokines such as IL-1 and IL-6. On the other hand, LTB5 (derivative from n-3) blocks the biosynthesis of the highly inflammatory LTB4 (Litwack 2022; Schmitz and Ecker 2008). While most AA-derived eicosanoids are proactive, EPA-derived eicosanoids have inhibitory effects. Therefore, the n-6/n-3 ratio in the cell membrane is important for membrane function and cell survival. Additionally, n-3 PUFAs are more potent ligands for nuclear receptors than n-6 PUFAs. n-3 PUFAs reduce inflammatory signaling and fatty acid synthesis through inhibition of NF-kB but may also increase lipid oxidation through activation of PPAR. Beyond all these examples, eicosanoids from n-6 or n-3 PUFAs are involved in numerous physiological activities such as chemotaxis, cell growth, and platelet aggregation, affecting inflammation, pain perception, hemostasis, and cell differentiation including cancer (Litwack 2022; Wang et al. 2021; Singh et al. 2023). Another important nutraceutical mechanism of action of PUFAs is anti-inflammation, and the effects of EPA and DHA in this regard are well characterized. The basis of the anti-inflammatory effect is the suppression of the pro-inflammatory NF-κB pathway, the production of anti-inflammatory mediators such as resolving and eicosanoids, and the decrease of pro-inflammatory cytokine and chemokine levels (Lenighan et al. 2019; Crupi and Cuzzocrea 2022; Serhan and Levy 2018). The anti-inflammatory effect is quite important for the management of atherosclerosis, obesity, insulin resistance, and ultimately NCDs, especially in the pathology in which low-grade systemic inflammation plays an important role (Crupi and Cuzzocrea 2022; Calder 2022). EPA also has a special place in the regulation of cardiovascular health and endothelial tone. As a matter of fact, according to in vitro evidence, EPA attenuates the formation of ROS, expression of adhesive molecules, pro-inflammatory cytokine release, and apoptotic cascade (Crupi and Cuzzocrea 2022; Lee et al. 2014).

8.4 Functional Aspects Regarding Diseases and Health

8.4.1 Cardiometabolic Health

When considering the relationship between fatty acid types and cardiometabolic health, the strongest evidence relates to replacing saturated fatty acids (SFA) with PUFA in the diet. There is a body of evidence from prospective studies that a change in dietary fat type, which involves a replacement from SFA to PUFA, reduces the risk of NCD, especially CVD and type 2 diabetes (T2DM) (Zong et al. 2016; Chen et al. 2016; Guasch-Ferré et al. 2015; Imamura et al. 2016; Billingsley et al. 2018; Sanders 2014). Moreover, a meta-analysis estimated a 29% reduction in coronary heart disease (CHD) events by changing dietary fat in this direction (Sacks et al. 2017). However, another important issue is that in most of these studies, total PUFA intake was considered, regardless of the distinction between the n-6 and n-3 series (Zong et al. 2016; Chen et al. 2016; Guasch-Ferré et al. 2015; Imamura et al. 2016; Billingsley et al. 2018; Sanders 2014). According to studies on the effects of dietary intake of PUFAs, the intake of PUFAs from both the n-6 and n-3 series is associated with a reduced risk of CVD. Systematic reviews and meta-analyses of cohort studies support that higher ALA intake protects against CHD mortality (Pan et al. 2012; Del Gobbo et al. 2016). However, according to the latest scientific advisory by the

AHA, current literature does not provide clear evidence that ALA reduces the overall incidence of CVD, although a higher intake of ALA may reduce fatal CHD (Sacks et al. 2017). It should be noted that another meta-analysis on the relationship between ALA consumption and CVD risk also suggests that the current consumption recommendation (2 g/d; 0.6–1% of energy) should be revised (Fleming and Kris-Etherton 2014). According to a meta-analysis on the effects of LA intake, the in-vivo level of LA in the circulation or tissue was associated with lower cardiovascular events, supporting its potential role in CVD prevention (Marklund et al. 2019). Higher dietary LA intake and higher serum LA levels are associated with a lower risk of T2DM, according to a meta-analysis involving prospective cohorts. According to the dose-response analysis performed in the same study, every 5% increase in the energy obtained from LA provides a 10% decrease in the risk of T2DM (Mousavi et al. 2021). According to another meta-analysis including 44 prospective studies, higher LA intake reduces all-cause and CVD mortality (Li et al. 2020). Also, in the science advisory published by the AHA in 2009, it was reported that the intake of n-6 as much as 5–10% of daily energy reduces the risk of CHD, based on existing observational and randomized clinical trial data (Harris et al. 2009). Non-alcoholic fatty liver disease (NAFLD), a disease closely related to cardiometabolic disorders, represents another potential area of the nutraceutical effects of PUFAs (El Hadi et al. 2019). NAFLD is associated with low serum PUFA levels, and it is thought that this condition may be associated with a defect in PUFA transfer from the periphery to the liver (Mocciaro et al. 2023). The nutraceutical potential of n-3 PUFAs (EPA and DHA) in NAFLD is based on the alleviation of hepatic steatosis and fibrosis through their anti-inflammatory, antioxidant, and lipid-lowering properties (Hegazi et al. 2023). There is significant promising evidence in the literature, regarding the effectiveness of n-3 PUFA supplements on CVD, T2DM, and NAFLD. In these studies, a decrease in major CVD events and mortality was reported in individuals with major cardiovascular risk factors such as hypercholesterolemia and previous myocardial infarction, following the use of EPA or DHA supplements in varying periods and doses (Liu 2021; Jacobson et al. 2014; Barry and Dixon 2021; Manson et al. 2019). According to a meta-analysis that included randomized clinical trials on the effectiveness of n-3 supplements in T2DM, n-3 supplements improve glycemic control and lipid profile in T2DM patients. Additionally, in the same study, a dose of 1000–2000 mg/d for at least 8 weeks was specified for the optimal use of n-3 supplementation in T2DM patients (Khalili et al. 2021). Similarly, according to a meta-analysis on the effectiveness of n-3 supplements in NAFLD, n-3 supplements may support treatment by improving lipid profile and body mass index (Lee et al. 2020).

8.4.2 Neurodegeneration

It is well known that n-3 PUFAs play important roles in neurotransmission, brain development, and neuromodulation, and DHA deficiency in the brain or retina can cause learning disabilities and visual dysfunction. The nutraceutical effectiveness of n-3 PUFAs is being investigated in neurodegenerative diseases such as Alzheimer's disease (AD) and Parkinson's disease (PD) (Singh et al. 2023; Shchepinov 2020; Sarparast et al. 2020). Preclinical evidence suggests that a higher brain DHA concentration is associated with better cognitive function, while n-3 PUFAs may be beneficial in neurodegeneration through neuroprotection and partial neuro-recovery (Kerdiles et al. 2017). According to recent meta-analyses, the evidence for the association of dietary fatty acids with AD risk is inconclusive (Zhu et al. 2021; Nwaru et al. 2022). However, it has also been previously reported that a higher intake of n-3 PUFA may be protective against mild cognitive impairments (MCI), and that marine-derived DHA reduces the risk of dementia and AD (Zhu et al. 2021; Zhang et al. 2016; Wei et al. 2023). The current scientific evidence regarding the potential nutraceutical role of PUFAs in neurodegeneration points to the efficacy of n-3 PUFAs, especially EPA and DHA, and more evidence is needed in this context.

8.4.3 Cancer

It is suggested that n-3 PUFA intake reduces the risk of cancer due to its anti-inflammatory and antiproliferative effects (Kim and Kim 2020). On the other hand, it is stated that the potential therapeutic role of PUFAs in cancer is complex, considering that n-3 PUFAs have an anti-apoptotic effect and n-6 PUFAs (especially AA) have a pro-apoptotic effect (Montecillo-Aguado et al. 2023). The results of meta-analyses based on prospective studies are also inconsistent at this point. According to a meta-analysis in this context, fish consumption and marine n-3 intake increase total cancer and especially prostate cancer survival (Wang et al. 2023). In another meta-analysis examining marine n-3 intake, marine n-3 PUFAs were protective against breast cancer, while ALA and n-6 intake did not have a significant relationship with breast or prostate cancers (Liu et al. 2021). In a meta-analysis focused on the effect of n-6 PUFA intake, serum n-6 level was found to be inversely associated with cancer risk, while n-6 intake was not found to be associated with cancer risk (Kim and Kim 2020). According to a meta-analysis in this context, a higher tissue LA level is associated with a lower risk of prostate cancer, and a 5% increase in tissue LA level corresponds to a 14% decrease in prostate cancer risk (Yousefi et al. 2023). Also, in a meta-analysis previously mentioned, higher LA intake was associated with decreasing cancer mortality (Li et al. 2020). On the other hand, a meta-analysis reported that an increase in total PUFA (including n-3 and n-6) intake slightly increases the risk of cancer (Hanson et al. 2020). Considering the current literature, it can be said that, apart from the protective effect of marine n-3 PUFAs, there is still limited data on the nutraceutical effectiveness of other dietary PUFAs in cancer. Among the points that remain unclear regarding the effectiveness of dietary PUFA intake in carcinogenesis or cancer treatment, attention is drawn to the optimal n-6/n-3 ratio, and further research is recommended (Liput et al. 2021). Another subject most frequently researched in the literature in this context is the use of n-3 PUFA supplements in cancer. According to meta-analyses on the effectiveness of n-3 supplements in cancer patients, although the supplements were not effective on CRP, they suppressed pro-inflammatory markers such as TNF-α and IL-6. Therefore, it can be said that there is a convincing level of evidence that n-3 supplements can be used to alleviate inflammation in the context of adjuvant treatment in cancer patients (Jamali et al. 2024; Mocellin et al. 2018). The relationship between the use of n-3 supplements in cancer cachexia and an increase in body weight was reported in a meta-analysis (Hosseini et al. 2024). On the other hand, it was noted in another recent meta-analysis that there is not yet sufficient evidence regarding the benefit of n-3 supplements on systemic inflammation and prognosis in cancer cachexia (de Castro et al. 2021).

8.5 Conclusion

There is no doubt that nutraceuticals will have an increasing importance in the management of NCDs in the aging population and developing healthcare system. In this context, scientific evidence for the efficacy of PUFAs as nutraceuticals is cumulatively increasing supported by various biochemical mechanisms. Among the proposed biochemical mechanisms, alleviation of low-grade inflammation in terms of managing NCDs attracts attention to the nutraceutical potential of PUFAs. Both dietary intake and supplementation of PUFAs have significant nutraceutical potential and clinical use in terms of cardiometabolic health, including CVD, T2DM, and NAFLD. Although there is promising nutraceutical activity in neurodegeneration and cancer, it can be said that only n-3 PUFAs (especially marine-derived ones) stand out. However, there are still some questions that need to be clarified, such as the ideal n-6/n-3 balance based on population or disease and specific dietary intake recommendations.

References

Abedi E, Sahari MA (2014) Long-chain polyunsaturated fatty acid sources and evaluation of their nutritional and functional properties. Food Sci Nutr 2(5):443–463

Barry AR, Dixon DL (2021) Omega-3 fatty acids for the prevention of atherosclerotic cardiovascular disease. Pharmacotherapy 41(12):1056–1065

Bassaganya-Riera J, Berry EM, Blaak EE, Burlingame B, le Coutre J, van Eden W et al (2020) Goals in nutrition science 2020-2025. Front Nutr 7:606378

Billingsley HE, Carbone S, Lavie CJ (2018) Dietary fats and chronic noncommunicable diseases. Nutrients 10(10)

Bjermo H, Iggman D, Kullberg J, Dahlman I, Johansson L, Persson L et al (2012) Effects of n-6 PUFAs compared with SFAs on liver fat, lipoproteins, and inflammation in abdominal obesity: a randomized controlled trial. Am J Clin Nutr 95(5):1003–1012

Burdge GC, Wootton SA (2002) Conversion of alpha-linolenic acid to eicosapentaenoic, docosapentaenoic and docosahexaenoic acids in young women. Br J Nutr 88(4):411–420

Burdge GC, Jones AE, Wootton SA (2002) Eicosapentaenoic and docosapentaenoic acids are the principal products of alpha-linolenic acid metabolism in young men*. Br J Nutr 88(4):355–363
Calder PC (2022) Dietary factors and low-grade inflammation in relation to overweight and obesity revisted. Br J Nutr 127(10):1455–1457
Chen M, Li Y, Sun Q, Pan A, Manson JE, Rexrode KM et al (2016) Dairy fat and risk of cardiovascular disease in 3 cohorts of US adults. Am J Clin Nutr 104(5):1209–1217
Chopra AS, Lordan R, Horbańczuk OK, Atanasov AG, Chopra I, Horbańczuk JO et al (2022) The current use and evolving landscape of nutraceuticals. Pharmacol Res 175:106001
Crupi R, Cuzzocrea S (2022) Role of EPA in inflammation: mechanisms, effects, and clinical relevance. Biomol Ther 12(2)
Daliu P, Santini A, Novellino E (2019) From pharmaceuticals to nutraceuticals: bridging disease prevention and management. Expert Rev Clin Pharmacol 12(1):1–7
de Castro GS, Andrade MF, Pinto FCS, Faiad JZ, Seelaender M (2021) Omega-3 fatty acid supplementation and its impact on systemic inflammation and body weight in patients with cancer cachexia-a systematic review and meta-analysis. Front Nutr 8:797513
Del Gobbo LC, Imamura F, Aslibekyan S, Marklund M, Virtanen JK, Wennberg M et al (2016) ω-3 polyunsaturated fatty acid biomarkers and coronary heart disease: pooling project of 19 cohort studies. JAMA Intern Med 176(8):1155–1166
El Hadi H, Di Vincenzo A, Vettor R, Rossato M (2019) Cardio-metabolic disorders in non-alcoholic fatty liver disease. Int J Mol Sci 20(9)
Elkhalifa AEO, Alshammari E, Adnan M, Alcantara JC, Awadelkareem AM, Eltoum NE et al (2021) Okra (Abelmoschus Esculentus) as a potential dietary medicine with nutraceutical importance for sustainable health applications. Molecules (Basel, Switzerland) [Internet] 26(3)
Fedacko J, Vargova V, Singh RB, Anjum B, Takahashi T, Tongnuka M et al (2012) Association of high w-6/w-3 fatty acid ratio diet with causes of death due to noncommunicable diseases among urban decedents in North India. Open Nutraceuticals J 5(1)
Fleming JA, Kris-Etherton PM (2014) The evidence for α-linolenic acid and cardiovascular disease benefits: comparisons with eicosapentaenoic acid and docosahexaenoic acid. Adv Nutr 5(6):863s–876s
Food and Agriculture Organization of the United Nations (2010) Fats and fatty acids in human nutrition. Report of an expert consultation. FAO Food Nutr Pap 91:1–166
Gibbs RA, Ian GD (2008) Current intakes of EPA and DHA in European populations and the potential of animal-derived foods to increase them: symposium on 'How can the n-3 content of the diet be improved?'. Proc Nutr Soc 67(3):273–280
Goyens PL, Spilker ME, Zock PL, Katan MB, Mensink RP (2005) Compartmental modeling to quantify alpha-linolenic acid conversion after longer term intake of multiple tracer boluses. J Lipid Res 46(7):1474–1483
Guasch-Ferré M, Babio N, Martínez-González MA, Corella D, Ros E, Martín-Peláez S et al (2015) Dietary fat intake and risk of cardiovascular disease and all-cause mortality in a population at high risk of cardiovascular disease. Am J Clin Nutr 102(6):1563–1573
Hall J, Harwood J. Brain lipids in health and disease. 2017
Hanson S, Thorpe G, Winstanley L, Abdelhamid AS, Hooper L (2020) Omega-3, omega-6 and total dietary polyunsaturated fat on cancer incidence: systematic review and meta-analysis of randomised trials. Br J Cancer 122(8):1260–1270
Harris WS, Mozaffarian D, Rimm E, Kris-Etherton P, Rudel LL, Appel LJ et al (2009) Omega-6 fatty acids and risk for cardiovascular disease: a science advisory from the American Heart Association Nutrition Subcommittee of the Council on Nutrition, Physical Activity, and Metabolism; Council on Cardiovascular Nursing; and Council on Epidemiology and Prevention. Circulation 119(6):902–907
Harwood JL (2019) Algae: critical sources of very long-chain polyunsaturated fatty acids. Biomol Ther 9(11):708

Hegazi OE, Alalalmeh SO, Alnuaimi GRH, Shahwan M, Jairoun AA, Alorfi NM et al (2023) NAFLD and nutraceuticals: a review of completed phase III and IV clinical trials. Front Med (Lausanne) 10:1227046

Hosseini F, Hemmati A, Takabi FS, Naeini F, Shab BS (2024) A dose-response meta-analysis of randomized clinical trials investigating the effects of omega-3 supplementation on body weight in patients with cancer cachexia. Clin Nutr ESPEN 59:378–386

Hussein N, Ah-Sing E, Wilkinson P, Leach C, Griffin BA, Millward DJ (2005) Long-chain conversion of [13C]linoleic acid and alpha-linolenic acid in response to marked changes in their dietary intake in men. J Lipid Res 46(2):269–280

Imamura F, Micha R, Wu JH, de Oliveira Otto MC, Otite FO, Abioye AI et al (2016) Effects of saturated fat, polyunsaturated fat, monounsaturated fat, and carbohydrate on glucose-insulin homeostasis: a systematic review and meta-analysis of randomised controlled feeding trials. PLoS Med 13(7):e1002087

Jacobson TA, Ito MK, Maki KC, Orringer CE, Bays HE, Jones PH et al (2014) National Lipid Association recommendations for patient-centered management of dyslipidemia: part 1 - executive summary. J Clin Lipidol 8(5):473–488

Jamali M, Zarezadeh M, Jamilian P, Ghoreishi Z (2024) The effect of n-3 polyunsaturated fatty acids on inflammation status in cancer patients: updated systematic review and dose- and time-response meta-analysis. PharmaNutr 27:100372

Johnson GH, Fritsche K (2012) Effect of dietary linoleic acid on markers of inflammation in healthy persons: a systematic review of randomized controlled trials. J Acad Nutr Diet 112(7):1029–1041, 41.e1–15

Kapoor B, Kapoor D, Gautam S, Singh R, Bhardwaj S (2021) Dietary polyunsaturated fatty acids (PUFAs): uses and potential health benefits. Curr Nutr Rep 10(3):232–242

Kerdiles O, Layé S, Calon F (2017) Omega-3 polyunsaturated fatty acids and brain health: preclinical evidence for the prevention of neurodegenerative diseases. Trends Food Sci Technol 69:203–213

Khalili L, Valdes-Ramos R, Harbige LS (2021) Effect of n-3 (Omega-3) polyunsaturated fatty acid supplementation on metabolic and inflammatory biomarkers and body weight in patients with type 2 diabetes mellitus: a systematic review and meta-analysis of RCTs. Meta 11(11)

Kim Y, Kim J (2020) N-6 polyunsaturated fatty acids and risk of cancer: accumulating evidence from prospective studies. Nutrients 12(9):2523

Kingsley Nwosu O, Ikechukwu UK (2020) Nutraceuticals: history, classification and market demand. In: Dable-Tupas G, Egbuna G (eds) Functional foods and nutraceuticals: bioactive components, formulations, and innovations. Springer, pp 13–23

Lee CH, Lee SD, Ou HC, Lai SC, Cheng YJ (2014) Eicosapentaenoic acid protects against palmitic acid-induced endothelial dysfunction via activation of the AMPK/eNOS pathway. Int J Mol Sci 15(6):10334–10349

Lee CH, Fu Y, Yang SJ, Chi CC (2020) Effects of Omega-3 polyunsaturated fatty acid supplementation on non-alcoholic fatty liver: a systematic review and meta-analysis. Nutrients 12(9)

Lenighan YM, McNulty BA, Roche HM (2019) Dietary fat composition: replacement of saturated fatty acids with PUFA as a public health strategy, with an emphasis on α-linolenic acid. Proc Nutr Soc 78(2):234–245

Li J, Guasch-Ferré M, Li Y, Hu FB (2020) Dietary intake and biomarkers of linoleic acid and mortality: systematic review and meta-analysis of prospective cohort studies. Am J Clin Nutr 112(1):150–167

Lichtenstein AH (2023) Lipids (fats and oils). In: Caballero B (ed) Encyclopedia of human nutrition, 4th edn. Academic, Oxford, pp 291–300

Liput KP, Lepczyński A, Ogłuszka M, Nawrocka A, Poławska E, Grzesiak A et al (2021) Effects of dietary n-3 and n-6 polyunsaturated fatty acids in inflammation and cancerogenesis. Int J Mol Sci 22(13)

Litwack G (2022) Chapter 8 - Eicosanoids. In: Litwack G (ed) Hormones, 4th edn. Academic, pp 195–212

Liu QK (2021) Triglyceride-lowering and anti-inflammatory mechanisms of omega-3 polyunsaturated fatty acids for atherosclerotic cardiovascular risk reduction. J Clin Lipidol 15(4):556–568
Liu J, Li X, Hou J, Sun J, Guo N, Wang Z (2021) Dietary intake of N-3 and N-6 polyunsaturated fatty acids and risk of cancer: meta-analysis of data from 32 studies. Nutr Cancer 73(6):901–913
Lordan R (2021) Dietary supplements and nutraceuticals market growth during the coronavirus pandemic—implications for consumers and regulatory oversight. PharmaNutr 18:100282
Manson JE, Cook NR, Lee IM, Christen W, Bassuk SS, Mora S et al (2019) Marine n-3 fatty acids and prevention of cardiovascular disease and cancer. N Engl J Med 380(1):23–32
Marklund M, Wu JHY, Imamura F, Del Gobbo LC, Fretts A, de Goede J et al (2019) Biomarkers of dietary Omega-6 fatty acids and incident cardiovascular disease and mortality. Circulation 139(21):2422–2436
Mocciaro G, Allison M, Jenkins B, Azzu V, Huang-Doran I, Herrera-Marcos LV et al (2023) Non-alcoholic fatty liver disease is characterised by a reduced polyunsaturated fatty acid transport via free fatty acids and high-density lipoproteins (HDL). Mol Metab 73:101728
Mocellin MC, Fernandes R, Chagas TR, Trindade E (2018) A meta-analysis of n-3 polyunsaturated fatty acids effects on circulating acute-phase protein and cytokines in gastric cancer. Clin Nutr 37(3):840–850
Montecillo-Aguado M, Tirado-Rodriguez B, Huerta-Yepez S (2023) The involvement of polyunsaturated fatty acids in apoptosis mechanisms and their implications in cancer. Int J Mol Sci 24(14):11691
Mousavi SM, Jalilpiran Y, Karimi E, Aune D, Larijani B, Mozaffarian D et al (2021) Dietary intake of linoleic acid, its concentrations, and the risk of type 2 diabetes: a systematic review and dose-response meta-analysis of prospective cohort studies. Diabetes Care 44(9):2173–2181
Neufeld LM, Ho E, Obeid R, Tzoulis C, Green M, Huber LG et al (2023) Advancing nutrition science to meet evolving global health needs. Eur J Nutr 62(Suppl 1):1–16
Nwaru BI, Dierkes J, Ramel A, Arnesen EK, Thorisdottir B, Lamberg-Allardt C et al (2022) Quality of dietary fat and risk of Alzheimer's disease and dementia in adults aged ≥50 years: a systematic review. Foog Nutr Res:66
Pan A, Chen M, Chowdhury R, Wu JH, Sun Q, Campos H et al (2012) α-Linolenic acid and risk of cardiovascular disease: a systematic review and meta-analysis. Am J Clin Nutr 96(6):1262–1273
Patel ARU, Christakopoulos P, Matsakas L (2020) Introduction to essential fatty acids. In: Pate AK, Matsakas L (eds) Nutraceutical fatty acids from oleaginous microalgae: a human health perspective. Scrivener Publishing LLC
Puri V, Nagpal M, Singh I, Singh M, Dhingra GA, Huanbutta K et al (2022) A comprehensive review on nutraceuticals: therapy support and formulation challenges. Nutrients 14(21)
Ramsden CE, Zamora D, Majchrzak-Hong S, Faurot KR, Broste SK, Frantz RP et al (2016) Re-evaluation of the traditional diet-heart hypothesis: analysis of recovered data from Minnesota Coronary Experiment (1968-73). BMJ 353:i1246
Rizzo G, Baroni L, Lombardo M (2023) Promising sources of plant-derived polyunsaturated fatty acids: a narrative review. Int J Environ Res Public Health 20(3)
Sacks FM, Lichtenstein AH, Wu JHY, Appel LJ, Creager MA, Kris-Etherton PM et al (2017) Dietary fats and cardiovascular disease: a presidential advisory from the American Heart Association. Circulation 136(3):e1–e23
Sanders TAB (2014) Protective effects of dietary PUFA against chronic disease: evidence from epidemiological studies and intervention trials. Proc Nutr Soc 73(1):73–79
Sarparast M, Dattmore D, Alan J, Lee KSS (2020) Cytochrome P450 metabolism of polyunsaturated fatty acids and neurodegeneration. Nutrients 12(11)
Schmitz G, Ecker J (2008) The opposing effects of n−3 and n−6 fatty acids. Prog Lipid Res 47(2):147–155
Serhan CN, Levy BD (2018) Resolvins in inflammation: emergence of the pro-resolving superfamily of mediators. J Clin Invest 128(7):2657–2669
Shchepinov MS (2020) Polyunsaturated fatty acid deuteration against neurodegeneration. Trends Pharmacol Sci 41(4):236–248

Singh RB, Watanabe S, Li D, Nakamura T, Juneja LR, Takahashi T et al (2022) Chapter 11 - Effects of polyunsaturated fatty acid–rich diets and risk of non-communicable diseases. In: Singh RB, Watanabe S, Isaza AA (eds) Functional foods and nutraceuticals in metabolic and non-communicable diseases. Academic, pp 165–185

Singh PK, Gupta MK, Nath R (2023) Chapter 16 - Omega-3 fatty acid as a protectant in lead-induced neurotoxicity. In: Martin CR, Patel VB, Preedy VR (eds) Treatments, nutraceuticals, supplements, and herbal medicine in neurological disorders. Academic, pp 285–301

Tortosa-Caparrós E, Navas-Carrillo D, Marín F, Orenes-Piñero E (2017) Anti-inflammatory effects of omega 3 and omega 6 polyunsaturated fatty acids in cardiovascular disease and metabolic syndrome. Crit Rev Food Sci Nutr 57(16):3421–3429

Trumbo P, Schlicker S, Yates AA, Poos M (2002) Dietary reference intakes for energy, carbohydrate, fiber, fat, fatty acids, cholesterol, protein and amino acids. J Am Diet Assoc 102(11):1621–1630

TURKOMP. Turkish National Food Composition Database (Türkomp-Ulusal Gıda Kompozisyon Veri Tabanı) 2023. [22.12.2023]. Available from: https://turkomp.tarimorman.gov.tr/main

Turpeinen A, Merimaa P (2011) 16 - Functional fats and spreads. In: Saarela M (ed) Functional foods. Woodhead Publishing, Second Edition, pp 383–400

USDA. FoodData Central: U.S. Department of Agriculture; 2024. [cited 2024 02.01.2024]. Available from: https://fdc.nal.usda.gov/fdc-app.html#/

Wang B, Wu L, Chen J, Dong L, Chen C, Wen Z et al (2021) Metabolism pathways of arachidonic acids: mechanisms and potential therapeutic targets. Signal Transduct Target Ther 6(1):94

Wang Y, Liu K, Long T, Long J, Li Y, Li J et al (2023) Dietary fish and omega-3 polyunsaturated fatty acids intake and cancer survival: a systematic review and meta-analysis. Crit Rev Food Sci Nutr 63(23):6235–6251

Wei BZ, Li L, Dong CW, Tan CC, Xu W (2023) The relationship of Omega-3 fatty acids with dementia and cognitive decline: evidence from prospective cohort studies of supplementation, dietary intake, and blood markers. Am J Clin Nutr 117(6):1096–1109

Yousefi M, Eshaghian N, Heidarzadeh-Esfahani N, Askari G, Rasekhi H, Sadeghi O (2023) Dietary intake and biomarkers of linoleic acid and risk of prostate cancer in men: a systematic review and dose-response meta-analysis of prospective cohort studies. Crit Rev Food Sci Nutr:1–17

Zhang Y, Chen J, Qiu J, Li Y, Wang J, Jiao J (2016) Intakes of fish and polyunsaturated fatty acids and mild-to-severe cognitive impairment risks: a dose-response meta-analysis of 21 cohort studies123. Am J Clin Nutr 103(2):330–340

Zhu RZ, Chen MQ, Zhang ZW, Wu TY, Zhao WH (2021) Dietary fatty acids and risk for Alzheimer's disease, dementia, and mild cognitive impairment: a prospective cohort meta-analysis. Nutrition 90:111355

Zong G, Li Y, Wanders AJ, Alssema M, Zock PL, Willett WC et al (2016) Intake of individual saturated fatty acids and risk of coronary heart disease in US men and women: two prospective longitudinal cohort studies. BMJ 355:i5796

Chapter 9
Probiotics and Prebiotics as Dietary Medicine

Taherah Mohammadabadi

Abstract Interest in functional foods containing probiotic bacteria or prebiotics by consumers has been increasing. Probiotics and prebiotics enhance colonizing microbiota, normalize the gastrointestinal function, strengthen the immune responses against infections, and improve autoimmune diseases and allergies toward human health. Prebiotics as food compounds that enhance beneficial gut microbes and the persistence of probiotic. Furthermore, the using probiotics prohibits pathogenic bacteria and their metabolites, reduces drug-resistant bacteria, controls blood cholesterol and pressure, and prevents cancerous conditions. Since probiotics contribute to the host's health, the interest in both scientific research and commercial uses is increasing. Traditional medicines can only relieve some diseases related to the microbiota of gut. Various clinical studies and evidences have proved that prebiotics and probiotics can prohibit diseases, but they are consider as dietary ingredients more than medicines. Although probiotics and prebiotics may be regarded as the future generation of medicine to control disease, more evidence that is scientific is needed on the benefits of prebiotics and probiotics on health complications.

Keywords Probiotics · Prebiotic · Dietary medicine · Lactobacillus · Bifidobacterium

9.1 Introduction

The interest in probiotic as 60% of bioactive foods, is increasing. Probiotics are recognized as live microorganisms' resistance to pancreatic secretions, biliary and gastric conditions, attaching and colonizing the intestine epithelial cells of human (Quin et al. 2018). Probiotic microorganisms must be safe and remain alive after food processing and passage through the gut (Gomaa 2020; Ballini et al. 2023).

T. Mohammadabadi (✉)
Faculty of Animal Science and Food Technology, Agricultural Sciences and Natural Resources University of Khuzestan, Ahvaz, Iran

S. Amir Ashraf, M. Adnan (eds.), *Food Bioactives and Nutraceuticals*,
https://doi.org/10.1007/978-981-97-9723-3_9

There are some yeasts and many genera of bacteria to be used as probiotic cultures, but Lactobacillus and Bifidobacterium are the most common, which are lactic acid producers and gram-positive, a major portion of intestinal microflora (Fujiya et al. 2014). Prebiotics as food components promote the proliferation and growth of certain probiotics in the large intestine and colon for human health. The most common prebiotics for stimulating probiotics are fructooligosaccharides, galactooligosaccharides, inulin, and soybean oligosaccharides (Hayat et al. 2014). Prebiotics promote the probiotic bacteria and inhibit unwanted and harmful microorganisms (Isacco et al. 2021). The studies proved that prebiotics affect insulin resistance, blood glucose, and immune factors such as IL-6 (Ballini et al. 2023). According to the research, probiotics prevent pathogenic bacteria, improve the function of beneficial microorganisms by altering environmental niche and pH in the gut, enhance the production of metabolites for the normal function of the gastrointestinal, and stimulate immune responses against infection. At the same time, prebiotics enhance the persistence of beneficial microbes in the gut (Ahmed et al. 2013). Probiotics have a key action on bacteria which are resistant to the drug, improve general health and activate the enzyme hydrolysis of undesirable proteins and lipids that cause allergic responses (Tomasik and Tomasik 2003). Probiotics stimulate the production of antimicrobial agents against infections. Some specific probiotic strains have competitive action on *H. pylori* and colonization in the epithelium cells of stomach (Saracino et al. 2020). The administration of certain probiotic for antibiotics, lactose intolerance and diarrhea caused by *Clostridium difficile* has been proven. Also, probiotics prevent ventilator-associated pneumonia inflammation and control the infections in the respiratory tract (Zeng et al. 2016; Alexandre et al. 2014; Campanella et al. 2018; Lee et al. 2021). Since probiotics and prebiotics can stimulate the beneficial gut microbiota and health (Li et al. 2021), they are regarded as health promoters rather than medicines. They are potentially important adjunct medicine or alternatives therapies to improve health, but more scientific evidence is needed on the medical benefits of probiotics and prebiotics (Ji et al. 2023).

9.2 Kinds of Probiotic Microorganisms

Currently, probiotics are live microorganisms usually of human origin and safe that resist pancreatic, gastric, and biliary secretions, attach and colonize in epithelial cells of the human intestine, and have competitive action against pathogens (Quin et al. 2018). Probiotic microorganisms must remain alive after the food processing and pass through the gastrointestinal tract (Quin et al. 2018). These microorganisms can resist antibiotic resistance and immobilization in the intestinal epithelium, and these bacteria can be alive in freeze-dried and fermented products (Santacroce et al. 2019) (Table 9.1).

Bifidobacteria spp., *Lactobacillaceae* family, and yeasts such as *Saccharomyces boulardii* are the most common probiotic species (Gomaa 2020). These bacteria, a major group of intestinal microflorae, are lactic acid-producing bacteria and

Table 9.1 Some major probiotic bacteria

Lactobacillaceae	*Bifidobacteriaceae*	Yeasts
Lactobacillus helveticus	*Bifidobacterium longum*	*Saccharomyces cerevisiae*
Lactobacillus johnsonii	*Bifidobacterium breve*	*Aspergillus niger*
Lactobacillus acidophilus	*Bifidobacterium bifidum*	*Saccharomyces boulardii*
Lactobacillus gasseri	*Bifidobacterium animalis*	*Aspergillus oryzue*
Lactobacillus crispatus	*Bifidobacterium adolescentis*	
Lacticaseibacillus rhamnosus	*Bifidobacterim lactis*	
Lacticaseibacillus casei	*Bifidobacterim infantis*	
Lactiplantibacillus plantarum		
Limosilactobacillus reuteri		
Ligilactobacillus salivarius		
Leuconostoc mesenteroides		
Enterococcus faecium		

Adapted from Santacroce et al. (2019)

gram-positive (Fujiya et al. 2014). Lactobacilli are mostly auto-fermentative, non-spore-forming bacteria, acidophilic, strictly anaerobic, and rod-shaped. Some bacteria uniquely produce exopolysaccharides with therapeutic properties (Senok et al. 2005). Bifidobacteria, as another major gut microbiota, can be seen in stools few days after birth, and 10^{10}–10^{11} CFU/g is the normal range in the colon of adult (Schrezenmeir and de Vrese 2001). Bifidobacteria mostly are rod-shaped, nonspore-forming bacteria, strictly anaerobic, and gram positive (Ahmad and Khalid 2018). Probiotics microorganisms are mostly *B. longum, S. thermophilus, S. boulardii, L. rhamnosus, L. kefiri, Enterococcus faecalis, B. pullorum, L. plantarum, Lactococcus lactis, Propionibacteria, L. pentosus, L. casei, L. fermentum, L. panis, L. delbrueckii subsp. Bulgaricus, L. delbrueckii subsp. lactis* and *B. subtilis* (Ahmad and Khalid 2018). Probiotics can be obtained from milk, yogurt, cheese, and fermented milk as dairy functional foods and non-dairy products. Various strains of *Lactobacillaceae* like *Lacticaseibacillus casei, Lactobacillus acidophilus, Lactobacillus helveticus* and *Lacticaseibacillus rhamnosus*, have been extensively explored for the inhibition of various diseases (Amara and Shibl 2015). Usually, the efficacy and safety of probiotic microorganisms are measured by the quantity of probiotics used. Probiotics in fermented foods improve immune responses and gut health (Ma et al. 2023).

9.3 Types and Sources of Prebiotics

Prebiotics are undigestible food components that enhance the development of probiotics (Ahmad and Khalid 2018) and the proliferation and function of gut proper bacteria, as well as inhibit harmful or unwanted microorganisms (Gibson et al. 2017) to improve health. The major criteria for selecting a substrate for probiotics are as follows: a prebiotic should resist hydrolysis and absorption in the gut, be a substrate for probiotics fermentation in the colon, and make microbiota healthier in the gut (Gibson and Roberfroid 1995). Prebiotics are generally divided into low-molecular-weight oligosaccharides and polysaccharides. They are a rich source of carbon for the growth of gut microbiota, and they are utilized by the large intestine microbes for fermentation and form Short-Chain Fatty Acids against harmful microbes (Rastall and Maitin 2002). Prebiotics generally are soluble alimentary fibers or carbohydrates (Ballini et al. 2023), as fructooligosaccharides, inulin, galactooligosaccharides, and soybean oligosaccharides regarded as major Prebiotics (Hayat et al. 2014). Galactooligosaccharides are monomers of galactose, fructose or glucose that are formed by glycosylic transferase enzyme and adding glucose units (Ahmad and Khalid 2018) that present in the livestock and human milk (about 3 mg/L) and yogurt (Isacco et al. 2021). Galactooligosaccharides can be degraded by Bifidobacterium in the large intestine and modulate the activity of Bifidobacterium (Marilyn et al. 2014). They can be hydrolyzed by probiotics to produce propionic acids and butyric acids as well as lactate in the large intestine that increase the function and growth of Bifidobacterium and Lactobacilli (Kolida et al. 2000). Galactooligosaccharides enhance the proliferation of the *Lactobacillaceae* and *Bifidobacteriaceae* families for the host's health (Isacco et al. 2021). These prebiotics can inhibit infection caused by pathogens because they prevent adhesion to epithelial cells by imitating binding sites (Ballini et al. 2023). Fructooligosaccharides are the most well-known prebiotics and contain β-d-fructans fructose units with β-2-1 glycosidic linkages that are dissolved at 80 °C and are only digestible by microbes in the large intestine (Ahmad and Khalid 2018). Fructooligosaccharides are fructans, a few fructose monomers with hydrolytic inulin derivatives (Slavin 2013). These prebiotics influence IL-6 production as immune factor and increase mucosal immunity, produce anti-inflammatory cytokines, improve insulin resistance and blood glucose, strengthen the intestinal mucosa integrity, reduce pH by the production of SCFAs (Slavin 2013; Ballini et al. 2023).

Xylooligosaccharides are formed from units of xylose with linkages of β-1-4. Xylooligosaccharides can be fermented in the large intestine by bifidobacteria and lactobacilli and enhance the Bifidobacterium counts more than fructooligosaccharides (Villamiel et al. 2014). Soybean oligosaccharides include raffinose and stachyose which can be hydrolyzed by microflora in the large intestine, improving the growth of bifidobacteria (Hayakawa et al. 2012). Lactulose is also formed from lactose with the structure Gal-β-1-4-Fructose. It cannot be hydrolyzed by gut microbes but fermented by large intestine bacteria. Lactulose is added to chocolate, cake, and yogurt to improve nutritional value (Ahmad and Khalid 2018). Lactosucrose is formed by using the β-fructofuranosidase enzyme by combining sucrose and lactose. Oral administration of lactosucrose increased Bifidobacterium by about 0.7 times and reduced the pathogenic bacteria by about 0.6 times (Bouhnik et al. 2004). Prebiotics can be found naturally in chicory, asparagus, milk, wheat, tomatoes, and human milk and synthetic forms are from carbohydrates. The health effects of prebiotics, such as 4 g/day inulin, are proven by increasing Bifidobacterium or reducing pathogenic microbes (McAllister et al. 1994) and found in plants, human milk and animal colostrum as free sugars (Ahmad and Khalid 2018). Fructose-containing oligosaccharides can be obtained from plants, enzymatic synthesis and hydrolysis of polysaccharides. Inulin β (Ahmad and Khalid 2018; Alexandre et al. 2014) fructans are found in leeks, garlic, onions, asparagus, shallots, artichokes, wheat, and bananas. Inulin originates from the radish roots by hot extraction, refining and spray drying (Ballini et al. 2023). Inulin and soybean oligosaccharides (raffinose and stachyose) can be extracted directly from chicory root and soybeans. Fructans include inulin and levan oligosaccharides naturally found in chicory, dahlia, Jerusalem artichoke, garlic, leek, salsify, onion, and wheat by-products (Callaway and Editors 2011).

9.4 Probiotics and Health

9.4.1 Effect on the Intestinal Microbiota

Probiotics have beneficial impacts on the gut microbiota, thus improving the harmful effects of antibiotics, corona viruses and infections (Santacroce et al. 2021). Oral administration of bacteria can improve intestinal microbiota balance (Quin et al. 2018). Foods's rich in probiotics like yogurt and kefir contain *Bifidobacteriaceae* and *Lactobacillaceae* families, compete with pathogens such as *C. perfringens* and *E. coli* that is important for the proper function of the intestine. The main effects of probiotics are competition for ingredients, fermentation of sugars into inhibitory components, stimulation of the activity and shielding of the epithelial cells, direct competition by bacteriocins, reduction of inflammation, and strengthening of the innate immune system (Inchingolo et al. 2019). Probiotics keep the Bifidobacterium species balance that disturbed by antibiotics (Seki et al. 2003). The major therapeutic actions of probiotics on the gastrointestinal tract are due to their activity and interaction with the mucosal microflora (Ashraf et al. 2022). Bifidobacterium and Lactobacillus, also producelactic acid, propionic acid, and acetic acid, which lower the pH and prevent the pathogens growth (Chaucheyras and Durand 2010).

9.4.2 Colon Cancer

Most cancer patients are in immune-compromised situations with various radiotherapy chemo therapy treatments, infection and bone marrow suppression (Zhu et al. 2011). The main reason for infections is pathogenic bacteria of the gut, which can be transferred through the mucosa. Probiotics which modulate the gut immune system, compete with pathogenic bacteria, and produce organic acids and some bacteriocins against pathogens infections. Probiotic strains and metabolites act as antitumor agents and reduce cancer risk. These strains' cancer-inhibiting functions and interactions with molecular signals and these metabolites prevent cancer (Maleki et al. 2016). Probiotic modified bacteria genes that express cytokines and are applied as a vector for the delivery of vaccines and drugs. Probiotics have antiproliferative effects on cancer cells, fight with microorganisms involved in the secretion and production of carcinogens and mutagens, protect DNA from oxidative stress, and regulate immune responses. Furthermore, they contribute to the expression of the genes involved in metastasis, infiltration, cell death and apoptosis, and control the cell cycle. Probiotics modulate signalling pathways that promote tumorigenesis (Śliżewska et al. 2020).

9.4.3 Diabetes

Probiotics enhance the gut microbiota effective in insulin-resistant and reduce glucose and other diabetes-related disorders (Sun and Buys 2016). The efficacy of probiotics on diabetes type I and 2 was proved by stimulating intestinal microbes (de Oliveira et al. 2014). The effect of probiotic yogurt in obese people compared with standard low-fat yogurt reported that there were no notable differences in fasting glucose, insulin, and lipoprotein in the *Lactobacillus salivarius* and the control (Gobel et al. 2010; Ahmad and Khalid 2018).

9.4.4 Diarrhea and Inflammatory Bowel Disease

Probiotics decrease the duration of diarrhea by increasing the immune system and producing antibodies against rotavirus and E. coli. Antibiotic-linked diarrhea happens due to the *Clostridium difficile* growth. Probiotics, including *Lactobacillus casei, Streptococcus thermophilus* and *Lactobacillus bulgaricus* reduce antibiotic-linked diarrhea by 21.6% (Hickson 2011). Probiotics are helpful in prohibiting diarrhea in children who suffer from undernutrition. Probiotics spp. For example, *L. rhamnosus* and *S. Boulardii* affect antibiotic-linked diarrhea and infections (Kopacz and Phadtare 2022; Awadelkareem et al. 2023). The oral administration of *L. reuteri* could save mice from inflammation of high-fat diet. Also, *S. Boulardii* interferes with the inflammatory cytokines production (Guslandi et al. 2000). Abdominal pain and changes in bowel habits are the signs of irritable bowel disease (Lovell et al. 2010). Probiotic bacterial types positively affect irritable bowel syndrome (Wasilewski et al. 2015). Butyric acid decreases gut inflammation and preventing mast cell degranulation (Parada et al. 2019; Sanders et al. 2019). Butyric acid's anti-inflammatory and immunoregulatory effects result in the healing of mucosal lesions, the improvement of electrolyte absorption and the mucus production which are important for the of the intestinal mucosa integrity and improve inflammatory bowel syndrome (Practice Guidelines et al. 2010).

9.4.5 *Lactose Intolerance*

Lactose intolerance is characterized by nausea, vomiting, abdominal pain, flatulence, bloating and watery diarrhea (Chitkara et al. 2005). Probiotics Lactobacillus and Bifidobacterium increase lactase enzyme production, improving lactose's digestibility. The administration of *Lactobacillus acidophilus* proved that lactose intolerance symptoms such as abdominal pain, diarrhea, and vomiting reached zero, with no side effects. Probiotics increased the lactose hydrolysis in dairy products which decrease malabsorption of lactose (Ahmad and Khalid 2018).

9.4.6 *Ulcerative Colitis and* H. pylori *Infections*

Ulcerative colitis is an inflammatory condition of the rectum mucous. The probiotics showed 1.35% remission rate in the ulcerative colitis patients in compared to 2% remission rate for control and Bifidobacterium showed 0.25%. Probiotic treatment is more effective in ulcerative colitis patients (Tsang et al. 2010). Infections of *H. pylori* impact about 50% of the world's population which is associated with gastric ulcers and carcinoma. In another study, probiotics had no any effect on eradicating *H. pylori* in school children (Pacifico et al. 2014).

9.4.7 *Crohn's Disease and Hypercholesterolemia*

Crohn's is an inflammatory disease known for chronic inflammation of the gastrointestinal and ileum walls. Administration of *Lactobacillus GG* by 1.2 × 1010 CFU to Crohn's disease patients for 1 year after surgery caused a 73% reduction in the disease activity index for 4 weeks. *Lactobacillus GG* could enhance barrier function of the gut in C. Crohn's disease children (Gupta et al. 2000). The consumption of probiotics leads to a significant increase in beneficial bacteria of gut and short-chain fatty acids and simultaneous reduction in harmful bacteria that improve the intestinal integrity and inhibit inflammatory responses in bowel inflammation, and consequently combat Crohn's disease (Ahmad and Khalid 2018). Probiotics bind bile salts to intestinal walls through the bile salt hydrolase enzyme. There was a significant decrease in total cholesterol and LDL-C, and probiotics were recommended to prevent hypercholesterolemia. Using lactobacillus for 4 weeks improve blood lipids and reduce the cardiovascular disease in hypercholesterolemic rats (Ahmad and Khalid 2018).

9.5 Prebiotics and Health

9.5.1 The Effect on the Gut Microbes

Dietary fibre, such as β-glucan, enhances the bioavailability of nutrients for supporting the growth of probiotics in the gut. Different non-digestible carbohydrates strengthen the development of gut microbes (Ahmad et al. 2010). Lactulose has effective effects on microbiota by stimulating fermentation and the production of lactic, butyric and acetic acid in the large intestine, which causes flatulence in the colon and removal of urease and ammonia-producing bacteria in the colon (Ahmad and Khalid 2018). A fiber-rich diet stimulates Bifidobacterium and Lactobacillus growth and the production of SCFAs for gut health (Xu et al. 2022).

9.5.2 Hypercholesterolemia and Obesity

Prebiotics oligo-fructose and inulin modify the hepatic lipid metabolism in rats and hamsters fed a diet with high fat and energy and lacking fiber, decreasing hypercholesterolemic activities. Triglyceride amounts and postprandial cholesterol were reduced about 15% and 50%, respectively, but the accumulation of alcohol triglyceride was prohibited. Furthermore, significant reduction of TC and TG proved using fiber and β-glucan in the diet (Ahmad and Khalid 2018). People with a low number of Bifidobacterium at birth have been associated with developing obesity after childhood. Moreover, obese mothers giving birth children with low counts of Bifidobacterium. Bifidobacterium has a key role in the metabolic system and obesity. Food products containing fructans and oligosaccharides increase the Bifidobacterium growth in the gut (Madjd 2016).

9.5.3 Gastric and Colorectal Cancer

Prebiotics increase the growth of beneficial bacteria in the gut, produce short-chain fatty acids as weak acids, modulate the gene expression in the cecum, colon and feces, increase absorption of micronutrients in the colon, enhance xenobiotic enzymes, and modify the immune system. Prebiotics change the gut microbiome and improve the growth of Lactobacillus and Bifidobacterium, which eliminate carcinogens in the gut. Therefore, prebiotics inhibit the risk of gut cancer. Prebiotics increases Bifidobacterium count and low pH, alter immune response, and decrease small intestine and colon tumours in the rodent (Ahmad and Khalid 2018).

9.5.4 Diabetes and Immunomodulation

Fruits, legumes, cereals and spices contain active ingredients and dietary fiber that can reduce glycemia amount and insulin immune responses. Inulin fructans are important prebiotics that can modulate the microflora in the gut. Similarly, arabinoxylan has a beneficial impact on hyperglycaemic levels (Bindels et al. 2015). Prebiotics produce SCFAs, which regulate hepatic lipogenic enzymes, modulate histone tail acetylation, and access many enzymes related to transcriptional factors. Prebiotics modify mucin and fructose production that are associated with enhancing lymphocytes in the peripheral blood. The IgA secretion is increased by GALT, which modulates the phagocytic activity of intraperitoneal macrophages (Liong 2007).

9.5.5 Uptake of Minerals

Prebiotic fructans can improve Ca absorption. Bone mineral density and ca absorption were increased in adults who utilized 8 g short-chain fructans/day. Insoluble Ca converts into ionic form because SCFAs produced from fiber fermentation in the colon lowers pH. SCFAs and low pH cause the intestine enlargement and elevated Ca absorption. Production of mucin due to prebiotic intake lowers bacterial transfer through gut barriers (Whisner et al. 2014). Prebiotics in fermented dairy products improve mineral absorption and bone density (Sanders et al. 2014).

9.5.6 Diarrhea, Irritable Bowel Syndrome, and Gut Diseases

Prebiotics protect Bifidobacterium and Lactobacillus from gastric acidic conditions and provide energy when they reach the colon. Also, it's confirmed that prebiotics and soluble fibers can be beneficial in decreasing the inflammations of irritable bowel syndrome and epithelial injury. Guar gum reduced abdominal pain and improved bowel habits. Prebiotics were proven to have remission effects in UC patients. Prebiotics have a useful impact on Crohn's disease. Patients suffering from CD fed 24 g inulin/day and decrease in the number of Bacteroides of feces was observed (Ahmad and Khalid 2018).

9.5.7 Allergies, Influenza and Atopic Dermititis

The studies concluded that allergies were decreased in infants fed GOS combined with *L. rhamnosus GG* and *LC705, B. breve Bb99* and *Propionibacterium freudenreichii ssp. Shermanii* for 6 months. Also, it has been proven that synbiotics can inhibit food allergies (Kuitunen et al. 2009). Prebiotics' useful effects on influenza infections have been proven in animal and clinical studies. There was a significant increase in probiotic and antibody count (H1N1, H3N2, and B) after using prebiotics for 8 weeks, (Ahmad and Khalid 2018). Over the last 30 years, Atopic dermatitis incidences have increased by 20% in young children and infants. Prebiotics and probiotics can decrease the severity of the disease. Feeding oligosaccharides and FOS to infants suffering from Atopic dermatitis at a ratio of 9:1 causes a significant decrease in the Atopic dermatitis in infants (Gruber et al. 2010) due to maintaining body health and intestinal homeostasis (Usami et al. 2011).

9.6 Prebiotic and Probiotic as Medicine

Dietary prebiotic can alter the gut microbe's function. Therapeutic benefits of probiotics include inhibiting mortality and morbidity linked with necrotizing enterocolitis in newborns with low birth weight and preterm (Jacobs et al. 2013), reduced incidence of atopic dermatitis and infections in the upper respiratory tract in the infants, reducing antibiotic diarrhea and the duration of infectious and traveler's diarrhea (Drakoularakou et al. 2010), regulate intestinal transit and improve blood lipid, alleviation of irritable bowel syndrome, improve calcium absorption, prevent specific allergies, reduction energy intake and insulin resistance, improve gut function and tolerance to antibiotics (Sanders et al. 2014). Prebiotic products obtained from prebiotic fermentation positively affect human health, such as antiapoptotic and anti-inflammatory properties and the prohibition of colon cancer and colitis (Kaur et al. 2022). *Bifidobacterium* spp. Ferment prebiotics produce butyrate, vital for the colonic mucosa's appropriate function and integrity and for stimulating immune system (Tidjani et al. 2016). Prebiotics as foods rich in fibers containing oligofructose and inulin, which modulate beneficial intestinal microorganisms. Dairy products and yogurt consumption lower diastolic blood pressure, diabetes risk, and cardiovascular issues, regulate immune responses, and reduce upper respiratory tract infections (Sanders et al. 2014). Prebiotics provide a fermentable food source for *the Lactobacillaceae* and *Bifidobacteriaceae* families. Administration of the fibers improves inflammatory bowel disease due to increasing beneficial metabolites for the microbiota (Ballini et al. 2023). Beneficial bacteria such as *Lactobacilli* and *Bifidobacteria* proliferate abundantly by prebiotics FOS and GOS that inhibits the pathogens colonization in the gut and keeps balance of gut microbiota (Karoline Ferreira Leite et al. 2023).

Prebiotics inulin fructans can impact immune functions and modulate dendritic cells, lead to maintaining intestinal homeostasis and the production of anti-inflammatory cytokines (Ma et al. 2023). Muramyl dipeptide notably decreases inflammation of the lung caused by influenza A virus infection, and elevates pulmonary function (Ma et al. 2023). Probiotics modulate intestinal immunity (Callaway and Editors 2011). Compared with some conventional medicines, probiotics are more suitable for long-term use and are considered medicines and live biotherapeutics (Ma et al. 2023). The probiotics combat infant malnutrition. *B. longum* subspecies, with oligosaccharides of human milk increase the counts of *Bifidobacteria* in the infant's gut with severe acute malnutrition. *B. longum* utilizes oligosaccharides present in human milk and maintains the gut microbiota due to produced SCFAs (Zabel et al. 2019). Probiotics such as *Limosilactobacillus fermentum* KBL374 and *B. longum* improve symptoms of IBS in mice by stimulating the immune response and enhancing the activity of the intestinal mucosa and microbiota (Jang et al. 2019; Zhou et al. 2020). *Lactobacillus reuteri* can modulate cytotoxic CD8+ T cells and prevent cancer (Bender et al. 2023). Disturbance in the microbiota can increase the inflammation, obesity and cancer (Rautava 2021). Therefore, using probiotics, prebiotics can regulate the immune responses and improve intestinal barrier function and gut microbiota (Cunningham et al. 2021). Probiotics modulate both innate and adaptive immunity by interacting with the host's immune system (Strisciuglio et al. 2023), which activates natural killer cells, boosts cytokine production, and improves the activity of antigen-presenting cells, leads to effective immune response. Probiotics produce antimicrobial components such as bacteriocin and organic acids and reduce the infection in the gut by restricting pathogen colonization (Chiumento et al. 2019; Wu et al. 2022). Probiotics break down harmful metabolic by-products and exogenous substances such as toxins, carcinogens, antibiotics, and heavy metals as well as break down uric acid and acetate to protect the liver (Wu et al. 2021). The researchers showed the effective properties of probiotics in pneumonia (Ma et al. 2023).

9.7 Conclusions

The administration of probiotics and prebiotics strengthens the microbiota balance in the gut. The prebiotics and vegetable fibers broken down by the microflora in the large intestine that led to increasing concentrations of the beneficial microbiota and reduction of inflammatory cytokines, which increases the risk of severe chronic diseases. The combinations of probiotic and prebiotic can act synergistically by increasing *Lactobacillus* and Bifidobacterium spp. in the colon. Probiotics can improve diseases by regulating the gut microbiota and removing exogenous substance and harmful metabolites. Prebiotics improve gut microbiota by modulating beneficial microorganisms and maintaining health. Studies show that probiotics will be vital in disease prevention and management and may become the future of representative medicines.

References

Ahmad A, Khalid S (2018) Therapeutic Aspects of Probiotics and Prebiotics. In: Diet, Microbiome and Health. Elsevier. https://doi.org/10.1016/B978-0-12-811440-7.00003-X

Ahmad A, Anjum FM, Zahoor T, Nawaz H, Ahmed Z (2010) Extraction and characterization of β-glucan from oat for industrial utilization. Int J Biol Macromol 46(3):304–309

Ahmed Z, Wang Y, Anjum N, Ahmad A, Khan ST (2013) Characterization of exopolysaccharide produced by Lactobacillus kefiranofaciens ZW3 isolated from Tibet kefir—part II. Food Hydrocoll 30:342–350

Alexandre Y et al (2014) Probiotics: a new way to fight bacterial pulmonary infections? Medecine et maladies infectieuses 44:9–17

Amara AA, Shibl A (2015) Role of probiotics in health improvement, infection control and disease treatment and management. Saudi Pharm J 23:107–114

Ashraf SA, Elkhalifa AE, Ahmad MF, Patel M, Adnan M, Sulieman AM (2022) Probiotic fermented foods and health promotion. In: African Fermented Food Products-New Trends. Springer, Cham, pp 59–88

Awadelkareem AM, Siddiqui AJ, Noumi E, Ashraf SA, Hadi S, Snoussi M, Badraoui R, Bardakci F, Ashraf MS, Danciu C et al (2023) Biosynthesized silver nanoparticles derived from probiotic Lactobacillus rhamnosus (AgNPs-LR) targeting biofilm formation and quorum sensing-mediated virulence factors. Antibiotics 12(6):986

Ballini A, Charitos IA, Cantore S, Topi S, Bottalico L, Santacroce L (2023) About functional foods: the probiotics and prebiotics state of art. Antibiotics (Basel) 12(4):635

Bender MJ, McPherson AC, Phelps CM et al (2023) Dietary tryptophan metabolite released by intratumoral Lactobacillus reuteri facilitates immune checkpoint inhibitor treatment. Cell 186(9):1846–1862

Bindels LB, Delzenne NM, Cani PD, Walter J (2015) Towards a more comprehensive concept for prebiotics. Nat Rev Gastroenterol Hepatol 12(5):303–310

Bouhnik Y, Raskine L, Simoneau G et al (2004) The capacity of nondigestible carbohydrates to stimulate fecal bifidobacteria in healthy humans: a double-blind, randomized, placebo-controlled, parallel-group, dose-response relation study2. Am J Clin Nutr 80(6):1658–1664

Callaway TR, Editors RSC (2011) Direct-fed microbials and prebiotics for animals. In: Science and mechanisms of action. Springer, New York Dordrecht Heidelberg London, p 215

Campanella V et al (2018) Oral probiotics influence oral and respiratory tract infections in pediatric population: a randomized double-blinded placebo-controlled pilot study. Eur Rev Med Pharmacol Sci 22:8034–8041

Chaucheyras F, Durand H (2010) Probiotics in animal nutrition and health. Benef Microbes 25:3–9

Chitkara DK, Montgomery RK, Grand RJ, Büller HA (2005) Lactose intolerance. Available from: www.uptodate.com

Chiumento S, Roblin C, Kieffer-Jaquinod S et al (2019) Ruminococcin C, a promising antibiotic produced by a human gut symbiont. Sci Adv 5(9):eaaw9969

Cunningham M, Azcarate-Peril MA, Barnard A et al (2021) Shaping the future of probiotics and prebiotics. Trends Microbiol 29(8):667–685

de Oliveira MEG, Garcia EF, de Oliveira CEV, Gomes AMP, Pintado MME, Madureira ARMF, da Conceicao ML, do Egypto Queiroga RCR, de Souza EL (2014) Addition of probiotic bacteria in a semi-hard goat cheese (coalho): survival to simulated gastrointestinal conditions and inhibitory effect against pathogenic bacteria. Food Res Int 64:241–247

Drakoularakou A, Tzortzis G, Rastall RA et al (2010) A double-blind, placebo-controlled, randomized human study assessing the capacity of a novel galacto-oligosaccharide mixture in reducing travellers' diarrhoea. Eur J Clin Nutr 64:146–152

Fujiya M, Ueno N, Kohgo Y (2014) Probiotic treatments for induction and maintenance of remission in inflammatory bowel diseases: a meta-analysis of randomized controlled trials. Clin J Gastroenterol 7(1):1–13

Gibson GR, Roberfroid MB (1995) Dietary modulation of the colonic microbiota: introducing the concept of prebiotics. J Nutr 125:1401–1412

Gibson GR, Hutkins R, Sanders ME et al (2017) Expert consensus document: the International Scientific Association for Probiotics and Prebiotics (ISAPP) consensus statement on the definition and scope of prebiotics. Nat Rev Gastroenterol Hepatol 14(8):491–502

Gobel R, Larsen N, Molgaard C, Jakobsen M, Michaelsen KF (2010) Probiotics to young children with atopic dermatitis: a randomized placebo-controlled trial. Int J Probiotics Prebiotics 5:53–60

Gomaa EZ (2020) Human gut microbiota/microbiome in health and diseases: a review. Antonie Leeuwenhoek 113:2019–2040

Gruber C, vas Stuijvenberg M, Mosca F, Moro G, Chirico G, Braegger CP, Riedler J, Boehm G, Wahn U, MIPS 1 Working Group (2010) Reduced occurrence of early atopic dermatitis because of immunoactive prebiotics among low-atopy-risk infants. J Allergy Clin Immunol 126(4):791–797

Gupta P, Andrew H, Kirschner BS, Guandalini S (2000) Is Lactobacillus GG helpful in children with Crohn's disease? Results of a preliminary, open label study. J Pediatr Gastroenterol Nutr 31:453–457

Guslandi M, Mezzi G, Sorghi M, Testoni PA (2000) Saccharomyces boulardii in maintenance treatment of Crohn's disease. Dig Dis Sci 45:1462–1464. https://doi.org/10.1023/A:1005588911207

Hayakawa M, Asahara T, Ishitani T, Okamura A, Nomoto K, Ganda S (2012) Synbiotic therapy reduces the pathological gram-negative rods caused by an increased acetic acid concentration in the gut. Dig Dis Sci 57(10):2642–2649

Hayat I, Ahmad A, Ahmed A, Khalil S, Gulfraz M (2014) Exploring the potential of red kidney beans (Phaseolus vulgaris L.) to develop protein based product for food applications. J Anim Plant Sci 24(3):860–868

Hickson M (2011) Probiotics in the prevention of antibiotic-associated diarrhea and Clostridium difficile infection. Ther Adv Gastroenterol 4:185–197

Inchingolo F, Santacroce L, Cantore S, Ballini A, Del Prete R, Topi S, Saini R, Dipalma G, Arrigoni R (2019) Probiotics and EpiCor® in human health. J Biol Regul Homeost Agents 33:1973–1979

Isacco CG, Ballini A, De Vito D, Nguyen KCD, Cantore S, Bottalico L, Quagliuolo L, Boccellino M, Di Domenico M, Santacroce L et al (2021) Rebalancing the oral microbiota as an efficient tool in endocrine, metabolic and immune disorders. Endocr Metab Immune Disord Drug Targets 21:777–784

Jacobs SE, Tobin JM, Opie GF et al (2013) Probiotic effects on late-onset sepsis in very preterm infants: a randomized controlled trial. Pediatrics 132:1055–1062

Jang YJ, Kim W-K, Han DH, Lee K, Ko G (2019) *Lactobacillus fermentum* species ameliorate dextran sulfate sodium-induced colitis by regulating the immune response and altering gut microbiota. Gut Microbes 10(6):696–711

Ji J, Jin W, Liu SJ, Jiao Z, Li X (2023) Probiotics, prebiotics, and postbiotics in health and disease. MedComm (2020) 4(6):e420

Karoline Ferreira Leite A, Vidal Fonteles T, Godoy Alves Filho E, da Silva A, Oliveira F, Rodrigues S (2023) Impact of orange juice containing potentially prebiotic ingredients on human gut microbiota composition and its metabolites. Food Chem 405:134706

Kaur H, Kaur G, Ali SA (2022) Dairy-based probiotic-fermented functional foods: an update on their health-promoting properties. Fermentation 8:425

Kolida S, Tuohy K, Gibson GR (2000) The human gut flora in nutrition and approaches for its dietary modulation. Nutr Bull 25:223–231

Kopacz K, Phadtare S (2022) Probiotics for the prevention of antibiotic-associated diarrhea. Healthcare 10:1450

Kuitunen M, Kukkonen K, Juntunen-Backman K, Korpela R, Poussa T, Tuure T, Haahtela T, Savilahti E (2009) Probiotics prevent IgE-associated allergy until age 5 years in cesarean-delivered children but not in the total cohort. J Allergy Clin Immunol 123(2):335–3341

Lee CH, Choi Y, Seo SY, Kim SH, Kim IH, Kim SW, Lee ST, Lee SO (2021) Addition of probiotics to antibiotics improves the clinical course of pneumonia in young people without comorbidities: a randomized controlled trial. Sci Rep 11(1):926

Li H-Y, Zhou D-D, Gan R-Y et al (2021) Effects and mechanisms of probiotics, prebiotics, synbiotics, and postbiotics on metabolic diseases targeting gut microbiota: a narrative review. Nutrients 13(9)

Liong MT (2007) Probiotics: a critical review of their potential role as antihypertensives, immune modulators, hypocholesterolemics, and perimenopausel treatments. Nutr Rev 65:316–328

Lovell DMW, Taylor J, Zwart A, Helliwell C (2010) Compositions can constraints on omics data lead analyses astray? CSIRO Report Number: EP10994. CSIRO, Dickson, ACT

Ma T, Shen X, Shi X et al (2023) Targeting gut microbiota and metabolism as the major probiotic mechanism—an evidence-based review. Trends Food Sci Technol 138:178–198

Madjd A (2016) Comparison of the effect of daily consumption of probiotic compared with low-fat conventional yogurt on weight loss in healthy obese women following an energy-restricted diet: a randomized controlled trial. Am J Clin Nutr 103(2):323–329

Maleki D, Homayouni A, Khalili L, Golkhalkhali B (2016) Probiotics in cancer prevention, updating the evidence. In: Preedy VR, Watson RR (eds) Probiotics, prebiotics, and synbiotics. Academic, London, UK.

Marilyn GF, Kuntz BSM, Trindade E (2014) A systematic review and meta-analysis of the prebiotics and synbiotics effects on glycaemia, insulin concentrations and lipid parameters in adult patients with overweight or obesity. Clin Nutr 1–14

McAllister TA, Bae HD, Jones GA, Cheng KJ (1994) Microbial attachment and feed digestion in the rumen. J Anim Sci 72:3004–3018

Pacifico L, Osborn JF, Tromba V, Romaggioli S, Bascetta S, Chiesa C (2014) Helicobacter pylori infection and extragastric disorders in children: a critical update. World J Gastroenterol 20:1379–1401

Parada VD, De la Fuente MK, Landskron G, González MJ, Quera R, Dijkstra G, Harmsen HJM, Faber KN, Hermoso MA (2019) Short chain fatty acids (SCFAs)-mediated gut epithelial and immune regulation and its relevance for inflammatory bowel diseases. Front Immunol 10:277

Practice Guidelines of the World Gastroenterology Organization (2010. [(Accessed on 7 December 2022)) Probiotics and prebiotics, nutritional therapy & metabolism—SINPE News, pp 1–22

Quin C, Estaki M, Vollman DM, Barnett JA, Gill SK, Gibson DL (2018) Probiotic supplementation and associated infant gut microbiome and health: a cautionary retrospective clinical comparison. Sci Rep 8:8283

Rastall R, Maitin V (2002) Prebiotics and synbiotics: towards the next generation. Curr Opin Biotechnol 13:490–496

Rautava S (2021) Chapter 6 – Early-life antibiotic exposure, the gut microbiome, and disease in later life. In: Koren O, Rautava S (eds) The human microbiome in early life. Academic, pp 135–153

Sanders ME, Lenoir-Wijnkoop I, Salminen S, Merenstein DJ, Gibson GR, Petschow BW, Nieuwdorp M, Tancredi DJ, Cifelli CJ, Jacques P, Pot B (2014 Feb) Probiotics and prebiotics: prospects for public health and nutritional recommendations. Ann N Y Acad Sci 1309:19–29

Sanders ME, Merenstein DJ, Reid G, Gibson GR, Rastall RA (2019) Probiotics and prebiotics in intestinal health and disease: from biology to the clinic. Nat Rev Gastroenterol Hepatol 16:605–616

Santacroce L, Charitos IA, Bottalico L (2019) A successful history: probiotics and their potential as antimicrobials. Expert Rev Anti-Infect Ther 17:635–645

Santacroce L, Inchingolo F, Topi S, Del Prete R, Di Cosola M, Charitos IA, Montagnani M (2021) Potential beneficial role of probiotics on the outcome of COVID-19 patients: an evolving perspective. Diabetes Metab Syndr 15:295–301

Saracino IM, Pavoni M, Saccomanno L, Fiorini G, Pesci V, Foschi C, Piccirilli G, Bernardini G, Holton J, Figura N et al (2020) Antimicrobial efficacy of five probiotic strains against *Helicobacter pylori*. Antibiotics 9:244

Schrezenmeir J, de Vrese M (2001) Probiotics, prebiotics, and synbiotics—approaching a definition. Am J Clin Nutr 73(2):361S–364S

Seki H, Shiohara M, Matsumura T, Miyagawa N, Tanaka M, Komiyama A, Kurata S (2003) Prevention of antibiotic-associated diarrhea in children by Clostridium butyricum MIYAIRI. Pediatr Int Off J Jpn Pediatr Soc 45:86–90. https://doi.org/10.1046/j.1442-200X.2003.01671.x

Senok AC, Ismaeel AY, Botta GA (2005) Probiotics: facts and myths. Clin Microbiol Infect 11(12):958–966

Slavin J (2013) Fiber and prebiotics: mechanisms and health benefits. Nutrients 5:1417–1435

Śliżewska K, Markowiak-Kopeć P, Śliżewska W (2020) The role of probiotics in cancer prevention. Cancers 13:20

Strisciuglio C, Vitale A, Perna F et al (2023) Bifidobacteria modulate immune response in pediatric patients with cow's milk protein allergy. Pediatr Res

Sun J, Buys N (2016) Glucose- and glycaemic factor-lowering effects of probiotics on diabetes: a meta-analysis of randomised placebo-controlled trials. Br J Nutr 7:1167–1177

Tidjani AM, Lagier JC, Raoult D (2016) Diet influence on the gut microbiota and dysbiosis related to nutritional disorders. Hum Microbiome J 1:3–11

Tomasik PJ, Tomasik P (2003) Probiotics and prebiotics. Cereal Chem 80(2):113

Tsang YH, Lamb A, Romero-Gallo J (2010) Helicobacter pylori CagA targets gastric tumor suppressor RUNX3 for proteasome-mediated degradation. Oncogene 29(41):5643–5650

Usami M, Miyoshi M, Kanbara Y, Aoyama M, Sakaki H, Shuno K, Hirata K, Takahashi M, Ueno K, Tabata S, Asahara T, Nomoto K (2011) Effects of perioperative synbiotic treatment on infectious complications, intestinal integrity, and fecal flora and organic acids in hepatic surgery with or without cirrhosis. J Parenter Enter Nutr 35(3):317–328

Villamiel M, Montilla A, Olano A, Corzo N (2014) Production and bioactivity of oligosaccharides derived from lactose. In: Moreno J, Sanz ML (eds) Food oligosaccharides: production, analysis and bioactivity. Wiley-Blackwell IFT Press, Chichester, pp 137–167

Wasilewski A, Zielin'ska M, Storr M, Fichna J (2015) Beneficial effects of probiotics, prebiotics, synbiotics, and psychobiotics in inflammatory bowel disease. Inflamm Bowel Dis 21:1674–1682

Whisner CM, Martin BR, Nakatsu CH, McCabe GP, McCabe LD, Peacock M, Weaver CM (2014) Soluble maize fibre affects short-term calcium absorption in adolescent boys and girls: a randomised controlled trial using dual stable isotopic tracers. Br J Nutr 112:446–456

Wu Y, Ye Z, Feng P et al (2021) *Limosilactobacillus fermentum* JL-3 isolated from "Jiangshui" ameliorates hyperuricemia by degrading uric acid. Gut Microbes 13(1):1897211

Wu M, Dong Q, Ma Y et al (2022) Potential antimicrobial activities of probiotics and their derivatives against *Listeria monocytogenes* in food field: a review. Food Res Int 160:111733

Xu T, Wu X, Liu J et al (2022) The regulatory roles of dietary fibers on host health via gut microbiota-derived short chain fatty acids. Curr Opin Pharmacol 62:36–42

Zabel B, Yde CC, Roos P et al (2019) Novel genes and metabolite trends in bifidobacterium longum subsp. infantis Bi-26 metabolism of human milk oligosaccharide 2′-fucosyllactose. Sci Rep 9(1):7983

Zeng J et al (2016) Effect of probiotics on the incidence of ventilator-associated pneumonia in critically ill patients: a randomized controlled multicenter trial. Intensive Care Med 42:1018–1028

Zhou C, Fang X, Xu J et al (2020) *Bifidobacterium longum* alleviates irritable bowel syndrome-related visceral hypersensitivity and microbiota dysbiosis via Paneth cell regulation. Gut Microbes 12(1):1782156

Zhu Y, Michelle Luo T, Jobin C, Young HA (2011) Gut microbiota and probiotics in colon tumorigenesis. Cancer Lett 309:119–127

Chapter 10
Food Bioactives in the Prevention and Clinical Management of Cardiovascular Diseases

Hmed Ben-Nasr, Bouzenna Hafsia, Arif J. Siddiqui, and Badraoui Riadh

Abstract Functional foods (FF) provide health promotion and benefits beyond their basic nutritive effects. FF contain biologically active compounds that have been used in traditional medicine including the prevention and clinical management of cardiovascular diseases (CVD). Furthermore, food is the most commonly used as effective biological products as it improves the normal physiological processes, compensates energy, development and growth as well as resolving cardiovascular pathological conditions or alleviating their symptoms. The potential effects of FF in both prevention and management of CDVs have gained high interest in the recent past. This chapter gives an overview of functional food with beneficial effects against CVD. It highlights the key role of FF in the neuroendocrine regulation of the cardiovascular system, the recent insights, applications, future challenges and perspectives about FF network in the health caring system.

Keywords Food bioactives · Nutraceuticals · Functional foods · Fenugreek · Flavonoids · Resveratrol · Angiotensin converting enzyme

---

H. Ben-Nasr
Department of Life Sciences, Faculty of Sciences of Gafsa, University of Gafsa, Gafsa, Tunisia

B. Hafsia
Laboratory of Biotechnology and Biomonitoring of the Environment and Oasis Ecosystems (LBBEEO), Faculty of Sciences of Gafsa, University of Gafsa, Gafsa, Tunisia

A. J. Siddiqui
Department of Biology, College of Science, University of Ha'il, Ha'il, Hail, Saudi Arabia

B. Riadh (✉)
Department of Biology, College of Science, University of Ha'il, Ha'il, Hail, Saudi Arabia

Histology-Cytology, Faculty of Medicine of Tunis, University of Tunis El Manar, Tunis, Tunisia
e-mail: riadh.badraoui@fmt.utm.tn

S. Amir Ashraf, M. Adnan (eds.), *Food Bioactives and Nutraceuticals*,
https://doi.org/10.1007/978-981-97-9723-3_10

10.1 Introduction

Along the mankind history, foods' acquisition was a central inquiry to sustain survival. As each individual needs appropriate basic nutrients such as carbohydrates, proteins, vitamins and minerals, all populations' socioeconomic and cultures opted various ways to satisfy these physiological demands. Since the discovery of vitamins and their role in preventing the development of several diseases by Funk (1912) in the last century, huge plethora of studies focused on defining what is a functional foods (FF), but neither universal nor conclusive definitions have been established yet (Alongi and Anese 2021). However, all scientific communities agree for the role of FF to improve the normal physiological processes. It also prevent and curing several pathological diseases (Ullah et al. 2020; Mutha et al. 2021; Scalbert et al. 2005). Herein, we will consider the first official Japanese definition postulating that a FF or a "Food for Specified Health Use" (FOSHU) designates a food that contains ingredient with functions for health and officially approved to claim their physiological effects on the human body (Martirosyan and Singh 2015). In addition to nutrients essential for compensating energy, development and growth, FF might be enriched in some bioactive molecules that resolve pathological conditions or alleviate their symptomatic tables. Outstandingly, this claims their potential to modulate subsets of pathological and/or physiological processes in favor of sustaining health. Among bioactive molecules with medicinal interest, phytochemicals derived from plants like polyphenols and flavonoids have been thoroughly studied and were proved to be effective in treating and preventing several diseases including cardiovascular diseases (CVD). The pharmacodynamics of FF or their active constituents might include direct effects and interactions with specific cytoplasmic or nuclear components of the targeted cell to modify its function and/or survival. For instance, the ascorbic acid, also known as vitamin C, constitutes a well- known nutrient that acts as co-factor for normal physiological activities of hydroxylase and monooxygenase, and used against many health disorders. Therefore, orange and lemon might be considered as FF to help treating several health disorders including covid-19 (Bae and Kim 2020; Jafari et al. 2019). Evenly, in addition to some toxic features that can be attributed to specific components or "posology", a FF combines multiple substances with probable synergistic or antagonistic interference for their targeted site of action (Erskine et al. 2022; Joshi et al. 2022). For that raison, the choice of a specific FF to counteract a health disorder should undertake a rough knowledge of its composition, bioavailability and pharmacology (Badraoui et al. 2023; Bédoui et al. 2024; Harrabi et al. 2024).

CVDs are multifactorial and remain a leading cause of death, worldwide. Hard works are spent to find adequate strategies to treat and prevent CVD (Vaduganathan et al. 2022). Dependently to the pathological nature of the CVD and the patient's status, the mechanistic scheme of the therapy includes inhibitors of angiotensin converting enzyme (ACE), antagonists of platelets and thrombois formation, ion channels' blocker; diuresis enhancers, stabilizer of cardiac rhythmicity and other forms of pharmacological pathways (Kesselheim et al. 2008; Manzoli et al. 2016).

In the approach of FF, the term "cardioceutical" has been coined to describe foods with interest in treating cardiovascular diseases (Table 10.1). They might contain some essential nutritional elements that offer protection against common CVD (Jain et al. 2015). This chapter gives an overview of functional food with beneficial effects against cardiovascular diseases.

10.2 Functional Food and Neuroendocrine Regulation of the Cardiovascular System

The cardiovascular system sustains survival by ensuring blood flow that supplies organs and tissues with necessary metabolites and respiratory gazes and permits, in association to kidneys, the removal of toxic metabolic wastes and xenobiotic. It comprises the heart and blood vessels that firmly interact with other organs, in particular the lungs and kidneys. The CVS responds to each minimal stressor to adapt metabolic exigence, and its function is under tight regulation by both nervous and endocrine networks. This fantastic hierarchized control involves, an autonomous neurogenic regulator (pacemaker), two antagonist innervations, namely the sympathetic and para-sympathetic neural pathways, the renin- angiotensin- aldosterone system, and many other endogenous regulators such as insulin and thyroxines. Several exogenous substances like dietary electrolytes (in particular, potassium and sodium ions) deeply affect the CVS function. Distortion of structures or functions of the CVS results in several related diseases collectively termed "cardiovascular diseases" (CVD). CVD encompasse atherosclerosis, myocardial infarction, stroke, rheumatic heart disorders, cardiac arrhythmias, venous blood clotting, pulmonary embolism and many other pathological forms and subtypes (Thiriet and Thiriet 2018). CVD were responsible of about one third of the overall global deaths, in the World in 2019 (Roth et al. 2020; Vaduganathan et al. 2022). Their management still causes heavy socioeconomic burden and remains a big challenge facing clinicians and scientific communities. Recent researches provided evidence for the important role of FF in preventing and treating subsets of CVDs. Bioactive components of FF are suggested to modify with one or more subsets of the regulatory processes of CVD.

10.3 Functional Food Network in Health-Caring Systems

In general, a single food contains several chemicals such as carbohydrates, proteins and minerals (Badraoui et al. 2023; Bardakci et al. 2023; Akacha et al. 2022). Such composition is necessary to sustain homeostasis and growth; and to compensate the used energy. Other elements, however, are of great interest such as enzymes' co-factors, stimulator or inhibitors of some physiological processes, and might also serve in preventing and healing many pathological situations. Because of that, the

Table 10.1 Some well-known functional foods, their major active phytochemicals, and the confirmed beneficial effects on the cardiovascular system

Food name	Image	Active compounds	Bioactivities	References
Tomatoes		β-carotenoids, lycopene, vitamins C and E, phenolic compounds, α-Tomatine, Tomatidine	↑ Antioxidant capacity, improve endothelial function, improve blood pressure, improve blood lipids	Friedman (2013)
Olive oil		Mono-unsaturated fatty acids, healthy fats, tocopherols, antioxidants	↓ Inflammation, healing, antioxidant, ↓ risk of developing CVD, ↓ risk of dying from CVD	Guasch-Ferré et al. (2022); Nocella et al. (2018)
Dates		Polyphenols, flavonoids	Maintain heart health, ↓ CVD, ↓ triglycerides, ↓ cholesterol levels	Al-Dashti et al. (2021)
Sea food: Salmon, Tuna, Sardines, Trout		Omega-3 fatty acids, unsaturated fats,	Support cardiovascular health, ↓ sudden cardiac death, ↓ Triglycerid levels, ↑ high density lipoprotein (HDL), ↓ low density lipoprotein (LDL), ↓ congestive heart failure, ↓ coronary heart disease, ↓ ischemic stroke	Rimm et al. (2018); Kraiem et al. (2024)
Nuts		Polyphenols, omega-3 essential fatty acid, α-linolenic acid, L-arginine, unsaturated fats, magnesium,	↓ LDL, ↓ stroke risk, support heart health, ↓ blood pressure, improve health of blood vessel	Ivey et al. (2021); Rajaram et al. (2021)

(continued)

Table 10.1 (continued)

Food name	Image	Active compounds	Bioactivities	References
Flax seeds		Omega-3 fatty acids, polyunsaturated fatty acids, hytoestrogens	Boost immunity, improve heart health, cardioprotective effects, ↓ hypertension, ↓ platelet aggregation, ↓ atherosclerosis, ↓ arrhythmias	Parikh et al. (2018)
Orange and lemon fruits: Oranges, papaya, cantaloupe, grapefruit		Antioxidants, vitamin C, bioflavonoids, hesperidin, potassium, magnesium	↓ Oxidation, anti-inflammatory, help the heart, boost immune system, ↓ CVD, ↓ ischemic stroke, improve blood pressure, better blood lipids	Aune et al. (2017); Tadros and Andrade (2021)
Oatmeal		Manganese, magnesium, phosphorous, iron, copper, zinc, vitamin B1	Maintain heart health, maintain vascular integrity	Hasler et al. (2000)
Lentils		Antioxidants, magnesium, potassium	Support heart health, ↓ CVD risks, ↓ stroke risk	Ganesan and Xu (2017)
Garlic		Sulfur-containing compounds, flavonols/ flavonones, flavonoids, terpenes	↑ Antioxidant capacity, healing, ↓ blood pressure, ↓ lipid levels, ↓ glycemia	Badraoui et al. (2020, 2023); Saoudi et al. (2021)
Avocado		Oleic acid, vitamins, monosaturated oils, folate, potassium, magnesium	↓ Risk of CVD, manage cholesterol, improve heart health	Pacheco et al. (2022)
Dark chocolate		Polyphenols, flavonoids, manganese, copper, procyanidin, catechin, epicatechin	↓ LDL, ↑ HDL, ↓ CVD, regulate blood pressure	Yang et al. (2024)

(continued)

Table 10.1 (continued)

Food name	Image	Active compounds	Bioactivities	References
Tea (true tea: green, black and oolong)		Polyphenols, catechins (like EGCG)	↓ LDL, slow the natural ↓ in HDL, ↓ CVD, ↓ atherosclerosis, ↓ inflammation, ↓ blood pressure, ↓ cardiovascular mortality rates	Huang et al. (2018); Wang et al. (2020); Bhardwaj and Khanna (2013); Soussi et al. (2022)
Sorghum and other whole grains		Magnesium, phenolic compounds	Anti-inflammatory, anti-atherogenic effects, improve total cholesterol, ↓ LDL, improve C-reactive protein and hemoglobin A1c	Satija and Hu (2018)
Berries		Antioxidants, vitamin C, vitamin K	↓ LDL, antioxidants, ↓ oxidation, ↓ blood pressure, ↓ arteries stiffness, ↓ risk factor of heart disease, maintain systolic blood pressure	Huang et al. (2016)

hallmark of an FF beneficial effects involves various mechanisms of actions and offers an important scaffold for the mechanistic schemes to counteract several desirable effects. Nevertheless, the detailed mechanisms of action are not- fully understood and might involve divergent pharmacological pathways, like antioxidant, anti-inflammatory, and antibacterial infections. Most of these bioactivities are commonly due to their rich content of chemical class of phenols.

10.3.1 ACE Inhibition by Food and It's Derived Bioactive Compounds

The cardiovascular function is under tight control of the renin- angiotensin system (RAS) that links it to kidneys, lungs and many other tissues. The renin is secreted by the juxtaglomural cells, and cleaves the angiotensinogen peptide generated by hepatic tissues into angiotensin I which is lately submitted to further sequential proteolytic action to generate other forms, namely angiotensin II to X. Angiotensin II is the most known active peptide resulting from such reaction through either enzymatic or non-enzymatic proteolysis by the angiotensin converting enzyme

(ACE). Angiotensin II presents a potent vasoconstrictor effect leading to levelling up of the blood pressure and heart rate (Campbell 1987; Velez 2009; Badraoui et al. 2021). Clinical investigations provided evidence for the significant role of the polymorphism and expression of ACE and its receptors in many pulmonary and CVDs like stroke and heart failure. (Badraoui et al. 2021, 2022; Goodfriend et al. 1996). ACE forms an active and well known target to offset CVDs. Several natural substances contained in foods have been proved to counteract hypertension through their actions on ACE and its vasoconstrictor pathways (Udenigwe and Mohan 2014; Lakhrem et al. 2024). In agreement to findings of many scientific contributors (Cho et al. 2015; Balaraman et al. 2006; Dib et al. 2017), in our previous studies we evaluated the effect of some culinary and medicinal plants like fenugreek (*Trigonella faenum- graecum*) and *Artemisia campestris* on hypertension. These two plants, which are used for their flavoring or medicinal benefits, showed relevant anti-hypertensive effect in both experimental and clinical studies (Ben-Nasr et al. 2014; Ben-Nasr et al. 2021). The intake of one or the other of these herbs results in a significant decrease of ACE release and activity (Hamden et al. 2017). While the pragmatic mechanistic pathway remains unclear, these anti-hypertensive effects might be attributed to their richness in several phytochemicals. Computational simulating studies of artemisinin and 4′-O-β-d-glucopyranoyl (E)-4-hydroxy-3-methylbut-2--enyl benzoate, two sesquiterpene extracted from Artemisia, revealed their interaction with ACE receptor, with varying binding energies dependently to its allelic polymorphism (Badraoui et al. 2022; Cho et al. 2015). These findings evidenced the potential for this sesquiterpene to competitively displace ACE and inhibit its binding to its specific receptor. Furthermore, polysaccharides isolated from the same genus were proven to inhibit ACE (Song et al. 2022). Diosgenin, is an active steroidal saponin, found in fenugreek seeds. It was evidenced to ameliorate treatment of CVD and other related pathological disorders such as dyslipidemia, renal failure and diabetes (Manivannan et al. 2013; Gan et al. 2020). It resuscitates atherosclerosis (Wu and Jiang 2019; Wang and Wang 2022) and hypertension (Manivannan et al. 2015). In addition to its inhibiting effect of inflammation (Tsukayama et al. 2021), diosgenin acts through depletion of ACE synthesis and activity, and enhancing the endogenous nitric oxide synthase expression (Manivannan et al. 2013). Perhaps, the potential of diosgenin to compete with ACE to its receptor constitutes a prominent form of its mechanism of action (Novianty et al. 2021). Further foods did also present such capacity such as the common barley (*Hordeum vulgare*) that contains various polyphenols; lutonarin, saponarin, isorientin and orientin (Ra et al. 2020) and several protein fractions (Gangopadhyay et al. 2016), which possessed ACE inhibiting activities.

10.3.2 Functional Food and Cardiovascular Electrophysiology

The blood-pumping function of the heart relies on ion channels activities that monitor myocardial contractility and rhythmicity. On the electrocardiograph, each phase of the cardiac contraction is a manifestation of different types of ions' currents through the membrane of the cardiac myocytes, such as sodium, potassium and calcium. These trans-membranous pores are of various pharmacological subtypes that differ in their localization, chronicity, signal operating gateways, and evenly function. Some of them, are targeted by pharmacological substances like calcium channels blockers and α and β blockers to resuscitate arrhythmia, atrial fibrillation and many other conditions of CVDs (Garcia-Elias and Benito 2018; Grant 2009; Priest and McDermott 2015). Ions' channels activities are involved in cardiac regulation, through two sets of modus operandi: the autonomic pacemaker and the peripheral innervations of the heart originating from the central nervous system. The cardiac automatism depends on voltage gated sodium and potassium channels and ryanodine receptor-coupled calcium channels. External innervation that responds to ion currents fluctuations, too, mediate the heart control through the liberation of synaptic neurotransmitters (acetylcholine, somatostatin, vasoactive intestinal peptide, nitric oxide; norepinephrine, calcitonin gene-related peptide and neuropeptide Y) (Palma and Benarroch 2014). Several scientific investigations revealed the importance of many substances, which are mainly contained in plant-originating foods, in modifying ion currents and the electrophysiological activity of injured heart. In particular, phenolic compounds have had gained much attention in such issue (Yan et al. 2023). Flavonoids have been thoroughly studied for their anti-hypertensive and cardiovascular protecting effects and were probed to modify calcium and potassium traffic bypassing cardio-myocyte membrane. However, their real mechanisms of action remain blurred. Carullo and her colleagues (2020) studied the potential of quercetin, a ubiquitous flavonoid in vegetable dietary, and its derivatives in potassium channels, and brought approval for these derivatives allosteric interactions with subunits of gated ionophores of the cardiovascular system. The natural flavonol was able to stimulate both the vasorelaxant calcium-activated potassium channel (Kca.1.) (Carullo et al. 2020) and the vasoconstrictor voltage-activated calcium channel (Cav1.2.) (Trezza et al. 2022) of the vascular bed. All of quercetin substitutes perform tight docking to the sixth segment of the pore-forming protein of the potassium channel, thus confirming their allosteric effects on the opening-closing status of the pore and the subsequent changes in the ion current (Carullo et al. 2020). Further experimental studies showed that galangin, quercetin, kaempferol, luteoli and many other flavonoids dock to the inward rectifying potassium channels and thereby inhibits smooth muscles contraction (Pettini et al. 2024). According to Oliván-Viguera et al. (2013), small to intermediate conductance of calcium activated potassium channels (Kca3.1 and Kca2.3) are also targeted by some polyphenolic compounds such as caffeic acid, vanillic acid and gallic acid. Ionophores-targeting by polyphenols was also proved for voltage-gated sodium channels by substances occurring in red grape. Evidences of such ionotropic have also been approved for the fenugreek isolated-diosgenin that protect against cardiac ischemia by modifying the ionic current through the ATP-sensitive potassium channel in mitochondria (Badalzadeh et al. 2015). Simultaneously, it also stimulates the neuronal calcium activated-potassium channels in the central nervous system (Wang et al. 2006).

Many authors used patch-clamp techniques to explore the action of phenolic derivatives on sodium channels and brought proof for their inhibitory potential of Na+ currents in both skeletal muscle and nerve tissues (Haeseler et al. 2001, 2002). While there was a consent that phenolic compounds modify cardiovascular function through indirect pathway by involving the neurogenic regulation, quercetin, catechin and resveratrol were proved to inhibit sodium current bypassing the Nav1.5 channel in cardio-myocytes. As these compounds are highly present in red wine and grapes, this would explain in part the antihypertensive effect of red wine and grapes consumption (Wallace et al. 2006).

10.3.3 Functional Food and Neurogenic Regulation of the Cardiovascular System

Many authors reviewed the importance of naturally occurring substances in dietetic in preventing neurological disruptions and/or diseases (Andriantsitohaina et al. 2012; Rebas et al. 2020; Saoudi et al. 2019; Allagui et al. 2014). Among these substances, quercetin, melatonin and caffeic acid were shown to inhibit acetylcholinesterase activity (Scozzafava et al. 2015; Roseiro et al. 2012; Allagui et al. 2014) and subsequently might prolong the "journey" of acetylcholine and its effect on heart and vascular tissues. The intake of grape wine at low doses resulted in re-establishing hypertension (Diebolt et al. 2001; Protić et al. 2013). The main effect was attributed to the polyphenolic constituents that stimulate the release of nitric oxide through enhancing the inducible nitric oxide synthase expression (Bernátová et al. 2002; Diebolt et al. 2001). It is well known that resveratrol, a polyphenolic chemical isolated from red wine, inhibits the norepinephrine and electrically evoked contraction of the portal vein in a dose dependent-manner (Protić et al. 2013). Likewise, curcumin was reported to prevent cardiac hypertrophy (Ahuja et al. 2011; Thandapilly et al. 2011). These results bring probe for the neurogenic-mediated action of these components on resuscitating CVD. Further molecules, such as catechin and its derivatives like epicatechin-3-gallate, which is an abundant flavonoid in tea, reduced the circulating noradrenaline (Churm et al. 2023) and thereby have direct effect on the blood pressure. In this context, some authors attributed the anti-hypertensive action of polyphenols to their molecular interactions with the muscarinic receptors (Byung-Sik et al. 2008; Ghayur and Gilani 2006; Rattmann et al. 2009).

10.3.4 Other Mechanisms of Action of Functional Foods

Further beneficial effects against CVD have been approved for FF. They include anti-inflammation, anti-thrombosis and antioxidant activities (Yahfoufi et al. 2018; Tangney and Rasmussen 2013; Ludovici et al. 2018; Zujko et al. 2015). Among active nutrients, long chain omega-3 fatty acids, like eicosapentaenoic acid (EPA) and docosahexaeinoic acids (DHA), which are found in sea foods, gained much attention for their protective effect against CVD and were recommended as nutritional supply to treat and prevent the development of CVD. According to Von Schacky and Harris (2007), ω-3 polyunsaturated fatty acids-enriched diet significantly reduced the risk of sudden cardiac death by 90%. These substances are intermediate metabolites for cyclooxygenases (COXs), a prominent non- steroidal pathway to treat inflammation (Calder and Yaqoob 2009). They increase the synthesis of vasodilator prostaglandin E1 (PGE1) and I2 (PGI2) and the endothelial nitric oxidase (eNO). EPA and DHA did also antagonize arrhythmia and thrombosis (Das 2005). Further experimental findings revealed that both polyunsaturated fatty acids inhibit the expression of the vasoconstrictor COX 2 enzyme through a mechanism involving the NF-κB signaling pathway, in endothelial cells of the umbilical vein (Lee et al. 2009).

10.4 Conclusion

A cardioceutical food, in addition to its nutritional value might provide elements with various potentials to prevent and treat CVD. Several bioactive molecules contained in FF present different pharmacological pathways to counteract CVD and sustain CVS function. Some of these substances modify ion currents that govern cardiac contractility and rhythmicity. Others, however, are capable of inhibiting ACE expression and binding to its receptor and therefore breakdown its hypertensive effect. A third class of cardioceutical foods affect the neuro-hormonal regulation of heart function. Interestingly, some molecules exert their action through diverse pharmacological pathways to counteract CVD. To valorize the concept of FF in clinical practices, in particular the cardiovascular ones, a detailed pharmacological study of each food bioactive compound should be undertaken to enrich the pharmaceutical medication/supplementation to improve the efficacy against CVD.

References

Ahuja S, Kohli S, Krishnan S, Dogra D, Sharma D, Rani V (2011) Curcumin: a potential therapeutic polyphenol, prevents noradrenaline-induced hypertrophy in rat cardiac myocytes. J Pharm Pharmacol 63(12):1604–1612

Akacha A, Badraoui R, Rebai T, Zourgui L (2022) Effect of Opuntia ficus indica extract on methotrexate-induced testicular injury: a biochemical, docking and histological study. J Biomol Struct Dyn 40(10):4341–4351

Al-Dashti YA, Holt RR, Keen CL, Hackman RM (2021) Date palm fruit (Phoenix dactylifera): effects on vascular health and future research directions. Int J Mol Sci 22(9):4665. https://doi.org/10.3390/ijms22094665

Allagui MS, Feriani A, Saoudi M, Badraoui R, Bouoni Z, Nciri R, Murat JC et al (2014) Effects of melatonin on aluminium-induced neurobehavioral and neurochemical changes in aging rats. Food Chem Toxicol 70:84–93

Alongi M, Anese M (2021) Re-thinking functional food development through a holistic approach. J Funct Foods 81:104466

Andriantsitohaina R, Auger C, Chataigneau T, Étienne-Selloum N, Li H, Martínez MC et al (2012) Molecular mechanisms of the cardiovascular protective effects of polyphenols. Br J Nutr 108(9):1532–1549

Aune D, Giovannucci E, Boffetta P et al (2017) Fruit and vegetable intake and the risk of cardiovascular disease, total cancer and all-cause mortality-a systematic review and dose-response meta-analysis of prospective studies. Int J Epidemiol 46(3):1029–1056. https://doi.org/10.1093/ije/dyw319

Badalzadeh R, Yousefi B, Tajaddini A, Ahmadian N (2015) Diosgenin-induced protection against myocardial ischaemia-reperfusion injury is mediated by mitochondrial KATP channels in a rat model. Perfusion 30(7):565–571

Badraoui R, Alrashedi MM, El-May MV, Bardakci F (2021) Acute respiratory distress syndrome: a life threatening associated complication of SARS-CoV-2 infection inducing COVID-19. J Biomol Struct Dyn 39(17):6842–6851

Badraoui R, Rebai T, Elkahoui S, Alreshidi M, Veettil VN, Noumi E et al (2020) *Allium subhirsutum* L. as a potential source of antioxidant and anticancer bioactive molecules: HR-LCMS phytochemical profiling, in vitro and in pharmacological study. Antioxidants 9(10):1003. https://doi.org/10.3390/antiox9101003

Badraoui R, Saoudi M, Hamadou WS, Elkahoui S, Siddiqui AJ, Alam JA et al (2022) Antiviral effects of Artemisinin and its derivatives against SARS-CoV-2 main protease: computational evidences and interactions with ACE2 allelic variants. Pharmaceuticals 15(2):129

Badraoui R, Siddiqui AJ, Bardakci F, Ben-Nasr H (2023) Ethnopharmacology and ethnopharmacognosy: current perspectives and future prospects. In: Ethnobotany and ethnopharmacology of medicinal and aromatic plants, pp 115–128. https://doi.org/10.1201/b22842-7

Bae M, Kim H (2020) The role of vitamin C, vitamin D, and selenium in immune system against COVID-19. Molecules 25(22):5346

Balaraman R, Dangwal S, Mohan M (2006) Antihypertensive effect of trigonella foenum-greacum. seeds in experimentally induced hypertension in rats. Pharm Biol 44(8):568–575

Bardakci F, Badraoui R, Saeed M, Siddiqui AJ, Hamadou WS, Alreshidi M et al (2023) Phytochemotaxonomy: role of phytochemicals in plant classification. In: Ethnobotany and ethnopharmacology of medicinal and aromatic plants, pp 165–194. https://doi.org/10.1201/b22842-10

Bédoui I, Nasr HB, Ksouda K, Ayadi W, Louati N, Chamkha M, Choura S et al (2024) Phytochemical composition, bioavailability and pharmacokinetics of Scorzonera undulata Methanolic extracts: antioxidant, anticancer, and apoptotic effects on MCF7 cells. Pharmacogn Mag 20(1):218–229

Ben-Nasr H, Harrabi B, Benabderrahim M-A, Turki M, Khaskhoussi A, Alaoui N, Hammami S, Ayedi F, Ksouda K (2021) Fenugreek nutritional supplement might improve diabetes mellitus associated to hypertension. Int J Diabetes 4(1):105–111

Ben-Nasr H, Salama M, Ksouda K, Zeghal KM, Hammami S (2014) Hemodynamic effects of aqueous extract of Artemisia campestris in adult men. J Appl Pharm Sci 4(4):038–042

Bernátová I, Pechánová O, Babál P, Kyselá S, Stvrtina S, Andriantsitohaina R (2002) Wine polyphenols improve cardiovascular remodeling and vascular function in NO-deficient hypertension. Am J Phys Heart Circ Phys 282(3):H942–H948

Bhardwaj P, Khanna D (2013) Green tea catechins: defensive role in cardiovascular disorders. Chin J Nat Med 11(4):345–353. https://doi.org/10.1016/S1875-5364(13)60051-5

Byung-Sik YU, Woo-Seok KO, Dong-Yoon LIM (2008) Inhibitory mechanism of polyphenol compounds isolated from red wine on catecholamine release in the perfused rat adrenal medulla. Biomol Ther 16(2):147–160

Calder PC, Yaqoob P (2009) Understanding omega-3 polyunsaturated fatty acids. Postgrad Med 121(6):148–157

Campbell DJ (1987) Circulating and tissue angiotensin systems. J Clin Invest 79(1):1–6

Carullo G, Ahmed A, Trezza A, Spiga O, Brizzi A, Saponara S et al (2020) Design, synthesis and pharmacological evaluation of ester-based quercetin derivatives as selective vascular KCa1. 1 channel stimulators. Bioorg Chem 105:104404

Cho JY, Park KH, Hwang DY, Chanmuang S, Jaiswal L, Park YK et al (2015) Antihypertensive effects of Artemisia scoparia Waldst in spontaneously hypertensive rats and identification of angiotensin I converting enzyme inhibitors. Molecules 20(11):19789–19804

Churm R, Williams LM, Dunseath G, Prior SL, Bracken RM (2023) The polyphenol epigallocatechin gallate lowers circulating catecholamine concentrations and alters lipid metabolism during graded exercise in man: a randomized cross-over study. Eur J Nutr 62(3):1517–1526

Das UN (2005) COX-2 inhibitors and metabolism of essential fatty acids. Med Sci Monit 11(7):RA233-7

Dib I, Tits M, Angenot L, Wauters JN, Assaidi A, Mekhfi H et al (2017) Antihypertensive and vasorelaxant effects of aqueous extract of Artemisia campestris L. from Eastern Morocco. J Ethnopharmacol 206:224–235

Diebolt M, Bucher B, Andriantsitohaina R (2001) Wine polyphenols decrease blood pressure, improve NO vasodilatation, and induce gene expression. Hypertension 38(2):159–165

Erskine E, Gültekin Subaşı B, Vahapoglu B, Capanoglu E (2022) Coffee phenolics and their interaction with other food phenolics: antagonistic and synergistic effects. ACS Omega 7(2):1595–1601

Friedman M (2013) Anticarcinogenic, Cardioprotective, and other health benefits of tomato compounds lycopene, α-Tomatine, and Tomatidine in pure form and in fresh and processed tomatoes. J Agric Food Chem 61:9534–9550. https://doi.org/10.1021/jf402654e

Funk C (1912) The etiology of the deficiency diseases. J State Med (1912–1937) 20(6):341–368

Gan Q, Wang J, Hu J, Lou G, Xiong H, Peng C et al (2020) The role of diosgenin in diabetes and diabetic complications. J Steroid Biochem Mol Biol 198:105575

Ganesan K, Xu B (2017) Polyphenol-rich lentils and their health promoting effects. Int J Mol Sci 18(11):2390. https://doi.org/10.3390/ijms18112390

Gangopadhyay N, Wynne K, O'Connor P, Gallagher E, Brunton NP, Rai DK, Hayes M (2016) In silico and in vitro analyses of the angiotensin-I converting enzyme inhibitory activity of hydrolysates generated from crude barley (Hordeum vulgare) protein concentrates. Food Chem 203:367–374

Garcia-Elias A, Benito B (2018) Ion channel disorders and sudden cardiac death. Int J Mol Sci 19(3):692

Ghayur MN, Gilani AH (2006) Radish seed extract mediates its cardiovascular inhibitory effects via muscarinic receptor activation. Fundam Clin Pharmacol 20(1):57–63

Goodfriend TL, Elliott ME, Catt KJ (1996) Angiotensin receptors and their antagonists. Drug Therapy 335(25):1649–1654

Grant AO (2009) Cardiac ion channels. Circ Arrhythm Electrophysiol 2(2):185–194

Guasch-Ferré M, Li Y, Willett WC et al (2022) Consumption of olive oil and risk of total and cause-specific mortality among U.S. adults. J Am Coll Cardiol 79(2):101–112. https://doi.org/10.1016/j.jacc.2021.10.041

Haeseler G, Maue D, Grosskreutz J, Bufler J, Nentwig B, Piepenbrock S et al (2002) Voltage-dependent block of neuronal and skeletal muscle sodium channels by thymol and menthol. Eur J Anaesthesiol 19(8):571–579

Haeseler G, Piepenbrink A, Bufler J, Dengler R, Aronson JK, Piepenbrock S, Leuwer M (2001) Structural requirements for voltage-dependent block of muscle sodium channels by phenol derivatives. Br J Pharmacol 132(8):1916–1924

Hamden K, Keskes H, Elgomdi O, Feki A, Alouche N (2017) Modulatory effect of an isolated triglyceride from fenugreek seed oil on of α-amylase, lipase and ACE activities, liver-kidney functions and metabolic disorders of diabetic rats. J Oleo Sci 66(6):633–645

Harrabi B, Ben Nasr H, Amri Z, Brahmi F, El Feki A, Zeghal K, Ghozzi H et al (2024) Chemical composition, nutritional value, antioxidative, and in vivo anti-inflammatory activities of Opuntia Stricta cladode. ACS Omega, vol 9, pp 26724–26734

Hasler CM, Kundrat S, Wool D (2000) Functional foods and cardiovascular disease. Curr Atheroscler Rep 2:467–475. https://doi.org/10.1007/s11883-000-0045-9

Huang S, Li J, Wu Y et al (2018) Tea consumption and longitudinal change in high-density lipoprotein cholesterol concentration in Chinese adults. J Am Heart Assoc 7(13):e008814. https://doi.org/10.1161/JAHA.118.008814

Huang H, Chen G, Liao D et al (2016) Effects of berries consumption on cardiovascular risk factors: a meta-analysis with trial sequential analysis of randomized controlled trials. Sci Rep 6:23625. (2016). https://doi.org/10.1038/srep23625

Ivey KL, Nguyen XT, Quaden RM et al (2021) Association of nut consumption with risk of stroke and cardiovascular disease: the million veteran program. Nutrients 13(9):3031. https://doi.org/10.3390/nu13093031

Jafari D, Esmaeilzadeh A, Mohammadi-Kordkhayli M, Rezaei N (2019) Vitamin C and the immune system. In: Nutrition and immunity, pp 81–102

Jain AP, Aggarwal KK, Zhang PY (2015) Omega-3 fatty acids and cardiovascular disease. Eur Rev Med Pharmacol Sci 19(3):441–445

Joshi T, Deepa PR, Sharma PK (2022) Effect of different proportions of phenolics on antioxidant potential: pointers for bioactive synergy/antagonism in foods and nutraceuticals. Proc Natl Acad Sci India Sect B Biol Sci 92(4):939–946

Kesselheim AS, Misono AS, Lee JL, Stedman MR, Brookhart MA, Choudhry NK, Shrank WH (2008) Clinical equivalence of generic and brand-name drugs used in cardiovascular disease: a systematic review and meta-analysis. JAMA 300(21):2514–2526

Kraiem M, Ben Hamouda S, Eleroui M, Ajala M, Feki A, Dghim A et al (2024) Anti-inflammatory and immunomodulatory properties of a crude polysaccharide derived from green seaweed Halimeda tuna: computational and experimental evidences. Mar Drugs 22(2):85

Lakhrem M, Eleroui M, Boujhoud Z, Feki A, Dghim A, Essayagh S, Hilali S et al (2024) Anti-vasculogenic, antioxidant, and anti-inflammatory activities of sulfated polysaccharide derived from Codium tomentosum: pharmacokinetic assay. Pharmaceuticals 17(6):672

Lee SA, Kim HJ, Chang KC, Baek JC, Park JK, Shin JK et al (2009) DHA and EPA down-regulate COX-2 expression through suppression of NF-κB activity in LPS-treated human umbilical vein endothelial cells. Korean J Physiol Pharmacol 13(4):301–307

Ludovici V, Barthelmes J, Nagele MP, Flammer AJ, Sudano I (2018) Polyphenols: anti-platelet nutraceutical? Curr Pharm Des 24(2):146–157

Manivannan J, Balamurugan E, Silambarasan T, Raja B (2013) Diosgenin improves vascular function by increasing aortic eNOS expression, normalize dyslipidemia and ACE activity in chronic renal failure rats. Mol Cell Biochem 384:113–120

Manivannan J, Shanthakumar J, Silambarasan T, Balamurugan E, Raja B (2015) Diosgenin, a steroidal saponin, prevents hypertension, cardiac remodeling and oxidative stress in adenine induced chronic renal failure rats. RSC Adv 5(25):19337–19344

Manzoli L, Flacco ME, Boccia S, D'Andrea E, Panic N, Marzuillo C et al (2016) Generic versus brand-name drugs used in cardiovascular diseases. Eur J Epidemiol 31:351–368

Martirosyan DM, Singh J (2015) A new definition of functional food by FFC: what makes a new definition of functional food unique? Funct Food Health Dis 5:209–223

Mutha RE, Tatiya AU, Surana SJ (2021) Flavonoids as natural phenolic compounds and their role in therapeutics: an overview. Future J Pharm Sci 7:1–13

Nocella C, Cammisotto V, Fianchini L, D'Amico A, Novo M, Castellani V, Stefanini L, Violi F, Carnevale R (2018) Extra virgin olive oil and cardiovascular diseases: benefits for human health. Endocr Metab Immune Disord Drug Targets 2018(18):4–13. https://doi.org/10.2174/1871530317666171114121533

Novianty R, Ananta S, Karim MA (2021) Herbal plants from Riau Province as inhibitors of COVID-19 binding to ACE2 receptor by computer aided molecular design an in-silico method. J Phys Conf Ser 1788(1):012019. IOP Publishing

Oliván-Viguera A, Valero MS, Murillo MD, Wulff H, García-Otín ÁL, Arbonés-Mainar JM, Köhler R (2013) Novel phenolic inhibitors of small/intermediate-conductance Ca2+-activated K+ channels, KCa3. 1 and KCa2. 3. PloS One 8(3):e58614

Pacheco LS, Li Y, Rimm EB, Manson JE, Sun Q, Rexrode K, Hu FB, Guasch-Ferré M (2022) Avocado consumption and risk of cardiovascular disease in US adults. J Am Heart Assoc 2022(11):e024014. https://doi.org/10.1161/JAHA.121.024014

Palma JA; Benarroch EE (2014) Neural control of the heart: recent concepts and clinical correlations. Neurology 83(3):261–271

Parikh M, Netticadan T, Pierce GN (2018) Flaxseed: its bioactive components and their cardiovascular benefits. Am J Phys Heart Circ Phys 314:2–H159. https://doi.org/10.1152/ajpheart.00400.2017

Pettini F, Spiga O, Furini S, Fusi F (2024) Electrophysiology, molecular modelling, and functional analysis of the effects of dietary quercetin and flavonoid analogues on Kir6. 1 channels in rat stomach fundus smooth muscle. Biochem Pharmacol 220:115969

Priest BT, McDermott JS (2015) Cardiac ion channels. Channels 9(6):352–359

Protić D, Beleslin-Čokić B, Novaković R, Kanjuh V, Heinle H, Šćepanović R, Gojković-Bukarica L (2013) Effect of wine polyphenol resveratrol on the contractions elicited electrically or by norepinephrine in the rat portal vein. Phytother Res 27(11):1685–1693

Ra JE, Woo SY, Jin H, Lee MJ, Kim HY, Ham H et al (2020) Evaluation of antihypertensive polyphenols of barley (Hordeum vulgare L.) seedlings via their effects on angiotensin-converting enzyme (ACE) inhibition. Appl Biol Chem 63:1–9

Rajaram S, Cofán M, Sala-Vila A et al (2021) Effects of walnut consumption for 2 years on lipoprotein subclasses among healthy elders: findings from the WAHA randomized controlled trial. Circulation 144(13):1083–1085. https://doi.org/10.1161/CIRCULATIONAHA.121.054051

Rattmann YD, Crestani S, Lapa FR, Miguel OG, Marques MC, da Silva-Santos JE, Santos AR (2009) Activation of muscarinic receptors by a hydroalcoholic extract of Dicksonia sellowiana Presl. HooK (Dicksoniaceae) induces vascular relaxation and hypotension in rats. Vasc Pharmacol 50(1–2):27–33

Rebas E, Rzajew J, Radzik T, Zylinska L (2020) Neuroprotective polyphenols: a modulatory action on neurotransmitter pathways. Curr Neuropharmacol 18(5):431–445

Rimm EB, Appel LJ, Chiuve SE et al (2018) Seafood long-chain n-3 polyunsaturated fatty acids and cardiovascular disease: a science advisory from the American Heart Association. Circulation 138(1):e35–e47. https://doi.org/10.1161/CIR.0000000000000574

Roseiro LB, Rauter AP, Serralheiro MLM (2012) Polyphenols as acetylcholinesterase inhibitors: structural specificity and impact on human disease. Nutr Aging 1(2):99–111

Roth GA, Mensah GA, Johnson CO, Addolorato G, Ammirati E, Baddour LM et al (2020) Global burden of cardiovascular diseases and risk factors, 1990–2019: update from the GBD 2019 study. J Am Coll Cardiol 76(25):2982–3021

Saoudi M, Badraoui R, Chira A, Saeed M, Bouali N, Elkahoui S, Alam JM et al (2021) The role of *Allium subhirsutum* L. in the attenuation of dermal wounds by modulating oxidative stress and inflammation in Wistar albino rats. Molecules 26(16):4875. https://doi.org/10.3390/molecules26164875

Saoudi M, Salem RBSB, Salem MB, Brahmi N, Badraoui R, Nasri M et al (2019) Beneficial effects of *Crataegus oxyacantha* extract on neurobehavioral deficits and brain tissue damages induced by an insecticide mixture of deltamethrin and chlorpyrifos in adult *wistar* rats. Biomed Pharmacother 114:108795. https://doi.org/10.1016/j.biopha.2019.108795

Satija A, Hu FB (2018) Plant-based diets and cardiovascular health. Trends Cardiovasc Med 28(7):437–441. https://doi.org/10.1016/j.tcm.2018.02.004

Scalbert A, Manach C, Morand C, Rémésy C, Jiménez L (2005) Dietary polyphenols and the prevention of diseases. Crit Rev Food Sci Nutr 45(4):287–306

Scozzafava A, Kalın P, Supuran CT, Gülçin İ, Alwasel SH (2015) The impact of hydroquinone on acetylcholine esterase and certain human carbonic anhydrase isoenzymes (hCA I, II, IX, and XII). J Enzyme Inhib Med Chem 30(6):941–946

Song Q, Gu L, Wu H, Ma S, Kong L, Zhang K (2022) Chemical structure and ACE inhibitory activity of polysaccharide from Artemisia vulgaris L. J Mol Struct 1250:131896

Soussi A, Gargouri M, Magné C, Ben-Nasr H, Kausar MA, Siddiqui AJ et al (2022) (−)-Epigallocatechin gallate (EGCG) pharmacokinetics and molecular interactions towards amelioration of hyperglycemia, hyperlipidemia associated hepatorenal oxidative injury in in alloxan induced diabetic mice. Chem Biol Interact 368:110230. https://doi.org/10.1016/j.cbi.2022.110230

Tadros FJ, Andrade JM (2021) Impact of hesperidin in 100% orange juice on chronic disease biomarkers: a narrative systematic review and gap analysis. Crit Rev Food Sci Nutr 62:8335–8354. https://doi.org/10.1080/10408398.2021.1927976

Tangney CC, Rasmussen HE (2013) Polyphenols, inflammation, and cardiovascular disease. Curr Atheroscler Rep 15:1–10

Thandapilly SJ, Louis XL, Yang T, Stringer DM, Yu L, Zhang S et al (2011) Resveratrol prevents norepinephrine induced hypertrophy in adult rat cardiomyocytes, by activating NO-AMPK pathway. Eur J Pharmacol 668(1–2):217–224

Thiriet M, Thiriet M (2018) Cardiovascular disease: an introduction. In: Vasculopathies: behavioral, chemical, environmental, and genetic factors, pp 1–90

Trezza A, Spiga O, Mugnai P, Saponara S, Sgaragli G, Fusi F (2022) Functional, electrophysiology, and molecular dynamics analysis of quercetin-induced contraction of rat vascular musculature. Eur J Pharmacol 918:174778

Tsukayama I, Mega T, Hojo N, Toda K, Kawakami Y, Takahashi Y, Suzuki-Yamamoto T (2021) Diosgenin suppresses COX-2 and mPGES-1 via GR and improves LPS-induced liver injury in mouse. Prostaglandins Other Lipid Mediat 156:106580

Udenigwe CC, Mohan A (2014) Mechanisms of food protein-derived antihypertensive peptides other than ACE inhibition. J Funct Foods 8:45–52

Ullah A, Munir S, Badshah SL, Khan N, Ghani L, Poulson BG et al (2020) Important flavonoids and their role as a therapeutic agent. Molecules 25(22):5243

Vaduganathan M, Mensah GA, Turco JV, Fuster V, Roth GA (2022) The global burden of cardiovascular diseases and risk: a compass for future health. J Am Coll Cardiol 80(25):2361–2371

Velez JCQ (2009) The importance of the intrarenal renin–angiotensin system. Nat Clin Pract Nephrol 5(2):89–100

Von Schacky C, Harris WS (2007) Cardiovascular benefits of omega-3 fatty acids. Cardiovasc Res 73(2):310–315

Wallace CH, Baczkó I, Jones L, Fercho M, Light PE (2006) Inhibition of cardiac voltage-gated sodium channels by grape polyphenols. Br J Pharmacol 149(6):657–665

Wang X, Liu F, Li J et al (2020) Tea consumption and the risk of atherosclerotic cardiovascular disease and all-cause mortality: the China-PAR project. Eur J Prev Cardiol 27(18):1956–1963. https://doi.org/10.1177/2047487319894685

Wang D, Wang X (2022) Diosgenin and its analogs: potential protective agents against atherosclerosis. In: Drug design, development and therapy, vol 16, pp 2305–2323

Wang YJ, Liu YC, Chang HD, Wu SN (2006) Diosgenin, a plant-derived sapogenin, stimulates Ca2+-activated K+ current in human cortical HCN-1A neuronal cells. Planta Med 72(05):430–436

Wu FC, Jiang JG (2019) Effects of diosgenin and its derivatives on atherosclerosis. Food Funct 10(11):7022–7036

Yahfoufi N, Alsadi N, Jambi M, Matar C (2018) The immunomodulatory and anti-inflammatory role of polyphenols. Nutrients 10(11):1618

Yan Z, Zhong L, Zhu W, Chung SK, Hou P (2023) Chinese herbal medicine for the treatment of cardiovascular diseases—targeting cardiac ion channels. Pharmacol Res 106765:106765

Yang J, Zhou J, Yang J et al (2024) Dark chocolate intake and cardiovascular diseases: a Mendelian randomization study. Sci Rep 14:968. https://doi.org/10.1038/s41598-023-50351-6

Zujko ME, Witkowska AM, Waśkiewicz A, Piotrowski W, Terlikowska KM (2015) Dietary antioxidant capacity of the patients with cardiovascular disease in a cross-sectional study. Nutr J 14:1–13

Chapter 11
Food Bioactives in the Prevention and Clinical Management of Diabetes

Seyashree Hazra, Najmun Nahar, and Runu Chakraborty

Abstract Food bioactives compounds are naturally present in different food sources and are advantageous to human health, particularly in the prevention and control of diabetes. Diabetes is a persistent medical illness defined by the presence of high amounts of glucose in the bloodstream, and if not treated properly, it can give rise to significant health issues. However, the inclusion of specific dietary bioactives in the diet has the potential to enhance blood sugar regulation and promote general well-being. Cereals, pulses, fruits, and vegetables are great sources of bioactive compounds like flavonoids, vitamins, phenolics, phytosterols, carotenoids, and non-digestible carbohydrates. This study will highlight the beneficial effects of various bioactive compounds on the enhancement of pancreatic β-cell function, improvement of glucose tolerance, reduction of apoptosis, oxidative stress reduction, insulin signalling activation, stimulation of insulin secretion, and protection against the development of diabetes. The mechanism behind this improvement includes the modification of the cellular redox state and the control of cell signalling activity. Bioactive substances in food can be affected positively or negatively by cooking. Roasting and steaming preserve vitamins and bioactive compounds better than blanching, microwaving, and boiling. To retain these chemicals, preparation at the right temperature and time and handling the food properly to reduce nutritional loss is recommended. This chapter focuses on the various bioactive compounds present in food and their effects on diabetes management. The impact of cooking on the bioactive compounds in food is also taken into consideration.

Keywords Diabetes mellitus · Bioactive compounds · Flavonoids · Carotenoids · Cooking

---

S. Hazra · R. Chakraborty (✉)
Department of Food Technology and Biochemical Engineering, Jadavpur University, Kolkata, West Bengal, India

N. Nahar
Department of Food Science, Maulana Abul Kalam Azad University of Technology, Nadia, West Bengal, India
e-mail: najmun.nahar@makautwb.ac.in

S. Amir Ashraf, M. Adnan (eds.), *Food Bioactives and Nutraceuticals*,
https://doi.org/10.1007/978-981-97-9723-3_11

11.1 Introduction

A bioactive compound may be defined as a chemical substance that possesses the capacity and aptitude to engage with one or more constituents of live tissue to exhibit a diverse array of potential effects such as anti-oxidant, anti-carcinogenic, anti-inflammatory, anti-diabetic, and protective against cardiovascular, and liver disease. The sources of these chemicals may include plants, animals, and other sources. On the other hand, synthetic sources can be either partially or entirely manufactured (Guaadaoui et al. 2014). Diabetes is among the most prevalent metabolic disorders worldwide, and its incidence among adults has increased over the past few decades. Urbanization has significantly altered ways of life, especially in developing nations (Guariguata et al. 2014). Many nations are turning to alternative treatments for diabetes. Furthermore, there were 463 million individuals between the ages of 20 and 79 who were afflicted with diabetes, a figure projected to increase to around 700 million by the year 2045 (Galicia-Garcia et al. 2020). The majority of individuals with diabetes reside in low-income and middle-income nations, and the incidence of the disease will increase most rapidly in these regions over the next 22 years (Guariguata et al. 2014). T2DM (Type 2 diabetes mellitus) is a prevalent metabolic condition characterized by: impaired production of insulin and the inadequate response of insulin-sensitive tissues to insulin (Galicia-Garcia et al. 2020). Weyer et al. (1999) elucidated that T2DM arises from abnormalities in the functioning of insulin, both in terms of its action and secretion, as well as in the production of glucose by the body. The transition from normal to impaired glucose tolerance (IGT) was linked to increased body weight, decreased insulin-stimulated glucose elimination, and decreased acute insulin secretory response (AIR) to intravenous glucose. Phenolics, tannins, flavonoids, vitamins C and E, and other natural antioxidants found in many foods can help to keep β-cells functioning properly and lower blood sugar levels (Kooti et al. 2016). Marine algae produce bioactive substances, including fucoxanthin, phlorotannins, siphonaxanthin, astaxanthin, neoxanthin, sulfated fucose (fucoidan), and alginate and also produce secondary bioactive chemicals such as Dieckol, Dioxinodehydroeckol, Diphlorethohydroxycarmalol, and L-Fucose. Marine algae slow carbohydrate digestion and metabolism by inhibiting digestive enzyme activity. Administration of marine algae extract improves energy utilization and fat breakdown (Magwaza and Islam 2023).

The chapter aimed to highlight the bioactive compounds present in many foods and their role in the management of diabetes. In addition to consumer preferences, the choice of cooking method plays a significant role in influencing both the chemical composition of food and the consumption of bioactive compounds within typical dietary contexts.

11.2 Bioactive Compounds Present in Foods, Both Plant and Animal

11.2.1 Cereals and Pulses

Cereals are enriched with several bioactive compounds. The predominant free and bound phenolic acids identified in millets were gallic, protocatechuic, caffeic acids, ferulic, and coumaric acids (Subba Rao and Muralikrishna 2002). The study revealed that the antioxidant activity (AOA) of a mixture of free phenolic acids was greater in comparison to a mixture of bound phenolic acids. According to (Przybylska-Balcerek et al. 2019), it was mentioned that sorghum is a gluten-free grain that is cultivated across the globe, mostly in warm temperate, tropical, and subtropical countries. White sorghum grain has been shown to contain higher levels of flavonoids and phenolic acids, particularly ferulic acid. Furthermore, it was observed that the white sorghum grain had a higher concentration of lutein. On the other hand, red-grain sorghum contains zeaxanthin and essential minerals, specifically iron and manganese. The study of (Donkor et al. 2012) aimed to evaluate bioactives in freeze-dried ground samples of several grains. The amounts of polyphenols in the chosen germinated grains and non-germinated methanolic extract were analyzed, revealing the presence of high levels of catechin, ferulic acid, and luteolin in all grain extracts including barley, brown rice, buckwheat, rye, oat, wheat, and sorghum. Ferulic acid and luteolin were found in brown rice at 1.13 ± 0.17 μg/ml and 3.66 ± 1.11 μg/ml respectively. The soluble dietary fiber known as native mixed linkage β-glucan, found in cereals, exhibits rheological properties that are typically comparable to those of guar gum and other polysaccharides (Wood 2007). The majority of these substances are often found on the cell wall of the endosperm in oat and barley grains, accounting for approximately 75% of their composition. Additionally, a smaller proportion of these substances, approximately 10.4%, can be found in the bran. Barley and oat grains have been shown to possess the greatest β-glucan levels among various grains, with values ranging from 2 to 20 g per 100 g of dry weight for barley and 3 to 8 g per 100 g of dry weight for oat. The molecular weight of the β-glucan found in oats is greater than that of the β-glucan present in barley (Bozbulut and Sanlier 2019). Pulses are rich in bioactive substances such as polyphenols, phytosterols, and non-digestible carbohydrates, which perform vital physiological and metabolic roles. The content of these chemicals varies among pulse species and variations. Pulse seed coverings contain water-insoluble fibers and antioxidant (polyphenols), while soluble fibers, oligosaccharides, and resistant starch are present in cotyledons. The main phytosterols found in pulses are stigmasterol, campesterol, and sitosterol (Singh et al. 2017). The maximum polyphenols were recorded in green lentils (284.32 ± 3.26) mg/100 g in household cooking followed by canned brown lentils (Margier et al. 2018). It can be suggested that pulses be included in the diet as staple foods to provide health to the human body.

11.2.2 Fruits and Vegetables

Fruits and vegetables have a wide variety of health-promoting compounds, making them an important subject of study for their bioactivity, whether consumed fresh or processed. Fruits, vegetables, nuts, seeds, and spices are all dietary sources of flavonoids. Citrus fruits like oranges, grapes, limes, and lemons are great source of flavanones (naringenin, hesperetin) (Kozłowska and Szostak-Wegierek 2014). Fruits and vegetables contain varying amounts of phenolic acids such as salicylic acid (SA) and benzoic acid (BA). The most significant recorded level of SA was found in fava beans, followed by blueberries, lemons, kiwi, oranges, and fruit juice (lemon and blueberry) samples (Aresta and Zambonin 2016). Pumpkin includes several phyto-constituents, which may be classified as flavonoids, alkaloids, palmitic, oleic, and linoleic acids. The significant medical qualities also include anti-diabetic, anti-inflammatory, anti-carcinogenic, and antioxidant activity (Adams et al. 2011). Tomato puree is a good source of provitamin A and high in lycopene. Carotenes were discovered to predominate, accounting for 90–95% of the total amount of carotenoids, with higher levels of lycopene (Borguini and da Silva Torres 2009). Vitamin C, also known as an ascorbate is mostly found in tomatoes, and citrus fruits. Reactive oxygen species (ROS), which are produced by cellular metabolism, are scavenged by ascorbate in the majority of cellular processes. Phytosterols are abundant in brussels, broccoli, sprouts, spinach, and cauliflower. Plant sterols may have anticancer, antiatherosclerosis, anti-inflammatory, and antioxidant properties (Shashirekha et al. 2015). Cruciferous vegetables, such as cabbage, broccoli, cauliflower, and other members of the Brassicaceae, are abundant in polyphenolic compounds like Vitamin C and E, dietary fiber, as well as glycosides of flavonoids such as quercetin and kaempferol. Nevertheless, the key characteristic that renders cruciferous vegetables highly advantageous for mankind is the presence of sulfur-containing glycosides, specifically glucosinolates (GSLs) (Miękus et al. 2020). The most prevalent families in the banana peel are free sterols and steryl esters. Among these, 31-norcyclolaudenone, cycloeucalenone, cycloeucalenyl, cycloartenyl, and 24-methylenecycloartanyl palmitates are identified as the primary components (Oliveira et al. 2008). The analysis of four readily accessible amaranth cultivars demonstrated the existence of the three primary phytosterols (β-sitosterol, campesterol, stigmasterol). Total sterol of amaranth cultivars presented at a higher range between 543 μg/100 g and 834 μg/100 g than other plants (Marcone et al. 2003).

11.2.3 Milk and Milk-Derived Peptides

Several dietary proteins form functionally and physiologically active peptides after digestion and fermentation with lactic acid bacteria. Once released, bioactive peptides (BPs) perform several physiological activities in the body. Most milk protein bioactivities are latent but are fully activated during proteolytic breakdown, releasing encrypted bioactive peptides. Native milk proteins contain bioactive peptides in their amino acid sequences. Milk peptides are highly valued for their versatility in enhancing health in dietary and medicinal uses (Park and Nam 2015). New technologies including membrane separation, nanofiltration, and ultrafiltration will introduce newly identified peptides to the food processing industry. New macroencapsulation and nanoencapsulation technologies enhance peptide stability in meals and digestion (Dziuba and Dziuba 2014). About 80% of milk proteins are casein and 20% are whey. At present, milk proteins constitute the primary source of bioactive peptides. Research suggests that whey proteins and caseins can reduce the physiological consequences of T2DM by increasing insulin production and controlling blood glucose levels (Patil et al. 2015). Oligosaccharides in goats' milk is higher than cow and sheep milk. Goats' milk has 5–8 g L-1 oligosaccharides, substantially lower than mature human breast milk (Martinez-ferez et al. 2006). Prebiotic and anti-infective qualities of human milk oligosaccharides may benefit infants. Human milk is comprised of a variety of low molecular weight peptides (LMWPs), which have the potential to impact on infant health and development. The low molecular weight peptides (LMWPs) was mostly obtained from three distinct areas of β-casein (Xiao et al. 2022).

11.2.4 Spices, Herbs, and Oilseeds

The volatile oils of cinnamon leaves, fruit, root bark, flowers, and blooms have been extracted via supercritical fluid extraction and hydro or steam distillation. The chemical compositions of volatile oils were detected using GC and GC-MS. Leaf oil is mostly eugenol. Cinnamaldehyde and camphor are essential compounds of volatile oils found in stem and root. Fruits, flowers, and stalks contain mainly trans-cinnamyl acetate. These volatile oils possess anti-oxidant, anti-bacterial, and anti-diabetic properties. Proanthocyanidins were identified in *C. zeylanicum* bark and fruits (Jayaprakasha and Rao 2011). The chemical composition of *Origanum vulgare* consisted primarily of thymol (58.31%), carvacrol (16.11%), p-cymene (13.45%), and γ-terpinene (4.64%). The compounds exhibited inhibitory effects on the enzymes acetylcholinesterase, butrylcholinesterase, tyrosinase, α-amylase, and α-glucosidase (Sarikurkcu et al. 2015). TPC, TFC, and AOA (measured with the DPPH test and the FRAP assay) were all analyzed for nine commonly used spices in Indian cooking (Gupta 2013). Cloves performed best according to the TPC and TFC result. Spices showed DPPH radical scavenging action, with cinnamon having the highest activity in FRAP assay, followed by cloves, cumin, and star anise. Spices with high phenolic content demonstrated significant AOA in DPPH and FRAP experiments, thereby confirming the correlation between TPC and AOA. The significant level of activity may be ascribed to the existence of eugenol, linalool, methyl chavicol, and β-caryophyllene in cinnamon, as well as eugenol, β-caryophyllene, vanillin, crategolic acid, tannins, and flavonoids in cloves. Methanolic extracts of various flavonoid and phenolic-rich spices from the family Apiaceae included chlorogenic acid (CGA), which was the most abundant component. Only ajmod and kaala zeera had detectable amounts of protocatechuic acid, while kaala zeera and sauf contained syringic acid. Rutin emerged as the predominant flavonoid compound, with the highest concentration within the dhania extract at 417 mg per 100 g. The compound was also found in zeera, ajwain, and dill. Apigenin was exclusively identified in the zeera extract, with a concentration of 249 ± 0.74 mg per 100 g of extract (Pandey et al. 2012). The identification of phytosterols in sea buckthorn (*Hippophae rhamnoides* L.) seed oil was carried out using GC-MS and flame ionization detection (FID) techniques (Li et al. 2007). The compounds that were identified include α-amyrin, β-amyrin, sitostanol, clerosterol, campesterol, lanosterol, Δ^5-avenasterol, $\Delta^{24(28)}$-stigmasta-en-ol, $\Delta^{5,24(25)}$-stigmastadienol, lupeol, gramisterol, Δ^7-sitosterol, cycloartenol, cycloeucalenol, Δ^7-avenasterol, 24-methylene-cycloartanol, 28-methylobtusifoliol, erythrodiol, citrostadienol, oleanol aldehyde, and uvaol. The most significant phytosterols in terms of quantity were sitosterol and Δ^5-avenasterol. The most common phytosterol was β-sitosterol (54–86.0% of total) among all imported nuts and seed except in fennel seed, where stigmasterol was more prevalent (Islam et al. 2017). Pine nuts (45.5 mg/100 g) had the highest Δ^5-avenasterol concentration, fennel seed (57.6 mg/100 g) contained the highest stigmasterol, and chia seed (29.4 mg/100 g) showed the highest campesterol level. The extraction method was effectively utilized for the analysis of sunflower, canola, maize, soybean, and olive oils. Among these oils, olive oil exhibited the lowest levels of total phytosterols, whilst corn oil demonstrated the highest concentrations (Almeida et al. 2020).

11.3 Mechanism of Bioactive Compounds in the Prevention and Management of Diabetes

11.3.1 Phenolics

According to Dal and Sigrist (2016) various endogenous enzymes like NADPH oxidase, lipo-oxygenases (LPOs), cyclo-oxygenases (COXs), xanthine oxidase, nitric oxide synthases (NOS) are capable of generating free radicals in normal quantities. Obesity and diabetes are commonly associated with a persistent condition of oxidative stress, which refers to an imbalance in the redox equilibrium. This imbalance has been linked to the emergence and advancement of problems such as micro- and macro-angiopathies. Glucose, along with its metabolites, can undergo a reaction with H_2O_2 in the presence of iron and copper ions, resulting in the formation of hydroxyl radicals by auto-oxidation and contributing to diabetes-related complications. The predominant polyphenols found in the diet are phenolic acids, including benzoic and cinnamic acids, as well as flavonoids (Garcia-Salas et al. 2010; Alara et al. 2021). Many phenolics present in the diet are antioxidants that can quench reactive oxygen species (ROS) and harmful free radicals from lipid peroxidation, resulting in anti-inflammatory and antioxidant effects (Shashirekha et al. 2015). Phloroglucinol (1,3,5-trihydroxybenzene) monomer units are polymerized to create algal polyphenols, also known as phlorotannins. Phlorotannins are found mostly in brown seaweeds. TPC of seaweed extracts correlated strongly with DPPH and peroxyl radical scavenging activity, highlighting algal polyphenols' potential as chain-breaking antioxidants (Wang et al. 2009). The green lentil crude extracts were analyzed using HPLC-PAD and HPLC-ESI-MS techniques, which provided data indicating the presence of different classes of phenolic chemicals. Trans-p-coumaric acid was the most abundant among the free phenolic acids (Amarowicz et al. 2010). The principal antioxidants, polyphenols, work by giving up hydrogen to free radicals and reducing the risk of chronic diseases. HPLC analysis was performed on freeze-dried, pulverized grain samples of barley, brown rice, buckwheat, rye, oat, wheat, and sorghum (Donkor et al. 2012). The concentrations of polyphenols found in specific germinated cereals and non-germinated methanolic extract contain phenolic acids such as ferulic acid (FA), gallic acid (GA), and p-coumaric acid (PC). Flavonoids like catechin, epicatechin, epigallocatechin gallate and leteolin were also identified. Furthermore, the inhibitory effect on α-glucosidase activity was observed to be significant ($p < 0.05$) in germinated sorghum, with germinated rye exhibiting a similar effect. In contrast, barley, wheat, buckwheat, brown rice, and oats exhibited limited inhibitory effects. The non-germinated samples exhibited the same pattern of inhibition. The activity of α-amylase in degrading starch was shown to be considerably decreased in germinated grains when compared to the non-germinated samples. The germinated forms of barley and sorghum exhibited inhibitory effects on the activity of α-amylase. The concurrent inhibition of α-amylase and α-glucosidase is regarded as an efficacious approach to managing diabetes through the reduction of glucose absorption. Inhibiting the digestive enzymes α-glucosidase and α-amylase can effectively decrease the rise in blood glucose after digestion, making it a valuable approach for controlling blood glucose levels in people with T2DM (Teng and Chen 2017). A study on CGA derived from coffee (Williamson 2020) found that 2 mM CGA stimulated glucose uptake in skeletal muscle cells and inhibited rat hepatic glucose-6-phosphatase. However, this inhibition may not be meaningful *in vivo*. Plants store CGAs instead of free phenolic acid due to the inactivation of phenolic activity by linked quinic acid. Hence, it showed limited biological activity. First, metabolism begins with the elimination of quinic acid through hydrolysis by digestive enzymes or gut microbiota in the colon. After being activated, ensuing free phenolic acids may have biological effects. A possible indirect action of CGA metabolites is through GLP-1. Coffee affects plasma GLP-1, which affects β-cells via cAMP to regulate IDX-1, maintaining the β-cell response to glucose. Chlorogenic acid, a common polyphenol, possesses immune-protective, anti-oxidant, and anti-inflammatory effects. CGA dramatically reduced diabetic kidney damage, known as diabetic nephropathy. CGA pre-treatment boosted nuclear translocation of Nrf2 and HO-1 expression while decreasing IκB phosphorylation and NF-κB nuclear translocation. HO-1, a Nrf2-regulated antioxidant enzyme, controlled intracellular ROS levels in response to various stimuli and HO-1 has anti-inflammatory properties. HBZY-1 cells were dramatically impacted by Nrf2 siRNA and the HO-1 inhibitor zinc protoporphyrin (ZnPPIX), which boosted NF-κB nuclear translocation and pro-inflammatory cytokine production that are responsible for kidney damage in diabetes. Additionally, CGA therapy reduces pro-inflammatory cytokines in kidney injury of animal models and inhibits NF-κB transcription (Bao et al. 2018). Scazzocchio et al. (2015) found that protocatechuic acid (PCA) enhances GLUT4 translocation and glucose absorption in human adipocytes by stimulating the insulin signalling system. PCA increased AMPK activity and activated the insulin pathway via targeting insulin receptor tyrosine kinase activity. Eugenol efficiently controls high blood sugar levels by inhibiting the production of glucose in the liver through the modulation of the CAMKK-AMPK-CREB signalling pathway (Jeong et al. 2014). The improvement of high blood sugar levels by sinapic acid is mostly linked to the increased expression of GLUT4 and/or enhanced absorption of glucose through the PLC-PKC pathway in diabetic rats (Cherng et al. 2013). Hence, phenolic acids have the potential to be utilized as a promising option for the creation of an innovative therapy aimed at controlling diabetes.

11.3.2 Flavonoids

Multiple studies indicate that flavonoids derived from medicinal plants exhibit advantageous impacts on diabetes through enhancements in glucose level management, lipid profile, and antioxidant status. Rutin, a flavonoid present in numerous plant species, has a diverse array of biological actions (Ghorbani 2017). Flavonoids are present in fruits, vegetables, seeds, and derived foods like cocoa, coffee, tea, red wine, and soy-based foods. Flavonoids reduce insulin resistance in insulin-sensitive cells by regulating insulin signalling pathways. Flavonoid helps to maintain normal glucose levels by initiating glucose absorption in insulin-sensitive tissues and regulating hepatic glucose output and release. Epigallocatechin-3-O-gallate (EGCG), a flavonoid, enhance glucose absorption in L6 skeletal muscle cells via promoting GLUT4 translocation to the plasma membrane (Manavi et al. 2021). Flavonoids have been identified as agents capable of inhibiting the generation of free radicals by the process of chelating iron and copper ions. This chelation process enables flavonoids to directly scavenge ROS and other harmful free radicals, thereby impeding lipid peroxidation (Shashirekha et al. 2015). Shobha et al. (2013) examined the inhibitory effect of lipid peroxidation and anti-diabetic activity of aniseed (*Pimpinella anisum* L.) fractions obtained through sequential solvents *in vitro* methods and model systems. The study found that ethyl acetate fraction contains potent bioactive compounds like flavonoids, and flavonols that effectively inhibit $FeSO_4$-induced lipid peroxidation in liver homogenate. Polyphenols are known for their ability to prevent fatty acid peroxidation and protect against oxidative stress caused by free radicals. It also showed anti-diabetic effects in inhibiting α-amylase and α-glucosidase. α-amylase hydrolyzes α-1,4-glucosidic links in starch, glycogen, and oligosaccharides, while α-glucosidase simplifies disaccharides for intestinal absorption. The inhibition of the activity of these enzymes diminishes the absorption of glucose, and slows down hydrolyzation from starch. Rutin was found in dhania, jeera, and ajwain spices (Pandey et al. 2012). Ghorbani (2017) proposed the potential mechanisms underlying the anti-hyperglycemic activity of rutin. Rutin has been found to have several effects on carbohydrate metabolism. It has been observed to reduce the absorption of carbohydrates, promote the release of insulin, provide protection to the Langerhans islet against cellular breakdown, enhance glucose uptake by tissues, and inhibit gluconeogenesis in the liver. Both chronic hyperglycemia and dyslipidemia can cause tissue damage and diabetes complications due to alterations in intracellular metabolic pathways. Diabetes problems involving kidney, ocular, and neural tissue damage have been linked to protein glycation caused by hyperglycemia (Laurean and Pashikanti 2007). Overproduction of AGE precursors leads to glycation of intracellular proteins and alteration of extracellular matrix (ECM) components. AGE precursors may diffuse from cells and alter plasma proteins. The interaction between AGE plasma proteins and AGE receptors can induce the synthesis of growth factors and inflammatory cytokines, resulting in the development of vascular disease. Rutin protects against metabolic alterations caused by hyperglycemia and dyslipidemia (Ghorbani 2017). In the rat, the production of N^{ε} -fructoselysine (FL), a biomarker for early-stage glycation, was suppressed ($p < 0.05$) by the addition of G-rutin, a water-soluble glucose derivative of rutin (Nagasawa et al. 2003). Rutin protects against STZ-induced hepatic and cardiac toxicity in adult male Wistar rats by preventing changes in ALT, AST, and LDH levels in the liver and heart. Rutin significantly reduced triacylglycerol levels in the liver and heart and increased glycogen levels (Fernandes et al. 2010). Oligonol is a phenolic compound obtained from lychee fruit extract. Oligonol effectively improved diabetic parameters, hindered the progression of diabetic renal disease, and safeguarded renal cells and their structural integrity. This was achieved by reducing oxidative stress caused by reduced nicotinamide adenine dinucleotide phosphate oxidase, inhibiting the formation of AGEs, and preventing cell death in db/db mice (Park et al. 2015a). Catechin exhibits therapeutic potential in preventing the development of vascular endothelial abnormalities (VED) in diabetes. This effect may be attributed to the stimulation of endothelial PI3K, which subsequently activates eNOS and leads to the creation of NO (Bhardwaj et al. 2014). Anthocyanins are a kind of flavonoid found in fruits and vegetables that give them a vibrant red to blue colour (He and Giusti 2010). Anthocyanin-activated adiponectin, produced from adipose tissue, has a necessary function in improving endothelial dysfunction associated with diabetes. Baicalein is abundant in natural food, which mitigates hypertensive and vascular damage in ID and IR models through anti-inflammatory and anti-glycation pathways. Baicalein reduced AGEs, TNF-α level, NF-κB activation, and minimized histopathological alterations in both models (El-bassossy et al. 2014). *In vitro* and *in vivo* investigations, strongly indicate that a significant number of flavonoids possess advantageous properties for combating diabetes problems.

11.3.3 Phytosterols

Phytosterols, also known as plant sterols (PS), serve as the counterparts to cholesterol found in animal products (Awad et al. 2000). Sterol-based bioactive substances are commonly found in plant-based foods. These chemicals have been shown to efficiently regulate insulin metabolism and cholesterol levels. Insulin resistance and hyperglycemia can increase oxidative stress and reactive oxygen species (ROS) production. These chemical alterations cause death in pancreatic cells, resulting in insulin deficiency. Plant sterols have been shown to reduce inflammation and oxidative stress damage related to DNA repair pathways (Jayaraman et al. 2021). T2DM increases oxidative stress, causing brain damage and perhaps contributing to cognitive impairments and neurodegenerative disorders. Phytosterol plays an important role in reducing oxidative stress which is crucial for cognitive function. A phytosterol extract from yellow corn (*Zea mays* L.) was administered daily at 100 mg/kg to diabetic mice for 8 weeks. The treatment reduced oxidized lipids and proteins in the cerebellum and frontal cortex of diabetic *db/db* mice (Salgado-Román et al. 2021). The process of normal insulin signalling originates with the binding of insulin to the insulin receptor, after phosphorylation of the tyrosine residue of IRS 1/2, and ultimately, the activation of the PI3K and Akt enzymes. The activation of Akt leads to the translocation of glucose for oxidation by activating the GLUT4 translocation proteins. Phytosterol exhibits strong anti-diabetic properties by augmenting the activity of insulin signalling molecules, including insulin receptor (IR), insulin receptor substrate 1/2 (IRS 1/2), phosphoinositol 3-kinase (PI3K), protein kinase B (Akt), and Akt substrate of 160 kDa (AS160), thereby facilitating the synthesis of glucose transporter type 4 (GLUT4) and enhancing glucose absorption. However, there remains a dearth of research investigating the impact of phytosterols on the GLUT4 translocation pathway (Prasad et al. 2022). The findings of the (Gao et al. 2017) investigation provide evidence that the phytosterol-enriched spread may have clinical efficacy in reducing the symptoms experienced by mothers with GDM and the difficulties observed in neonates.

11.3.4 Non-Digestible Carbohydrate

Legumes and whole grains may show good glycemic control because of the slow release of carbohydrates into the bloodstream after digestion. The process of finely grinding both grains and legumes results in the disruption of cell structures, hence increasing the accessibility of starch for digestion (Venn and Mann 2004). Arabinoxylan (AX) is a prominent dietary fiber found in several cereal grains. Following a period of 5 days of germination, it was observed that the overall content of arabinoxylans (AX) in barley, oats, and rye grains exhibited a reduction. Conversely, the AX content in buckwheat, sorghum wheat, and brown rice grains displayed an increase when compared to the AX content in non-germinated grains

(Donkor et al. 2012). Cereals, such as barley or portions of oat bran, contain a significant amount of soluble fiber known as β-glucan. Bozbulut and Sanlier (2019) revealed that β-glucan has been shown to enhance glycaemic control and improve the antioxidant profile. The effect is significant as it helps reduce oxidative stress associated with diabetes by upregulating the levels of catalase (CAT) and superoxide dismutase (SOD) in the liver and kidneys. Additionally, β-glucan increases Akt kinase levels and decreases the degradation of pro-caspase-3. Additionally, there have been reports indicating that β-glucans can activate the survival pathway, leading to systematic recovery. Dong et al. (2011) found that the high dosage of oat β-glucan in streptozotocin-induced diabetic mice showed the potential to impede the activities of intestinal disaccharidases. This inhibition occurs through a mechanism involving the reduction of glucose digestion in the small intestine, ultimately leading to a decrease in blood glucose levels. Hence, the administration of a substantial dosage of β-glucan in oats has the potential to effectively regulate blood glucose levels and mitigate symptoms associated with diabetes through the inhibition of α-glycosidase activities. Additionally, it has been suggested that oat β-glucan may have the capacity to block α-glycosidase (Grundy et al. 2016).

11.3.5 Carotenoids and Bioactive Peptides

Carotenoids are found in all red, yellow, and orange-coloured fruits and vegetables. Both beta-carotene and lycopene are powerful antioxidants that can effectively suppress the production of peroxide (Shashirekha et al. 2015). Carotenoids including fucoxanthin, astaxanthin, and bixin lower blood sugar via promoting PPARγ expression. Alternatively, research indicates a direct correlation between plasma carotenoid levels and insulin sensitivity. Two endogenous regulators of serine phosphorylation of IRS-1 are JNK and IKKb. They promote insulin resistance by phosphorylating IRS-1. Inhibiting these targets may help to overcome insulin resistance (Roohbakhsh et al. 2017). Crocetin is a carotenoid constituent found in saffron. Crocetin and bixin decrease insulin resistance by suppressing JNK and IKKb and inducing PPARγ expression (Yang et al. 2010). Oxidative stress reduces insulin secretion and raises insulin resistance. Carotenoids like bixin and astaxanthin, acts as antioxidants. By mitigating oxidative stress, they improves the consequences of DM (Park et al. 2015b; Roehrs et al. 2014). Milk protein-derived peptides possess inhibitory properties against α-glucosidase and DPP IV, which can regulate postprandial hyperglycemia in both healthy individuals and those with type 2 diabetes. This regulation is achieved by preventing the inactivation of incretin hormones as well as carbohydrate hydrolyzing enzymes (Patil et al. 2015). GIP and GLP-1 are two gut hormones (incretins), released by enteroendocrine cells within minutes of eating. DPP4 inhibitors raise endogenous blood levels of active incretins, which results in extended incretin activity. Oral glucose, but not intravenous glucose, promotes the production of incretin hormones, which in turn boosts insulin secretion (Kim and Egan 2008; Deacon and Ahrén 2011). Ali and Agha (2009) investigated exogenous 90 mg/kg lycopene treatment to STZ-induced hyperglycemic rats, reduced glucose levels and increased insulin concentration via reducing free radical activity. Further study is needed to determine the relationship between lycopene-rich diets and diabetes incidence. Meanwhile a summarized table presented here with detailed listing of food bioactive compounds and its role in human health (Table 11.1).

Table 11.1 Different bioactive compounds and its therapeutic effect

Bioactive compounds	Sources	Role/effect	References
Phenolic acid			
Ferulic acid	Banana, broccoli, rice bran, citrus fruits	Ferulic acid effectively reduces blood glucose levels and inhibits the activities of G6pase and PEPCK. Additionally, it increases glycogen and insulin concentrations, as well as the activity of glucokinase (GK), in C57BL/6 N mice fed a high-fat diet	Son et al. (2011)
Gallic acid	*Terminalia bellirica* fruit, Fresh matured beans of *C. tetragonoloba* (L.) Taub. (Fabaceae)	The islets of gallic acid treated rats exhibited regeneration of beta-cells, in contrast to the untreated diabetic rats	Latha and Daisy (2011)
		GA decreased insulin resistance by partially activating PPARγ, specifically in adipose tissue. It also increased glucose absorption by activating GLUT4 through the PI3k/p-Akt pathway in the epididymal adipose tissue of an insulin-resistant animal model. GA enhanced the condition of hyperlipemia and maintained the density of β-cells by reducing the free radicals	Gandhi et al. (2014)
Cinnamic acid	Cinnamon	In type 2 diabetic rats, it enhances glucose tolerance dose-dependently and increases glucose-induced insulin secretion in isolated islets	Hafizur et al. (2015)
Caffeic acid	Blueberry, plum, apple, cherry, pear, chicory, artichoke, kiwi, cider, potato, and coffee	Caffeic acid, Naringenin, and quercetin seem to have a positive affect on the expression of β-cell survival and regulatory genes, and the likelihood of survival in INS-1E cells exposed to glucotoxicity	Bhattacharya et al. (2014), Vinayagam et al. (2016)
Flavonoids			
Proanthocyanidins	Grape seed	Proanthocyanidins have been shown to enhance the condition of diabetic peripheral neuropathy in type 1 diabetic rats and reduced endoplasmic reticulum stress in the pancreas, and skeletal muscles	Shi et al. (2013)

(continued)

Table 11.1 (continued)

Bioactive compounds	Sources	Role/effect	References
Quercetin	Tea, red wine, fruits, vegetables, and nuts	Evidences showed that dorsal root ganglion neuron death increased in hyperglycemic rats, but was reduced by quercetin. Further research suggests quercetin's protective properties originate from its impact on the Nrf-2/HO-1 pathway, ROS scavenging, and NF-κB suppression. Inhibiting local NF-κB activity may prevent or improve diabetic neuropathy problems	Ding et al. (2014), Testa et al. (2016)
Epicatechin	Cocoa, green tea, fruits and vegetables	*In vivo*, epicatechin has been observed to reduce the accumulation of AGE products in retinas in a dose-dependent way, as well as break down preformed glycated human serum albumin *in vitro*. An enhanced retinal vascular apoptosis was seen in rats treated with a (−)-epicatechin after they were given an exogenous AGE injection	Kim et al. (2015)
Rutin	Apples, tea, red wine, and onion	The findings of the study demonstrated that rutin exhibits anti-diabetic properties, as evidenced by a reduction in blood glucose levels and an insulin levels enhancement in diabetic rats	Ola et al. (2015)
Polysaccharide			
Polysaccharide	Seaweeds	Post-meal glucose levels are lowered because soluble fibers delay glucose availability and absorption in the proximal small intestine	Holdt and Kraan (2011)
Polysaccharide: Arabinoxylan (AX)	Cereal grains	Improves metabolic regulation in type 2 diabetes	Donkor et al. (2012), Lu et al. (2004)
β-glucan	Cereal grains	High viscosity causes a delay in the process of absorption	Grundy et al. (2016)

11.4 Cooking

Food nutrients are greatly affected by cooking. Positive effects of cooking include softening vegetable tissues and enabling bioactive component extraction. However, heating can destroy these chemicals, which can be unfavourable also (Murador et al. 2016). Heating and cooking reduce nutritional value and phytochemicals present in foods. Canned foods showed a reduction in protein, total dietary fiber, magnesium, and phytate than home-cooked foods. The impact of the heating method on the vitamin B1, B5, and B6 contents was found to be negligible. However, the canning procedure significantly decreased the levels of vitamin B2. In contrast to the culinary practice of household cooking, the process of canning has been found to result in a notable decrease in the vitamin B9 concentration of white beans and chickpeas (Margier et al. 2018). Microwave heating and stir-frying had no major effect on radicle scavenging activity (RSA), total polyphenol content (TP), ascorbic acid content analysis (AsA), or total carotenoid levels, but boiling partially destroyed them. Prolonged boiling time resulted in increased loss of AOA in cooked tissues. Research suggests that microwave and stir-fried peppers are more effective in retaining their bioactive components (Nahar et al. 2022; Kumari et al. 2023). It is advisable to reduce the amount of water used and the cooking time when simmering is an absolute necessity to preserve the maximum medicinal properties of the bioactive compounds found in peppers (Chuah et al. 2008). Lee et al. (2019) examined how several cooking methods (boiling, blanching, microwaving, roasting, and steaming) affect the preservation of bioactive compounds and micronutrients in shiitake mushrooms (*Lentinus edodes*). Mineral retention was lowest in boiled samples. Microwaved and roasted samples retained minerals better than other cooking methods. Boiling produced lower retention of vitamin and bioactive components, while roasting provided better values. The antioxidant activities of blanched, boiled, and steamed samples were reduced compared to raw samples. However, antioxidant activity in microwaved and roasted samples was preserved or enhanced. Murador et al. (2016), analysed that raw kale had the highest carotenoid content, while steamed red cabbage had the highest anthocyanin concentration. Steaming minimizes excessive use of oil, prevents leaching, and yields high bioactive chemical content and antioxidant activity in studied veggies. A similar conclusion was observed in the study investigating how different cooking procedures affect the phytochemicals and antioxidant activity of fresh kale, and it was found that boiling drastically reduced water-soluble phytochemicals. Stir-frying displayed the highest degradation rate for all phytochemicals while steaming had the lowest (Akdaş and Bakkalbaşı 2017). López-Berenguer et al. (2007) indicated an overall decline in the concentrations of all the examined constituents, with the exception of mineral nutrients, which remained consistent throughout all cooking settings in the microwave. In contrast to phenolic chemicals and glucosinolates, which mostly showed losses from leaching into water, vitamin C showed the most significant losses from degradation and leaching. To prevent nutrient losses, it is generally recommended to avoid prolonged microwave cooking durations and the use of excessive amounts of cooking water.

Traditional methods are commonly utilized in gastronomy, particularly in community catering. Lobefaro et al. (2021) used a new technique in their study. The Niko Romito Food Processing Technique (NR-FPT) is a low-aggressive culinary method that combines four cooking practices and three processing methods for various dishes. This approach shares a decreased food aggression due to less intense and indirect heat treatments. NR-FPT, a new food processing approach that uses lower cooking temperatures and least invasive processes, has been shown to preserve the nutritional integrity of foods. Implementing a suitable cooking method may contribute to enhanced uptake of functional molecules, hence augmenting their efficacy and minimizing the likelihood of chronic ailments.

11.5 Conclusion

Currently, the management of metabolic syndrome and the prevention of diabetes entails making changes to one's lifestyle, such as engaging in more physical activity and maintaining a healthy weight through the reduction of calorie consumption. The dietary guidelines for those at risk of diabetes are placing more emphasis on consuming plant-based foods including whole grains, berries, fruits, and vegetables as well as animal foods. These foods are known to be rich sources of bioactive compounds like dietary phenols, flavonoids, and fibre, which can affect glucose metabolism by slowing down the rate of carbohydrate digestion, stimulating insulin secretion, regulating liver glucose release, and activating insulin receptors and uptake in insulin-sensitive tissues. This chapter focuses on polyphenolic compounds present in foods that have the potential to act as anti-diabetic agents. Cooking can have both positive and negative effects on polyphenolic compounds in food, depending on factors such as temperature, cooking time, and the type of food. Dietary components that might influence diabetes by interacting with and regulating molecular pathways have the potential to significantly transform the methods of treating diabetes and its consequences in the near future.

References

Adams GG, Imran S, Wang S, Mohammad A, Kok S, Gray DA et al (2011) The hypoglycaemic effect of pumpkins as anti-diabetic and functional medicines. Food Res Int 44(4):862–867

Akdaş ZZ, Bakkalbaşı E (2017) Influence of different cooking methods on color , bioactive compounds , and antioxidant activity of kale. Int J Food Prop 20(4):877–887

Alara OR, Abdurahman NH, Ukaegbu CI (2021) Extraction of phenolic compounds: a review. Curr Res Food Sci 4:200–214

Ali MM, Agha FG (2009) Amelioration of streptozotocin-induced diabetes mellitus, oxidative stress and dyslipidemia in rats by tomato extract lycopene. Scand J Clin Lab Invest 69(3):371–379

Almeida CAS, Baggio SR, Mariutti LRB, Bragagnolo N (2020) One-step rapid extraction of phytosterols from vegetable oils. Food Res Int 130:108891

Amarowicz R, Estrella I, Hernández T, Robredo S, Troszyńska A, Kosińska A et al (2010) Free radical-scavenging capacity, antioxidant activity, and phenolic composition of green lentil (Lens culinaris). Food Chem 121(3):705–711

Aresta A, Zambonin C (2016) Simultaneous determination of salicylic, 3-methyl salicylic, 4-methyl salicylic, acetylsalicylic and benzoic acids in fruit, vegetables and derived beverages by SPME–LC–UV/DAD. J Pharm Biomed Anal 121:63–68

Awad AB, Chan KC, Downie AC, Fink CS (2000) Peanuts as a source of β-sitosterol, a sterol with anticancer properties. Nutr Cancer 36(2):238–241

Bao L, Li J, Zha D, Zhang L, Gao P, Yao T et al (2018) Chlorogenic acid prevents diabetic nephropathy by inhibiting oxidative stress and in fl ammation through modulation of the Nrf2 / HO-1 and NF- κ B pathways. Int Immunopharmacol 54:245–253

Bhardwaj P, Khanna D, Balakumar P (2014) Catechin averts experimental diabetes mellitus-induced vascular endothelial structural and functional abnormalities. Cardiovasc Toxicol 14:41–51

Bhattacharya S, Oksbjerg N, Young JF, Jeppesen PB (2014) Caffeic acid, naringenin and quercetin enhance glucose-stimulated insulin secretion and glucose sensitivity in INS-1E cells. Diabetes Obes Metab 16(7):602–612

Borguini RG, da Silva Torres EAF (2009) Tomatoes and tomato products as dietary sources of antioxidants. Food Rev Int 25(4):313–325

Bozbulut R, Sanlier N (2019) Promising effects of β-glucans on glyceamic control in diabetes. Trends Food Sci Technol 83:159–166

Cherng Y, Tsai C, Chung H, Lai Y, Kuo S, Cheng J (2013) Antihyperglycemic action of sinapic acid in diabetic rats. J Agric Food Chem 61(49):12053–12059

Chuah AM, Lee YC, Yamaguchi T, Takamura H, Yin LJ, Matoba T (2008) Effect of cooking on the antioxidant properties of coloured peppers. Food Chem 111(1):20–28

Dal S, Sigrist S (2016) The protective effect of antioxidants consumption on diabetes and vascular complications. Diseases 4(3):24

Deacon CF, Ahrén B (2011) Physiology of incretins in health and disease. Rev Diabet Stud 8(3):293–306

Ding Y, Dai X, Zhang Z, Jiang Y, Ma X, Cai X et al (2014) Proanthocyanidins protect against early diabetic peripheral neuropathy by modulating endoplasmic reticulum stress. J Nutr Biochem 25(7):765–772

Dong J, Cai F, Shen R, Liu Y (2011) Hypoglycaemic effects and inhibitory effect on intestinal disaccharidases of oat beta-glucan in streptozotocin-induced diabetic mice. Food Chem 129(3):1066–1071

Donkor ON, Stojanovska L, Ginn P, Ashton J, Vasiljevic T (2012) Germinated grains - sources of bioactive compounds. Food Chem 135(3):950–959

Dziuba B, Dziuba M (2014) Milk proteins-derived bioactive peptides in dairy products: molecular, biological and methodological aspects. Acta Sci Pol Technol Aliment 13(1):5–25

El-bassossy HM, Ahmed N, Fouad M, Fahmy A (2014) Baicalein protects against hypertension associated with diabetes: effect on vascular reactivity and stiffness. Eur J Integr Med 21(12):1742–1745

Fernandes AAH, Novelli ELB, Okoshi K, Okoshi MP, Di MBP, Guimarães JFC et al (2010) Influence of rutin treatment on biochemical alterations in experimental diabetes. Biomed Pharmacother 64(3):214–219

Galicia-Garcia U, Benito-Vicente A, Jebari S, Larrea-Sebal A, Siddiqi H, Uribe KB et al (2020) Pathophysiology of type 2 diabetes mellitus. Int J Mol Sci 21(17):1–34

Gandhi GR, Jothi G, Antony PJ, Balakrishna K, Paulraj MG, Ignacimuthu S et al (2014) Gallic acid attenuates high-fat diet fed-streptozotocin-induced insulin resistance via partial agonism of PPARγ in experimental type 2 diabetic rats and enhances glucose uptake through translocation and activation of GLUT4 in PI3K/p-Akt signaling pathway. Eur J Pharmacol 745:201–216

Gao F, Wang G, Wang L, Guo N (2017) Phytosterol nutritional supplement improves pregnancy and neonatal complications of gestational diabetes mellitus in a double-blind and placebo-controlled clinical study. Food Funct 8(1):424–428

Garcia-Salas P, Morales-Soto A, Segura-Carretero A, Fernández-Gutiérrez A (2010) Phenolic-compound-extraction systems for fruit and vegetable samples. Molecules 15(12):8813–8826

Ghorbani A (2017) Mechanisms of antidiabetic effects of flavonoid rutin. Biomed Pharmacother 96:305–312

Grundy MM, Edwards CH, Mackie AR, Gidley MJ, Butterworth PJ, Ellis PR (2016) Re-evaluation of the mechanisms of dietary fibre and implications for macronutrient bioaccessibility , digestion and postprandial metabolism. Br J Nutr 116(5):816–833

Guaadaoui A, Benaicha S, Elmajdoub N, Bellaoui M, Hamal A (2014) What is a bioactive compound? A combined definition for a preliminary consensus. Int J Food Sci Nutr 3(3):17–179

Guariguata L, Whiting DR, Hambleton I, Beagley J, Linnenkamp U, Shaw JE (2014) Global estimates of diabetes prevalence for 2013 and projections for 2035. Diabetes Res Clin Pract 103(2):137–149

Gupta D (2013) Comparative analysis of spices for their phenolic content , flavonoid content and antioxidant capacity. Am Int J Res Formal, Appl Nat Sci 4(1):38–42

Hafizur RM, Hameed A, Shukrana M, Raza SA, Chishti S, Kabir N et al (2015) Cinnamic acid exerts anti-diabetic activity by improving glucose tolerance *in vivo* and by stimulating insulin secretion *in vitro*. Phytomedicine 22(2):297–300

He J, Giusti MM (2010) Anthocyanins: natural colorants with health-promoting properties. Annu Rev Food Sci Technol 1:163–187

Holdt SL, Kraan S (2011) Bioactive compounds in seaweed: functional food applications and legislation. J Appl Phycol 23:543–597

Islam MA, Jeong BG, Jung J, Shin EC, Choi SG, Chun J (2017) Phytosterol determination and method validation for selected nuts and seeds. Food Anal Methods 10:3225–3234

Jayaprakasha GK, Rao LJM (2011) Chemistry, biogenesis, and biological activities of Cinnamomum zeylanicum. Crit Rev Food Sci Nutr 51(6):547–562

Jayaraman S, Roy A, Vengadassalapathy S, Sekar R, Veeraraghavan VP, Rajagopal P et al (2021) An overview on the therapeutic function of foods enriched with plant sterols in diabetes management. Antioxidants 10(12):1903

Jeong KJ, Kim DY, Quan HY, Jo HK, Kim GW, Chung SH (2014) Effects of eugenol on hepatic glucose production and AMPK signaling pathway in hepatocytes and C57BL/6J mice. Fitoterapia 93:150–162

Kim W, Egan JM (2008) The role of incretins in glucose homeostasis and diabetes treatment. Pharmacol Rev 60(4):470–512

Kim J, Kim CS, Moon MK, Kim JS (2015) Epicatechin breaks preformed glycated serum albumin and reverses the retinal accumulation of advanced glycation end products. Eur J Pharmacol 748:108–114

Kooti W, Farokhipour M, Asadzadeh Z, Ashtary-Larky D, Asadi-Samani M (2016) The role of medicinal plants in the treatment of diabetes: a systematic review. Electron Physician 8(1):1832–1842

Kozłowska A, Szostak-Wegierek D (2014) Flavonoids--food sources and health benefits. Rocz Państwowego Zakładu Hig 65(2):79–85

Kumari P, Hazra S, Nahar N (2023) Effect of different drying techniques on the quality attributes and antioxidant properties of green banana pulp. J Adv Zool 44:2048–2056

Latha RCR, Daisy P (2011) Insulin-secretagogue, antihyperlipidemic and other protective effects of gallic acid isolated from Terminalia bellerica Roxb. in streptozotocin-induced diabetic rats. Chem Biol Interact 189(1–2):112–118

Laurean DC, Pashikanti S (2007) Rutin derivatives as potent AGE inhibitors and its relevance in diabetes complications. FASEB J 21(5):A633

Lee K, Lee H, Choi Y, Kim Y, Jeong HS, Lee J (2019) Effect of different cooking methods on the true retention of vitamins, minerals, and bioactive compounds in shiitake mushrooms (Lentinula edodes). Food Sci Technol Res 25(1):115–122

Li TSC, Beveridge THJ, Drover JCG (2007) Phytosterol content of sea buckthorn (Hippophae rhamnoides L.) seed oil: extraction and identification. Food Chem 101(4):1633–1639

Lobefaro S, Piciocchi C, Luisi F, Miraglia L, Romito N, Luneia R et al (2021) Cooking techniques and nutritional quality of food: a comparison between traditional and innovative ways of cooking. Int J Gastron Food Sci 25:100381

López-Berenguer C, Carvajal M, Moreno DA, García-Viguera C (2007) Effects of microwave cooking conditions on bioactive compounds present in broccoli inflorescences. J Agric Food Chem 55(24):10001–10007

Lu ZX, Walker KZ, Muir JG, O'Dea K (2004) Arabinoxylan fibre improves metabolic control in people with type II diabetes. Eur J Clin Nutr 58(4):621–628

Magwaza STN, Islam MS (2023) Roles of marine macroalgae or seaweeds and their bioactive compounds in combating overweight, obesity and diabetes: a comprehensive review. Mar Drugs 21(4):258

Manavi SP, Amiri T, Mozafaryan MJ (2021) Role of flavonoids in diabetes. J Rev Med Sci [Internet] 1(3):149–161

Marcone MF, Kakuda Y, Yada RY (2003) Amaranth as a rich dietary source of β-sitosterol and other phytosterols. Plant Foods Hum Nutr 58:207–211

Margier M, Georgé S, Hafnaoui N, Remond D, Nowicki M, Du Chaffaut L et al (2018) Nutritional composition and bioactive content of legumes: characterization of pulses frequently consumed in France and effect of the cooking method. Nutrients 10(11):1–12

Martinez-ferez A, Rudloff S, Guadix A, Henkel CA, Pohlentz G, Boza JJ et al (2006) Goats ' milk as a natural source of lactose-derived oligosaccharides: isolation by membrane technology. Int Dairy J 16(2):173–181

Miękus N, Marszałek K, Podlacha M, Iqbal A, Puchalski C, Świergiel AH (2020) Health benefits of plant-derived sulfur compounds, glucosinolates, and organosulfur compounds. Molecules 25(17):3804

Murador DC, Mercadante AZ, De Rosso VV (2016) Cooking techniques improve the levels of bioactive compounds and antioxidant activity in kale and red cabbage. Food Chem 196:1101–1107

Nagasawa T, Tabata N, Ito Y, Nishizawa N, Aiba Y, Kitts DD (2003) Inhibition of glycation reaction in tissue protein incubations by water soluble rutin derivative. Mol Cell Biochem 249:3–10

Nahar N, Hazra S, Raychaudhuri U, Adhikari SP (2022) Effect of different drying methods on drying kinetics, antioxidant of lotus rhizomes (Nelumbo nucifera). Food Process Preserv 46:e–16842

Ola MS, Ahmed MM, Ahmad R, Abuohashish HM, Al-Rejaie SS, Alhomida AS (2015) Neuroprotective effects of rutin in streptozotocin-induced diabetic rat retina. J Mol Neurosci 56:440–448

Oliveira L, Freire CSR, Silvestre AJD, Cordeiro N (2008) Lipophilic extracts from banana fruit residues: a source of valuable phytosterols. J Agric Food Chem 56(20):9520–9524

Pandey MM, Vijayakumar M, Rastogi S, Rawat AKS (2012) Phenolic content and antioxidant properties of selected Indian spices of Apiaceae. Int J Geogr Inf Syst 18(3):246–256

Park YW, Nam MS (2015) Bioactive peptides in milk and dairy products: a review. Korean J Food Sci Anim Resour 35(6):831–840

Park CH, Noh JS, Fujii H, Roh SS, Song YO, Choi JS et al (2015a) Oligonol, a low-molecular-weight polyphenol derived from lychee fruit, attenuates gluco-lipotoxicity-mediated renal disorder in type 2 diabetic db/db mice. Drug Discov Ther 9(1):13–22

Park CH, Xu FH, Roh SS, Song YO, Uebaba K, Noh JS et al (2015b) Astaxanthin and corni fructus protect against diabetes-induced oxidative stress, inflammation, and advanced glycation end product in livers of streptozotocin-induced diabetic rats. J Med Food 18(3):337–344

Patil P, Mandal S, Tomar SK, Anand S (2015) Food protein-derived bioactive peptides in management of type 2 diabetes. Eur J Nutr 54:863–880

Prasad M, Jayaraman S, Eladl MA, El-sherbiny M (2022) A comprehensive review on therapeutic perspectives of phytosterols in insulin resistance: a mechanistic approach. Molecules 27(5):1595

Przybylska-Balcerek A, Frankowski J, Stuper-Szablewska K (2019) Bioactive compounds in sorghum. Eur Food Res Technol 245:1075–1080

Roehrs M, Figueiredo CG, Zanchi MM, Bochi GV, Moresco RN, Quatrin A et al (2014) Bixin and norbixin have opposite effects on glycemia, lipidemia, and oxidative stress in streptozotocin-induced diabetic rats. Int J Endocrinol 2014:839095

Roohbakhsh A, Karimi G, Iranshahi M (2017) Carotenoids in the treatment of diabetes mellitus and its complications: a mechanistic review. Biomed Pharmacother 91:31–42

Salgado-Román JM, Loza-Cornejo S, Hernández-Soto R, Hernández-Maldonado JA, Márquez-Villalobos FA, Pérez-Vázquez V et al (2021) Phytosterol extract decreases the oxidative damage in the brains of diabetic db/db mice. Metab Syndr Relat Disord 19(5):305–311

Sarikurkcu C, Zengin G, Oskay M, Uysal S (2015) Composition , antioxidant , antimicrobial and enzyme inhibition activities of two Origanum vulgare subspecies (subsp . vulgare and subsp . hirtum) essential oils. Ind Crop Prod 70:178–184

Scazzocchio B, Varì R, Filesi C, Del Gaudio I, D'Archivio M, Santangelo C et al (2015) Protocatechuic acid activates key components of insulin signaling pathway mimicking insulin activity. Mol Nutr Food Res 59(8):1472–1481

Shashirekha MN, Mallikarjuna SE, Rajarathnam S (2015) Status of bioactive compounds in foods, with focus on fruits and vegetables. Crit Rev Food Sci Nutr 55(10):1324–1339

Shi Y, Liang X, Zhang H, Wu Q, Qu L, Sun Q (2013) Quercetin protects rat dorsal root ganglion neurons against high glucose-induced injury *in vitro* through Nrf-2 / HO-1 activation and NF-κB inhibition. Acta Pharmacol Sin 34(9):1140–1148

Shobha R, Rajeshwari C, Andallu B (2013) Anti-peroxidative and anti-diabetic activities of aniseeds (Pimpinella anisum L.) and identification of bioactive compounds. Am J Phytomedicine Clin Ther 1(5):516–527

Singh B, Singh JP, Shevkani K, Singh N, Kaur A (2017) Bioactive constituents in pulses and their health benefits. J Food Sci Technol 54(4):858–870

Son MJ, Rico CW, Nam SH, Kang MY (2011) Effect of oryzanol and ferulic acid on the glucose metabolism of mice fed with a high-fat diet. J Food Sci 76(1):H7–H10

Subba Rao MVSST, Muralikrishna G (2002) Evaluation of the antioxidant properties of free and bound phenolic acids from native and malted finger millet (ragi, Eleusine coracana Indaf-15). J Agric Food Chem 50(4):889–892

Teng H, Chen L (2017) α-Glucosidase and α-amylase inhibitors from seed oil: a review of liposoluble substance to treat diabetes. Crit Rev Food Sci Nutr 57(16):3438–3448

Testa R, Bonfigli AR, Genovese S, De Nigris V, Ceriello A (2016) The possible role of flavonoids in the prevention of diabetic complications. Nutrients 8(5):1–13

Venn BJ, Mann JI (2004) Cereal grains, legumes and diabetes. Eur J Clin Nutr 58(11):1443–1461

Vinayagam R, Jayachandran M, Xu B (2016) Antidiabetic effects of simple phenolic acids: a comprehensive review. Phyther Res 30(2):184–199

Wang T, Jónsdóttir R, Ólafsdóttir G (2009) Total phenolic compounds , radical scavenging and metal chelation of extracts from Icelandic seaweeds. Food Chem 116(1):240–248

Weyer C, Bogardus C, Mott DM, Pratley RE (1999) The natural history of insulin secretory dysfunction and insulin resistance in the pathogenesis of type 2 diabetes mellitus. J Clin Invest 104(6):787–794

Williamson G (2020) Protection against developing type 2 diabetes by coffee consumption: assessment of the role of chlorogenic acid and metabolites on glycaemic responses. Food Funct 11(6):4826–4833

Wood PJ (2007) Cereal β-glucans in diet and health. J Cereal Sci 46(3):230–238

Xiao H, Jiang H, Tu H, Jia Y, Wang H, Lü X et al (2022) Extraction, isolation and identification of low molecular weight peptides in human milk. Foods 11(13):1836

Yang L, Qian Z, Ji H, Yang R, Wang Y, Xi L et al (2010) Inhibitory effect on protein kinase CΘ by Crocetin attenuates palmitate-induced insulin insensitivity in 3T3-L1 adipocytes. Eur J Pharmacol 642(1–3):47–55

Chapter 12
Food Bioactives as Natural Anticancer Therapeutic Agents

Thuane Passos Barbosa Lima, Pedro Paulo Saldanha Coimbra, Carlos Fernando Araujo-Lima, and Anderson Junger Teodoro

Abstract Cancer, a disease characterized by high incidence, morbidity, and mortality rates, represents a significant challenge in global public health, requiring continuous efforts from the scientific community. A promising perspective in the battle against this disease emerges with food bioactives. When found in vegetables, these compounds are known as phytochemicals or phytonutrients, being secondary metabolites produced by plants as part of their self-defense. They exhibit physiologically active properties that go beyond basic nutrition, providing benefits to human health. Numerous food bioactives have already demonstrated to be useful in cancer prevention and therapy due to their antioxidant, anti-inflammatory, and chemopreventive properties. *In vivo* and *in vitro* studies suggest that these compounds can regulate the carcinogenesis process by targeting several molecules and signaling pathways. Some suggested mechanisms of action include inhibiting cancer cell growth, modulating the cell cycle, inducing apoptosis, reducing angiogenesis, and modulating redox imbalance that damages DNA. In addition to their preventive role, some bioactives have shown promise in combined therapy with conventional treatments or as alternative therapeutic approaches. In this chapter, we delve deeper into the prospecting of bioactives from plant matrices, with focus in phytochemicals, as well as their metabolic and genetic modulation effects related to anticancer therapeutic properties.

T. P. B. Lima · P. P. S. Coimbra
Federal University of State of Rio de Janeiro (UNIRIO), Graduate Program in Food and Nutrition (PPGAN), Rio de Janeiro, Brazil
e-mail: pedro.coimbra@edu.unirio.br

C. F. Araujo-Lima
Federal University of State of Rio de Janeiro (UNIRIO), Biomedical Institute, Department of Genetics and Molecular Biology, Rio de Janeiro, Brazil
e-mail: araujo.lima@unirio.br

A. J. Teodoro (✉)
Fluminense Federal University (UFF), College of Nutrition, Department of Nutrition and Dietetics, Niterói, Brazil

S. Amir Ashraf, M. Adnan (eds.), *Food Bioactives and Nutraceuticals*,
https://doi.org/10.1007/978-981-97-9723-3_12

Keywords Phytochemicals · Nutrigenomics · Flavones · Antitumor · Bioactive compounds

12.1 Introduction

Cancer stands out as a significant contributor to mortality in both economically disadvantaged nations and affluent countries, imposing a substantial burden on societies from scientific, economic, and political perspectives, being considered a worldwide public health problem (Koh et al. 2020). According to the International Agency Research Cancer, the global incidence of all cancer types summed was 19.2 million in 2020 and the number of deaths reached 9.9 million. Moreover, projections indicate a substantial increase in incidence by the year 2040, with an estimated 30.2 million new cases (IARC 2020). Cancer is characterized by the uncontrolled growth of cells in tissues, forming a tumor that can potentially spread to entire organs or other tissues, known as metastasis. This abnormal cell behavior may result from hereditary genetic factors or epigenetic changes in key genes (oncogenes) related to the cell cycle and apoptosis. Additionally, cancerous cells show dysregulation in programmed cell death (apoptosis) and abnormal behavior of microtubules, crucial in the mitotic process (Garcia-Oliveira et al. 2021). More than 500 different genes can undergo modifications during the development of cancer (Khatoon et al. 2022). Mutations arise from various sources like DNA damage (infections, UV-induced tissue damage, reactive oxygen species), and DNA adducts linked to altered phase I (overexpression) and II (downregulation) enzymes in chemical metabolic activation. Epigenetic modifications (DNA methylation, histone modification, miRNA) also contribute. The cellular microenvironment is crucial in cancer development, with inappropriate inflammatory responses playing a pivotal role. Pro-inflammatory macrophages accelerate tumor initiation, progression, and promotion, while anti-inflammatory macrophages support tumor growth, angiogenesis, and metastasis (Koh et al. 2020). Due to disease profile, the treatment of cancer has been a constant combat with relatively less success. Current approaches involve surgical removal and radiation therapy for the substantial cancer mass, followed by systemic chemotherapy for maintenance. Key chemotherapeutic agents include antimetabolites (e.g., methotrexate), DNA-interactive agents (e.g., cisplatin, doxorubicin), anti-tubulin agents (taxanes), hormones, and molecular targeting agents (Nussbaumer et al. 2011; Choudhari et al. 2020). However, chemotherapy has major drawbacks, including cancer recurrence, drug resistance, and toxic effects on non-targeted tissues, limiting the use of anticancer drugs and impacting the patient's quality of life (Choudhari et al. 2020).

The increasing interest in complementary and alternative medicines, driven by the limitations of conventional cancer chemotherapies and the perceived benefits of natural treatment options has prompted research into bioactive compounds derived from various food matrices (Molassiotis et al. 2005). When found in vegetables, these compounds are known as phytochemicals and they exhibit physiologically active properties that go beyond basic nutrition, providing benefits to human health (Dillard and German 2000). Extracted from pulp, roots, bulbs, barks, leaves, seeds, and other parts, these phytochemicals show promising potential as anticancer drugs, as demonstrated in both *in vitro* and *in vivo* tests (Chinembiri et al. 2014). Over 3000 plant species with hundreds of bioactive compounds exhibit anticancer properties (Haque et al. 2021). Accumulating evidence indicates that many of these phytochemicals can target a diverse range of signaling molecules involved in cellular growth, proliferation, differentiation, and death across various cancer types (Rizeq et al. 2020). These activities modulate pathways, including proinflammatory cytokines, oxidative stress markers, cell cycle regulators, apoptotic/autophagy regulators, proliferative signaling pathways, invasion/migration/metastasis/angiogenesis regulators, stemness markers, and tumor-suppressive miRNA expression (De et al. 2023). The low toxicity and diverse modulation pathways of these phytochemicals make them a promising alternative or adjunct to conventional chemotherapy, offering a novel approach to enhance cancer treatment efficacy while minimizing adverse effects and costs (Gao et al. 2022). In this chapter, we delve deeper into the prospecting of bioactives from vegetable matrices, as well as it is discussed concerns about their metabolic and genetic modulation effects related to anticancer therapeutic properties and the safety of the food matrix used as source.

12.2 Food Bioactives with Anticancer Bioactivity

As the bioactives reach the cells, different pathways may be activated or inhibited, leading to the modulation of the energy production, cell cycle and cell death. Regardless the lack of information about the molecular mechanisms of some specific bioactives, in fact, some others had their activities already clarified or partially clarified as their influence in the redox potential, energy production and DNA expression. Those mechanisms are discussed below.

The literature brings evidence of anticancer properties of food bioactives against a diverse number of cancers, with different outcomes. However, some bioactives may not be efficient against a type of cancer as it was to another (Su et al. 2023). These differences may be related to the cancer cells metabolism variations and evasions mechanisms that may be active or not. With more than 3000 publications only in 2023, the prospection of bioactive compounds with anticancer properties advances. Thus, the Table 12.1 below brings a summary of some recent publications that demonstrated the anticancer activity as well as the possible mechanism of action of food bioactives in different *in vitro* studies.

The *in vitro* studies are a good way to initiate the prospection of the bioactives once it turns possible to identify molecular pathways and the regulation (up or down) of gene expression (Mirabelli et al. 2019). However, *in vivo* studies bring the possibility to identify other features related with the anticancer activity as the bioaccessibility and bioavailability of the compounds, possibility of influence of the

Table 12.1 Overview of different cancer types and bioactives with anticancer activity in *in vitro* assays published in 2023

Cancer type	Cell line	Bioactive or extract	Concentration	Main outcomes and conclusion	References
Breast cancer	MCF-7 and MDA-MB-231	Beetroot peel flour aqueous extract	2.5–25 mg/mL, 24–72 h treatment	Dose and time dependent effects were observed in cell viability. LC50 value stated as 20 mg/mL	Coimbra et al. (2023)
Breast cancer	MDA-MB-231	Apple peel phenolic extract	1–80 mg/mL for 24 h	No cytotoxicity was observed in doses below 50.2 mg/ mL. Suppression of IGF-1R/PI3K/ Akt pathway was observed.	Sair et al. (2023)
Lung cancer	A549	Hesperidin methyl chalcone (HMC- flavonoid)	6.25, 12.5, 25, 50, and 100 µM for 24 h	Action comparable to hesperetin. IC50 stablished at 51.12 µM.	Rizvi et al. (2023)
Lung cancer	A549 and LLC	*Pericarpium Citri Reticulatae* 'Chachi' flavonoid rich extracts	500 µg/mL for 24 h	Inhibition of cell proliferation and cell cycle arrest at G2/M phase via ROS-dependent pathway.	Tao et al. (2023)
Endometrial carcinoma	Ishikawa cells	Apigenin	0, 30, 50, and 70 µmol/L for 48 h	IC50 of 45.55 µmol/L in absence of caspases inhibitors and 141.44 µmol/Lin the presence of caspase inhibitor. Dose dependence and induction of autophagy were observed.	Liang et al. (2023)
Tongue carcinoma	HNO-97	Pomegranate peel (PPE) and blueberry (BE) extracts	0.1, 1.0, 10, 100, and 1000 µg/mL 24–72 h.	IC50 of 73.35 µg/mL for PPE and 525.38 µg/mL for BE. Significant inhibition of cell migration ability and induction of apoptosis.	Ahmed et al. (2023b)
Liver cancer	HepG2	Chokeberry polyphenol and anthocyanins' rich extract from different varieties	1, 2, 3, 4 or 5 mg/ mL for 1 h	Dose dependency was observed in cytotoxicity assay, neochlorogenic acid and anthocyanins were pointed as main source of cellular antioxidant activity.	Gao et al. (2023)

Liver cancer	HepG2	Gold nanoparticle synthesized using leaf extract of *Citrus medica*	1, 10, 50, 100, and 200 µg/mL for 24 h	Growth inhibition via generation of ROS and MDA, reduction of GSH level and SOD activity, and over expression of Bax/Bcl-2 and Caspase-3 mRNA.	Liu et al. (2023)
Pancreatic cancer	PANC-1 and AsPC-1	Anthocyanins and metabolites isolated from plasma after consumption of anthocyanins' rich juice	Plasma extracts 1:1 with culture media for 36 h	Significant reduction of PANC-1 migration capacity. Improvement in AsPC-1 intercellular adhesion	Mostafa et al. (2023)
Lung cancer, liver cancer and colon cancer	A549, HepG2 and SW620	Six partially purified fractions containing 2,4′-dihydroxy-6′-methoxy-3′,5′-dimethylchalcone (DMC) from *Cleistocalyx nervosum* var. paniala seeds	0.01, 0.1, 1, 10, 100, 200, 400, and 800 µg/mL for 48 h	All fractions present anticancer activity, but fraction MK4 present higher anticancer activity than other fractions, especially against colon cancer cells.	Vachiraarunwong et al. (2023)
Breast adenocarcinoma	Ehrlich ascites carcinoma (EAC) cells	β-carotene (BC) and lupeol (LUP)	5–200 µg/mL of BC, LUP or a combination for 24 h	Synergism was observed in BC + LUP (IC50 = 54.8 ± 1.84) compared to BC and LUP isolated (IC50 = 101.5 ± 2.66 and 84.08 ± 3.56, respectively).	Jana et al. (2023)
Ovarian cancer, breast cancer, cervix adenocarcinoma, lung cancer	SKOV3, MCF-7, HeLa and A549	Raspberries water and ethanol extracts	0–1000 µg/mL for 24 h	Growth inhibition of SKOV3 and MCF-7. Dose dependency was observed in SKOV3 cells. No activity against A549 and HeLa was observed.	Su et al. (2023)

(continued)

Table 12.1 (continued)

Cancer type	Cell line	Bioactive or extract	Concentration	Main outcomes and conclusion	References
Colon adenocarcinoma	SW480	Ethanolic extract of *P. edulis* leave. (Luteolin, chlorogenic acid, moupinamide, isoorientin, glucosyl passionflower, cyclopasifloic acid E and saponarin)	50, 100, 200, 500, 1000, 2000, and 3000 μg/mL for 24 and 48 h	Reduced cell viability. Possible inhibitory effects by coupling on Acetyl-CoA Carboxylase 1 and 2 and 3-hydroxy-3-methyl-glutaryl--CoA reductase by bioactive compounds	Villada Ramos et al. (2023)
Glioblastoma multiforme	LN229 and SNB19	Purified polyphenolic fraction from *Ipomoea pes-caprae*	0, 5, 10 and 15 μg/mL for 48 h	Cell death via ROS-mediated caspase dependent pathway. Regulation of cell proliferation through mitogen-activated protein kinase signaling pathway.	Murugesan et al. (2023)
Melanoma	MM418C1 and D24	*Garcinia humilis* peel extract	15–500 μg/mL for 48 h	Dose-dependent effects were observed in cell viability with IC_{50} ranging of 130–200 μg/mL (MM418C1) and 122–176 μg/mL (D24). Morphological changes indicating apoptosis.	Poojary et al. (2023)
Cervical cancer	HeLa, SiHa and C33A	3-hydroxymorindone isolated from *Knoxia roxburghii*	0, 10, 20, 40, 60, 80 and 160 μmol/L for 24 h	Apoptosis s by up-regulating ROS-mediated mitochondrial dysfunction and down-regulating PI3K/AKT/NF-κB signaling pathways.	Chen et al. (2023)
Colorectal cancer	Caco-2	Hydroxytyrosol from extra virgin olive oil	10 μM for 7 days	Decreases endothelin receptor type A gene expression through epigenetic modification	del Saz-Lara et al. (2023)
Leiomyoma	ELT-3	Manzamine A	0.5–10 μM for 24 and 48 h	Inhibition of cell proliferation by arresting cell cycle progression and inducing caspase-mediated apoptosis.	Lin et al. (2023)

Cholangiocarcinoma	KKU213A, KKU100, and KKU055	Ethanolic extracts from *Chlorella sp* (EE), Lutein (LU) and gallic acid (GA)	50–500 μg/mL (EE), 100–500 μg/mL (LU) and 25–125 μg/mL (GA) for 24 and 48 h	EE promote cancer cell death via inhibition of AKT/mTOR pathway. GA and LU caused a significant reduction of cell viability.	Sawasdee et al. (2023)
Breast, lung, cervical adenocarcinoma, and rhabdomyosarcoma	MCF-7, A549, HeLa and RD	*Uvaria chamae* and *Dicliptera paniculata* extracts	200–0.064 μg/mL for 24, 48 and 72 h	Cytotoxicity across all cell lines in a time a nd dose-dependent manner. Results indicated that the intrinsic apoptotic pathway was caspase-dependent.	Asenye et al. (2023)

liver, as well as their excretion (Ruggeri et al. 2014). Thus, the Table 12.2 below will demonstrate the main outcomes of some *in vivo* studies with food bioactives in the prospection of anticancer compounds.

12.3 Molecular Mechanisms

As the bioactives reach the cells, different pathways may be activated or inhibited, leading to the modulation of the energy production, cell cycle and cell death. Regardless the lack of information about the molecular mechanisms of some specific bioactives, in fact, some others had their mechanisms of action already clarified or partially clarified as their influence in the redox potential, energy production and DNA expression. Those mechanisms are discussed below.

12.3.1 Influence Over Cancer Energy Metabolism and Redox Imbalance

The cellular energy production is directly associated with the oxidative stress that the cell undergoes. Oxidative stress or redox imbalance is an unbalance of the antioxidant apparatus capacity of scavenging free radicals and the concentration of free radicals formed in cell metabolism, mostly originated from the cell respiration in mitochondria and mainly represented by reactive oxygen species (ROS), but also by reactive nitrogen species (RNS), from nitrogen compounds degradation and hydroxyls (OH•) and metals from Fenton-like reactions (Thomas et al. 2009; Kwon et al. 2021; Hayes et al. 2020).

Cancer cells may choose between the aerobic (mitochondrial respiration) and anaerobic metabolism according to the oxygen offer in their tissue microenvironments or due to adaptations (Samec et al. 2021). In anaerobic conditions, the glycolysis is active leading to the production of lactate and low concentrations of adenosine triphosphate (ATP). Alongside the ATP and lactate production, low concentrations of ROS are also generated (Sosa et al. 2013).

On the other hand, the cancer cell may choose to follow the anaerobic metabolism even when the oxygen offer is high and the mitochondria is perfect, an effect known as Warburg effect (Liberti and Locasale 2016), that firstly suggested that the mitochondria of these cells were not able to produce enough energy to their development and mitosis, but now recognized as a voluntary adaptation of the cell, modifying the mitochondria shape and function and even inactivating it, bringing mitochondrial dysfunction (Kheirandish-Rostami et al. 2020). The lower production of ROS by increasing anaerobic metabolism and reducing the mitochondrial respiration may be enough to generate ROS levels able to promote adaptations avoiding oxidative stress triggers of cell cycle arrest and apoptosis keeping the cell

Table 12.2 Overview of different cancer types and bioactives with anticancer activity in *in vivo* studies published in 2023

Cancer type	Animal model	Bioactive or extract and methodology	Concentration	Main outcomes and conclusion	References
Breast cancer	Murine Erlich ascites carcinoma in Swiss albino mouse	Hesperidin methyl chalcone (HMC- flavonoid)	30 mg/kg of HMC or hesperetin 24 h after tumor inoculation followed by daily administration for 15 days.	Significant inhibition of tumor growth, improvement of the expression of antioxidant enzymes.	Rizvi et al. (2023)
Lung cancer	male C57BL/6 mice	*Pericarpium Citri Reticulatae* 'Chachi' flavonoid rich extracts	Pre-treatment with 400 mg/kg for 5 days, followed by tumor inoculation and 14 days of treatment with 400 mg/kg/day	Reduction of tumor volume, induction of apoptosis *in vivo*. No toxicity was found in heart, liver, spleen, lung and kidney.	Tao et al. (2023)
Colon cancer	BALB/c mice	Black rice bran rich in flavonoids, anthocyanins and phenolics.	Concentration adjusted to match AIN-93 M ration standard. Animals were fed for 17 weeks after inoculation.	Modification in intestine microbiome, inhibition of tumor growth, upregulation of caspase-3 and caspase-8.	Budijanto et al. (2023)
Endometrial cancer	BALB/c nude mice	Apigenin	Intraperitoneal injection of 50 mg/kg for 21 days	Inhibition of tumor growth with reduction of 50% of the weight and volume. Apoptotic cells were observed in tumor area.	Liang et al. (2023)
Hepatocarcinoma and colorectal cancer	Wistar rats induced with diethylnitrosamine (liver cancer) and 12-dimethylhydrazine (colon cancer)	Six partially purified fractions containing 2,4′-dihydroxy-6′-methoxy-3′,5′-dimethylchalcone (DMC) from *Cleistocalyx nervosum* var. paniala seeds	20 mg/kg bw of DMC for 10 weeks	Inhibition of the formation of neoplasic colonies in treated rats. Upregulated expression of GST enzyme and reduced the aberrant crypt foci formation.	Vachiraarunwong et al. (2023)

(continued)

Table 12.2 (continued)

Cancer type	Animal model	Bioactive or extract and methodology	Concentration	Main outcomes and conclusion	References
Breast adenocarcinoma	Murine Erlich ascites carcinoma in Swiss albino mouse	β-carotene (BC) and lupeol (LUP)	Injection and oral administration of BC (10 mg/kg) and LUP (200 mg/kg) or a combination of them for 9 days.	BC and Lupeol decreased the tumor's volume, weight, and viable cell count while enhancing the life span of mice bearing EAC tumors. No toxicity was observed in doses until 2000 mg/kg.	Jana et al. (2023)
Glioma	Sprague-Dawley rats transfected with C6 rat glioma cells	*Brassica oleracea* var. *italica* seeds, sprout and inflorescences freeze-dried methanol extracts	Pre-treatment of animals with 100 mg/kg daily for 30 days.	Treatment reduced tumor growth and progression. Effects were partially attributed to isothiocyanate sulforaphane.	Baenas et al. (2023)
Hepato-renal carcinoma	Male rats with oral administration of rhodamine B	*Saurauia vulcani* ethanol extract (EES)	Rats received EES at dose 500 mg/kg for 22 days	Improvement of severe hepatorenal histological abnormalities, reduction of liver damage and repair of renal tubular damage.	Sinaga et al. (2023)
Leiomyoma	BALB/c mice with transplanted ELT-3 cancer cells	Manzamine A (Manz A)	Intraperitoneainjection of Manz A (2 mg/kg/day) once daily for 10 days	The inhibitory effect of ManzA on cell proliferation was partially restored by PERK knockdown and eliminated by tauroursodeoxycholic acid, suggesting oxidative stress plays a critical role in the mechanism of action of Manz A	Lin et al. (2023)

proliferation by upregulating the nuclear erythroid-related factor 2 (NRF2), leading to a hyper expression of antioxidant enzymes as glutathione, thioredoxin reductase, peroxiredoxins and NADPH (Hayes et al. 2020; Park et al. 2020). Additionally, a reduced expression of p53 protein, related to the cell cycle arrest and modulated by the redox balance, is also observed (Hayes et al. 2020). The redox imbalance may harm DNA, leading to new mutations related with the initiation, progression and metastasis of the tumor (Gonzalez-Hunt et al. 2018; Tafani et al. 2016) or the activation of oncogenes (Garcia-Oliveira et al. 2021). The preference for the anaerobic metabolism may explain the increased glucose uptake from cancer cells when compared with non-tumor counterparts. With a lesser production of energy by glycolysis when compared with tri carboxylic acid cycle, a higher substrate concentration is required to produce the same amount of energy and carbon skeletons needed in mitosis. This, alongside to 10–100 times faster degradation of glucose by anaerobic metabolism compared with oxidative mitochondrial metabolism, results in a faster energy production and higher glucose uptake (Liberti and Locasale 2016).

The preference of the glycolysis for energy production alongside a reduction in mitochondria activity creates a therapeutic target against cancer cells, once only a few enzymes are involved in this metabolism, with the lactate dehydrogenase (LDH) enzymes over expressed (Hayes et al. 2020; Park et al. 2020). These enzymes are responsible for the conversion of pyruvate to lactate and backwards, with the LDHa exhibiting preference for the lactate production pathway (Park et al. 2020). Unfortunately, information about food bioactives with LDH direct inhibition capacity is scarce. Vulgaxanthin I, a betaxanthin present in beetroot (Sawicki and Wiczkowski 2018) is reported as possible LDH inhibitor by molecular docking (Kusumawati et al. 2018), and may be involved in beetroot anticancer activity (Coimbra et al. 2023). Recently, other three phytochemicals from *Oroxylum indicum* (Chrysin-7-O-glucuronide, Oroxindin and Oroxin A) were presented as potential inhibitors of the LDH enzyme (Ahmed et al. 2023a). Despite this lack of information about direct interactions of food bioactives with the LDH, there is some information about the influence over the hypoxia-induced factor 1 (HIF-1). The HIF-1 is a regulator of different enzymes involved in Warburg phenotype with flavonoids, present in a plenty of edible vegetables, reported as potential inhibitors (Samec et al. 2021). The redox imbalance also is related with poor protein expression by endoplasmic reticulum (ER), leading to a poor capacity of protein folding, resulting in ER stress, which is associated with tumorigenesis due to the activation of response mechanisms to avoid apoptosis (Kim and Kim 2018). The antioxidant activity of bioactive compounds may improve the redox imbalance of the tumor cells, bringing it to a reduced state. Once back to normal redox conditions, the tumor cell redox environment may favor an adjustment in NRF-2 expression and the pathway follow-up, which may contribute to the cell apoptosis (Ezzati et al. 2020).

12.3.2 Epigenetic Influence

Epigenetics originates from the interaction between DNA and histone proteins (H2A, H2B, H3, H4, and various variant), two complex molecules intricately linked within chromatin. Modifications to DNA and histone proteins play a role in regulating accessibility and function. An anomaly in these modifications is a recognized characteristic of cancer. Epigenetic therapies aim to restore normal patterns of DNA methylation and post-translational modifications on histones that encourage or sustain a malignant phenotype (Bates 2020). DNA methylation, the initial epigenetic modification identified in humans, serves to determine gene activation or inhibition (Chen et al. 2022). DNA methylation is the covalent addition of a methyl group to the 5′ carbon of cytosinenucleotides adjacent to guanines, also known as CpG dinucleotides, creating 5-methylcytosine nucleotides (Zhao et al. 2020). This event is mainly catalyzed by the DNA methyltransferase (DNMT) family, including DNMT1, DNMT3A, and DNMT3B. S-adenosylmethionine (SAM) is the methyl donor provided by the methionine cycle (Chen et al. 2022). Guanine-rich regions, called CpG islands, are mostly unmethylated in normal cells and located in regulatory regions of housekeeping genes, tissue-specific genes and tumour suppressors (Stefanska et al. 2012). Aberrant DNA methylation patterns play a crucial role in genomic instability, activation of oncogenes and silencing of tumor suppressor genes involved in cell proliferation, cell cycle, DNA repair, stress response and apoptosis (Carlos-Reyes et al. 2019). Numerous genes are linked to apoptosis, either promoting or impeding the process. For instance, proteins from proto-oncogenes MYC and BAX, the tumor-suppressor gene p53, and the E2F transcription factor activate apoptosis. Conversely, oncoproteins such as BCL2, ABL, and RAS act to inhibit apoptosis. Additionally, RB has the potential to inhibit the growth of human cancer cells (Kontomanolis et al. 2020). Histone methylation consists of three components; writers (Histone Methyl Transferases: HMTs), readers (Histone Methylation Recognizing Proteins), and erasers (Histone Demethylases: HDMs). It can occur in arginine and lysine and is catalyzed by HMTs (KMTs or RMTs), that transfer the methyl group from SAM to arginine and lysine residues. Lysine residues can be mono-, di- or tri-methylated, while arginine residues can only be mono- or di-methylate (Bouyahya et al. 2022). The methylation of histones is associated with activation, as well as repression of gene expression, depending on the position of amino acid residues in a protein and the level of methylation (Nowacka-Zawisza and Wiśnik 2017). In general, H3K4, H3K36, and H3K79 are considered as active marks that occupy the actively transcribed gene regions in chromatin. H3K9, H3K27, and H4K20 are known as repressive marks that are usually associated with the silenced gene expression and condensed chromatin (Fang and Han 2021). The methylation isa key regulators of cellular proliferation, differentiation, division and plasticity (Carlos-Reyes et al. 2019).

Histone acetylation and deacetylation are correlated with the active/open chromatin conformation and inactive/condensed chromatin form, respectively. This modification occurs at N-terminal lysine residues through the actions of histone acetyl transferases (HATs) and histone deacetylases (HDACs), contributing to epigenetic regulation (Bouyahya et al. 2022; Yen et al. 2016). HATs and HDACs modulate the acetylation of a wide range of cellular proteins in addition to histones. Non-histone targets for acetylation control include transcription factors (e.g., p53, HIF-1), chaperones (e.g. HSP90), structural proteins (e.g. tubulin), signaling molecules (e.g., Smad7) and DNA damage regulators (e.g., Ku70) (Spange et al. 2009; Ganesan et al. 2009). Through these targets, HDACs influence diverse biological pathways such as cell division, survival, differentiation, motility, and angiogenesis (Ganesan et al. 2009). Non-coding RNAs significantly influence epigenetic modification, regulating gene and chromosome expression to control cell differentiation. They're not translated into proteins, categorized into maintenance and regulatory types, further divided into long and small RNAs, such as microRNAs (miRNA) (Wei et al. 2017; Ramassone et al. 2018). MicroRNAs influence gene expression by inhibiting translation of their bound messenger RNA or facilitating degradation of the target messenger RNA (Sevinç 2021). Studies revealed that DNA methylation governs miRNA expression. Moreover, miRNAs have the ability to regulate histone modifications and the expression of DNMT, in addition to influencing DNA methylation itself (Sevinç 2021). The miRNA genes exhibit dual regulation in cancer cells, either being up-regulated as oncogenes to down-regulate tumor-suppressor genes or down-regulated as tumor suppressors to inhibit oncogenes. The specific function of a miRNA gene is determined by its targets in a particular tissue. In one cell type, a miRNA gene may act as a tumor suppressor if its crucial target is an oncogene, while in a different cell type, it can function as an oncogene if its target is a tumor-suppressor gene (Croce 2008). For example miR-17, miR-20a, miR-19b-1, miR-18a, miR-19a, miR-92-1, miR-21, and miR-155 may have critical roles as oncogenic miRNAs (oncomiRs), while miR-15, miR-16, let-7, miR-34, miR-48, miR-84, miR-241 may have anticancer activities as tumor suppressor miRNAs (Gezici and Sekeroglu 2017). Therefore, the regulation of miRNA levels by bioactive compounds holds promise as a strategy for cancer treatment. Given the substantial advantages in terms of miRNA modulation by bioactive dietary agents and their potential to down-regulate oncomiRs (such as miR-21, miR-17-92 cluster, miR-92, etc.), activate tumor suppressor genes (including p53, PDCD4, and PTEN), and sensitize treatment-resistant cells, it presents a compelling avenue for therapeutic intervention (Gavrilas et al. 2016). Recent reviews have highlighted the significant therapeutic role of bioactive compounds against cancer through epigenetic pathways. Alkaloids, terpenoids, quinones, and isothiocyanates have demonstrated the ability to modulate epigenetic modifications in human cancer cells, involving various mechanisms such as DNA inhibition and/or methylation, as well as histone modifications (El Omari et al. 2021). Similarly, polyphenols, in general, appear to sensitize tumor cells through different epigenetic targets, including oncogenes, tumor suppressor genes, as well as DNMTs and HDACs (Bouyahya et al. 2022). Epigenetic studies of various individual flavonols (quercetin, kaempferol, myricetin), flavones (luteolin, apigenin), flavan-3-ols (fisetin, EGCG), isoflavones (genistein, daidzein), flavanones (silybin, hesperetin, naringenin), and anthocyanidins (cyanidin, delphinidin, malvidin, and pelargonidin) have shown strong preclinical potential (Khan et al. 2021). Specifically, kaempferol, gallic acid, curcumin, resveratrol, EGCG, and quercetin negatively regulate HDAC expression, primarily HDAC 1 and HDAC 3. Furthermore, combining these polyphenols with chemotherapeutic agents demonstrates a synergistic effect in cancer therapy (Rajendran et al. 2022). The efficacy of red fruits in inhibiting epigenetic modulations involved in cancer has been studied, especially blackberries and their derived anthocyanin extracts. These modifications involve the reactivation of tumor suppressor genes by demethylation of their promoters and suppression of DNA methyltransferase 1. Anthocyanin metabolites megal, gal, and 2,4,6-trihydroxybenzene exert anticarcinogenic effects through modulation of multiple signaling pathways, but it is not yet established whether these metabolites can alter epigenetic modulations in cancer (Abo El-Ella and Bishayee 2019). Regarding organic acids, vitamin C stands out for its DNA demethylation mechanism mediated by activation of the ten-eleven translocation enzyme (Shenoy et al. 2018). Recent studies suggest that vitamin C, at physiological concentrations, associated with hypomethylating agents, can act synergistically to promote DNA demethylation (Gillberg et al. 2018). Some compounds such as selenium, isothiocyanates, curcumin, tea polyphenols, genistein, withaferin A, and indole-3-carbinol show promise by acting through DNA methylation, histone modification, and non-coding RNAs (Özyalçin and Sanlier 2020). Compounds that are components of one-carbon metabolism, such as folate, riboflavin, pyridoxine, cobalamin, choline, betaine, and methionine, affect DNA methylation by regulating the levels of SAM. Other natural compounds target histone modifications and levels of non-coding RNAs, such as resveratrol, which activates the deacetylase sirtuin and regulates oncogenic and tumor suppressor micro-RNAs; or vitamin D, which recruits histone acetylases (Stefanska et al. 2012). Lastly, coumarin, hydroxamic acid, and their derivatives act through inhibition of histone deacetylase (Khai and Vu 2024). Furthermore, a pictorial depiction were made (Fig. 12.1) to understand the molecular pathways of food bioactives and its anticancer activity.

12.4 Safety Concerns with Food Matrix in Bioactives Prospection

Regarding the prospection of bioactive compounds with anticancer activity in vegetables, naturally the focus becomes the obtaining of these compounds and the evaluation of the anticancer capacity against different types of cancer. However, leaving aside the safety of obtaining these compounds from the food matrix is a common mistake once food contamination continues to be a global problem (Thompson and Darwish 2019). The main food contaminants are the presence and accumulation of heavy metals, the use of pesticides in cultivation or accumulation in groundwater, and the presence of natural toxins alongside to the possibility of the presence of pharmaceutical residues (Thompson and Darwish 2019). The main contaminant of the food may be different according to the food part. While roots are

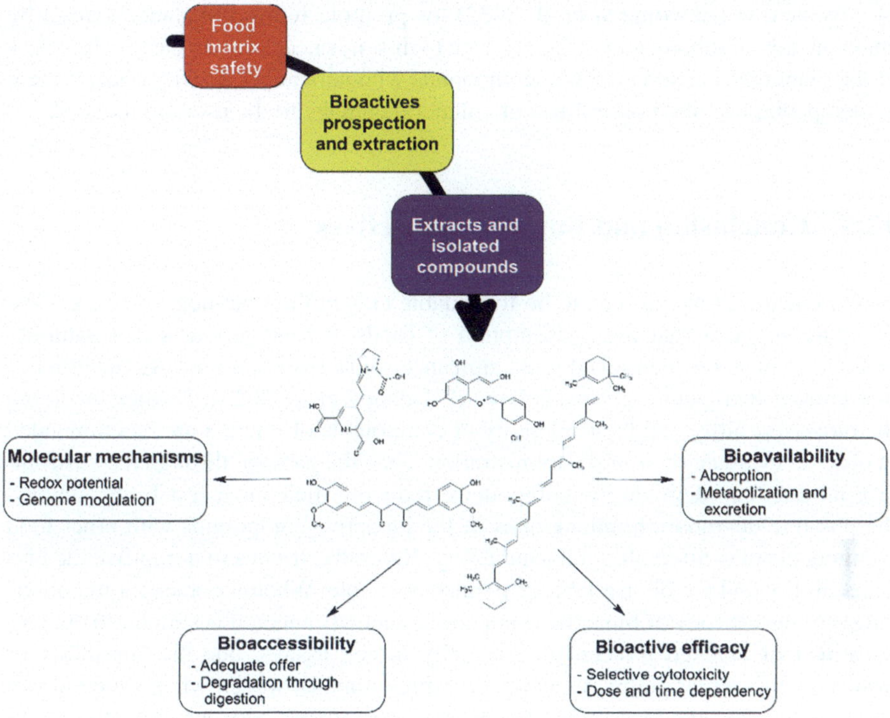

Fig. 12.1 Prospection of food bioactives with anticancer activity

more susceptible to ground contaminants as heavy metals, pesticides, and pharmaceutical residues (Knez et al. 2022), leaves and fruits are more susceptible to pesticides accumulation by direct exposition (Wahab et al. 2022). In addition, grains may also present relevant mycotoxins as aflatoxin and ochratoxin (Ekwomadu et al. 2022). Once the main vegetable residues are leaves, peels and seeds (Nguyen et al. 2020), it is recommended to pay special attention to food contaminants when prospecting anticancer bioactives in these food matrixes. The concern with the presence of contaminants in food matrices is related to the extraction processes of the bioactive compounds of interest from each matrix, using conventional extraction methods, with the use of organic solvents and water (Brglez Mojzer et al. 2016), and also adjuvant methods for greater release of compounds, such as the use of ultrasound (probe, bath, etc.) (Bhargava et al. 2021), microwaves (Guzik et al. 2022), enzymes (Gligor et al. 2019), or combinations of these techniques. However, in addition to the release of bioactive compounds, it is possible that there will also be a greater release of contaminants, previously fixed in the plant matrix but which are now released due to changes in the polymeric structure resulting from the use of these adjuvant technologies. The presence of contaminants in the extract can create false positives for anticancer activity, since these contaminants are harmful to human health and can induce cell apoptosis, such as by promoting DNA damage, as done by mycotoxins (Ekwomadu et al. 2022), or promote redox imbalance caused by heavy metals (Paithankar et al. 2021). Due to this, it is necessary to ensure the safety of the plant matrix used to obtain compounds with anticancer activity once, if there is therapeutic use, the possibilities of collateral damage to the user are reduced.

12.5 Conclusion and Future Perspectives

Food bioactives have proven to be formidable aids in the treatment of cancer. The literature suggests that the consumption of foods or food products that naturally contain, or have been enriched with, anticancer bioactives can improve the efficacy that current therapeutic protocols have (Elfadadny et al. 2023). Studies involving the bioaccessibility and bioavailability of compounds of interest are recommended in order to evaluate their oral administration, once the passage through the gastrointestinal tract can degrade the compounds, reducing their bioaccessibility. Beyond the possible degradation, interactions of the bioactive compounds with other food components may affect their bioavailability. Naturally, studies that improve the bioaccessibility and the bioavailability are also desirable. When it comes for the observation of the efficacy of bioactive compounds against cancer, it is simpler to observe the effect of isolated molecules, especially in cell assays, due the possibility of observing molecular pathways and gene expressions. However, when studying isolated molecules, the possibility to observe synergisms between bioactive compounds obtained from complex extracts of food matrices is lost. On the other hand, evaluating food extracts becomes an arduous task due to the diversity of compounds that can be extracted, their different concentrations and possibilities of molecular interaction with each other and with therapeutic targets, which brings the necessity of the identification of the compounds present in the extract. The quantification of identified compounds is also a relevant task because, associated with bioaccessibility and bioavailability studies, it will guide the establishment of therapeutic doses that can be further studied in animal models as pre-clinical trials. Additionally, the clearance of the bioactives after the absorption and possible related toxicity must also be addressed. It is suggested that these concerns should be taken into consideration in the prospecting of bioactive compounds with anticancer activity in food matrices. Therefore, bioactive food compounds can be recognized for cancer therapeutic and can be added to official protocols.

References

Abo El-Ella DM, Bishayee A (2019) The epigenetic targets of berry anthocyanins in cancer prevention. In: Epigenetics of cancer prevention. Elsevier, pp 129–148

Ahmed SS, Rahman MO, Alqahtani AS, Sultana N, Almarfadi OM, Ali MA, Lee J (2023a) Anticancer potential of phytochemicals from Oroxylum indicum targeting Lactate Dehydrogenase a through bioinformatic approach. Toxicol Rep 10:56–75

Ahmed RM, Esmaeil DAM, El-Nagdy SY, El-Sisi NA (2023b) Antitumoral properties of the pomegranate peel and blueberry extracts against tongue carcinoma (in vitro study). Saudi Dent J 35:985–995

Asenye GE, Akinleye TE, George UE, Mgbeje BI, Adewumi OM, Adeniji JA, Ogbole OO, Ebong PE (2023) Cytotoxic action of the leaves of Uvaria chamae P.Beauv. and Dicliptera paniculata (Forssk.) I.Darbysh. From Nigeria mediated through intrinsic apoptotic pathway induction in four cancer cell lines. Phytomed Plus 3:100423

Baenas N, Vega-García A, Manjarrez-Marmolejo J, Moreno DA, Feria-Romero IA (2023) The preventive effects of broccoli bioactives against cancer: evidence from a validated rat glioma model. Biomed Pharmacother 168:115720

Bates SE (2020) Epigenetic therapies for cancer. N Engl J Med 383:650–663

Bhargava N, Mor RS, Kumar K, Sharanagat VS (2021) Advances in application of ultrasound in food processing: a review. Ultrason Sonochem 70:105293

Bouyahya A, Mechchate H, Oumeslakht L et al (2022) The role of epigenetic modifications in human cancers and the use of natural compounds as epidrugs: mechanistic pathways and pharmacodynamic actions. Biomol Ther 12:367

Brglez Mojzer E, Knez Hrnčič M, Škerget M, Knez Ž, Bren U (2016) Polyphenols: extraction methods, antioxidative action, bioavailability and anticarcinogenic effects. Molecules 21:901

Budijanto S, Nurtiana W, Muniroh A et al (2023) Dietary supplementation of black rice bran to colon carcinogen-induced mice: examining the development of colorectal cancer by improving environmental colon conditions. Heliyon 9:e18528

Carlos-Reyes Á, López-González JS, Meneses-Flores M, Gallardo-Rincón D, Ruíz-García E, Marchat LA, Astudillo-de la Vega H, Hernández de la Cruz ON, López-Camarillo C (2019) Dietary compounds as epigenetic modulating agents in cancer. Front Genet 10. https://doi.org/10.3389/fgene.2019.00079

Chen C, Wang Z, Ding Y, Wang L, Wang S, Wang H, Qin Y (2022) DNA methylation: from cancer biology to clinical perspectives. Front Biosci (Landmark Ed) 27:326

Chen X-J, Liu Z-B, Li X et al (2023) 3-Hydroxymorindone from Knoxia roxburghii (Spreng.) M. A. Rau induces ROS-mediated mitochondrial dysfunction cervical cancer cells apoptosis via inhibition of PI3K/AKT/NF-κB signaling pathway. J Funct Foods 103:105498

Chinembiri T, du Plessis L, Gerber M, Hamman J, du Plessis J (2014) Review of natural compounds for potential skin cancer treatment. Molecules 19:11679–11721

Choudhari AS, Mandave PC, Deshpande M, Ranjekar P, Prakash O (2020) Phytochemicals in cancer treatment: from preclinical studies to clinical practice. Front Pharmacol 10. https://doi.org/10.3389/fphar.2019.01614

Coimbra PPS, da Silva-e-Silva ACAG, da S Antonio A, Pereira HMG, da Veiga-Junior VF, Felzenszwalb I, Araujo-Lima CF, Teodoro AJ (2023) Antioxidant capacity, antitumor activity and metabolomic profile of a beetroot peel flour. Metabolites 13. https://doi.org/10.3390/metabo13020277

Croce CM (2008) Oncogenes and cancer. N Engl J Med 358:502–511

De S, Paul S, Manna A et al (2023) Phenolic phytochemicals for prevention and treatment of colorectal cancer: a critical evaluation of in vivo studies. Cancers (Basel) 15:993

del Saz-Lara A, Boughanem H, López de las Hazas M-C, Crespo C, Saz-Lara A, Visioli F, Macias-González M, Dávalos A (2023) Hydroxytyrosol decreases EDNRA expression through epigenetic modification in colorectal cancer cells. Pharmacol Res 187:106612

Dillard CJ, German JB (2000) Phytochemicals: nutraceuticals and human health. J Sci Food Agric 80:1744–1756

Ekwomadu T, Mwanza M, Musekiwa A (2022) Mycotoxin-linked mutations and cancer risk: a global health issue. Int J Environ Res Public Health 19:7754

El Omari N, Bakrim S, Bakha M, Lorenzo JM, Rebezov M, Shariati MA, Aboulaghras S, Balahbib A, Khayrullin M, Bouyahya A (2021) Natural bioactive compounds targeting epigenetic pathways in cancer: a review on alkaloids, terpenoids, quinones, and isothiocyanates. Nutrients 13:3714

Elfadadny A, Ragab RF, Hamada R, Al Jaouni SK, Fu J, Mousa SA, El-Far AH (2023) Natural bioactive compounds-doxorubicin combinations targeting topoisomerase II-alpha: anticancer efficacy and safety. Toxicol Appl Pharmacol 461:116405

Ezzati M, Yousefi B, Velaei K, Safa A (2020) A review on anti-cancer properties of Quercetin in breast cancer. Life Sci 248:117463. https://doi.org/10.1016/j.lfs.2020.117463

Fang D, Han J (2021) Histone mutations and cancer. Adv Exp Med Biol. https://doi.org/10.1007/978-981-15-8104-5

Ganesan A, Nolan L, Crabb S, Packham G (2009) Epigenetic therapy: histone acetylation, DNA methylation and anti-cancer drug discovery. Curr Cancer Drug Targets 9:963–981

Gao Q, Feng J, Liu W et al (2022) Opportunities and challenges for co-delivery nanomedicines based on combination of phytochemicals with chemotherapeutic drugs in cancer treatment. Adv Drug Deliv Rev 188:114445

Gao N, Si X, Han W, Gong E, Shu C, Tian J, Wang Y, Zhang J, Li B, Li B (2023) The contribution of different polyphenol compositions from chokeberry produced in China to cellular antioxidant and antiproliferative activities. Food Sci Human Wellness 12:1590–1600

Garcia-Oliveira P, Otero P, Pereira AG, Chamorro F, Carpena M, Echave J, Fraga-Corral M, Simal-Gandara J, Prieto MA (2021) Status and challenges of plant-anticancer compounds in cancer treatment. Pharmaceuticals 14:157

Gavrilas L, Ionescu C, Tudoran O, Lisencu C, Balacescu O, Miere D (2016) The role of bioactive dietary components in modulating miRNA expression in colorectal cancer. Nutrients 8:590

Gezici S, Sekeroglu N (2017) Regulation of MicroRNAs by natural products and bioactive compounds obtained from common medicinal plants: novel strategy in cancer therapy. Indian J Pharm Educ Res 51:s483–s488

Gillberg L, Ørskov AD, Liu M, Harsløf LBS, Jones PA, Grønbæk K (2018) Vitamin C—a new player in regulation of the cancer epigenome. Semin Cancer Biol 51:59–67

Gligor O, Mocan A, Moldovan C, Locatelli M, Crişan G, Ferreira ICFR (2019) Enzyme-assisted extractions of polyphenols—a comprehensive review. Trends Food Sci Technol 88:302–315

Gonzalez-Hunt CP, Wadhwa M, Sanders LH (2018) DNA damage by oxidative stress: measurement strategies for two genomes. Curr Opin Toxicol 7:87–94

Guzik P, Kulawik P, Zając M, Migdał W (2022) Microwave applications in the food industry: an overview of recent developments. Crit Rev Food Sci Nutr 62:7989–8008

Haque A, Brazeau D, Amin AR (2021) Perspectives on natural compounds in chemoprevention and treatment of cancer: an update with new promising compounds. Eur J Cancer 149:165–183

Hayes JD, Dinkova-Kostova AT, Tew KD (2020) Oxidative stress in cancer. Cancer Cell 38:167–197

IARC (2020) International Agency Research on Cancer/WHO, GloboCan 2020. In: International Agency Research on Cancer/WHO. https://gco.iarc.fr/. Accessed 16 Nov 2023

Jana S, Sarkar N, Chakraborty M, Kar A, Banerjee S, Bhattacharya S, Mukherjee PK, Haldar PK (2023) Combination synergy between β-carotene and lupeol against breast adenocarcinoma in vitro and in vivo using combination index. Phytomed Plus 3:100392

Khai NQ, Vu TK (2024) Coumarin-derived hydroxamic acids as histone deacetylase inhibitors: a review of anti-cancer activities. Anti Cancer Agents Med Chem 24:18–29

Khan H, Belwal T, Efferth T et al (2021) Targeting epigenetics in cancer: therapeutic potential of flavonoids. Crit Rev Food Sci Nutr 61:1616–1639

Khatoon E, Banik K, Harsha C, Sailo BL, Thakur KK, Khwairakpam AD, Vikkurthi R, Devi TB, Gupta SC, Kunnumakkara AB (2022) Phytochemicals in cancer cell chemosensitization: current knowledge and future perspectives. Semin Cancer Biol 80:306–339

Kheirandish-Rostami M, Roudkenar MH, Jahanian-Najafabadi A, Tomita K, Kuwahara Y, Sato T, Roushandeh AM (2020) Mitochondrial characteristics contribute to proliferation and migration potency of MDA-MB-231 cancer cells and their response to cisplatin treatment. Life Sci 244:117339

Kim C, Kim B (2018) Anti-cancer natural products and their bioactive compounds inducing ER stress-mediated apoptosis: a review. Nutrients 10. https://doi.org/10.3390/nu10081021

Knez E, Kadac-Czapska K, Dmochowska-Ślęzak K, Grembecka M (2022) Root vegetables—composition, health effects, and contaminants. Int J Environ Res Public Health 19:15531

Koh Y-C, Ho C-T, Pan M-H (2020) Recent advances in cancer chemoprevention with phytochemicals. J Food Drug Anal 28:14–37

Kontomanolis EN, Koutras A, Syllaios A et al (2020) Role of oncogenes and tumor-suppressor genes in carcinogenesis: a review. Anticancer Res 40:6009–6015

Kusumawati R, Nasrullah AH, Pesik RN, Muthmainah, Indarto D (2018) Secondary metabolites of Mirabilis jalapa structurally inhibit Lactate Dehydrogenase A in silico: a potential cancer treatment. IOP Conf Ser Mater Sci Eng 333:012078. https://doi.org/10.1088/1757-899X/333/1/012078

Kwon N, Kim D, Swamy KMK, Yoon J (2021) Metal-coordinated fluorescent and luminescent probes for reactive oxygen species (ROS) and reactive nitrogen species (RNS). Coord Chem Rev 427:213581

Liang Y, Zhong Q, Ma R, Ni Z, Thakur K, Zhang J, Wei Z (2023) Apigenin, a natural flavonoid, promotes autophagy and ferroptosis in human endometrial carcinoma Ishikawa cells in vitro and in vivo. Food Sci Human Wellness 12:2242–2251

Liberti MV, Locasale JW (2016) The Warburg effect: how does it benefit cancer cells? Trends Biochem Sci 41:211–218

Lin L-C, Chang H-Y, Kuo T-T, Chen H-Y, Liu W-S, Lo Y-J, Hsia S-M, Huang T-C (2023) Oxidative stress mediates the inhibitory effects of Manzamine A on uterine leiomyoma cell proliferation and extracellular matrix deposition via SOAT inhibition. Redox Biol 66:102861

Liu B, Li C, Han J, Chen Y, Zhao Z, Lu H (2023) Biosynthesized gold nanoparticles using leaf extract of Citrus medica inhibit hepatocellular carcinoma through regulation of the Wnt/β--catenin signaling pathway. Arab J Chem 16:104800

Mirabelli P, Coppola L, Salvatore M (2019) Cancer cell lines are useful model systems for medical research. Cancers (Basel) 11:1098

Molassiotis A, Fernadez-Ortega P, Pud D et al (2005) Use of complementary and alternative medicine in cancer patients: a European survey. Ann Oncol 16:655–663

Mostafa H, Behrendt I, Meroño T, González-Domínguez R, Fasshauer M, Rudloff S, Andres-Lacueva C, Kuntz S (2023) Plasma anthocyanins and their metabolites reduce in vitro migration of pancreatic cancer cells, PANC-1, in a FAK- and NF-kB dependent manner: results from the ATTACH-study a randomized, controlled, crossover trial in healthy subjects. Biomed Pharmacother 158:114076

Murugesan M, Kandhavelu M, Thiyagarajan R, Natesan S, Rajendran P, Murugesan A (2023) Marine halophyte derived polyphenols inhibit glioma cell growth through mitogen-activated protein kinase signaling pathway. Biomed Pharmacother 159:114288

Nguyen TT, Rosello C, Bélanger R, Ratti C (2020) Fate of residual pesticides in fruit and vegetable waste (FVW) processing. Food Secur 9:1468

Nowacka-Zawisza M, Wiśnik E (2017) DNA methylation and histone modifications as epigenetic regulation in prostate cancer. Oncol Rep 38:2587–2596

Nussbaumer S, Bonnabry P, Veuthey J-L, Fleury-Souverain S (2011) Analysis of anticancer drugs: a review. Talanta 85:2265–2289

Özyalçin B, Sanlier N (2020) The effect of diet components on cancer with epigenetic mechanisms. Trends Food Sci Technol 102:138–145

Paithankar JG, Saini S, Dwivedi S, Sharma A, Chowdhuri DK (2021) Heavy metal associated health hazards: an interplay of oxidative stress and signal transduction. Chemosphere 262:128350

Park JH, Pyun WY, Park HW (2020) Cancer metabolism: phenotype, signaling and therapeutic targets, vol 9. Cells. https://doi.org/10.3390/cells9102308

Poojary MM, Nguyen TD, Dekiwadia C, Dias DA, Huynh T (2023) Phenolic compounds-containing fruit peel extracts of Garcinia humilis exhibit anti-melanoma activity. Food Biosci 52:102428

Rajendran P, Abdelsalam SA, Renu K, Veeraraghavan V, Ben Ammar R, Ahmed EA (2022) Polyphenols as potent epigenetics agents for cancer. Int J Mol Sci 23:11712

Ramassone A, Pagotto S, Veronese A, Visone R (2018) Epigenetics and MicroRNAs in cancer. Int J Mol Sci 19:459
Rizeq B, Gupta I, Ilesanmi J, AlSafran M, Rahman MM, Ouhtit A (2020) The power of phytochemicals combination in cancer chemoprevention. J Cancer 11:4521–4533
Rizvi SMD, Mudagal MP, Boregowda SS, Hussain T, Al Hagbani T, Abdallah MH, Khafagy E-S, Hussain A, Yousif Adam FA, Abu Lila AS (2023) The flavonoid hesperidin methyl chalcone as a potential therapeutic agent for cancer therapy: molecular docking, in vitro cytotoxicity, and in vivo antitumor activity. Arab J Chem 16:104769
Ruggeri BA, Camp F, Miknyoczki S (2014) Animal models of disease: pre-clinical animal models of cancer and their applications and utility in drug discovery. Biochem Pharmacol 87:150–161
Sair AT, Li Y, Zhao W, Li T, Liu RH (2023) Anticancer activity of apple peel extracts against human breast cancer cells through insulin-like growth factor-1 signal transduction pathway. J Agric Food Res 11:100507
Samec M, Liskova A, Koklesova L et al (2021) Flavonoids targeting HIF-1: implications on cancer metabolism. Cancers (Basel) 13:1–27
Sawasdee N, Jantakee K, Wathikthinnakon M, Panwong S, Pekkoh J, Duangjan K, Yenchitsomanus P, Panya A (2023) Microalga chlorella sp. extract induced apoptotic cell death of cholangiocarcinoma via AKT/mTOR signaling pathway. Biomed Pharmacother 160:114306
Sawicki T, Wiczkowski W (2018) The effects of boiling and fermentation on betalain profiles and antioxidant capacities of red beetroot products. Food Chem 259:292–303
Sevinç S (2021) Effects of DNA methylation on cancer and aging. J Exp Basic Med Sci 1:126–130
Shenoy N, Creagan E, Witzig T, Levine M (2018) Ascorbic acid in cancer treatment: let the phoenix fly. Cancer Cell 34:700–706
Sinaga E, Hasanah U, Sipahutar FRP (2023) Chemopreventive potential of Saurauia vulcani korth in improving Rhodamine B induced hepato-renal carcinoma in rats. Pharmacol Res Mod Chin Med 9:100336
Sosa V, Moliné T, Somoza R, Paciucci R, Kondoh H, LLeonart ME (2013) Oxidative stress and cancer: an overview. Ageing Res Rev 12:376–390
Spange S, Wagner T, Heinzel T, Krämer OH (2009) Acetylation of non-histone proteins modulates cellular signalling at multiple levels. Int J Biochem Cell Biol 41:185–198
Stefanska B, Karlic H, Varga F, Fabianowska-Majewska K, Haslberger A (2012) Epigenetic mechanisms in anti-cancer actions of bioactive food components—the implications in cancer prevention. Br J Pharmacol 167:279–297
Su J, Jin L, Yang R, Liang Y, Nile SH, Kai G (2023) Comparative studies on selection of high polyphenolic containing Chinese raspberry for evaluation of antioxidant and cytotoxic potentials. J Agric Food Res 12:100603
Tafani M, Sansone L, Limana F, Arcangeli T, De Santis E, Polese M, Fini M, Russo MA (2016) The interplay of reactive oxygen species, hypoxia, inflammation, and sirtuins in cancer initiation and progression. Oxidative Med Cell Longev 2016. https://doi.org/10.1155/2016/3907147
Tao Y, Li R, Xiang L, Zhang X, Liu R, Wu T, Pan S, Xu X (2023) Inhibitory effects of Pericarpium Citri Reticulatae 'Chachi' flavonoid extracts during storage against NSCLC in vivo and in vitro. J Funct Foods 110:105840
Thomas C, Mackey MM, Diaz AA, Cox DP (2009) Hydroxyl radical is produced via the Fenton reaction in submitochondrial particles under oxidative stress: implications for diseases associated with iron accumulation. Redox Rep 14:102–108
Thompson LA, Darwish WS (2019) Environmental chemical contaminants in food: review of a global problem. J Toxicol 2019:1–14
Vachiraarunwong A, Tuntiwechapikul W, Wongnoppavich A, Meepowpan P, Wongpoomchai R (2023) 2,4′-dihydroxy-6′-methoxy-3′,5′-dimethylchalcone from Cleistocalyx nervosum var. paniala seeds attenuated the early stage of diethylnitrosamine and 1,2-dimethylhydrazine-induced colorectal carcinogenesis. Biomed Pharmacother 165:115221

Villada Ramos JA, Aguillón Osma J, Restrepo Cortes B, Loango Chamarro N, Maldonado Celis ME (2023) Identification of potential bioactive compounds of Passiflora edulis leaf extract against colon adenocarcinoma cells. Biochem Biophys Rep 34:101453

Wahab S, Muzammil K, Nasir N, Khan MS, Ahmad MF, Khalid M, Ahmad W, Dawria A, Reddy LKV, Busayli AM (2022) Advancement and new trends in analysis of pesticide residues in food: a comprehensive review. Plan Theory 11:1106

Wei J-W, Huang K, Yang C, Kang C-S (2017) Non-coding RNAs as regulators in epigenetics. Oncol Rep 37:3–9

Yen C-Y, Huang H-W, Shu C-W, Hou M-F, Yuan S-SF, Wang H-R, Chang Y-T, Farooqi AA, Tang J-Y, Chang H-W (2016) DNA methylation, histone acetylation and methylation of epigenetic modifications as a therapeutic approach for cancers. Cancer Lett 373:185–192

Zhao SG, Chen WS, Li H et al (2020) The DNA methylation landscape of advanced prostate cancer. Nat Genet 52:778–789

Chapter 13
Food Bioactives in the Prevention and Clinical Management of Obesity

Alphonse Laya, Eugenia Carvalho, Iva Fernandes, Joana Oliveira, Fabrice Bruno Siewe, Honoré Wangso, and Raphael Djakba

Abstract Obesity is a major public health concern worldwide, which has been increasing at exponential rates over the last decades. In the recent era, conventional drugs, such as orlistat, used to treat obesity present significant side effects such as, nausea, vomiting, and gastrointestinal discomfort on patients, leading the scientific community across the world to find alternative compounds or develop new treatment approaches with no adverse effects. Food bioactives are natural products used as basic nutritional constituents and for their beneficial health effects which include polyunsaturated fatty acids, bioactive peptides, polysaccharides, carotenoids, polyphenolics, vitamins, microminerals, probiotics, phytosterols, and fibers. The consumption of these natural functional foods, have been linked to the reduction of the incidence and progression of obesity in animal models and in clinical trials. These

A. Laya (✉)
Department of Biological Sciences, Faculty of Science, University of Maroua, Maroua, Cameroon

Center for Neuroscience and Cell Biology, Faculdade de Medicina, University of Coimbra, Coimbra, Portugal

E. Carvalho
Center for Neuroscience and Cell Biology, Faculdade de Medicina, University of Coimbra, Coimbra, Portugal

Instituto de Investigação Interdisciplinar, University of Coimbra, Casa Costa Alemão, Rua Dom Francisco de Lemos, Coimbra, Portugal
e-mail: ecarvalh@cnc.uc.pt

I. Fernandes · J. Oliveira
Laborato'rio Associado para a Qu'ımica Verde-REQUIMTE, Departamento de Qu'ımica e Bioqu'ımica, Faculdade de Ciências, Universidade do Porto, Porto, Portugal
e-mail: iva.fernandes@fc.up.pt; jsoliveira@fc.up.pt

F. B. Siewe
Department of Food Science and Technology, Faculty of Agriculture and Veterinary Medicine, University of Buea, Buea, Cameroon

H. Wangso · R. Djakba
Department of Chemistry, Faculty of Science, University of Maroua, Maroua, Cameroon

S. Amir Ashraf, M. Adnan (eds.), *Food Bioactives and Nutraceuticals*,
https://doi.org/10.1007/978-981-97-9723-3_13

food bioactives have anti-obesity activity through mechanisms that include reducing energy intake, lipid synthesis and promote lipolysis by affecting lipase activity. In addition, these compounds regulate genes and pathways related to lipid synthesis and catabolism, and modulate advanced glycation end product accumulation, AMP-activated protein kinase (AMPK) signaling pathway, inflammation, free radical scavenging and immunomodulatory effects. This chapter examines in details food bioactives for effective prevention and clinical management of obesity.

Keywords Anti-obesity · Bioactive compounds · Clinical management · Enzyme · Functional foods · Obesity

13.1 Introduction

Obesity is defined as a chronic disease. It is characterized by abnormal fat accumulation and the degrees in body fat accumulation are measured using the body mass index (BMI) as recommended by the World Health Organization (WHO) as, underweight (<18.5 kg/m^2), normal weight (18.5 to <25 kg/m^2), overweight (25 to <30 kg/m^2) or obese (≥30 kg/m^2), (WHO 2023). It is also normally measured using the waist/hip circumference, that for male is 0.90 and for female is 0.85. It is a multifactorial disease with both environmental and genetic factors involved in its genesis. Obesity is considered as one of the top major public health concerns worldwide according to WHO (2023). The growing rates of obesity are linked with excess calorie intake and reduced levels of physical activity. Importantly, the intake of ultra-processed foods and/or foods highly rich in salt, sugar, as well as saturated fat usually very poor in essential vitamins, minerals, polyphenolic compounds and fibers, have become a real health problem and have worsened the obesity pandemic rates worldwide. According to the organization for economic co-operation and development (OECD), the number of people with obesity has reached more than 60 percent in the European Union (UE) and about 70 percent in the United States of America (USA) (OECD 2019). The high incidence of obesity is an important key risk factors for diabetes, cardiovascular diseases, hypertension, heart diseases, dementia, stroke, fatty liver disease, Alzheimer's and Parkinson's disease as well as, some types of cancers that include breast, colon and prostate cancer (Hachuła et al. 2023; WHO 2024). The OECD has reported that European Union countries spent over 8 billion per year for obesity treatments and the United States of America have spent over 700 billion (OECD 2019). According to World Obesity (WO 2023), the obesity pandemic rates in low- and middle-income countries, have tripled in comparison to 2010 (WO 2023). To date, more than two-thirds of deaths from non-communicable diseases take place in developing countries, where also two-thirds of people with overweight or obesity live (World Bank 2017a, b). However, it was recently reported that more than 51% or over 4 billion people of the global population will have overweight or obesity by 2035, and this pandemic will happen over all regions and continents of the world (World Obesity Atlas, WOA 2023). Due to this epidemic increase in obesity, people will live on average 2.7 years less in the OECD countries (OECD 2019).

In the last decades, several classes of drugs have been developed and utilized for the management of obesity, as shown on Table 13.1. However, these conventional drugs such as, orlistat, phentermine, sibutramine, oleoyl-estrone, liraglutide, naltrexone/bupropion, and lorcaserin as older drugs and liraglutide and semaglutide, tirzepatide, Cagrilintide, and Setmelanotide, phentermine–topiramate as newer drugs have high costs and present serious side effects such as nausea, vomiting, dysfunction of the central nervous system, increased blood pressure, fatigue, headache, sleeping disorders, and gastrointestinal discomfort in the patients (Frohmader et al. 2010; Fidler et al. 2011; Kumar et al. 2022; Kelly 2023; Coutinho and Halpern 2024) as shown on Table 13.1.

As a result, this has led to the common use of plant bioactive product based remedies that lack adverse effects, as described on Table 13.2 below. Obesity can be prevented by changing lifestyle such as, eating a healthy diet and increasing the levels of daily physical activity. People should eat a healthy and balanced diet, and practice regular daily exercise for the basic prevention of obesity related chronic diseases. Otherwise, to tackle the obesity pandemic worldwide, natural food bioactives can be used as excellent alternative strategies as safe and effective ways to reduce the overweight and obesity. Plant-based diets are rich in bioactive compounds including polyphenolics, saponins, carotenoids, vitamins, bioactive peptides as well as polysaccharides, polyunsaturated fatty acids, phytosterols, minerals, short-chain fatty acids (SCFAs), gamma-aminobutyricacid (GABA), and fibers which play key roles in promoting better health benefits. Their synergistic intake and modes of action increase the bioavailability and action of various targets for single treatment or prevention of obesity (Hu et al. 2018; Rahman et al. 2017; Câmara et al. 2021; Duque-Soto et al. 2023; Laya et al. 2023). Polyphenolics found in plants are classified in two groups based on their chemical structure including, flavonoids and non-flavonoids including lignans, stilbenes, xantones and diarlyheptanoids (Tain and Hsu 2022). Bioactive compounds are found in fruits, vegetables, leafy vegetables, some roots and/or tubers as well as other plant-based foods, as described on Table 13.2 below. Seafoods including fish, selfish, crabs, and seaweeds, also contain a large amount of these natural bioactive compounds especially proteins, lipids, phenolics, phlorotanins, vitamins and fibers (Ford et al. 2019; Abeyrathne et al. 2022). These natural bioactive food components combat obesity through the modulation of mechanisms and pathways including antioxidant, antimicrobial, anti-inflammatory, as well as immunomodulatory effects. These food bioactives can reduce the physiological weight gain by modulating/regulating or inhibiting specific cellular properties via signaling pathways such as, leptin, adiponectin, PI3K/AKT, NF-κB, Nrf2, PPARγ, MC4 RA, and AMPK that occurs over time among the global population. Emerging evidence has shown that food bioactives as shown on Table 13.2 below contained such as, phenolics can increase thermogenesis by suppressing fatty acids synthase, energy expenditure by inhibiting adipogenesis, and reduce oxidative stress by inhibiting reactive oxygen species.

Table 13.1 Drugs used for obesity treatment

Drugs	Class of drug	Potential sides effects	Studies	Placebo subtracted Mean weight loss (%)	References
Orlistat[a]	LI	Flatulence, diarrhea, bloating, kidney stones, stool incontinence, oily stool, fatty stool, new-onset heart failure and renal failure	RCT	2.9–3.4	Padwal et al. (2004); Ardissino et al. (2022)
Phentermine-topiramate[a]	SAA	Increase anxiety, depression and cognitive impairment	RCT	10.5	Gadde et al. (2011)
Naltrexone-bupropion[a]	AD/OA	Nausea, constipation, headache, vomiting, dizziness, insomnia, dry mouth, increase blood pressure and pulse	RCT	4.8	Nissen et al. (2016); Wadden et al. (2011); Alobaida et al. (2021)
Liraglutide[b]	GLP-1RA	Nausea, diarrhea, constipation, vomiting, and dyspepsia, cholelithiasis, cholecystitis and resting heart rate	RCT	5.4	Pi-Sunyer et al. (2015); le Roux et al. (2017)
Semaglutide[b]	GLP-1RA	Nausea, constipation, diarrhea, vomiting, and abdominal pain, headache, dizziness, and fatigue	RCT	15	Wilding et al. (2021); Garvey et al. (2022); Rubino et al. (2021); Blüher et al. (2023)
Cagrilintide[b]	GLP-1RA	Hypoglycemia, nausea, constipation and diarrhea	RCT	10.8	Lau et al. (2021)
Cagrilintide-Semaglutide[b]	GLP-1RA	Nausea, diarrhea, hypoglycemia, decreased appetite, abdominal pain and suicidal ideation	RCT	15.7–17.1	Enebo et al. (2021)
Tirzepatide[b]	GIPR	Nausea, diarrhea, vomiting, constipation and dyspepsia	RCT	20.9	Jastreboff et al. (2022); Coutinho and Halpern (2024)
Setmelanotide	MC4RA	Hyperpigmentation, nausea and vomiting	RCT	27.1–43.7	Blüher et al. (2023)
CagriSema[c]	GLP-1RA	Gastrointestinal disorders	RCT	17.1	Hay et al. (2015); Jeon et al. (2023)

(continued)

Table 13.1 (continued)

Drugs	Class of drug	Potential sides effects	Studies	Placebo subtracted Mean weight loss (%)	References
LY3502970 (orforglipron)[c]	GLP-1RA	Vomiting, nausea, constipation, and headache	RCT	9.4–14.7	Jeon et al. (2023)
BI 456906[c]	GIPR/ GLP-1RA	Nausea, eructation, dyspepsia, vomiting, diarrhea, abdominal pain and headache	RCT	13.8	Jungnik et al. (2023); Yazawa et al. (2023)
LY3437943 (retartrutide)[c]	GIPR/ GLP-1RA and GLP-1RA/ GCGR	Diarrhea, nausea and vomiting	RCT	10	Coskun et al. (2022); Urva et al. (2022)

[a]Older drugs
[b]Most recent approved by FDA
[c]Waiting approval by FDA
RCT Randomised Controlled Trial, *FDA* Food and Drug Administration, *AD/OA* Antidepressant/opioid antagonist, *SAA* Sympathomimetic amine anorectic/antiepileptic, *LI* Gastric and pancreatic lipase inhibitor, *GLP-1RA* glucagon-like peptide-1 receptor agonist, *MC4RA* melanocortin 4 receptor (MC4R) agonists, *GIPR* Glucose-dependent insulinotropic polypeptide receptor, *GIPR* Glucose-dependent insulinotropic polypeptide, *GCGR GLP-1RA* glucagon receptor

In addition, these bioactives can promote immunomodulatory effects through antioxidant and anti-inflammatory activities as well as regulate gut microbiota by increasing lactic bacteria and SCFAs production thereby supporting weight loss, and decreasing metabolic disorders (Zhang et al. 2023). For example, bioactive fermented foods can increase the number of beneficial gut bacteria that maintain symbiosis between the host and its microbiome (Jalili et al. 2023), hence increasing the production of SCFAs via fermentation of prebiotic and GABA. SCFAs can inhibit the harmful bacteria and modulate immunity of the host, therefore modifying adipose tissue, a primary risk factor in type 2 diabetes development. GABA may suppress weight gain via alteration, such as thermogenesis, energy expenditure, in adipose tissues (Hwang et al. 2019). In addition, these bioactive food components can modulate mitochondrial activity to maintain metabolic homeostasis (Kusminski and Scherer 2012).

The present chapter presents natural food bioactives and its mechanisms as excellent alternative strategy for safe and effective prevention and clinical management of obesity.

Table 13.2 Anti-obesity mechanisms of food bioactives

Foods	Source	Model	Bioactive components	Effects	References
Hibiscus genera	Plant	*In vitro*	Phenolic acids; quercetin; anthocyanins; flavonoids	Oxidative stress and weight gain reduction Modulation of obesity-associated microbiota	Herranz-López et al. (2020); Silva et al. (2022)
Eggs	Animal	*In vivo*	Peptide	Prevented weight gain; increased the concentration of adiponectin; reduced high-density lipoprotein	Liu et al. (2022)
Zebrafish	Animal	*In vivo*	Kaempferol, quercetin	Reduction of triglyceride and lipid droplets accumulation	Seo et al. (2015); Lee et al. (2015)
Fish	Animal	*In vivo*	Lipoprotein	Reduced body weight, abdominal fat tissue and pericardial fat	Carrera et al. (2023)
Edible mushrooms	Human/animal	*In vivo*	Extract	Fatty oil levels, and fat tissue	Harada et al. (2020)
Krill	Animal	*In vivo*	n-3 PUFAs	Inhibited lipid synthesis through the modification of gene expressions in the liver and adipose tissues and adipokine-mediated pathways	Hwang et al. (2022)
Seeds of black cumin	Animal	*In vivo*	Thymoquinone	Inhibited fat metabolism by regulating the 5'-AMP-activated protein kinase (AMPK) expression.	Bashir et al. (2023)
Insect, (*Allomyrina dichotomy*)	Animal	*In vivo*	Peptide	Inhibited fat deposition in the liver and reduced oxidative stress	Fan et al. (2021)
Dioscorea oppositifolia	Plant	*In vivo*	3,5-dimethoxy-2,7phenanthrenediol	Suppressed feeding efficiency and fat absorption	Jeong et al. (2016)
Fruits, cherries	Plant	*In vivo*	Anthocyanins	Enhanced redox status and alleviated inflammation	Nemes et al. (2019)

(continued)

Table 13.2 (continued)

Foods	Source	Model	Bioactive components	Effects	References
Fruits, grape-seed	Plant	*In vivo*	Proanthocyanidin	Reduced body weight, total adiposity, and liver steatosis	Sierra-Cruz et al. (2021)
Green tea	Plant	*In vivo*	Polyphenols	Reduced body weight gain and visceral fat	Tan et al. (2017)
Soy products, grains and legumes	Plant	*In vivo*	Genistein	Reduced body weight gain; downregulation of superoxide dismutase	Smith et al. (2006)
Red alga (*Gelidium amansii*)	Plant	*In vivo*	Polyphenolics	Decreased adipose tissue weight and liver weight	Kang et al. (2017)
Root (*Pachyrhizus erosus*)	Plant	*In vivo*	Extract	Prevented fat tissue weight gain, and prevented liver weight gain by blocking lipid droplet accumulation	Lee et al. (2023a, b)
Garlic	Plant	*In vivo*	Extract	Reduced adipose tissue, TG, free FA, weight gain	Kim et al. (2012); Ha et al. (2015a, b)
Soy bean	Plant	*In vivo*	Isoflavones (soy proteins)	Reduced adipose tissue, body weight, BMI, adiposity, fat pad	Panneerselvam et al. (2016); Ruscica et al. (2016)
Fish	Animal	*In vivo*	Collagen peptide	Decreased the inflammatory cytokine	Astre et al. (2018)
Sea cucumber	Animal	*In vivo*	Collagen peptide, saponins, dregs, polysaccharides	Suppressed adipose accumulation, and reduced serum and hepatic lipids	Hu et al. (2018)
Sea cucumber	Animal	*In vivo*	Sulfated polysaccharide	Inhibited fat accumulation, modulated the gut microbiota and improved microbial metabolites and gut tissue.	Zhu et al. (2018)
Bovine milk casein	Animal	*In vitro*	Peptide	Increased phosphorylation of Akt and AMPK	Song et al. (2017)
Black soy	Plant	*In vivo*	Isoflavone-free peptide	Decreased appetite	Jang et al. (2008)

(continued)

Table 13.2 (continued)

Foods	Source	Model	Bioactive components	Effects	References
Tuna	Animal	*In vivo*	Peptide	Inhibited C/EBPs and PPARγ expression	Kim et al. (2015)
Soybean	Plant	*In vivo*	Aglycin	Maintained IR and IRS1 expression	Lu et al. (2012)
Milk	Animal	*In vitro*	Peptide	Upregulated PPARγ and adiponectin	Chakrabarti and Wu (2015)
Grape	Plant	*In vitro*	Resveratrol	Decreased lipogenesis and increased fatty acids oxidation	Mercader et al. (2011)
Tart cherry	Plant	*In vivo*	Anthocyanins	Decreased NF-kB and adiposity	Seymour et al. (2009)
Green alga	Plant	*In vitro*	Xanthophyll	Inhibited lipase activity	Kurniawan et al. (2023)
Shrimp/crabs	Animal	*In vivo*	Chitooligosaccharide	Promote thermogenesis and β-oxidation	Zhou et al. (2023)
Milk	Animal	*In vitro*	Peptide	Inhibited dipeptidyl peptidase-IV and lipase	Mudgil et al. (2018)
Brown alga	Plant	*In vivo*	Extract	Decreased body weight and fat accumulation; induced downregulation of PPARγ in white adipose tissue, and upregulation of the thermogenic genes in adipose tissue	Kang et al. (2020)
Onion	Plant	*In vivo*	Extract	Inhibited lipid accumulation and reduced the hypertrophy of adipose tissue.	Yu et al. (2021)
Edible fungi	Plant	*In vivo*	Carotenoids	Decreased body weight, alleviated hepatic steatosis and adipocyte hypertrophy, reversed obesity-induced dysbiosis, increased benefits bacteria	Jiang et al. (2021)
Green alga	Plant	*In vitro*	Extract/phenolics	Inhibited lipase activity	Nurkolis et al. (2023)

13.2 Natural Bioactive Components in Food Bioactives

13.2.1 Polyhenolics

Phenolics are classified as flavonoids, non-flavonoids and others. Plant and animal–based foods contain these bioactive compounds that are affected significantly by the time of sampling, geographical distribution and species. Epidemiological, clinical and nutritional trials demonstrated strong evidence that dietary phenolics promote health benefits for human body. Previous studies found diverse phenolics in marine organisms include benzoic, cinnamic acids, catechins, and flavonoids in algae. Bromophenolic compounds have been found in red algae as the major source, however green and brown algae also contained which are ingested by fish, crabs and shrimps via food chain (Malmvärn et al. 2005; Mateos et al. 2020). According to chemical structure, the phlorotanins in brown algae are classified in fuhalols, halogenation, sulfonation, eckols, fucophlorethols, phlorethols, and Fucols and they various mechanisms to combat obesity (El-Sheekh et al. 2023; Zheng et al. 2022). The quantity and quality of these phenolics in algae varied between and within families and genuses. Example, Aminina et al. (2020) found in *Thalassiophyllum clathrus* ethyl acetate extract gallic acid, chlorogenic acid, coumaric acid, epicatechin, catechin and caffeic acid, however others phenolics did not found. Fish include salmon, tuna and sardines are also rich in phenolics such as catechins, and others phenolic acids (Ashraf et al. 2020). Curcumin is another phenolics found in some fish such as tambaqui (Saccol et al. 2017). In addition, quercetin and fisetin are found in some fish such as salmon and trout (Pereira and Cotas 2023). Sponges are also rich in phenolics such as bisabolenes (Pereira and Cotas 2023) therefore consumption of several bioactive foods may be benefits.

Plant foods including fruit, vegetable, tea, ginger, garlic, onion, coffee, and other herbs contain high amounts of different classes of phenolics. They can be categorized into phenolics acids, flavonoids, terpenes, saponins, stilbenes, lignans and others. Flavonoids are sub-divided into flavone, flavanol, flavanone, flavanonol, flavonone, flavononol, flavanol (catechin), isoflavone, and anthocyanidin (Abourashed 2013). However, benzoic acids and cinnamic acids, and salicylic acid, gentisic acid, p-Hydoxybenzoic acid, protocatechuic acid, vanillic acid, syringic acid, gallic acid, p-coumaric acid, ferulic acid, caffeic acid, and sinapic acid are phenolic acids (Abeyrathne et al. 2022). The major stilbenes found in plant foods are piceid, resveratrol, piceatannol, and pterostilben (Abeyrathne et al. 2022).

13.2.2 Fatty Acids and Carotenoids

Fatty acids (FAs) are organic molecules composed by a hydrocarbon chain and a carboxyl functional group. Based on their number of double bonds carbon-carbon FAs are categorized into three classes namely saturated fatty acids (SFAs), monounsaturated fatty acids (MFAs), and polyunsaturated fatty acids (PUFAs). Fish is particularly source of polyunsaturated fatty acids (PUFAs) include ω-3 PUFAs and ω-6 PUFAs. *Oviductus ranae* is also a source of ω-3 PUFAs and ω-6 PUFAs (Zhang et al. 2019). In addition, chicken and milk are reported to be rich in PUFAs (Mancinelli et al. 2022; Wanderley et al. 2022). For marine organisms, edible seaweeds are known as an important source of PUFAs include linoleic acid, linolenic acid, arachidonic acid, eicosapentaenoic acid (EPA) and docosahexaenoic acid (DHA) and these PUFAs ranged from 43.47% to 48.19% in brown seaweeds (Lorenzo et al. 2017; Ford et al. 2019). Apple seed oil was reported to be a great source of PUFAs (Berwal et al. 2022). Other plant-based foods include soybean, canola, peanuts, sunflower, flaxseed oil, palm oil and seed sesame are amongst another source of PUFAs that offer multiple health benefits. All these bioactive foods are considered the key factors in the prevention of some undesirable systemic reactions, metabolic syndrome include obesity, diabetes, cardiovascular diseases, hypertension and certain cancers.

Carotenoids are tetraterpenoid derivatives constituted of eight isoprenoid units constituting a linear chain possessing numerous conjugated double bonds and an inverse symmetry in the center of the molecule. At present, approximately 1100 carotenoid structures, divided into carotenes and xanthophylls (Pérez-Gálvez et al. 2020). Carotenes are hydrocarbon carotenoids (e.g., α-, β-, γ- and δ-carotene, lycopene, and phytoene) while xanthophylls are oxygenated derivatives of carotenes in which oxygen atom can be present as a hydroxyl (e.g., lutein, zeaxanthin, and β-cryptoxanthin), epoxide (e.g., neoxanthin, violaxanthin, and fucoxanthin), carbonyl (e.g., astaxanthin, canthaxanthin, and capsanthin), or another functional group (Saini et al. 2020). Red, orange, and yellow fruits and vegetables, as well as dark green leafy vegetables, constitute the core sources of carotenoids for humans, but cannot be ignored the contribution of animal-derived foods, such as dairy products, egg yolk, and seafood (Maiani et al. 2009).

13.2.3 Proteins/Peptides

Peptides are responsible for the regulation of essential physiological natural functions, particularly the endocrinology of human body. Food proteins could be hydrolyzed during processing (fermentation, and cooking) and/or through the gastro-intestinal tract releasing a large number of bioactive peptides (BPs). Some foods BPs are similar in structure to those endogenous active peptides, hence present high tissue affinity, specificity, and efficiency, and can interact with receptors, enzymes, and others biomolecules in the organism, thereby conferring health promoting effects including anti-obesity activity (Hernández-Ledesma et al. 2014). Despite, their anti-obesity potentiality, discrepancies have been noted in findings from in vitro or ex vivo studies, when compared to in vivo findings. *Ex vivo* studies present the best-case scenarios whereby most BPs interact with their cellular targets in situ to elicit bioactivity. Contrariwise, FBPs administered in vivo face many challenges, such as susceptibility to hydrolysis by proteases and peptidases present in the stomach, intestinal lumen, and brush border. In addition, the interaction with other components of the food matrix, and GI tract components such as mucin, limits the bio-accessibility of FBPs. FBPs are found in seafood including fish, shellfish and seaweeds. Meat, egg and milk are also a major source of FBPs and are produced during processing, especially fermentation that hydrolyze proteins into polypeptides and then into smaller peptides with high bioactivities (Abeyrathne et al. 2022). Honey is another dietary source of natural bioactive peptides (Noori et al. 2014). Other peptides are produced from microorganisms include *Bacillus* spp. and, *Lactobacillus* spp. during fermentation action.

13.2.4 Dietary Vitamins and Minerals

Vitamins play various important functions for the healthy living cell. Animal and plant-based foods both are source of diverse vitamins. Vitamin B12 complex, vitamin D, and vitamin A, especially found in poultry, fish, egg, milk, meat and other dairy products. And plant-based foods are potentially rich in vitamins C, A, and carotenoids include tomatoes, carrots, lemons, mangoes, cereals, herbs, and leafy vegetables or vegetables. These plant-based foods are also rich in essential minerals, especially trace elements such as iron, zinc, selenium, and copper. Example, numerous edible mushrooms contained vitamin B complex and green algae had high contents of vitamins and carotenoids. Vitamin D and A may have the ability to inhibit adipogenesis and inflammation thus improve the status of obese patients (de Carvalho Gomes et al. 2021; Lee 2023). Vitamin B1 prevent hyperglycaemia-induced and the lipid peroxidation thus reducing adiposity and ameliorate lipid profile (Lee 2023). Minerals have many functions in human body include for boosting enzymes actions, absorbing nutrients, formation of blood, tissue, bones and optimal antioxidant and inflammatory activities. Example, edible mushrooms rich in minerals such as K, P, Na, Ca, and Mg, Cu, Zn, Fe, Mo, and Cd that are essential for health boosting of the consumers (Malinowska et al. 2004; Niego et al. 2021). Algae are also rich in macro and trace elements that combat obesity via various signaling pathways. Edible fungi are another source of diverse vitamins and minerals (Jiang et al. 2021). Furthermore, high calcium foods can inhibit adipocyte expansion and suppress weight gain or increase lipolysis thereby preventing obesity.

13.2.5 Non-digestible Oligosaccharides (NDOs)

Natural NDOs (fructooligosaccharides (FOS), galactooligosaccharides (GOS), isomaltooligosaccharides (IMO), and xylooligosaccharides (XOS)) are saccharide polymers encompassing a small number (generally 3–10) of monosaccharides. In NDOs, the anomeric C atoms (C1 or C2) of the monosaccharide units have a configuration that makes their glycosidic bonds nondigestible to human digestive carbohydrases (Fontana et al. 2017). Due to their fermentability they have prebiotic properties. When NDO are ingested, the undigested fraction serves as food for probiotic. Food oligosaccharides can be obtained by extraction from natural sources or via enzymatic/chemical processing. Natural NDOs are typically edible constituents of whole grains (rye, barley, wheat), fruits (banana, yacon), vegetables (onion, tomato, asparagus, sugar beet, artichoke, chicory, garlic, bamboo shoots), marine algae and other foods (milk, honey, sugarcane juice, pulses) as well as some plant-based materials, with concentrations ranging between 0.3% and 20% (Kaur and Sharma 2019; Fan et al. 2022; Qiu et al. 2022; Hu et al. 2023). Amongst the marine-derived bioactive polysaccharides, oyster is nutritious shellfish that contained a huge amount of polysaccharides (Ma et al. 2022). In addition, brown algae are a source of bioactive polysaccharides with a potential health benefits include anti-oxidant, anti-inflammatory, anti-microbial and immunoregulatory or immunomodulatory activities (Priya and Khora 2023). Edible mushrooms contain also an important quantity of bioactive polysaccharides with a potential health benefits including anti-obesity properties (Chen et al. 2020; El-Ramady et al. 2022). Besides, edible mushrooms contain an important quantity of NDOs with a potential health benefit including anti-obesity (Chen et al. 2020; El-Ramady et al. 2022). Moreover, edible fungi are rich in NDOs (Jiang et al. 2021).

13.2.6 Probiotics, Paraprobiotic and Postbiotic

Probiotics are group of live microorganisms when administered in adequate amounts confer numerous health benefits to the consumer (Pacheco-Martínez et al. 2023), while paraprobiotics are nonviable microbial cells (intact or broken) or crude cell extracts which, when administered in adequate amounts, confer a benefit on the host (Martyniak et al. 2021). Postbiotics are soluble fractions secreted by probiotic's metabolic activity or released after the rupture of its cells, which cause any physiological benefit to the host. The postbiotics may be products or by-products of microbial metabolism, as well as polysaccharides, enzymes, proteins, organic acids, vitamins, peptides, plasmalogens, and SCFAs (Zendeboodi et al. 2020). These health benefits include suppression of pathogenic infection, improvement of the immune system, and regulation of fatty metabolism. The traditional strains of probiotics and parabiotics belong to *Bifidobacterium* and *Lactobacillus*. The primary *Lactobacillus* strains found in fermented foods are *Lactobacillus reuteri, Lactobacillus plantarum, Lactobacillus paracasei, Lactobacillus acidophilus, Lactobacillus paracasei, Lactobacillus casei*, and *Lactobacillus fermentum* (Martyniak et al. 2021; Pacheco-Martínez et al. 2023). *Bifidobacterium* genus is an important commensal member of gut microbiota encompasses *Bifidobacterium bifidum, Bifidobacterium breve, Bifidobacterium longum,* and *Bifidobacterium lactis* (Pacheco-Martínez et al. 2023). Fermented foods such as yogurt, dairy milk, fish, meat, fruits, and vegetables or their derived are the main sources of probiotic and parabiotics (Laya et al. 2023; Lee et al. 2023a, b).

13.3 Anti-obesity Evidences and Control Mechanisms of Food Bioactives

Currently, obesity is considered a global pandemic that affects people of any age group, gender, sex or ethnicity, with related disease complications including diabetes, reproductive disorders, cardiovascular diseases, respiratory, gastro-intestinal, cancers and so on. Many scientific studies show potential evidence that consumption of natural bioactive compounds from plants and animals can prevent and or reverse obesity and/or the related complications, via various control mechanisms (Mudgil et al. 2018; Nemes et al. 2019 Yu et al. 2021). These studies have been carried out in, *in vitro*, *in vivo* pre-clinical and animal and in clinical studies or both using various food bioactives.

13.3.1 Evidence of Anti-obesity in In Vitro *Studies*

Driven by the prospect of positive epidemiological data, mechanistic research evidences that food bioactives may promote anti-obesity effects via diverse mechanisms and these may act via multiple molecular targets. Selective examples of *in vitro* studies evidencing food bioactives' activities in the control of obesity are presented in this section. Table 13.2 presents various in *in vitro* studies that show the anti-obesity effects of natural plant-based bioactive foods, that include *Hibiscus* spp., grapes and green algae (Mercader et al. 2011; Herranz-López et al. 2020; Silva et al. 2022; Kurniawan et al. 2023; Nurkolis et al. 2023). The anti-obesity effects of bioactive foods are linked to their bioactive compounds including carotenoids, xanthophyll, resveratrol, polyphenolics, anthocyanins and flavonoids as shown on Table 13.2.

Polyphenolics may reduce the oxidative stress and inhibit the weight gain. Ofosu et al. (2020) reported that flavonoids in decorticated sorghum grains exerted anti-obesity properties by inhibiting pancreatic lipases thus reducing intestinal fat absorption. It was also reported that resveratrol decreased body mass in primates by suppressing appetite and increasing satiety (Dal-Pan et al. 2010). It has also been reported that resveratrol inhibits adipogenesis by itself or in synergy with other compounds such as quercetin, kaempferol, saponins and as a result, food bioactives controlled obesity (Mohamed et al. 2014). Polyphenolics are also known as excellent prebiotics that have the ability to modulate the obesity-associated microbiota via the increase of bacteria that produce high amounts of SCFAs. SCFAs exert anti-obesity effects by inhibiting the harmful bacteria that can induce inflammation and modulate host immunity via their antioxidant and anti-inflammatory activities. In addition, SCFAs can inhibit inflammatory pathways and lower appetite hormones such as leptin (Adalsteinsdottir et al. 2018). Its anti-obesity control is also effective via a central modulation of the hypothalamus and GABAergic neurons to regulate appetite (Frost et al. 2014). In addition, these types of bacteria also exert anti-obesity properties by reducing lipopolysaccharide production and regulating the super pathway of heme biosynthesis, associated with inflammation (Jiang et al. 2021). On the other hand, some polyphenolics such as resveratrol decrease lipogenesis and increase fatty acids oxidation. This compound also increases insulin sensitivity by regulating visceral fat derived adipokine levels such as leptin, resistin, and visfatin (Hoca et al. 2023). Polyphenolics may control obesity by suppressing appetite via hormonal processes including adiponectin, leptin and apelin, as seen in Fig. 13.1 below, which regulate energy balance and glucose homeostasis, in the human body. As an example, adiponectin enhances the activity of AMP-Kinase that increases the fatty acid oxidation, boosting energy expenditure (Song et al. 2020). Besides, polyphenolics act as effective thermogenic regulators by interfering with the gene expression of adenosine monophosphate-activated protein kinase (AMPK) and sirtuins such as SIRT1, SIRT2, SIRT3, SIRT4, and SIRT5 thus regulating adipose tissue metabolism (Mele et al. 2017). Polyphenolics also suppress fatty acid synthase, thereby inhibiting adipocyte differentiation and adipogenesis, and improving insulin sensitivity. Kim et al. (2020) found in their *in vitro* and *in vivo* studies that anthocyanins and carotenoids from sweet potatoes control obesity by reducing triglyceride content and peroxisome proliferator-activated receptor gamma (PPARγ) protein expression during adipogenesis. Yet another important mechanism is the stimulation of lipolysis and adipocyte apoptosis, particularly found with genistein (Cizmarova et al. 2023). Polyphenolics may also act by preventing mitochondrial dysfunction (Fig. 13.1) as the main factor of systemic metabolic alterations (Cavaliere et al. 2023). Carotenoids or xanthophylls inhibited lipase activity (Table 13.2). These compounds act by stimulating energy expenditure via the increase of fatty acid oxidation and thermogenesis (Xu et al. 2021). Furthermore, polyphenolics, peptides, vitamins and polysaccharides may exert a strong synergistic effects on obesity control via different signaling pathways (Fig. 13.1). Lee et al. (2023a, b) reported the anti-obesity effect of vegetable juice fermented by lactic acid bacteria isolated from kimchi. Similarly, the probiotic *Lactiplantibacillus plantarum* MGEL20154 from fermented food demonstrated the anti-obesity potential through the regulation of carbohydrate uptake by upregulating gene expression such as zo-1, pparα, and erk2 in intestinal epithelial cells (Fig. 13.1) (Park et al. 2023).

13.3.2 Anti-obesity in Animal Models

Utilization of animal models is a key step in understanding how food bioactives act and better understand the mechanisms underlying the specific regulating the anti-obesity activities. These data are important to guide novel treatment approaches or

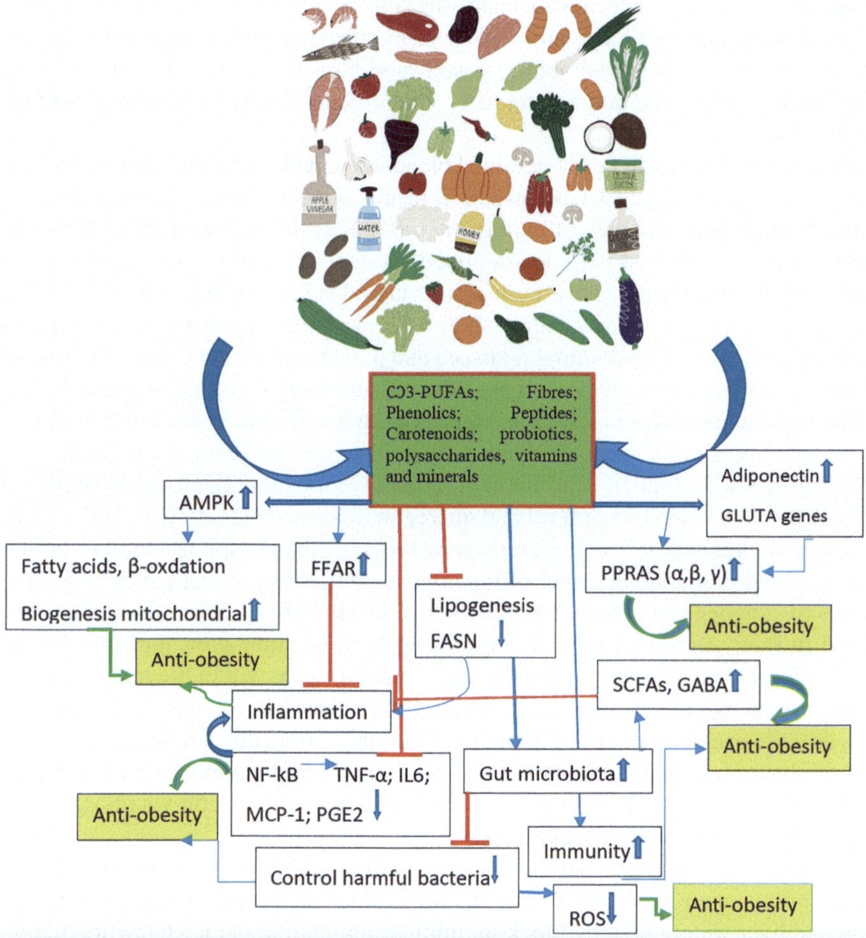

Fig. 13.1 Summarized signaling mechanisms mediating effects of food bioactives. AMPK AMP-activated protein kinase, NF-kB Nuclear factor kappa B; TNFα Tumor necrosis factor alpha; FFAR Free fatty acid receptor, ROS Reactive oxygen species, MCP-1 Monocyte chemoattractant protein-1, PGE2 Pro-inflammatory eicosanoids, PPRAS (α, β, γ) Peroxisome proliferator-activated receptor (alpha, beta et gamma), FASN Fatty acid synthase, Ɯ3-PUFAs Omega −3- polyunsaturated fatty acids, GLUTA Glucose transporter A, IL6 Interleukin-6, SCFAs Short-chain fatty acid, GABA gamma -amino butyric acid

even to elucidate the protective effects against obesity observed in populations with high consumption of foods rich in bioactive compounds. Consumption of natural foods has been widely studied for its anti-obesity benefits, in part due to its effective prominent bioactive components. In the last few decades, natural food bioactives are considered for safe and effective alternative natural products to prevent and manage obesity and its complications. Recent studies carried out in animal models demonstrated that food bioactives possess strong anti-obesity effects. These natural bioactive foods act via various mechanisms (Fig. 13.1). Table 13.2 presents studies on food bioactives that have anti-obesity properties, performed in pre-clinical animal model. These plant and animal based food bioactives contain peptides, polyphenolics, carotenoids, polysaccharides, vitamins, minerals, prebiotics, probiotics bacteria and others.

A study carried out in eggs contained bioactive peptides that increased adiponectin as observed in Fig. 13.1 below. As a result, bioactive peptides reduced high-density lipoprotein and weight gain (Liu et al. 2022). In one another study carried out in insects, contained peptides inhibited fat deposition in the liver and reduced oxidative stress in obese mice after feeding via oral gavage (100 mg/kg) for 5 weeks (Fan et al. 2021). Similarly, other studies showed that lipoprotein and peptides reduced body weight, abdominal fat tissue and pericardial fat, inhibited C/EBPs and PPARγ expression as well as decreased the inflammatory cytokine in obese mice in supplementing peptides in drinking water (4 g/kg b w/d) for 20 weeks by inhibiting de novo adipogenesis and increasing fatty acid oxidation (Kim et al. 2015; Astre et al. 2018; Hu et al. 2018; Carrera et al. 2023). On the other hand, other studies on zebrafish showed that kaempferol and quercetin increased reduction of triglycerides and lipid accumulation in obese animals by blocking the phosphorylation of protein kinase B and mammalian target of rapamycin (mTOR) signaling pathway at early stages of adipogenesis (Seo et al. 2015; Lee et al. 2015). Choi et al. (2015) also found in their study that dieckol suppressed lipid accumulation high-fat diet-fed zebrafish and mice by inhibiting the early adipogenesis via cell-cycle arrest and AMPKα activation. Similarly, other molecules such as 5-bromo-3,4-dihydroxybenzaldehyde in red algae inhibited the intracellular lipid accumulation and triglycerides by lowering protein expression of adipogenic genes including PPARγ, C/EBPα, SREBP-1, FABP4, FAS, leptin, and adiponectin via phosphorylation of AMPK and ACC (Ko et al. 2019).

The extracts of various natural foods including edible mushrooms, root plants, garlic, brawn algae, and onions reduced fatty oil levels, prevented fat tissue weight gain and liver weight gain by blocking lipid accumulation, induced downregulation of PPARγ in white adipose tissue, and upregulation of the thermogenic genes in adipose tissue thereby controlling the obesity in obese individuals (Kim et al. 2012; Ha et al. 2015a, b; Harada et al. 2020; Kang et al. 2020; Yu et al. 2021; Lee et al. 2023a, b). Several studies reported that ginger is safe and effective in the prevention and management of obesity (Suk et al. 2017; Mao et al. 2019). The ginger extracts contain many bioactive compounds thus their synergic action can be greater. In line with this, it has been demonstrated in many studies that polyphenolics including proanthocyanidin, flavonoids, genistein, catechins and resveratrol can induce apoptosis of adipocytes, increase insulin production and sensitivity, inhibit adipogenesis and the expression of fatty acid synthase and suppress the expression of adipocyte differentiation. In addition, these compounds upregulate AMP-activated protein kinase signaling thereby protecting cells from oxidative stress and preserving mitochondrial activities hence promoting anti-obesity properties (Feng et al. 2016; van den Eynde et al. 2018; Saad et al. 2022; Mahboob et al. 2023) as shown on Table 13.2. Polyphenolics and fibres can also act by other mechanisms, example they can interfere with nutrient digestion, leading to lower energy intake and increasing satiety in obese animals. Besides, flavonoids can enhance reactive oxygen species (ROS) scavenging activities thus preventing inflammatory activities as seen in Fig. 13.1. Furthermore, garlic can lower SREBP-1C mRNA expression, which causes downregulation of lipid and cholesterol metabolism, thus anti-obesity control (Ha et al. 2015a, b). Mustafa et al. (2022) reported also that edible mushrooms reduced plasma triglyceride, total cholesterol, low-density lipoprotein, and plasma glucose levels.

Another study using krill, that contains high levels of (n-3) PUFAs, inhibited lipid synthesis through the modification of gene expressions in the liver and adipose tissues and adipokine-mediated pathways in obese mice after receiving an oral gavage in a volume of 10 mL/kg once a day for 12 weeks (Hwang et al. 2022). Similarly, another study reported that (n-3) PUFAs including EPA and DHA increased adiponectin in adipose tissue and lowered body weight via PPAR-γ activation in obese mice (Kalupahana et al. 2011), as shown in Fig. 13.1. By other mechanisms such as gene regulation, (n-3) PUFAs can stimulate the AMPK pathway and can induce lipid oxidation, inhibit lipogenesis, and glucose uptake by adipose tissue (Wu et al. 2003). In addition, (n-3) PUFAs have the ability to increase anti-inflammatory activities in adipose tissue, thus downregulating the expression of toll-like receptors. Many other studies reported that (n-3) PUFAs such DHA and EPA can control obesity by increasing mitochondrial biogenesis thereby increasing energy expenditure (Hensler et al. 2011). Similarly, these compounds may be able to increase fatty acids beta oxidation and reduce lipogenesis thereby upregulating mitochondrial biogenesis (Hein et al. 2010). Sea cucumber rich in sulfated polysaccharides inhibited fat accumulation, modulated the gut microbiota and improved microbial metabolites and gut tissue (Zhu et al. 2018). Another study on shrimps/crabs contained chitooligosaccharide promoted thermogenesis and β-oxidation (Zhou et al. 2023). Another study demonstrated also that polysaccharides of edible mushrooms suppressed mRNA expression for cholesterol 7-α-hydroxylase 1 (CYP7A1) genes in 3 T3-L1 adipocytes in mice fed for 4 weeks, thus contributing to their anti-obesity effects (Mustafa et al. 2022). Tart cherry and soybean contained anthocyanins and aglycin decreased NF-kB and adiposity, maintained insulin receptors (IR) and insulin receptor substrate 1 (IRS-1) expression (Seymour et al. 2009; Lu et al. 2012) as shown on Table 13.2.

Black soy and soybean rich in isoflavone-free peptide and isoflavones-soy proteins decreased appetite, reduced adipose tissue, body weight, and adiposity in subjects (Jang et al. 2008; Panneerselvam et al. 2016; Ruscica et al. 2016). It was reported in another study that these compounds lowered cholesterol and triglyceride levels, reduced adiposity, and improved insulin sensibility (Ørgaard and Jensen 2008). Similarly, soy isoflavones were reported to have protective effect on adipose tissue oxidative stress and anti-inflammatory activity on obese animals (Panneerselvam et al. 2016). On the other hand, it was reported that fish proteins such as tuna protein hydrolysate increased fatty acid oxidation in mitochondria, promoted lipid excretion, and suppressed lipid deposition in adipocytes tissue (Maeda et al. 2017). Furthermore, Lee et al. (2017) found that amino-acids such as eicosapentaenoic (EPA) and docosahexaenoic (DHA) acids in fish had the ability to downregulate adipogenesis by upregulating mitochondrial biogenesis and inducing beta oxidation in fat thus improving obesity in mice. In another study, casein peptides improved the production of cholecystokinin that induced the sensation of satiety, and increased weight loss thereby controlling obesity (Erkkila et al. 2004; Paddon-Jones et al. 2008).

Furthermore, thymoquinone of seeds of black cumin inhibited fat metabolism by regulating the 5'-AMP-activated protein kinase (AMPK) expression (Bashir et al. 2023). The compound 3,5-dimethoxy-2,7phenanthrenediol found in *Dioscorea oppositifolia* suppressed feeding efficiency and fat absorption in obese mice after feeding with high-fat diet containing 60% kcal fat (5.24 kcal/g) for 8 weeks thereby preventing obesity in subjects (Jeong et al. 2016). Several studies on fruits, grapeseeds and cherries contained proanthocyanidin, and anthocyanins reduced significantly body weight, total adiposity, and liver steatosis, enhanced redox status and alleviated inflammation (Nemes et al. 2019; Sierra-Cruz et al. 2021) as seen on Table 13.2. Similar results have been reported in many experimental studies (Pan et al. 2016; Nowicka et al. 2018; Peng et al. 2022; Yu et al. 2021). In addition, edible fungi rich in carotenoids decreased body weight, alleviated hepatic steatosis and adipocyte hypertrophy, reversed obesity-induced dysbiosis by reducing lipopolysaccharide production responsible of inflammation, and increased the benefits bacteria that inhibited harmful bacteria responsible of inflammation in obese mice (Jiang et al. 2021) (Table 13.2). Thus, different bioactive foods compounds positively affect metabolism via several mechanisms affecting body weight and adipocytes biology to help curb obesity.

13.3.3 Anti-obesity in Human Trials

Several studies have demonstrated that natural bioactive foods components had potential power to prevent and/or to treat obesity and overweight via various signaling pathways. These bioactive rich foods contain many bioactive components that modulate metabolism in adipose tissue and decrease gene expression of fatty acid synthase genes that impact adipocyte differentiation and adiopogenesis. Although clinical trials are limited, there are some promising studies that demonstrated clearly the evidence of bioactive foods with prominent anti-obesity activities. For example, one randomized control trial showed that curcumin (containing 800 mg/dose/diet of Curcuma longa extract) supplemented twice per day reduced weight gain and adiposity in 30 and 60 days among 44 overweight or obese individuals (DiPierro et al. 2015). In addition, another study showed that curcumin reduced the levels of cytokines IL-1β and IL-4 in 37 obese individuals aged 18–65 years after receiving daily dose of 1 g for a 30-day treatment (Ganjali et al. 2014). Another study reported that curcumin supplementation for 30-days reduced significantly oxidative stress in obese individuals (Sahebkar et al. 2013). Furthermore, Timmers et al. (2011) found that 30 days of resveratrol supplementation treatment (150 mg/day) of 11 obese men induced energy expenditure, reduced cytokines such as leptin and decreased adiposity by reducing number and size of cells, plasma fatty acids and glycerol levels. The reduction of abdominal fat, cholesterol, LDL, TNF-α and MCP-1 levels after treatment with black soybean (2.5 g/day), in randomized clinical study in 63 overweight or obese subjects (Lee et al. 2016). Other randomized crossover design trial, in 7 men consumed 1.4 litre for 2 days, it showed that whole milk and fermented milk lowered plasma cholesterol concentration and decreased the triacylglycerol contents in all lipoprotein fractions and of CCK, GIP, GLP-1, and PYY (Sanggaard et al. 2004). Besides, drinking kefir supplementation reduced TC, LDL-C, TG non-HDLC, TC/HDLC LDLC/HDLC in 75 obese women aged 25–45 years, after receiving 2 servings per day for 8 weeks oral intake, as reported by Fathi et al. (2017). Moreover, the consumption of tempeh by 13 obese females aged 18–20 years for 1 week during a crossover study showed an increase of acyl-ghrelin, arginine levels, as well as the insulin response (Noer et al. 2021). Another randomized control study demonstrated that the consumption of quinoa seeds (*Chenopodium quinoa*) by 50 overweight and obese women aged 20–60 years consumed 1 quinoa sachet per day for 12 weeks significantly reduced serum triglyceride levels (Mohsin et al. 2023). Similarly, another human clinical study, oral uptake of 2 grams of Oolong tea four times a day for 6 weeks, reduced weight gain, fat accumulation, plasma triglyceride levels and total cholesterol in 102 obese individuals aged 35–65 years (He et al. 2009). Furthermore, green tea (491 mg catechins containing 302 mg EGCG) consumed by 100 obese women aged 16–60 years received three times per day for 12 weeks induced weight loss, decreased triglycerides, LDL-cholesterol, and increased adiponectin and ghrelin (Hsu et al. 2008). Decaffeinated green tea consumed for 8 weeks, increased fat oxidation, and energy expenditure, but decreased LDL-cholesterol, in 27 obese individuals aged in average 35–45 years (20 females and 7 males) (Roberts et al. 2021). Green tea consumed for 6 weeks showed that LDL-cholesterol decreased, and leptin increased in 73 overweight or obese individuals aged 18–65 years (Huang et al. 2018). These results reveal that natural bioactive compound-containing foods are able to ameliorate metabolism in obese individuals rendering them healthier.

13.4 Conclusions

Obesity is a pandemic disease across the world and its incidence is increasing rapidly. It is a preventable disease by the moderate ingestion of natural foods containing bioactive compounds as a potential obesity treatment, shifting from the current surgical or pharmacological treatments. These natural bioactive foods are rich in essential bioactive components include polyphenolics, fibres, bioactive peptides and polysaccharides, carotenoids, fatty acids, vitamins, and minerals that possess antioxidant, anti-inflammatory, immunomodulatory and anti-microbial activities. There are considered as safe and efficient agents to prevent and to treat obesity and its complications. The mechanism of these bioactives may be via various signaling pathways. Food bioactives can inhibit fat metabolism by regulating the 5'-AMP-activated protein kinase expression. Some components in bioactive foods can modulate adipose tissue metabolism and decrease gene expression of adipocyte differentiation and adiopogenenic genes. Example, PUFAs such DHA and EPA can control obesity by increasing mitochondrial biogenesis thereby increasing energy expenditure. Lipoproteins and peptides may reduce body weight, abdominal fat tissue and pericardial fat, inhibit C/EBPs and PPARγ expression as well as decrease the inflammatory cytokine such as tumor necrosis factor-α (TNF-α), interleukin (IL)-1, IL-6 in obese patients. Bioactive components can also induce apoptosis of adipocytes, increase insulin production and sensitivity, inhibit adipogenesis and expression of fatty acid synthase and suppress the expression of adipocyte differentiation, upregulate AMP-activated protein kinase signaling thereby protecting cells from oxidative stress and preserving mitochondrial activities. These mechanisms regulate energy homeostasis in individuals and preserve obesity. Thus food bioactives are considered as safe and efficient modern obesity treatment without side effects in comparison with conventional drugs. Currently, there are only a few clinical trials on food bioactives, and also the quantity to be injected daily is scarcely determined thus more clinical studies are needed to determine dose, verify and certify their safe and efficient state.

References

Abeyrathne EDNS, Nam K, Huang X, Ahn DU (2022) Plant- and animal-based antioxidants' structure, efficacy, mechanisms, and applications: a review. Antioxidants. 11:1025. https://doi.org/10.3390/antiox11051025

Abourashed EA (2013) Bioavailability of plant-derived antioxidants. Antioxidants 2:309–325

Adalsteinsdottir SA, Magnusdottir OK, Halldorsson TI, Birgisdottir BE (2018) Towards an individualized nutrition treatment: role of the gastrointestinal microbiome in the interplay between diet and obesity. Curr Obes Rep 7:289–293

Alobaida M, Alrumayh A, Oguntade AS, Al-Amodi F, Bwalya M (2021) Cardiovascular safety and superiority of anti-obesity medications. Diabetes Metab Syndr Obes 14:3199–3208

Aminina NM, Karaulova EP, Vishnevskaya TI, Yakush EV, Kim Y-K, Nam K-H et al (2020) Characteristics of polyphenolic content in brown algae of the Pacific coast of Russia. Molecules 25:3909. https://doi.org/10.3390/molecules25173909

Ardissino M, Vincent M, Hines O, Amin R, Eichhorn C, Tang AR et al (2022) Long-term cardiovascular outcomes after orlistat therapy in patients with obesity: a nationwide, propensity-score matched cohort study. Eur Heart J Cardiovasc Pharmacother 8:179–186

Ashraf SA, Adnan M, Patel M, Siddiqui AJ, Sachidanandan M, Snoussi M et al (2020) Fish-based bioactives as potent nutraceuticals: exploring the therapeutic perspective of sustainable food from the sea. Mar Drugs 18:265

Astre G, Deleruyelle S, Dortignac A, Bonnet C, Valet P, Dray C (2018) Diet-induced obesity and associated disorders are prevented by natural bioactive type 1 fish collagen peptides (Naticol®) treatment. JPhysiolBiochem. https://doi.org/10.1007/s13105-018-0650-0

Bashir KMI, Kim JW, Kim JK, Chun YS, Choi JS, Ku SK (2023) Efficacy confirmation test of black cumin (Nigella sativa L.) seeds extract using a high-fat diet mouse model. Meta 13:501. https://doi.org/10.3390/metabo13040501

Berwal MK, Ram C, Gurjar PS, Gora JS, Kumar R, Verma AK et al (2022) The bioactive compounds and fatty acid profile of bitter apple seed oil obtained in hot, arid environments. Horticulturae 8:259. https://doi.org/10.3390/horticulturae8030259

Blüher M, Aras M, Aronne LJ, Batterham RL, Giorgino F, Ji L et al (2023) New insights into the treatment of obesity. Diabetes Obes Metab:1–15. https://doi.org/10.1111/dom.15077

Câmara JS, Albuquerque BR, Aguiar J, Corrêa RCG, Gonçalves JL, Granato D et al (2021) Food bioactive compounds and emerging techniques for their extraction: polyphenols as a case study. Food Secur 10:37. https://doi.org/10.3390/foods10010037

Carrera I, Corzo L, Naidoo V, Martínez-Iglesias O, Cacabelos R (2023) Cardiovascular and lipid-lowering effects of a marine lipoprotein extract in a high-fat diet-induced obesity mouse model. Int JMedSci 20(3):292–306. https://doi.org/10.7150/ijms.80727

Cavaliere G, Cimmino F, Trinchese G, Catapano A, Petrella L, D'Angelo M (2023) From obesity-induced low-grade inflammation to lipotoxicity and mitochondrial dysfunction: altered multi-crosstalk between adipose tissue and metabolically active organs. Antioxidants 12:1172. https://doi.org/10.3390/antiox12061117

Chakrabarti S, Wu JP (2015) Milk-derived tripeptides IPP (Ile-Pro-Pro) and VPP (Val-Pro-Pro) promote adipocyte differentiation and inhibit inflammation in 3T3-F442A cells. PLoS One 10:e0117492

Chen L, Zhang L, Wang W, Qiu W, Liu L, Ning A et al (2020) Polysaccharides isolated from Cordyceps sinensis contribute to the progression of NASH by modifying the gut microbiota in mice fed a high-fat diet. PLoSONE 15:e0232972

Choi HS, Jeon HJ, Lee OH, Lee BY (2015) Dieckol, a major phlorotannin in Ecklonia cava, suppresses lipid accumulation in the adipocytes of high-fat diet-fed zebrafish and mice: inhibition of early adipogenesis via cell-cycle arrest and AMPKα activation. Mol Nutr Food Res 59:1458–1471

Cizmarova B, Tomeckova V, Hubkova B, Birkova A (2023) Anti-obesity properties and mechanism of action of genistein. Food Funct Food Sci Obesity 1(7):26–35. https://doi.org/10.31989/fffso.v1i7.1111

Coskun T, Urva S, Roell WC, Qu H, Loghin C, Moyers JS et al (2022) LY3437943, a novel triple glucagon, GIP, and GLP-1 receptor agonist for glycemic control and weight loss: from discovery to clinical proof of concept. Cell Metab 34:1234–1247. https://doi.org/10.1016/j.cmet.2022.07.013

Coutinho W, Halpern B (2024) Pharmacotherapy for obesity: moving towards efficacy improvement. Diabetol Metab Syndr 16:6. https://doi.org/10.1186/s13098-023-01233-4

Dal-Pan A, Blanc S, Aujard F (2010) Research article Resveratrol suppresses body mass gain in a seasonal non-human primate model of obesity. BMC Physiol 10:11. https://doi.org/10.1186/1472-6793-10-11

DiPierro F, Bressan A, Ranaldi D, Rapacioli G, Giacomelli L, Bertuccioli A (2015) Potential role of bioavailable curcumin in weight loss and omental adipose tissue decrease: preliminary data of a randomized, controlled trial in overweight people with metabolic syndrome. Preliminary study. Eur Rev Med Pharmacol Sci 19:4195–4202

Duque-Soto C, Expósito-Almellón X, García P, Pando ME, Borrás-Linares I, Lozano-Sánchez J (2023) Extraction, characterization, and bioactivity of phenolic compounds-a case on hibiscus genera. Food Secur 12:963. https://doi.org/10.3390/foods12050963

El-Ramady H, Abdalla N, Badgar K, Llanaj X, Törős G, Hajdú P et al (2022) Edible mushrooms for sustainable and healthy human food: nutritional and medicinal attributes. Sustain For 14:4941. https://doi.org/10.3390/su14094941

El-Sheekh M, Fathy AA, Saber H, Saber AS (2023) Medicinal and pharmaceutical applications of seaweeds. Egypt J Bot 63(1):1–29

Enebo LB, Berthelsen KK, Kankam M, Lund MT, Rubino DM, Satylganova A et al (2021) Safety, tolerability, pharmacokinetics, and pharmacodynamics of concomitant administration of multiple doses of cagrilintide with semaglutide 2·4 mg for weight management: a randomised, controlled, phase 1b trial. Lancet 397(10286):1736–1748

Erkkila AT et al (2004) Fish intake is associated with a reduced progression of coronary artery atherosclerosis in postmenopausal women with coronary heart disease. Am J Clin Nutr 80:626

Fan M, Choi YJ, Tang Y, Kim JH, Bg K, Lee B et al (2021, 1973) AGL9: a novel hepatoprotective peptide from the larvae of edible insects alleviates obesity-induced hepatic inflammation by regulating AMPK/Nrf2 signaling. Food Secur:10

Fan S, Guo D, Zhang J, Yang Y, Xue H, Xue T, Bai B (2022) Structure, physicochemical properties, antioxidant, and hypoglycemic activities of water-soluble polysaccharides from millet bran. JFood Sci 87(12):5263–5275. https://doi.org/10.1111/1750-3841.16362

Fathi Y, Ghodrati N, Zibaeenezhad MJ, Faghih S (2017) Kefir drink causes a significant yet similar improvement in serum lipid profile, compared with low-fat milk, in a dairy-rich diet in overweight or obese premenopausal women: a randomized controlled trial. J Clin Lipidol 11:136–146

Feng S, Reuss L, Wang Y (2016) Potential of natural products in the inhibition of adipogenesis through regulation of PPARγ expression and/or its transcriptional activity. Molecules 21:1278. https://doi.org/10.3390/molecules21101278

Fidler MC, Sanchez M, Raether B, Weissman NJ, Smith SR, Shanahan WR et al (2011) One-year randomized trial of lorcaserin for weight loss in obese and overweight adults: the BLOSSOM trial. J Clin Endocrinol Metab 96:3067–3077

Fontana JD, Tiboni M, Koop HS, Grzybowski A, Chalcoski BM, Spirandelli R et al (2017) Importance of phosphoric acid for functional foods: prebiotics oligosaccharides. In: Food bioconversion, pp 433–465

Ford L, Theodoridou K, Sheldrake GN, Walsh PJ (2019) A critical review of analytical methods used for the chemical characterisation and quantification of phlorotannin compounds in brown seaweeds. Phytochem Anal 30:587–599

Frohmader KS, Pitchers KK, Balfour ME, Coolen LM (2010) Mixing pleasures: review of the effects of drugs on sex behavior in humans and animal models. Horm Behav 58:149–162

Frost G, Sleeth ML, Sahuri-Arisoylu M, Lizarbe B, Cerdan S, Brody L et al (2014) The short-chain fatty acid acetate reduces appetite via a central homeostatic mechanism. Nat Commun 5:3611

Gadde KM, Allison DB, Ryan DH, Peterson CA, Troupin B, Schwiers ML et al (2011) Effects of low-dose, controlled-release, phentermine plus topiramate combination on weight and associated comorbidities in overweight and obese adults (CONQUER): a randomised, placebo-controlled, phase 3 trial. Lancet 377:1341–1352

Ganjali S, Sahebkar A, Mahdipour E, Jamialahmadi K, Torabi S, Akhlaghi S et al (2014) Investigation of the effects of curcumin on serum cytokines in obese individuals: a randomized controlled trial. Sci World J:898361

Garvey WT, Batterham RL, Bhatta M, Buscemi S, Christensen LN, Frias JP et al (2022) Two-year effects of semaglutide in adults with overweight or obesity: the STEP 5 trial. Nat Med 28:2083–2091. https://doi.org/10.1038/s41591-022-02026-4

de Carvalho Gomes C, Passos TS, AHA M (2021, 1921) Vitamin A status improvement in obesity: findings and perspectives using encapsulation techniques. Nutrients 13. https://doi.org/10.3390/nu13061921

Ha AW, Ying T, Kim WK (2015a) The effects of black garlic (Allium satvium) extracts on lipid etabolism in rats fed a high fat diet. Nutri Res Practice 9(1):30–36

Ha AW, Ying T, Kim WK (2015b) The effects of black garlic (Allium satvium) extracts on lipid metabolism in rats fed a high fat diet. Nutr Res Pract 9:30–36

Hachuła M, Kosowski M, Zielanska K, Basiak M, Okopien B (2023) The impact of various methods of obesity treatment on the quality of life and mental health—a narrative review. Int J Environ Res Public Health 20:2122. https://doi.org/10.3390/ijerph20032122

Harada E, Morizono T, Kanno T, Saito M, Kawagishi H (2020) Medicinal mushroom, grifola gargal (Agaricomycetes), lowers triglyceride in animal models of obesity and diabetes and in adults with prediabetes. Int J Med Mushrooms 22:79–91

Hay DL, Chen S, Lutz TA, Parkes DG, Roth JD (2015) Amylin: pharmacology, physiology, and clinical potential. Pharmacol Rev 67:564–600. https://doi.org/10.1124/pr.115.010629

He RR, Chen L, Lin BH, Matsui Y, Yao XS, Kurihara H (2009) Beneficial effects of oolong tea consumption on diet-induced overweight and obese subjects. Chin J Integr Med 15:34–41

Hein GJ, Bernasconi AM, Montanaro MA, Pellon-Maison M, Finarelli G, Chicco A et al (2010) Nuclear receptors and hepatic lipidogenic enzyme response to a dyslipidemic sucrose-rich diet and its reversal by fish oil n-3 polyunsaturated fatty acids. Am JPhysiol Endocrinol Metab 298:E429

Hensler M, Bardova K, Jilkova ZM, Wahli W, Meztger D, Chambon P et al (2011) The inhibitionoffat cellproliferation byn-3fattyacidsin dietaryobesemice. Lipids Health Dis 10:128–135

Hernández-Ledesma B, García-Nebot MJ, Fernández-Tomé S, Amigo L, Recio I (2014) Dairy protein hydrolysates: peptides for health benefits. Inter Dairy J 38(2):82–100

Herranz-López M, Olivares-Vicente M, Rodríguez-Gallego E, Encinar JA, Pérez-Sánchez A, Ruiz-Torres V et al (2020) Quercetin metabolites from Hibiscus sabdariffa contribute to alleviate glucolipotoxicity-induced metabolic stress in vitro. Food Chem Toxicol 144:111606

Hoca M, Becer E, Vatansever HS (2023) The role of resveratrol in diabetes and obesity associated with insulin resistance. Arch Physiol Biochem 129:2. https://doi.org/10.1080/13813455.2021.1893338

Hu I, Matthews A, Shmueli-Blumberg D, Killeen TK, Tai B, Van Veldhuisen P (2018) Prevalence of obesity for opioid- and stimulant-dependent participants in substance use treatment clinical trials. Drug and Alcohol Dependence- 190:255–262. https://doi.org/10.1016/j.drugalcdep.2018.06.014

Mohsin RH, Allami RH, Alnabi DIA, Mouhamad RS, Tarkan EA (2023) Effects of quinoa seed on some biochemical parameters in overweight and Obese Iraqi Women. J Hunan Univer (Nat Sci) 50(03)

Hsu CH, Tsai TH, Kao YH, Hwang KC, Tseng TY, Chou P (2008) Effect of green tea extract on obese women: a randomized, double-blind, placebo-controlled clinical trial. Clin Nutr 27:363–370

Hu Y-C, Hu J-L, Li J, Wang J, Zhang X-Y, Wu X-Y et al (2023) Physicochemical characteristics and biological activities of soluble dietary fibers isolated from the leaves of different quinoa cultivars. Food Res Int 163:112166. https://doi.org/10.1016/j.foodres.2022.112166

Huang LH, Liu CY, Wang LY, Huang CJ, Hsu CH (2018) Effects of green tea extract on overweight and obese women with high levels of low density-lipoprotein-cholesterol (LDL-C): a randomised, double-blind, and cross-over placebo-controlled clinical trial. BMC Complement Altern Med 18:294

Hwang I, Jo K, Shin KC, Kim JI, Ji Y, Park YJ, Park J et al (2019) GABA-stimulated adipose-derived stem cells suppress subcutaneous adipose inflammation in obesity. Proc Natl Acad Sci 116:11936–11945

Hwang SM, Kim YU, Kim JK, Chun YS, Kwon YS et al (2022) Preventive and therapeutic effects of krill oil on obesity and obesity-induced metabolic syndromes in high-fat diet-fed mice. Mar Drugs 20:483. https://doi.org/10.3390/md20080483

Jalili M, Nazari M, Magkos F (2023) Fermented foods in the management of obesity: mechanisms of action and future challenges. Int J Mol Sci 24:2665. https://doi.org/10.3390/ijms24032665

Jang E, Moon J, Ko J, Ahn C, Lee H, Shin J et al (2008) Novel black soy peptides with antiobesity effects: activation of leptin-like signaling and AMP-activated protein kinase. Int J Obes 32:1161

Jastreboff AM, Aronne LJ, Ahmad NN, Wharton S, Connery L, Alves B et al (2022) Tirzepatide once weekly for the treatment of obesity. N Engl J Med 387:205–216. https://doi.org/10.1056/NEJMoa2206038

Jeon E, Lee KY, Kim K-K (2023) Approved anti-obesity medications in 2022 KSSO guidelines and the promise of phase 3 clinical trials: anti-obesity drugs in the sky and on the horizon. J Obes Metab Syndr 32:106–120. https://doi.org/10.7570/jomes23032

Jeong EJ, Jegal J, Ahn J, Kim J, Yang MH (2016) Anti-obesity effect of Dioscorea oppositifolia extract in high-fat diet-induced obese mice and its chemical characterization. Biol Pharm Bull 39:409–414

Jiang X, Hao J, Liu Z, Ma X, Feng Y, Teng L et al (2021) Anti-obesity effects of Grifola frondosa through the modulation of lipid metabolism via ceramide in mice fed a high-fat diet. Food Funct 12:6725–6739. https://doi.org/10.1039/d1fo00666e

Jungnik A, Arrubla Martinez J, Plum-Mörschel L, Kapitza C, Lamers D, Thamer C et al (2023) Phase I studies of the safety, tolerability, pharmacokinetics and pharmacodynamics of the dual glucagon receptor/glucagon-like peptide-1 receptor agonist BI 456906. Diabetes Obes Metab 25:1011–1023. https://doi.org/10.1111/dom.14948

Kalupahana NS, Claycombe KJ, Moustaid-Moussa N (2011) (n-3) fatty acids alleviate adipose tissue inflammation and insulin resistance: mechanistic insights. Adv Nutr 2:304–316

Kang JH, Lee HA, Kim HJ, Han JS (2017) Gelidium Amansii extract ameliorates obesity by downregulating Adipogenic transcription factors in diet-induced obese mice. Nutr Res Pract 11:17

Kang M-C, Lee H-G, Kim H-S, Song K-M, Chun Y-G, Lee MH et al (2020) Anti-obesity effects of Sargassum thunbergii via downregulation of Adipogenesis gene and upregulation of thermogenic genes in high-fat diet-induced obese mice. Nutrients 12:3325. https://doi.org/10.3390/nu12113325

Kaur R, Sharma M (2019) Cereal polysaccharides as sources of functional ingredient for reformulation of meat products: a review. J Funct Foods 62:103527. https://doi.org/10.1016/j.jff.2019.103527

Kelly AS (2023) Current and future pharmacotherapies for obesity in children and adolescents. Nat Rev Endocrinol 19:534–541. https://doi.org/10.1038/s41574-023-00858-9

Kim EJ, Lee DH, Kim HJ, Lee SJ, Ban JO, Cho MC et al (2012) Thiacremonone, a sulfur compound isolated from garlic, attenuates lipid accumulation partially mediated via AMPK activation in 3T3-L1 adipocytes. J Nutr Biochem 23:1552–1558

Kim H-J, Koo KA, Park WS, Kang D-M, Kim HS, Lee BY et al (2020) Anti-obesity activity of anthocyanin and carotenoid extracts from color-fleshed sweet potatoes. J Food Biochem 44:e13438. https://doi.org/10.1111/jfbc.13438

Kim YM, Kim IH, Choi JW, Lee MK, Nam TJ (2015) The anti-obesity effects of a tuna peptide on 3T3-L1 adipocytes are mediated by the inhibition of the expression of lipogenic and adipogenic genes and by the activation of the Wnt/β-catenin signaling pathway. Int J Mol Med 36:327–334

Ko SC, DingY KJ, Ye BR, Kim EA, Jung WK et al (2019) Bromophenol (5-bromo-3,4-dihydroxybenzaldehyde) isolated from red alga Polysiphonia morrowii inhibits adipogenesis by regulating expression of adipogenic transcription factors and AMP-activated protein kinase activation in 3T3-L1 adipocytes. Phyther Res 33:737–744

Kumar M, Kaushik D, Kaur J, Proestos C, Oz F, Oz E et al (2022) A critical review on obesity: herbal approach, bioactive compounds, and their mechanism. Appl Sci 12:8342. https://doi.org/10.3390/app12168

Kurniawan R, Nurkolis F, Taslim NA, Subali D, Surya R, Gunawan WB et al (2023) Carotenoids composition of green algae Caulerpa racemosa and their antidiabetic, anti-obesity, antioxidant, and anti-inflammatory properties. Molecules 28:3267. https://doi.org/10.3390/molecules28073267

Kusminski CM, Scherer PE (2012) Mitochondrial dysfunction in white adipose tissue. Trends Endocrinol Metab 23:435–443

Lau DCW, Erichsen L, Francisco AM, Satylganova A, le Roux CW, McGowan B et al (2021) Once-weekly cagrilintide for weight management in people with overweight and obesity: a multicentre, randomised, double-blind, placebocontrolled and active-controlled, dose-finding phase 2 trial. Lancet 398(10317):2160–2172

Laya A, Wangso H, Fernandes I, Djakba R, Oliveira J, Carvalho E (2023) Bioactive ingredients in traditional fermented food condiments: emerging products for prevention and treatment of obesity and type 2 diabetes. J Food Qual. https://www.hindawi.com/journals/jfq/2023/5236509/

le Roux CW, Astrup A, Fujioka K, Greenway F, Lau DC, Van Gaal L et al (2017) 3 Years of liraglutide versus placebo for type 2 diabetes risk reduction and weight management in individuals with prediabetes: a randomised, double-blind trial. Lancet 389:1399–1409. https://doi.org/10.1016/S0140-6736(17)30069-7

Lee CY (2023) Effects of dietary vitamins on obesity-related metabolic parameters. J Nutr Sci 12:e47. https://doi.org/10.1017/jns.2023.30

Lee EJ, Hur J, Ham SA, Jo Y, Lee SY, Choi MJ et al (2017) Fish collagen peptide inhibits the adipogenic differentiation of preadipocytes and ameliorates obesity in high fat diet-fed mice. Int J Biol Macromol 104:281–286

Lee M, Yun Y-R, Choi EJ, Song JH, Kang JY, Kim D et al (2023b) Anti-obesity effect of vegetable juice fermented with lactic acid bacteria isolated from kimchi in C57BL/6J mice and human mesenchymal stem cells. Food Funct 14:1349–1356. https://doi.org/10.1039/d2fo02998g

Lee M, Sorn SR, Park Y, Park HK (2016) Anthocyanin rich-black soybean Testa improved visceral fat and plasma lipid profiles in overweight/obese Korean adults: a randomized controlled trial. J Med Food 19:995–1003

Lee SM, Bok SH, Kim MH, Lee KS, Park DH (2023a) Pachyrhizus erosus inhibits Adipogenesis via the leptin-PPARγ-FAS pathway in a high-fat diet-induced mouse model. PRO 11:735. https://doi.org/10.3390/pr11030735

Lee YJ, Choi HS, Seo MJ, Jeon HJ, Kim KJ, Lee BY (2015) Kaempferol suppresses lipid accumulation by inhibiting early adipogenesis in 3T3-L1 cells and zebrafish. Food Funct 6:2824–2833

Liu Z, Ding S, Jiang H, Fang J (2022) Egg protein transferrin-derived peptides Irw (Lle-Arg-Trp) and Iqw (Lle-Gln-Trp) prevent obesity mouse model induced by a high-fat diet via reducing lipid deposition and reprogramming gut microbiota. Int JMol Sci 23:11227. https://doi.org/10.3390/ijms231911227

Lorenzo JM, Agregán R, Munekata PES, Franco D, Carballo J, Sahin S et al (2017) Proximate composition and nutritional value of three macroalgae: Ascophyllum nodosum, Fucus vesiculosus and Bifurcaria bifurcata. Mar Drugs 15:360

Lu J, Zeng Y, Hou W, Zhang S, Li L, Luo X et al (2012) The soybean peptide aglycin regulates glucose homeostasis in type 2 diabetic mice via IR/IRS1 pathway. J Nutr Biochem 23:1449–1457

Ma Y, Zhu L, Ke H, Jiang S, Zen M (2022) Oyster (Crassostrea gigas) polysaccharide ameliorates obesity in association with modulation of lipid metabolism and gut microbiota in high-fat diet fed mice. Int J Biol Macromol 216:916–926. https://doi.org/10.1016/j.ijbiomac.2022.07.100

Maeda H, Hosomi R, Fukuda M, Ikeda Y, Yoshida M, Fukunaga K et al (2017) Dietary tuna dark muscle protein attenuates hepatic steatosis and increases serum high-density lipoprotein cholesterol in obese type-2 diabetic/obese KK-Ay mice. J Food Sci 82:1231–1238

Mahboob A, Samuel SM, Mohamed A, Wani MY, Ghorbel S, Miled N et al (2023) Role of flavonoids in controlling obesity: molecular targets and mechanisms. Front Nutr 10:1177897. https://doi.org/10.3389/fnut.2023.1177897

Maiani G, Periago Castón MJ, Catasta G, Toti E, Cambrodón IG, Bysted A et al (2009) Carotenoids: actual knowledge on food sources, intakes, stability and bioavailability and their protective role in humans. Mol Nutr Food Res 53(S2). https://doi.org/10.1002/mnfr.200800053

Malinowska E, Szefer P, Falandysz J (2004) Metals bioaccumulation by bay bolete, Xerocomus badius, from selected sites in Poland. Food Chem 84:405–416

Malmvärn A, Marsh G, Kautsky L, Athanasiadou M, Bergman Å, Asplund L (2005) Hydroxylated and methoxylated brominated diphenyl ethers in the red algae Ceramiumtenuicorne and blue mussels from the Baltic Sea. Environ Sci Technol 39:2990–2997

Mancinelli AC, Veroli AD, Mattioli S, Cruciani G, Bosco AD, Castellini C (2022) Lipid metabolism analysis in liver of different chicken genotypes and impact on nutritionally relevant polyunsaturated fatty acids of meat. Sci Rep 12:1888. https://doi.org/10.1038/s41598-022-05986-2

Martyniak A, Medyńska-Przęczek A, Wędrychowicz A, Skoczen S, Tomasik PJ (2021) Prebiotics, probiotics, synbiotics, paraprobiotics and postbiotic compounds in IBD. Biomol Ther 11:1903. https://doi.org/10.3390/biom11121903

Mao Q-Q, Xu X-Y, Cao S-Y, Gan R-Y, Corke H, Trust Beta T et al (2019) Bioactive Compounds and Bioactivities of Ginger (Zingiber officinale Roscoe). Foods 8:185. https://doi.org/10.3390/foods8060185

Mateos R, Pérez-Correa JR, Domínguez H (2020) Bioactive properties of marine phenolics. Mar Drugs 18:501. https://doi.org/10.3390/md18100501

Mele L, Bidault G, Mena P, Crozier A, Brighenti F, Vidal-Puig A, Del Rio D et al (2017) Dietary (Poly)phenols. Brown Adipose Tissue Activation, and Energy Expenditure: A Narrative Review. Adv Nutr 8(5):694–704. https://doi.org/10.3945/an.117.015792

Mercader J, Palou A, Bonet ML (2011) Resveratrol enhances fatty acid oxidation capacity and reduces resistin and Retinol-Binding Protein 4 expression in white adipocytes. J Nutr Biochem 22:828–834

Mudgil P, Kamal H, Yuen GC, Maqsood S (2018) Characterization and identification of novel antidiabetic and antiobesity peptides from camel milk protein hydrolysates Food Chem, 259:46–54. https://doi.org/10.1016/j.foodchem.2018.03.082

Mustafa F, Chopra H, Baig AA, Avula SK, Kumari S, Mohanta TK et al (2022) Edible mushrooms as novel myco-therapeutics: effects on lipid level, obesity, and BMI. J Fungi 8:211. https://doi.org/10.3390/jof8020211

Mohamed GA, Ibrahim SRM, Elkhayat ES, El Dine RS (2014) Natural anti-obesity agents. Bul Fac Pharmacy, Cairo Univ 52:269–284. https://doi.org/10.1016/j.bfopcu.2014.05.001

Nemes A, Homoki JR, Kiss R, Hegedűs C, Kovács D, Peitl B et al (2019) Effect of anthocyanin-rich tart cherry extract on inflammatory mediators and adipokines involved in type 2 diabetes in a high fat diet induced obesity mouse model. Nutrients 11:1966. https://doi.org/10.3390/nu11091966

Niego AG, Rapior S, Thongklang N, Raspé O, Jaidee W, Lumyong S et al (2021) Macrofungi as a nutraceutical source: promising bioactive compounds and market value. JFungi 7:397. https://doi.org/10.3390/jof7050397

Nissen SE, Wolski KE, Prcela L, Wadden T, Buse JB, Bakris G et al (2016) Effect of naltrexone-bupropion on major adverse cardiovascular events in overweight and obese patients with cardiovascular risk factors: a randomized clinical trial. JAMA 315:990–1004

Noer ER, Dewi L, Kuo CH (2021) Fermented soybean enhances post-meal response in appetite-regulating hormones among Indonesian girls with obesity. Obes Res Clin Pract 15:339–344

Noori S, Faiza S, Mohammed A, Amjed A, Khelod YS et al (2014) Effects of natural honey on polymicrobial culture of various human pathogens. Arch Medi Sci 10:246–250

Nowicka P, Wojdyło A, Laskowski P (2018) Inhibitory potential against digestive enzymes linked to obesity and type 2 diabetes and content of bioactive compounds in 20 cultivars of the peach fruit grown in Poland. Plant Foods H Nutrit 73:314–320. https://doi.org/10.1007/s11130-018-0688-8

Nurkolis F, Taslim NA, Qhabibi FR, Kang S, Moon M, Choi J et al (2023) Ulvophyte Green Algae Caulerpa lentillifera: Metabolites Profile and Antioxidant, Anticancer, Anti-Obesity, and In Vitro Cytotoxicity Properties. Molecules 28:1365. https://doi.org/10.3390/molecules28031365

OECD (Organisation for Economic Co-operation and Development). The heavy burden of Obesity; 2019. Visitoe.cd/obesity2019. Visited on 7/11/2023

Ofosu FK, Elahi F, Daliri EB-M, Yeon S-J, Ham HJ, Kim J-H et al (2020) Flavonoids in decorticated sorghum grains exert antioxidant, antidiabetic and antiobesity activities. Molecules 25:2854. https://doi.org/10.3390/molecules25122854

Ørgaard A, Jensen L (2008) The effects of soy isoflavones on obesity. Exp Biol Med 233: 1066–1080

Pacheco-Martínez K, Saucedo-Castañeda CG, Rodríguez-Durán V (2023) Luis; Pérez-Chabela, María de Lourdes. Características de microorganismos utilizados como probióticos tradicionales y nuevos probióticos. Actualidades Biológicas 45:119. https://doi.org/10.17533/udea.acbi/v45n119a05

Paddon-Jones D, Westman E, Mattes RD, Wolfe RR, Astrup A, Westerterp-Plantenga M (2008) Protein, weight management, and satiety. Am J Clin Nutr 87:1558–1561

Padwal R, Li SK, Lau DC (2004) Long-term pharmacotherapy for obesity and overweight. Cochrane Database Syst Rev 2003:CD004094

Pan M-H, Tung Y-C, Yang G, Lia S, Ho C-T (2016) Molecular mechanisms of the anti-obesity effect of bioactive compounds in tea and coffee. Food Funct 7:4481. https://doi.org/10.1039/c6fo01168c

Panneerselvam S, Packirisamy RM, Bobby Z, Sridhar MG (2016) Protective effect of soy isoflavones (from Glycine max) on adipose tissue oxidative stress and inflammatory response in an experimental model of post-menopausal obesity: the molecular mechanisms. Biochem Anal Biochem 5:266. https://doi.org/10.4172/2161-1009.1000266
Park SY, Choi JW, Oh DN, Lee EJ, Kim DP, Yoon SJ et al (2023) Anti-obesity potential through regulation of carbohydrate uptake and gene expression in intestinal epithelial cells by the probiotic Lactiplantibacillus plantarum MGEL20154 from fermented food. J Microbiol Biotechnol 33(5):621–633. https://doi.org/10.4014/jmb.2212.12005
Peng H, Sun M, Hu X, Han H, Su J, Peng E et al (2022) Prevalence, awareness, and associated factors of high blood pressure among female migrant workers in Central South China. PeerJ 10:e13365. https://doi.org/10.7717/peerj.13365. eCollection 2022
Pereira L, Cotas J (2023) Therapeutic potential of polyphenols and other micronutrients of marine origin. Mar Drugs 21:323. https://doi.org/10.3390/md210
Pérez-Gálvez A, Viera I, Roca M (2020) Carotenoids and chlorophylls as antioxidants. Antioxidants 9:505. https://doi.org/10.3390/antiox9060505
Pi-Sunyer X, Astrup A, Fujioka K, Greenway F, Halpern A, Krempf M et al (2015) A randomized, controlled trial of 3.0 mg of liraglutide in weight management. N Engl J Med 373:11–22
Priya RR, Khora SS (2023) Antioxidant potentials of polysaccharides derived from marine brown algae. In: Marine antioxidants preparations, syntheses, and applications, pp 433–448. https://doi.org/10.1016/B978-0-323-95086-2.00012-6
Qiu Z, Qiao Y, Zhang B, Sun-Waterhouse D, Zheng Z (2022) Bioactive polysaccharides and oligosaccharides from garlic (Allium sativum L.): production, physicochemical and biological properties, and structure–function relationships. Comp Revi Food Scie Food Saf 21:3033–3095. https://doi.org/10.1111/1541-4337.12972
Rahman HA, Saari N, Abas F, Ismail A, Mumtaz MW, Hamid AA (2017) Anti-obesity and antioxidant activities of selected medicinal plants and phytochemical profiling of bioactive compounds. InterJFood Prop 20(11):2616–2629. https://doi.org/10.1080/10942912.2016.1247098
Roberts JD, Willmott AGB, Beasley L, Boal M, Davies R, Martin L (2021) The impact of decaffeinated green tea Extracton fat oxidation, body composition and cardio metabolic healthin overweight, recreationally active individuals. Nutrients 13:764
Rubino D, Abrahamsson N, Davies M, Hesse D, Greenway FL, Jensen C et al (2021) Effect of continued weekly subcutaneous semaglutide vs placebo on weight loss maintenance in adults with overweight or obesity: the STEP 4 randomized clinical trial. JAMA 325:1414–1425
Ruscica M, Pavanello C, Gandini S, Gomaraschi M, Vitali C, Macchi C et al (2016) Effect of soy on metabolic syndrome and cardiovascular risk factors: a randomized controlled trial. Eur J Nutr 27:1–13
Saad B, Kmail A, Haq SZH (2022) Anti-diabesity middle eastern medicinal plants and their action mechanisms. Evid Based Complement Alternat Med:2276094. https://doi.org/10.1155/2022/2276094
Saccol EMH, Toni C, Pês TS, Ourique GM, Gressler LT, Silva LVF et al (2017) Anaesthetic and antioxidant effects of Myrcia Sylvatica (G. Mey.) DC. and Curcuma Longa L. essential oils on Tambaqui (Colossoma Macropomum). Aquac Res 48:2012–2031
Sahebkar A, Mohammadi A, Atabati A, Rahiman S, Tavallaie S, Iranshahi M et al (2013) Curcuminoids modulate pro-oxidant-antioxidant balance but not the immune response to heat shock protein 27 and oxidized LDL in obese individuals. Phytother Res 27:1883–1888
Saini RK, Keum Y-S, Daglia M, Rengasamy KRR (2020) Dietary carotenoids in cancer chemoprevention and chemotherapy: a review of emerging evidence. Pharmacol Res 157:104830. https://doi.org/10.1016/j.phrs.2020.104830
Sanggaard KM, Holst JJ, Rehfeld JF, Sandström B, Raben A, Tholstrup T (2004) Different effects of whole Milk and a fermented Milk with the same fat and lactose content on gastric emptying and postprandial Lipaemia, but not on Glycaemic response and appetite. Br J Nutr 92:447–459
Seo MJ, Lee YJ, Hwang JH, Kim KJ, Lee BY (2015) The inhibitory effects of quercetinon obesity and obesity-induced inflammation by regulation of MAPK signaling. J Nutr Biochem 26:1308–1316

Seymour EM, Lewis SK, Urcuyo-Llanes DE, Tanone II, Kirakosyan A, Kaufman PB et al (2009) Regular tart cherry intake alters abdominal adiposity, adipose gene transcription, and inflammation in obesity-prone rats fed a high fat diet. J Med Food 12:935–942

Sierra-Cruz M, Miguéns-Gómez A, Grau-Bové C, Rodríguez-Gallego E, Blay M, Pinent M et al (2021) Grape-Seed Proanthocyanidin Extract Reverts Obesity-Related Metabolic Derangements in Aged Female Rats. Nutrients 13:2059. 10.3390/nu13062059

Silva M, Cueva C, Alba C, Rodriguez JM, de Pascual-Teresa S, Jones J et al (2022) Gut microbiome-modulating properties of a polyphenol-enriched dietary supplement comprised of hibiscus and lemon verbena extracts. Monitoring of phenolic metabolites. J Funct Foods 91:105016

Smith JT, Acohido BV, Clifton DK, Steiner RA (2006) KiSS-1 Neurones are direct targets for leptin in the Ob/Ob mouse. J Neuroendocrinol 18:298–303

Song JJ, Wang Q, Du M, Li TG, Chen B, Mao XY (2017) Casein glycomacropeptide-derived peptide IPPKKNQDKTE ameliorates high glucose-induced insulin resistance in HepG2 cells via activation of AMPK signaling. Mol Nutr Food Res 61:1600301

Song M, Yang G, Hoa TQ, Hieu HD, Amin ASM, Choe W et al (2020) Anti-obesity effect of fermented persimmon extracts via activation of AMP-activated protein kinase. Biol Pharm Bull 43:440–449

Suk S, Kwon GT, Lee E, Jang WJ, Yang H, Kim JH et al (2017) Gingerenone A, a polyphenol present in ginger, suppresses obesity and adipose tissue inflammation in high-fat diet-fed mice. Mol Nutr Food Res 61:1700139

Tain YL, Hsu CN (2022) Novel insights on dietary polyphenols for prevention in early-life origins of hypertension: a review focusing on preclinical animal models. Int J Mol Sci 23:6620. https://doi.org/10.3390/ijms23126620

Tan Y, Kim J, Cheng J, Ong M, Lao WG, Jin XL et al (2017) Green tea polyphenols ameliorate non-alcoholic fatty liver disease through upregulating AMPK activation in high fat fed Zucker fatty rats. World J Gastroenterol 23:3805

Timmers S, Konings E, Bilet L, Houtkooper RH, van de Weijer T, Goossens GH et al (2011) Calorierestriction-likeeffectsof30daysofresveratrolsupplementation on energy metabolism and metabolic profile in obese humans. Cell Metab 14:612–622

Urva S, Coskun T, Loh MT, Du Y, Thomas MK, Gurbuz S et al (2022) LY3437943, a novel triple GIP, GLP-1, and glucagon receptor agonist in people with type 2 diabetes: a phase 1b, multicentre, double-blind, placebo-controlled, randomised, multiple-ascending dose trial. Lancet 400:1869–1881. https://doi.org/10.1016/S0140-6736(22)02033-5

van den Eynde MDG, Geleijnse JM, Scheijen J et al (2018) Quercetin, but not epicatechin, decreases plasma concentrations of methylglyoxal in adults in a randomized, double-blind, placebo-controlled, crossover trial with pure flavonoids. J Nutr 48:1911–1916. https://doi.org/10.1093/jn/nxy236

Wadden TA, Foreyt JP, Foster GD, Hill JO, Klein S, O'Neil PM et al (2011) Weight loss with naltrexone SR/bupropion SR combination therapy as an adjunct to behavior modification: the COR-BMOD trial. Obesity (Silver Spring) 19:110–120

Wanderley AM, Ítavo LCV, dos Santos GT, Ferreira Ítavo CCB, Dias AM, dos Santos DG et al (2022) Sources of polyunsaturated fatty acids from oilseeds in dairy cows diets can alter yield of and fatty acid profile in milk. Food Sci Technol 42:e11902. https://doi.org/10.1590/fst.119021

Wilding JP, Batterham RL, Calanna S, Davies M, Van Gaal LF, Lingvay I et al (2021) Once-weekly semaglutide in adults with overweight or obesity. N Engl J Med 384:989–1002

WO. One billion people globally estimated to be living with Obesity by 2030. https://time.com/6264865/global-obesity-rates-increasing/. Visited on 7/11/2023

WOA (World Obesity Atlas). More Than Half of the World Will Be Obese By 2035, Report Says. https://edition.cnn.com/2023/03/03/world/world-obesity-2023-wellness-intl-scli/index.html. Visited on 7/11/2023

World Bank (2017a) An overview of links between obesity and food systems implications for the food and agriculture global practice agenda. 65p. Visited on 7/11/2023

World Bank (2017b) Food and agriculture global practice, an overview of links between obesity and food systems implications for the food and agriculture global practice agenda. p65
WHO- Malnutrition in women-Nutrition Landscape Information System (NLiS)- WHO, 2023-Visited on 7/11/2024
Wu X, Motoshima H, Mahadev K, Stalker TJ, Scalia R, Goldstein BJ (2003) Involvement of AMP-activated protein kinase in glucose uptake stimulated by the globular domain of adiponectin in primary rat adipocytes. Diabetes 52:1355–1363
Xu H, Wang E, Chen F, Xiao J, Wang M (2021) Neuroprotective phytochemicals in experimental ischemic stroke: mechanisms and potential clinical applications. Oxidative Med Cell Longev:6687386
Yazawa R, Ishida M, Balavarca Y, Hennige AM (2023) A randomized phase I study of the safety, tolerability, pharmacokinetics and pharmacodynamics of BI 456906, a dual glucagon receptor/glucagon-like peptide-1 receptor agonist, in healthy Japanese men with overweight/obesity. Diabetes Obes Metab 25:1973–1984. https://doi.org/10.1111/dom.15064
Yu S, Li H, Cui T, Cui M, Piao C, Wang S et al (2021) Onion (Allium cepa L.) peel extract effects on 3T3-L1 adipocytes and high-fat diet-induced obese mice. Food Biosci. https://doi.org/10.1016/j.fbio.2021.101019
Zendeboodi F, Khorshidian N, Mortazavian AM, da Cruz AG (2020) Probiotic: conceptualization from a new approach. Cur Opi Food Sci 32:103–123
Zhang D, Jian Y-P, ZhangY-N LY, Gu L-T, Sun H-H et al (2023) Short-chain fatty acids in diseases. Cell Commun Signals 21:212. https://doi.org/10.1186/s12964-023-01219-9
Zhang Y, Wang Y, Li M, Liu S, Yu J, Yan Z et al (2019) Traditional uses, bioactive constituents, biological functions, and safety properties of Oviductus ranae as functional foods in China. Oxid Med Cell Longev:4739450. https://doi.org/10.1155/2019/4739450
Zheng H, Zhao Y, Guo L (2022) A bioactive substance derived from brown seaweeds: Phlorotannins. Mar Drugs 20:742. https://doi.org/10.3390/md20120742
Zhou M, Huang J, Zhou J, Zhi C, Bai Y, Che Q et al (2023) Anti-obesity effect and mechanism of Chitooligosaccharides were revealed based on Lipidomics in diet-induced obese mice. Molecules 28:5595. https://doi.org/10.3390/molecules28145595
Zhu Z, Zhu B, Sun Y, Ai C, Wang L, Wen C et al (2018) Sulfated Polysaccharide from Sea Cucumber and its Depolymerized Derivative Prevent Obesity in Association with Modification of Gut Microbiota in High-Fat Diet-Fed Mice. Mol Nutr Food Res 62(23):e1800446. https://doi.org/10.1002/mnfr.201800446

Chapter 14
Role of Food Bioactives in Promoting Bone Health

Mohamad Taleuzzaman, Anupam, Manjari Verma, Mamta Bhatia, and Raju Singh

Abstract Dietary bioactive compounds are additional nutritious elements found in small amounts in foods. Different bioactive compounds are essential for bone health. The biomineral framework of bone is composed of approximately 70% inorganic calcium hydroxyapatite (HA) crystals Ca_{10} $(PO_4)_6(OH)_2$. The remaining 30% is made up of organic materials. Certain peptides, including ECM-Derived Peptides, BMP-Derived Peptides, and Other Peptides, can promote and accelerate the bone recovery process. Peptide structure alteration may enhance antibacterial, proangiogenic, and immune-modulating activities. Proteins are constructed from amino acids. Polar and charged amino acids (AAs) are ubiquitous in non-collagenous peptides and play a role in the activation of bone. The importance of growth variables in bone regeneration has long been acknowledged. Growth factors are a type of polypeptide hormone that stimulates a variety of biological activities, including cell growth, chemotaxis differentiation, and ECM protein creation. They also control the creation and function of local factors that have an immediate impact on cellular metabolism (autocrine and paracrine actions). Because of their antioxidant and anti-inflammatory effects, phytochemicals, particularly flavonoids, may benefit bone health. It's possible that phytosterols influence osteoblast growth and differentiation. Oxysterols are tiny, cholesterol-derived compounds found in both human and animal tissues and blood flow which have been shown to be osteo inductive. Liposomes are biocompatible lipid-based vesicles that are commonly employed in

M. Taleuzzaman (✉)
Department of Pharmaceutical Chemistry, Faculty of Pharmacy, Maulana Azad University, Jodhpur, Rajasthan, India

Anupam · M. Verma
Rameesh Institute of Vocational & Technical Education, Greater Noida, U.P., India

M. Bhatia
Department of Pharmacognosy, Faculty of Pharmacy, Maulana Azad University, Jodhpur, Rajasthan, India

R. Singh
School of Pharmacy, Noida International University, Uttar Pradesh, Noida, India
e-mail: raju.singh@niu.edu.in

S. Amir Ashraf, M. Adnan (eds.), *Food Bioactives and Nutraceuticals*,
https://doi.org/10.1007/978-981-97-9723-3_14

bone repair therapy. Statins are popular drugs that are used to decrease LDL cholesterol levels to avoid atherosclerosis. Bacterial cellulose is an exciting biopolymer which can be employed in the creation of modern biomaterials for the directed rebuilding of bone tissue due to its biocompatibility, biodegradability, high crystallinity, porosity, and tensile strength with mechanical resilience.

Keywords Bone health · Dietary bioactive compounds · Oxysterols · LDL cholesterol

14.1 Introduction

Plant-based foods are eaten all over the globe due to their substantial amount of energy and high nutritious value. Their dietary intake has increased due to the rise in metabolic issues linked to non-vegetarian diets. In addition to being high in vitamins and minerals, plant meals contain several bioactive components that have been scientifically shown to have positive health effects. Particular functional groups that emerge as a result of food metabolism and their molecular variations give functional foods their therapeutic potential. Functional food ingredients are typically found in a variety of forms, including hydroxylated, thiolylated, esterified, and glycosylated forms (Samtiya et al. 2021). Components that are physiologically and biologically active and offer health advantages exceeding simple nourishment are what make a diet functioning. The components of these foods that give them their medicinal properties have traditionally been referred as "bioactive compounds" (Banwo et al. 2021). Bioactive compounds, which are non-nutritive food ingredients may have positive effects on one's health. Numerous kinds of bioactive chemicals can be found in plant-based foods such fruits, vegetables, cereals, red meat, chicken, and seafood as well as in algae, seaweeds, and other aquatic goods (Miyashita and Hosokawa 2018). According to Dr. Martirosyan and Pisarski, Food bioactive compounds, which prevent or alleviate chronic disease or its symptoms, are the main and secondary metabolites of nutritive and non-nutritive natural components (Martirosyan and Miller 2018). Based on the vast differences in their molecular structures and functions, food bioactive chemicals are classified into important types such as polyphenols, carotenoids, peptides and amino acids, lipids and fatty acids, polysaccharides, glucosinolates, terpenoids, alkaloids, and furocoumarins as shown in Fig. 14.1 (Shahidi et al. 2019).

The dynamic nature of bone allows for constant remodeling through the balancing and synchronization of osteoclast and osteoblast cell production and resorption processes (Teodoro 2019). Normal bone structure and functions maintained by a elusive balance between osteoblast and osteoclast activity (Waugh and Grant 2014). Bone contains abundant extracellular matrix that surround widely separated cells 55% of the ECM is made up of crystalline mineral salts, 30% collagen fibers, and 15% water. The mineral salt calcium phosphate [$Ca_3(PO_4)_2$] is the most ubiquitous. Calcium hydroxide [$Ca(OH)_2$], another mineral salt, interacts with it to generate

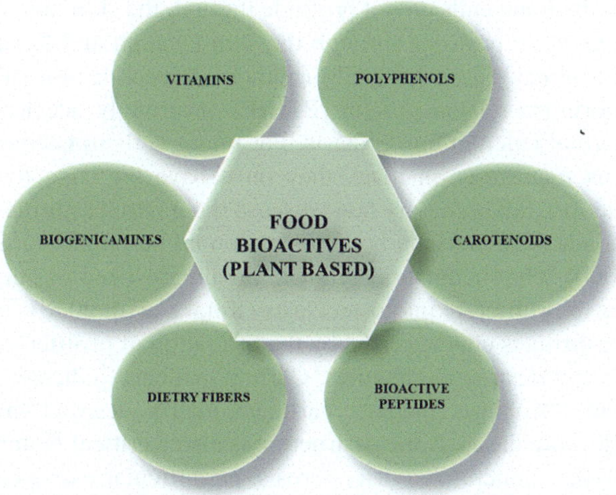

Fig. 14.1 Different classes of plant-based food bioactive compounds.

hydroxyapatite [$Ca_{10}(PO_4)_6(OH)_2$] crystals. Together with other mineral salts like calcium carbonate ($CaCO_3$) and ions like magnesium, fluoride, potassium, and sulfate, the crystals mix as they form. The tissue hardens as an outcome of the mineral salts crystallizing as they are deposited into the extracellular matrix's collagen fiber framework. Osteoblasts (a type of cells that help build bones) start the process, which is known as calcification (Tortora and Derrickson 2018).

An imbalance between bone growth and resorption can lead to a variety of bone disorders. Osteoporosis, a localized or systemic form of bone loss, caused by bone breakdown exceeding bone synthesis after the age of 40 (Sobh et al. 2022). A condition known as osteoporosis is brought on by increased bone resorption during menopause as a result of estrogen deficiency (Teodoro 2019). Significant increases in bone resorption and calcium loss from the bone (hypercalcemia of malignancy) are skeletal problems linked to bone metastases of breast and prostate malignancies as well as other cancers. There are over 30 varieties of bone cancer. The most frequent ones are chordoma, Ewing sarcoma, chondrosarcoma, and osteosarcoma. Osteosarcoma is a malignant primary bone cancer that typically affects the arm and leg bones. It originates from the bone cells. More often than not, chondrosarcoma is a secondary sarcoma originating from cartilage cells. In joints, cartilage is a type of connective tissue. Cartilage is not hard; it is flexible, unlike bone. In contrast, Ewing sarcoma is a cancerous growth that originates from a primitive nerve cell. Typically, children and teenagers are affected (Boulehmi et al. 2018). An increase in osteoclast activity and quantity characterizes Paget's disease of the bone, which frequently affects several sites across the skeleton. Increased local bone turnover and bone creation occur in response to higher bone resorption, resulting in woven bone—a dense, fragile material that is prone to breaking and bowing. There are three types of bone fractures: pathological, which occurs when a disease weakens a bone;

simple, in which the bone ends do not protrude through the skin; and compound, in which the bone ends do protrude through the skin (Waugh and Grant 2014). The most frequent hormonal reason for HHM (humoral hypercalcemia of malignancy) is parathyroid hormone-related protein (PTHrP), it strongly accelerates turnover and overrides normal calcium homeostasis. For tumor cells to become established and proliferate as metastases in bone, they must have the capacity to stimulate osteoclast production. Inflammatory bone disease rheumatoid arthritis is typified by extensive subchondral osteoclastic bone resorption and articular cartilage damage (Romas et al. 2000). The rheumatoid synovial membrane becomes inflamed, causing macrophages to proliferate into osteoclasts (Kotake et al. 2020). The pathology of Rheumatoid arthritis is characterized by the formation of proliferative tissue that exhibits invasive and destructive features. Current evidence indicates that synovial macrophages and fibroblasts mediate cartilage matrix destruction through the release of metalloproteinases. Bone destruction is also a critical feature of RA, and it consists of 3 types: generalized osteoporosis, periarticular osteopenia, and focal erosion of marginal and subchondral bone (Romas et al. 2000). After ossification is complete, adults have osteomalacia and children develop rickets. They are brought on by a vitamin D shortage, which encourages calcium absorption in the small intestine and bone calcification. Vitamin D deficiency can arise from various causes, including inadequate food, malabsorption, gastrointestinal surgeries, lack of solar exposure, excessive loss of vitamin D or its precursors, and medications that cause vitamin D to be broken down. Bone infections are known as osteomyelitis. Staphylococcus aureus is the most frequent pathogen, usually affecting the growing sections of children's long bones. Elderly diabetics frequently have soft tissue infections in their foot, which can eventually extend to the bones. A genetic anomaly causes achondroplasia and osteogenesis imperfecta, often known as brittle bone syndrome.. Achondroplasia, characterized by unusual cartilage growth, particularly in the epiphyseal cartilage of long bones, results in inequality of the base of the skull bones and characteristic dwarfism. Osteogenesis imperfect is a group of conditions characterized by a congenital defect of osteoblasts that causes ossification to fail. The brittle nature of the bones makes them prone to breaking, either on their own or after very minor impact (Waugh and Grant 2014).

The two main phytochemicals which are thought to be bioactive and potentially support improved well-being for humans are phenolics and carotenoids. The majority of fruits and vegetables with orange and yellow hues are rich in lipophilic compounds called carotenoids. Polyphenols include, among other things, anthocyanins, flavonoids, and phenolic acids (Samtiya et al. 2021). One of the most common phenolic acids is caffeine, which makes up between 75% and 100% of the total hydroxycinnamic acids in some fruits (Manach et al. 2004). Numerous foods, such as berries, grapes, peanuts, vine, and wine, contain small amounts of resveratrol, also known as stilbene, that is made by stressed or pathogenically infected plants. The highest lignan-containing foods are sesame, flax, and linseeds. It is well known that lignans have insecticidal, antiviral, and antifeedant qualities. Barley, red finger millets, and sorghum with a pigmented testa layer are examples of foods that have an astringent quality due to the presence of tannins. Avenanthramide, ferulic acid, dihydroxybenzoic acid, sinapic acid, and p-coumaric acids are the most significant bioactive compounds found in grain and oat brans. Naturally occurring oat phenolic compounds, which have strong antioxidant activity, are the source of avenanthramides. Several studies and numerous other investigations, demonstrate the efficacy of dietary supplements and other bioactives in preventing bone loss, there is a scarcity of research focused on illustrating their potential for promoting bone formation. The prevailing belief is that once bone loss, especially in the trabecular region, occurs, restoring bone to its normal state becomes challenging. Therefore, discovering natural compounds with the ability to reverse this process would be highly intriguing (Arjmandi et al. 2010). These recommendations were made with regard to the bone-related nutrients calcium, magnesium, phosphorus, fluoride, and vitamin D in order to support the health of bones and preserve a healthy diet for people at various phases of life. Nevertheless, a number of other minerals and vitamins, such as as manganese, copper, boron, iron, zinc, vitamin A, vitamin K, vitamin C, and the B vitamins, are required for chemical reactions connected to bone. The goal of prescribed dietary intake for bone health is to increase bone mass and strength (Palacios 2006).

14.2 Different Sources of Food Bioactives

Plants are abundant in bioactive substances that work in concert or additively to prevent disease and promote good health (Adefegha et al. 2022). Extra nutritional ingredients known as "food bioactive compounds" are usually found in trace amounts in food. Furthermore, a wide range of bioactive substances seem to have positive health impacts (Kussmann et al. 2023). Food bioactives include polyphenols, alkaloids, flavonoids, terpenoids, saponins, polyunsaturated fatty acids, carotenoids, peptides, and omega-3 fatty acids, vitamins, phytoestrogens etc. (Table 14.1).

14.3 Factors Affecting Bone Health

Several factors are responsible for the bone health. A balance approach is essential to maintain the bone health. Basic knowledge of determinants for prevention and treatment of bone disorder are helpful for bone life. The following are general information that must be care for the health of bone (Proia et al. 2021; Shin et al. 2013; Hernlund et al. 2013; Gardinier et al. 2015; Malik et al. 2022).

- Individual bone mass is influenced by hereditary factors. A person's strong or fragile skeleton depends on their diet and level of exercise.
- Since calcium is so vital to bone health, it is being singled out as a key cause of concern today.

Table 14.1 Bioactive compounds in different foods

S. No.	Bioactive compounds	Food sources	Inference	References
A	Flavonoids: Favones			Hamzalıoğlu and Gökmen (2024)
1.	Luteolin	Broccoli, bell pepper, parsley, oregano, carrots, rosemary, spinach, squash, sweet potato,	Provide more effective therapeutic implication by progressive bone loss including osteoporosis	Kim et al. (2009)
2.	Apigenin	Many fruits and vegetables, parsley, celery, celeriac, and chamomile tea	API could promote osteogenesis; API may be a promising therapeutic candidate for bone fracture repair.	Pan et al. (2021)
3.	Tangeritin & Nobiletin	Citrus peels	Regulate adipocyte differentiation and adipocytokine secretion &suppresses bone resorption and restore the bone loss	Tominari et al. (2012)
4.	Quercetin	Fruits and vegetables, especially onions, citrus, apples, dark berries, grapes, olive oil, green tea, and red wine	Anti-osteoporosis effects	Cao et al. (2022)
5.	Kaempferol	Apples, grapes, tomatoes, green tea, potatoes, onions, broccoli, Brussels sprouts, squash, cucumbers, lettuce, green beans, peaches, blackberries, raspberries, and spinach	It achieves the bone-protective effects by inhibiting adipogenesis, inflammation, oxidative stress, osteoclastic autophagy and osteoblastic apoptosis	Adhikary et al. (2018)
6.	Myricetin	Vegetables, fruits, nuts, berries, tea, and red wine	Osteoporosis protection, anti-inflammatory, osteoclastogenesis.	Imran et al. (2021)
7.	Galangin	Propolis, lesser galangal	Galangin protects human rheumatoid arthritis fibroblast-like synoviocytes	Fu et al. (2018)
B	Flavanone			

(continued)

Table 14.1 (continued)

S. No.	Bioactive compounds	Food sources	Inference	References
1.	Hesperitin & Neohesperidin	Citrus fruits	Used to control bone diseases, such as osteoporosis and osteoarthritis	Ortiz et al. (2022)
2.	Naringenin	Grapefruits, oranges, and tomatoes	Naringenin also inhibited osteoclastogenesis and bone resorption by inhibiting inflammation and the RANKL pathway	Nor Muhamad et al. (2022)
C	Phytoestrogens			
1.	Glycitein, & Daidzein	Soy products such as; Soybeans, soy flour, soy milk, and tofu	Increased BMD through increasing bone formation markers, reducing bone resorption markers, rescuing deteriorated bone structure, quality and bone resistance to fracture.	Sarker and Nahar (2020)
2.	Genistein	Soybeans and soy-derived foods	Genistein exhibits estrogen-like activity in protecting against osteoporosis and osteoarthritis	Wu and Liu (2022), Elkhalifa et al. (2023)
D	Vitamins			
1.	Pyridoxal	Whole-grain products, vegetables, nuts, and bananas	To increase bone mineral density (BMD)	Welan (2023)
2.	Niacin	Cereal, peanuts, legumes, avocados, tomatoes, leaf vegetables, carrots, sweet potatoes, asparagus, nuts, whole grain.	Niacin is a promising agent for preventing bone loss in iron overload conditions	Tao et al. (2024)
3.	Riboflavin	Leafy vegetables, legumes, mushrooms, and almonds, some grains, such as quinoa	Decrease risk of osteoporosis	Wan et al. (2023)
4.	Nicotinamide	Nuts, mushrooms, and some vegetables	Promotes osteogenesis Enhance osteoporotic bone fracture healing & prevents osteoporosis	Lu et al. (2023), Song et al. (2019)
5.	Ascorbic acid	Citrus fruits, tomatoes, tomato juice, potatoes, red and green peppers, kiwifruit, broccoli, strawberries, Brussels sprouts, and cantaloup	Essential for collagen formation and normal bone development, promotes vitamin D metabolism and binding, affects markers of osteoblast activity	Simon and Hudes (2001)

(continued)

Table 14.1 (continued)

S. No.	Bioactive compounds	Food sources	Inference	References
E	Others			
1.	Punicalagin	Pomegranates	Shows antitumor effect on osteosarcoma (Bone tumor), treats postmenopausal osteoporosis by promoting osteogenesis	Huang et al. (2020)
2.	Curcumin	Turmeric	To treat musculoskeletal disorders, such as osteoarthritis, osteoporosis, musculocartilaginous disorders, and sarcoma	Peddada et al. (2015)

- Calcium absorption is enhanced by vitamin D, which is why it is very important for bone health.
- Bone function is dependent on physical exercise all throughout life. It lowers the chance of injury and assists to maintain or grow bone mass. Exercise of all kinds, though in various manners, is beneficial for bone health.
- Underweight people are more likely to fracture and lose bone. While an appropriate diet and regular exercise can help to prevent this loss, weight reduction is also linked to bone loss.
- A number of particular strategies have shown promise, such as expert home risk analysis and modification, retraining of muscle strength and balance, and the reduction or cessation of psychiatric medication.
- Menorrhea, or the cessation of menstruation, following puberty but before menopause poses a major risk to bone health and should be treated by patients and doctors who treat them.
- Numerous medical disorders and prescription drugs can impact bone health through different ways.
- Excessive consumption of alcohol has been linked to decreased bone density and elevated risk of fracture.
- Hormonal factors, such as estrogen, testosterone, thyroid hormones, and parathyroid hormone, profoundly influence bone health. Estrogen and testosterone support bone density, while thyroid hormones and parathyroid hormone regulate calcium balance
- Deficiencies in estrogen or testosterone, as seen in menopause or hypogonadism, accelerate bone loss, leading to osteoporosis. Hyperthyroidism or hyperparathyroidism can increase bone turnover, weakening bones. Conversely, excess estrogen from hormone therapy can elevate fracture risk. Maintaining hormonal balance through appropriate medical management is essential for optimal bone health.
- Medical conditions such as osteoporosis, diabetes, and chronic kidney disease can compromise bone health by impairing bone density and strength. Medications like glucocorticoids and anticonvulsants may also contribute to bone loss.

Managing underlying conditions and medications is crucial for preserving bone health.

- Some medications, such as glucocorticoids, anticonvulsants, and certain cancer treatments, can adversely affect bone density and metabolism.
- Lack of sunlight exposure limits the synthesis of vitamin D, affecting calcium absorption and potentially compromising bone health.
- Environmental pollutants, heavy metals, and toxins can interfere with bone metabolism and contribute to bone-related diseases.

14.4 Essential Requirements for Bone Health

Maintaining good bones must be emphasized through activity and a nutritious diet. Calcium and vitamin D are essential for the postmenopausal women to reduce the risk of fractures, that it recommended dietary supplements. For good bone health precaution is also very important that always required for the patients of weak bone, elders and postmenopausal like lifestyle changes, home safety modifications, hip protectors, and fall prevention. Exercise will also be noted as a crucial—yet typically inadequate—factor for maintaining healthy bones (Sahni et al. 2015; Ahsan et al. 2023). Fish that are high in oil, like swordfish and salmon, and low in tuna are good dietary sources of vitamin D. Obtaining adequate quantities of vitamin D alone through diet is challenging. Milk and milk products are the main dietary sources of calcium, while leafy green vegetables like spinach, kale, and turnip greens, salmon, and almonds are other minor sources (Jakobsen et al. 2019). Hip fracture risk is correlated with low calcium intake; however, hip fracture risk does not decrease when one boosts calcium consumption over 750 mg/day. Since slightly greater calcium consumption may have extra health benefits, an acceptable supplemental dose is 400–800 mg/day to reach the National Institutes of Health's recommended daily intake of 12,000 mg/day. Excessive intake of calcium concentrations has been linked to a higher risk of myocardial infarction and kidney stones. As a result, calcium supplements ought to meet the dietary intake that is advised not going beyond. As exercise increases muscle tone, balance, and coordination, it also lowers the chance of falling (Wang et al. 2015). The importance of magnesium in maintaining healthy bones is becoming more well acknowledged. A few dietary sources of magnesium are peanuts, cashews, and almonds. Lentils, kidney beans, black-eyed peas, brown rice, potato skins, and raisin bran cereal are among more sources. A serving of eight ounces of milk contains about 25 milligrams of magnesium. For optimal health, an intake of 320–420 mg/day is advised. It makes sense to take 250 mg of magnesium daily as dietary supplements to promote healthy bones and other elements of good health in general (Rondanelli et al. 2021).

An additional important component that supports healthy bones is silicon. A deficit in silicon is linked to impaired skeletal growth since silicon is an important nutrient. One component involved in the start of the mineralization process is silicon. A part of the mineralization process is initiated by silicon. Due to silicon's distinct piezoresistive qualities as a semi-conductor element, it is also known that silicon is utilized in computer electronics and micro-pressure transducers. Carrots, green beans, whole grains, and cereals are dietary sources of silicon (Jugdaohsingh 2007). Another little-known vitamin that is crucial for healthy bones is vitamin K. There are various types of vitamin K, but the naturally occurring forms are K_1 and K_2. Kale, collard greens, fresh spinach, Brussels sprouts, iceberg lettuce, and prunes are some helpful food sources of vitamin K. More people are becoming aware of boron's numerous medical benefits, include its ability to support healthy bones. Although boron's exact mode of action for the maintenance of bones is uncertain, it maintains and lengthens the half-lives of estrogen and vitamin D. With about 3–4 mg of boron per three-ounce serving, prunes are an excellent source of boron (Khalil et al. 2021). Even while the more prevalent deficiencies of calcium, vitamin D, and exercise are becoming more widely acknowledged as factors in bone health, doctors are rarely informed about these less commonly recognized deficiencies of magnesium, silicon, vitamin K, and Boron. Vitamin C, inositol, and L-arginine, along with to these necessary nutrients, are excellent for bone health. (Dominguez et al. 2021). The synthesis of collagen and the rehabilitation of fractures depend on vitamin C. L-arginine is a building block for the synthesis of nitric oxide (NO), which enhances endothelial function, lowers vascular resistance, encourages angiogenesis, and affects many metabolic processes. Combining arginine, inositol, and silicon has been shown through research to boost the durability of bones and density of minerals. Adding physiological doses of these three nutrients to the diet may help maintain bone health (Albaugh et al. 2017).

14.5 Bone Therapy with Herbal Extracts

Novel extract of *Boswellia serrata* significantly reduced the levels of high-sensitive C-reactive protein which is potential inflammatory marker concern with the osteoarthritis of the knee. The bioactive compounds of the extract which are responsible for the therapeutic effects are 3-acetyl-11-keto-β-boswellic acid and β-boswellic acid. Both compounds have synergetic effects to exert anti-inflammatory/antiarthritic activity (Siemoneit et al. 2010). Anti-inflammatory and antioxidant effects of turmeric extract have reported. The extract has polyphenol compounds which strongly inhibit the production of inflammatory mediators, like interleukin-1, tumor necrosis factor-alpha (TNF-α), IL-8, NO, and a variety of MMPs, via diminishing the activation of NF-κB, protein kinase B (Akt), and MAPK signaling pathways. Also, inhibit COX-2 that reduced the synthesis of prostaglandin (Chin 2016). The aqueous extract of *Eremostachys laciniata* has shown antioxidant and anti-inflammatory activity. The application of 0.5% *E. laciniata* ointment is effective for the joints pain in arthritis. The bioactive molecules which are mainly responsible for the therapeutic effects are iridoid glycosides (Erdemoglu et al. 2006). *Eucommia ulmoides* extract have potential to reverse the LPS-induced production of IL-1β, IL-6, TNF-α, inducible nitric oxide synthase (iNOS), and COX-2 via modulating the activation of toll-like receptor (TLR) 4 in murine macrophages. In Osteoarthritis the aqueous extract of this plant is very effective, Aucubin, a bioactive molecule has been reported for to reduce reactive oxygen species (ROS) production, caspase-3 activity, and cell apoptosis (Xia et al. 2019). Paeoniflorin bioactive compound is present in Water/ethanol extracts of *Radix Paeoniae.* This molecule has potential inhibit the production of PGE2, leukotriene B4, NO, ROS, and other pro-inflammatory mediators. It has been reported in research. The paeoniflorin component of TGP may also inhibit osteoclast differentiation and TNF-α-induced apoptosis via inhibition of NF-κB. Bioactive (Zhang and Dai 2012). The phenolic compounds like apigenin, quercetin, patuletin, luteolin, and glucosides are available in *Matricaria Chamomile* extracts. The inflammatory effects of this extract have reported in research. These bioactive molecules have shown anti-inflammatory action by reducing cytokines and PGE2 (Shoara et al. 2015). Macrophages, lymphocytes, synovial fibroblasts, and chondrocytes are all inhibited in their expression of pro-inflammatory cytokines and mediators, adhesion molecules, and matrix metalloproteinases by extracts of the root of *Tripterygium wilfordii Hook F.* Moreover, stop the growth of synovial fibroblasts and lymphocytes by causing apoptosis in them. Extracts have been shown to have immunosuppressive, cartilage-protective, and anti-inflammatory properties (Pyatt et al. 2000). It has been discovered that *W. somnifera* extract reduces the activation of NF-κB and activator protein 1 (AP-1) signaling pathways, which in turn inhibits the production of TNF-α, IL-1β, and IL-12. By suppressing collagenase activity, extract also reduced the rate at which type I collagen in the bovine Achilles tendon degraded. In a rat model of collagen-induced arthritis, treatment with *W. somnifera* reduced edema, redness, deformity, and ankylosis (Gupta and Singh 2013). *Zingiber offcinale* extract has a strong correlation with the reduction of inflammatory symptoms. Ginger's anti-inflammatory properties have been well studied in both in vitro and in vivo models as well as in humans. Ginger extract formulation has resulted in a decrease in the generation of PGE2, NO, IL-1β, IL-12, TNF-α, and controlled on activation, normal T cell expressed and secreted (RANTES). Additionally, ginger has been shown to suppress the expression of MHC class II molecules, interferon-gamma (IFN-γ), and IL-2, which has been shown to suppress T cell function and macrophage antigen-presenting ability (Tripathi et al. 2008).

14.6 Conclusion

In this chapter, we tried to summarized the role of different bioactive compounds in bone health. The different classes of bioactive compounds have different mechanism to improve and maintain the bone health. Apart from calcium and vitamin D the different bioactive compounds like flavonoid, flavones, phytoestrogens and vitamins have importance in bone life.

References

Adefegha SA, Oboh G, Oluokun OO (2022) Food bioactives: the food image behind the curtain of health promotion and prevention against several degenerative diseases. Stud Nat Prod Chem 72:391–421

Adhikary S, Choudhary D, Ahmad N, Karvande A, Kumar A, Banala VT, Mishra PR, Trivedi R (2018) Dietary flavonoid kaempferol inhibits glucocorticoid-induced bone loss by promoting osteoblast survival. Nutrition 53:64–76

Ahsan N, Imran M, Mohammed Y, Al Anouti F, Khan MI, Banerjee T, Adnan M et al (2023) Mechanistic insight into the role of vitamin D and zinc in modulating immunity against COVID-19: a view from an immunological standpoint. Biol Trace Elem Res 201(12):5546–5560

Albaugh VL, Mukherjee K, Barbul A (2017) Proline precursors and collagen synthesis: biochemical challenges of nutrient supplementation and wound healing. J Nutr 147(11):2011–2017. https://doi.org/10.3945/jn.117.256404

Arjmandi BH, Johnson CD, Campbell SC, Hooshmand S, Chai SC, Akhter MP (2010) Combining fructooligosaccharide and dried plum has the greatest effect on restoring bone mineral density among select functional foods and bioactive compounds. J Med Food 13(2):312–319

Banwo K, Olojede AO, Adesulu-Dahunsi AT, Verma DK, Thakur M, Tripathy S et al (2021) Functional importance of bioactive compounds of foods with Potential Health Benefits: a review on recent trends. Food Biosci 43:101320

Boulehmi H, Mahersia H, Hamrouni K (2018) Bone cancer diagnosis using GGD analysis. In 2018 15th International Multi-Conference on Systems, Signals and Devices (SSD), pp. 246–251. https://doi.org/10.1109/SSD.2018.8570658

Cao L, Wang J, Zhang Y, Tian F, Wang C (2022) Osteoprotective effects of flavonoids: evidence from in vivo and in vitro studies. Mol Med Rep 25(6):1–9

Chin KY (2016) The spice for joint inflammation: anti-inflammatory role of curcumin in treating osteoarthritis. Drug Des Dev Ther 10:3029–3042

Dominguez LJ, Farruggia M, Veronese N, Barbagallo M (2021) Vitamin D sources, metabolism, and deficiency: available compounds and guidelines for its treatment. Meta 11(4):255. https://doi.org/10.3390/metabo11040255

Elkhalifa AEO, Al-Shammari E, Kuddus M, Adnan M, Sachidanandan M, Awadelkareem AM, Qattan MY et al (2023) Structure-based multi-targeted molecular docking and dynamic simulation of soybean-derived isoflavone genistin as a potential breast cancer signaling proteins inhibitor. Life 13(8):1739

Erdemoglu N, Turan NN, Cakõcõ I, Şener B, Aydõn A (2006) Antioxidant activities of some Lamiaceae plant extracts. Phytother Res 20:9–13. https://doi.org/10.1002/ptr.1816

Fu Q, Gao Y, Zhao H, Wang Z, Wang J (2018) Galangin protects human rheumatoid arthritis fibroblast like synoviocytes via suppression of the NF κb/NLRP3 pathway. Mol Med Rep 18(4):3619–3624

Gardinier J, Mohamed F, Kohn D (2015) PTH signaling during exercise contributes to bone adaptation. J Bone Miner Res 30(6):1053–1063

Gupta A, Singh S (2013) Evaluation of anti-inflammatory effect of Withania somnifera root on collagen-induced arthritis in rats. Pharm Biol 52:308–320

Hamzalıoğlu A, Gökmen V (2024) Interaction between bioactive carbonyl compounds and asparagine and impact on acrylamide. Acrylamide in Food Academic Press, pp 433–455

Hernlund E, Svedbom A, Ivergård M, Compston J, Cooper C, Stenmark J et al (2013) Osteoporosis in the European Union: Medical management, epidemiology and economic burden. A Report Prepared in Collaboration with the International Osteoporosis Foundation (IOF) and the European Federation of Pharmaceutical Industry Associations (EFPIA). Arch Osteoporos 8:136

Huang T, Zhang X, Wang H (2020) Punicalagin inhibited proliferation, invasion and angiogenesis of osteosarcoma through suppression of NF κb signaling. Mol Med Rep 22(3):2386–2394

Imran M, Saeed F, Hussain G, Imran A, Mehmood Z, Gondal TA, El-Ghorab A, Ahmad I, Pezzani R, Arshad MU, Bacha U (2021) Myricetin: a comprehensive review on its biological potentials. Food Sci Nutr 9(10):5854–5868

Jakobsen J, Smith C, Bysted A, Cashman KD (2019) Vitamin D in wild and farmed Atlantic Salmon (Salmo Salar)-what do we know? Nutrients 11(5):982. https://doi.org/10.3390/nu11050982

Jugdaohsingh R (2007) Silicon and bone health. J Nutr Health Aging 11(2):99–110

Khalil Z, Alam B, Akbari AR, Sharma H (2021) The medical benefits of vitamin K2 on calcium-related disorders. Nutrients 13(2):691. https://doi.org/10.3390/nu13020691

Kim SY, Jung JW, Kim TH, Hong JM, Kim HJ, Park EK (2009) Effect of luteolin on bone resorption, bone loss and microarchitecture in ovariectomized mice. In: Proceedings of the 55th Annual Meeting of the Orthopaedic Research Society, Las Vegas, p. 1

Kotake S, Sato K, Kim KJ, Takahashi N, Udagawa N, Nakamura I et al (2020) Interleukin-6 and soluble Interleukin-6 receptors in the synovial fluids from rheumatoid arthritis patients are responsible for osteoclast-like cell formation. J Bone Miner Res 11:88–95

Kussmann M, Cunha DH, Berciano S (2023) Bioactive compounds for human and planetary health. Front Nutr 10. https://doi.org/10.3389/fnut.2023.1193848

Lu Z, Jiang L, Lesani P, Zhang W, Li N, Luo D, Li Y, Ye Y, Bian J, Wang G, Dunstan CR (2023) Nicotinamide mononucleotide alleviates osteoblast senescence induction and promotes bone healing in osteoporotic mice. J Gerontol 78(2):186–194

Malik MA, Jan Y, Al-Keridis LA, Haq A, Ahmad J, Adnan M, Alshammari N, Ashraf SA, Panda BP (2022) Effect of Vitamin-D-Enriched edible Mushrooms on Vitamin D status, bone health and expression of CYP2R1, CYP27B1 and VDR gene in Wistar rats. J Fungi 8(8):864

Manach C, Scalbert A, Morand C, Rémésy C, Jiménez L (2004) Polyphenols: food sources and bioavailability. Am J Clin Nutr 79(5):727–747. https://academic.oup.com/ajcn/article-abstract/79/5/727/4690182

Martirosyan D, Miller E (2018) Bioactive compounds: the key to functional foods. Bioact Compd Health Dis 1:36–39

Miyashita K, Hosokawa M (2018) Health impact of marine carotenoids. J Food Bioact 1:31–40

Nor Muhamad ML, Ekeuku SO, Wong SK, Chin KY (2022) A scoping review of the skeletal effects of Naringenin. Nutrients 14(22):4851

Ortiz AD, Fideles SO, Reis CH, Bellini MZ, Pereira ED, Pilon JP, de Marchi MÂ, Detregiachi CR, Flato UA, Trazzi BF, Pagani BT (2022) Therapeutic effects of citrus flavonoids neohesperidin, hesperidin and its aglycone, hesperetin on bone health. Biomol Ther 12(5):626

Palacios C (2006) The role of nutrients in bone health, from A to Z. Crit Rev Food Sci Nutr 46(8):621–628

Pan FF, Shao J, Shi CJ, Li ZP, Fu WM, Zhang JF (2021) Apigenin promotes osteogenic differentiation of mesenchymal stem cells and accelerates bone fracture healing via activating Wnt/β--catenin signaling. Am J Physiol Endocrinol Metab 320(4):E760–E771

Peddada KV, Peddada KV, Shukla SK, Mishra A, Verma V (2015) Role of curcumin in common musculoskeletal disorders: a review of current laboratory, translational, and clinical data. Orthop Surg 7(3):222–231

Proia P, Amato A, Drid P, Korovljev D, Vasto S, Baldassano S (2021) The impact of diet and physical activity on bone health in children and adolescents. Front Endocrinol 12:704647. https://doi.org/10.3389/fendo.2021.704647

Pyatt DW, Yang Y, Mehos B, Le A, Stillman W, Irons RD (2000) Hematotoxicity of the Chinese herbal medicine Tripterygium wilfordii hook f in CD34-positive human bone marrow cells. Mol Pharm 57:512–518

Romas E, Bakharevski O, Hards DK, Kartsogiannis V, Quinn JM, Ryan PF et al (2000) Expression of osteoclast differentiation factor at sites of bone erosion in collagen-induced arthritis. Arthritis Rheum 43:821–826

Rondanelli M, Faliva MA, Tartara A, Gasparri C, Perna S, Infantino V, Riva A, Petrangolini G, Peroni G (2021) An update on magnesium and bone health. Biometals 34(4):715–736. https://doi.org/10.1007/s10534-021-00305-0

Sahni S, Mangano KM, McLean RR, Hannan MT, Kiel DP (2015) Dietary approaches for bone health: lessons from the Framingham osteoporosis study. Curr Osteoporos Rep 13(4):245–255. https://doi.org/10.1007/s11914-015-0272-1

Samtiya M, Aluko RE, Dhewa T, Moreno-Rojas JM (2021) Potential health benefits of plant food-derived bioactive components: an overview. Food Secur 10:839. https://doi.org/10.3390/foods10040839

Sarker SD, Nahar L (2020) Medicinal natural products: a disease-focused approach. Academic Press

Shahidi F, Ramakrishnan VV, Oh WY (2019) Bioavailability and metabolism of food bioactives and their health effects: a review. J Food Bioact 8. https://doi.org/10.31665/JFB.2019.8204

Shin S, Hong K, Kang S, Joung H (2013) A milk and cereal dietary pattern is associated with a reduced likelihood of having a low bone mineral density of the lumbar spine in Korean adolescents. Nutr Res 33(1):59–66. https://doi.org/10.1016/j.nutres.2012.11.003

Shoara R, Hashempur MH, Ashraf A, Salehi A, Dehshahri S, Habibagahi Z (2015) Efficacy and safety of topical Matricaria chamomilla L. (chamomile) oil for knee osteoarthritis: a randomized controlled clinical trial. Complement Ther Clin Pract 21:181–187

Siemoneit U, Koeberle A, Rossi A, Dehm F, Verhoff M, Reckel S, Maier TJ, Jauch J, Northoff H, Bernhard F et al (2010) Inhibition of microsomal prostaglandin E2 synthase-1 as a molecular basis for the anti-inflammatory actions of boswellic acids from frankincense. Br J Pharm 162:147–162

Simon JA, Hudes ES (2001) Relation of ascorbic acid to bone mineral density and self-reported fractures among US adults. Am J Epidemiol 154(5):427–433

Sobh MM, Abdalbary M, Elnagar S, Nagy E, Elshabrawy N, Abdelsalam M, Asadipooya K, El-Husseini A (2022) Secondary osteoporosis and metabolic bone diseases. J Clin Med 11(9):2382

Song J, Li J, Yang F, Ning G, Zhen L, Wu L, Zheng Y, Zhang Q, Lin D, Xie C, Peng L (2019) Nicotinamide mononucleotide promotes osteogenesis and reduces adipogenesis by regulating mesenchymal stromal cells via the SIRT1 pathway in aged bone marrow. Cell Death Dis 10(5):336

Tao Z, Tao M, Zhou M, Wu XJ (2024) Niacin treatment prevents bone loss in iron overload osteoporotic rats via activation of SIRT1 signaling pathway. Chem Biol Interact 388:110827

Teodoro AJ (2019) Bioactive compounds of food: their role in the prevention and treatment of diseases. Oxidative Med Cell Longev 2019:1–4. https://doi.org/10.1155/2019/3765986

Tominari T, Hirata M, Matsumoto C, Inada M, Miyaura C (2012) Polymethoxy flavonoids, nobiletin and tangeretin, prevent lipopolysaccharide-induced inflammatory bone loss in an experimental model for periodontitis. J Pharmacol Sci 119(4):390–394

Tortora GJ, Derrickson BH (2018) Principles of anatomy and physiology, 15th edn. John Wiley & Sons

Tripathi S, Bruch D, Kittur DS (2008) Ginger extract inhibits LPS induced macrophage activation and function. BMC Complement Altern Med 8:1–7. https://doi.org/10.1186/1472-6882-8-1

Wan M, Wu H, Wang X, Gu Y, Meng G, Zhang Q, Liu L, Zhang J, Sun S, Jia Q, Song K (2023) There is a significantly inverse relationship between dietary riboflavin intake and prevalence of osteoporosis in women but not in men: results from the TCLSIH cohort study. Front Nutr 10:1112028

Wang D, Chen XH, Fu G, Gu LQ, Zhu QT, Liu XL, Qi J, Xiang JP (2015) Calcium intake and hip fracture risk: a meta-analysis of prospective cohort studies. Int J Clin Exp Med 8(8):14424–14431

Waugh A, Grant A (2014) Ross & Wilson Anatomy and physiology in health and illness E-book. Elsevier Health Sciences

Welan R (2023) Effect of vitamin B6 on osteoporosis fracture. J Bone Metabol 30(2):141–147

Wu Z, Liu L (2022) The protective activity of genistein against bone and cartilage diseases. Front Pharmacol 13:1016981

Xia T, Gao R, Zhou G, Liu J, Li J, Shen J (2019) Trans-Cinnamaldehyde inhibits IL-1β-stimulated inflammation in chondrocytes by suppressing NF-κB and p38-JNK pathways and exerts chondrocyte protective effects in a rat model of osteoarthritis. Biomed Res Int 2019:4039472–4039412

Zhang W, Dai SM (2012) Mechanisms involved in the therapeutic effects of Paeonia lactiflora Pallas in rheumatoid arthritis. Int Immunopharmacol 14:27–31

Chapter 15
Immune-Boosting Potential of Food Bioactives

Soubhagya Tripathy, Deepak Kumar Verma, Alok Kumar Gupta, Prem Prakash Srivastav, Mónica L. Chávez-González, and Cristobal N. Aguilar

Abstract A robust immune system is essential for maintaining good health. The characteristics of a robust immune system is to safeguard the body against invasion of outer microscopic materials and subsequent infection. Food ingredients have long been acknowledged to play a crucial role in the growth and development of the human body, providing protective functions against foreign matter that can pose serious public health issues. According to numerous studies, natural products are rich in bioactive substances like vitamins such as vitamins A, C, and E, minerals, and phytochemicals such as beta-carotene, flavonoids, tannins, and phenolics. These substances can improve our immune system by promoting lymphocyte proliferation, scavenging free radicals, lowering oxidative stress, enhancing anti-inflammatory and immunomodulatory pathways, and promoting platelet aggregation. Through encapsulations, bioactive substances can safely deliver inside the body. The advantages of encapsulated bioactives include their high bioavailability and stability, prolonged release of functional compounds, capacity to prevent undesirable interactions, and excellent antibacterial and antioxidant activity. In summary, this chapter will present an in-depth review of the human immune system and the immune-boosting potential of food bioactives, emphasising their potential mechanism and the techniques employed for applications in food matrices.

S. Tripathy (✉) · D. K. Verma (✉) · P. P. Srivastav
Agricultural and Food Engineering Department, Indian Institute of Technology Kharagpur, Kharagpur, West Bengal, India
e-mail: pps@agfe.iitkgp.ac.in

A. K. Gupta
Division of Post-harvest Management, ICAR-Central Institute for Subtropical Horticulture (Ministry of Agriculture and Farmers Welfare, Government of India), Lucknow, Uttar Pradesh, India

M. L. Chávez-González · C. N. Aguilar
Bioprocesses Research Group, Food Research Department, School of Chemistry, Universidad Autonoma de Coahuila, Saltillo, Mexico
e-mail: monicachavez@uadec.edu.mx; cristobal.aguilar@uadec.edu.mx

S. Amir Ashraf, M. Adnan (eds.), *Food Bioactives and Nutraceuticals*,
https://doi.org/10.1007/978-981-97-9723-3_15

Keywords Bioactive compounds · Immune system · Functional foods · Nutrition · Vitamins

15.1 Introduction

The immune system is one of the body's many defense mechanisms against pathogenic invasion. It is an intricate network of cells, tissues, and organs that protect the body from potential injury (Singh et al. 2023). The immune system protects against external and internal pathogenic microorganisms and other threats. An immune protection system against foreign chemicals is provided by a number of anatomical physical barriers, including the skin, ciliated epithelial cells, mucous blankets, and mucous membranes (Mitra et al. 2022). It never stops being proactive and can discriminate between "self" and "non-self." The immune system responds quickly to protect the body from foreign chemicals that can slip through these barriers as "non-self" particles (Maheshwari et al. 2022). Innate or nonspecific immunity and acquired or specific immunity are the two categories into which immunity can be divided (Singh et al. 2022). Innate immunity, present from birth and serves as a defense against harmful substances, is distinct from acquired immunity, developed through interaction with the invader and specific to that invader (Singh et al. 2023). An innate immune response is elicited by infectious agents regardless of how frequently they are encountered. Pathogen-associated molecular patterns (PAMPs), another name for receptors that recognize foreign particles, are essential for the prompt onset of innate immunity (Elalouf et al. 2023). The innate immune system is a generic, flexible, and instantaneous defensive system. It comprises leukocyte responses against pathogens, cell-intrinsic responses against viral infections, and soluble mediators, including complementary proteins. T and B cells are specific to the adaptive immune response. The cells generate antibodies to remove infections and help attract other immune cells to combat the pathogen by recognizing specific antigens from the invading microorganism (Cai et al. 2020; Pecora et al. 2020).

It is commonly known, that nutrition and food greatly influence immunological function. A lack of certain nutrients or undernutrition can interfere with immune system function and increase the risk of many illnesses in our bodies. Thus, eating a balanced diet with the appropriate proportions of each nutrient is crucial for our immune system to operate appropriately (Maheshwari et al. 2022). Eating a diet rich in foods that have immunomodulatory qualities can help to fortify the immune system. Many foods contain components that support the health of both our acquired immune system (T cells and B cells) and innate immunity (macrophages, NK cells, and neutrophils) (Singh et al. 2023). Conversely, poor diet can hinder the development of the immune system and result in immunological incompetence, making the body more vulnerable to allergies, chronic inflammation, and infections (Cooper and Ma 2017; Singh et al. 2023). This emphasizes the possible benefits of including particular foods to boost immunity and prevent illness. Consumers now choose certain foods over others to attain desired health outcomes due to a greater understanding of the intimate relationship between food and human health in developed nations (Tripathy et al. 2021a, b; Banwo et al. 2021). Because they aim to improve quality of life by reducing diet-related disorders, functional foods are great alternatives in this context (Díaz et al. 2020; Verma and Thakur 2021). Foods that provide health advantages over their nutritional content, such as dietary fiber, vitamins, essential oils, probiotics, and prebiotic substances, are referred to as functional foods. The body receives nutrients from immune-boosting foods when they are consumed daily. These nutrients strengthen biological defense mechanisms, prevent and treat certain diseases, slow aging, and manage physical and mental illnesses (Stephen et al. 2023). Eating a diet high in foods that stimulate the immune system can help prevent and treat several diseases and health issues, including inflammation. Certain foods—like broccoli, spinach, and tomatoes—and green vegetables—like chili peppers and mushrooms—help the body become more disease-resistant. A diet high in plant-based fats could help boost immunity (Mishra and Patel 2020; Tripathy et al. 2021b). Flavonoids, carotenoids and Vitamins A, C, E are naturally occurring functional dietary elements found to be potent antioxidants and anti-inflammatory agents that support immune system function (Singh et al. 2022).

Numerous studies have demonstrated a clear link between a higher intake of bioactive substances and a lower risk of developing chronic illnesses. According to reports, several biological factors could influence the prevention of chronic illnesses by consuming bioactive chemicals (Maheshwari et al. 2022). According to various experimental investigations, using vitamins and minerals to enhance immune functioning has shown promising results (Bora 2023; Pedroza-García et al. 2023; Stefanache et al. 2023). Specific vitamins, minerals, and trace elements are critical for immune system maintenance; this has been shown in animal models and, in rare instances, in malnourished humans. Immunological function recovery and infection resistance can be improved by supplementing or changing one's diet to include lacking nutrients, such as vitamins (Mitra et al. 2022). Antioxidants also improve the acquired (antigen-specific) immune response and the natural (antigen-independent) immune response. Antioxidant molecules, in particular, are able to withstand the B and T lymphocytes, two "weapons" of acquired immunity (Pangrazzi 2019). Fruits contain a variety of phytochemicals, vitamins, and minerals that shield our bodies against the damaging effects of reactive oxygen species (ROS), including flavonoids, quercetin, kaempferol, ellagic acid, catechins, ascorbic acid, folic acid, zinc, and selenium (Maheshwari et al. 2022). Furthermore, immune cells' antioxidant balance must be preserved to shield them from oxidation, and an abundance of micronutrients may benefit the antioxidant system functionally (Elmadfa and Meyer 2019).

In light of this, this review aims to outline the characteristics of the immune system and discuss the application of food-bioactive substances as a prophylactic strategy to lower morbidities in people with autoimmune disorders or immune system invasions. To date, various investigations have been conducted to offer details regarding the immunomodulatory properties of bioactive substances. Nevertheless, a thorough analysis was lacking that covered the various widely eaten bioactive substances and the function of phytochemicals in bolstering particular immune system components and their mechanism. Furthermore, a novel idea in this field—delivering these bioactive substances into the human body by encapsulation—is discussed. In addition, the creation of various functional foods using encapsulated bioactive chemicals has been thoroughly examined for each category of food products. Future researchers may find it helpful in devising techniques such as selecting powerful bioactive components for preparing immune-enhancing functional food.

15.2 Human Immune System

The human immune system is an invulnerable mechanism for disease resistance made up of tissues, immune cells, and organs that work together to shield the body from foreign substances called antigens. These foreign components include parasites, fungus, viruses, and bacteria that invade the body and make people sick. According to Stephen et al. (2023), compounds that infect humans can multiply if a person's immunity is weakened and spread to other individuals by inflicting disease, serious infection, or even death. The term "immunity" describes the host's defense against infections, which is provided by the immune system. Immunity is classified into two categories based on the way the human body responds to an immune stimulus: passive immunity involves receiving antibodies from external sources, while active immunity involves the host body producing antibodies for the immune response (Singh et al. 2023). While active immunity plays a major role in the immunological responses that individuals experience, passive immunity is transient and less effective in the battle against infections (Jafari et al. 2023). Based on its source, active immunity is further separated into two groups: innate immunity and acquired immunity (Fig. 15.1) (Tripathy et al. 2021a).

Since the human body produces antibodies that will instantly combat any kind of antigen, innate immunity is not antigen-specific. Dendritic cells, mast cells, basophils, eosinophils, neutrophils, and macrophages are among the antibodies involved in this kind of immunization. First and foremost, this innate immunity begins to function as soon as the host comes into contact with the virus, which can happen minutes or hours later (Park et al. 2023). The initial line of defense is innate immunity, which is triggered as soon as an antigen is encountered is shown in Fig. 15.1. There is no immunological memory in innate immunity. It implies that immunization does not produce immunological capacity.

Alternatively, the presence of many pathogens might reduce innate immunity's effectiveness, leading to adaptive immunity's activation. Adaptive immunity, involving B and T cells, is an added advantage and a sophisticated defense mechanism (Stephen et al. 2023). B cells produce antibodies specific to each antigen produced by a disease. Parija (2023) state that these antibodies, part of the immunoglobulin protein family, have several functions inside the immune system. There are five main groups of immunoglobulins: IgA, IgD, IgE, IgG, and IgM. There are three

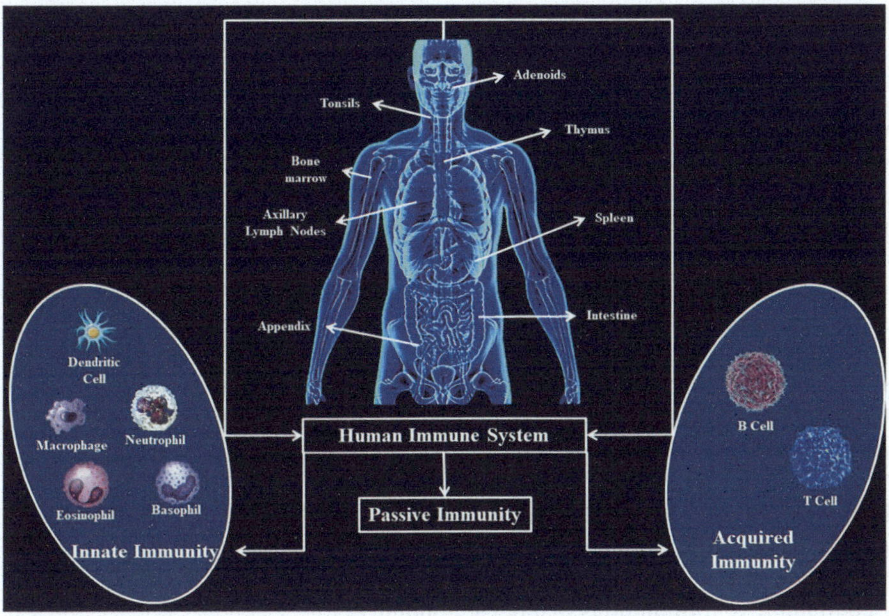

Fig. 15.1 Human immunity system with different categories. (Tripathy et al. 2021b; Singh et al. 2023)

different types of T lymphocytes, each with a specific function: helper T cells, killer T cells, and suppressor T cells (Singh et al. 2023). The job of helper T cells is to stimulate B cells to produce more antibodies and to coordinate with other cells. As their name suggests, killer T cells, sometimes referred to as cytotoxic T lymphocytes, attack other cells. These work by identifying and concentrating on small viral fragments that are present on the surface of infected cells, which helps in the detection and removal of viruses. Finally, suppressor T cells provide other T cells and B cells the signal to cease operations when the enemy is eliminated entirely from the body (Matsuzaka and Yashiro 2023; Razizadeh et al. 2023; Zhu et al. 2023). Upon pathogen recognition, the adaptive immune response is triggered, activating the lymphocytes in the body's system. Adaptive immunity is specific to antigens and requires time for the body to generate antibodies through vaccination (Gray and Gibbs 2022). Immunological memory is a key component of adaptive immunity, which is important for defense should the same infection recur in the future (Domínguez-Andrés et al. 2023). Adaptive immunity, the basis of vaccination, can recognize and target specific entities like viruses by producing antibodies through immunologic memory—adaptive immunity functions as the second line of protection by targeting pathogenic organisms (Marshall et al. 2018).

15.3 Major Bioactive Compounds Responsible for Immunological Activity and Their Mechanism

Nutrition has a major role in human immune development. Antioxidants, vitamin D, zinc, iron, dietary fiber, and omega-3 fatty acids are essential in meals that help improve immunity and maintain a healthy immune system. Adding nutrients from natural sources like fruits and vegetables enhances the immune system by lowering viral infections, cancer, cardiovascular disease, blood pressure, memory loss, and neurological disorders (Jayawardena et al. 2020; Tripathy et al. 2021a; Stephen et al. 2023). Table 15.1 illustrates the role of nutrients in diets that strengthen the immune system. Strong immunomodulatory effects have been reported for a number of plant species and plant-based products, including several Tinospora and Phyllanthus species, *Camptotheca accuminata*, *Catharanthus roseus*, *Centella asiatica*, *Annona muricata*, *Zingiber zerumbet Viscum album*, *Panax ginseng*, *Asparagus racemosus*, and *Neolitsea* spp. (Jantan et al. 2019; Haque et al. 2020; Tripathy et al. 2022, 2023). Similarly, natural ingredients are increasingly being used in the food, nutraceutical, and pharmaceutical industries to make dietary supplements, herbal medications, spices, and other condiments. They are also being used to improve formulations that boost immunity.

15.3.1 Vitamin A

Vitamin A (VitA) is a group of unsaturated monohydric alcohols with an alicyclic ring. Vitamin A is fat soluble but insoluble in water. Natural killer cells with antiviral defenses are activated more when vitamin A is present. Immune responses that are T cell-mediated (Fig. 15.2) and antibody-dependent are both compromised by vitamin A deficiency (Huang et al. 2018). One way to include vitamin A in the diet is to eat carotene, which is its precursor. Provitamin A, or carotenoids, is a diverse group of compounds characterized by a polyisoprenoid carbon chain with several conjugated double bonds (Olatunde et al. 2020). The body metabolizes carotenoids into two molecules of pre-vitamin A, also known as retinol. The enzyme beta-carotene 15,15′-dioxygenase in the gut mucosal layer facilitates this process. There are two groups of carotenoids: beta-carotene and alpha-carotene. Beta-carotene is the primary carotenoid converted into vitamin A (Stephen et al. 2023). Beta-carotene was used as a marker to identify populations that consumed a lot of fruits and vegetables. Carrots and other orange veggies are abundant in it. These carotenoids promote macrophage production of cytokines and increase the population of lymphocytes (Mishra et al. 2024). The body produces cytokines, which are hormone regulators required for cell signaling. When there is a vitamin A shortage, the inflammatory immune system is assisted by the rise of cytokines (Ghahramanipour et al. 2023).

Table 15.1 Immunological activity of different class of bioactive compounds

Classes	Compounds	Mode of Action	References
Phenols	Caffeic acid, chlorogenic acid, cinnamic acid, ferulic acid, coumaric acid, benzoic acid, rosmarinic acid, catechin, gallic acid, dihydroxyl phenolic acid, and curcumin	Enhanced production of immunological proteins, including IgG, IgM, and IgA, occurred Suppresses the production of IL-6, TNF-α, and IL-1β from PBMCs	Barakat et al. (2023), Gairola et al. (2023), Hooda et al. (2023), Alkhalaf (2021)
Flavonoids	Quercetin, kaempferol, phyllantidine, isoquercetin, astragaline, guajaverine, gallocatechin, epigallocatechin gallate, gallic acid, and tannic acid	Activate Th1 cells Increase in B-cells and lymphocytic T-cell proliferation Release the cytokines such as IL-4, TNF-α, and IFN-γ Increase the release of NO from macrophages Improve phagocytic index and increased antibody response Increased expression of the IL-8 gene, indicating immune-boosting action, enhance the mitogenic activity of B-lymphocytes	Suja et al. (2024), Masad et al. (2021), Sutoyo et al. (2018), Muthulakshmi et al. (2016)
Essential oils	Eucalyptol, eugenol, β-caryophyllene, terpinen-4-ol, γ-terpinene, linalool, camphor, thymol, resveratrol, and carvacrol	In the spleen, the Augmentation of IFNγ-producing CD4+ and CD8+ T cells is accompanied by CD8+ central memory T lymphocytes Th1 cytokines IFNγ and IL-2 levels rose, but levels of Th2 cytokines IL-4 and IL-10 dropped in the serum	Grazul et al. (2023), Ghafarifarsani et al. (2022), de Lima Nunes et al. (2021), Firmino et al. (2021), Sandner et al. (2020), Parker (2018)
Vitamins	V-A, V-B, V-C, V-D, V-E	Decreases CPR, demonstrating an immunomodulatory action Protected against oxidative stress Enhanced immunological response in contagious illness Increased DTH reaction and expansion of T cells	Stephen et al. (2023), Maheshwari et al. (2022), Trasino (2020), Yamamoto and Jørgensen (2019), Yoshii et al. (2019)
Minerals	Iron, Selenium, Zinc	Enhance immunity against illnesses Increased T cell count, reduced infection rates, enhanced natural killer cell activity It enhances cell-mediated immunity and strengthens immunological response in individuals with deficits	Renu et al. (2023), Stephen et al. (2023), Phipps et al. (2021), Lange and Nakamura (2020), Read et al. (2019), Avery and Hoffmann (2018)

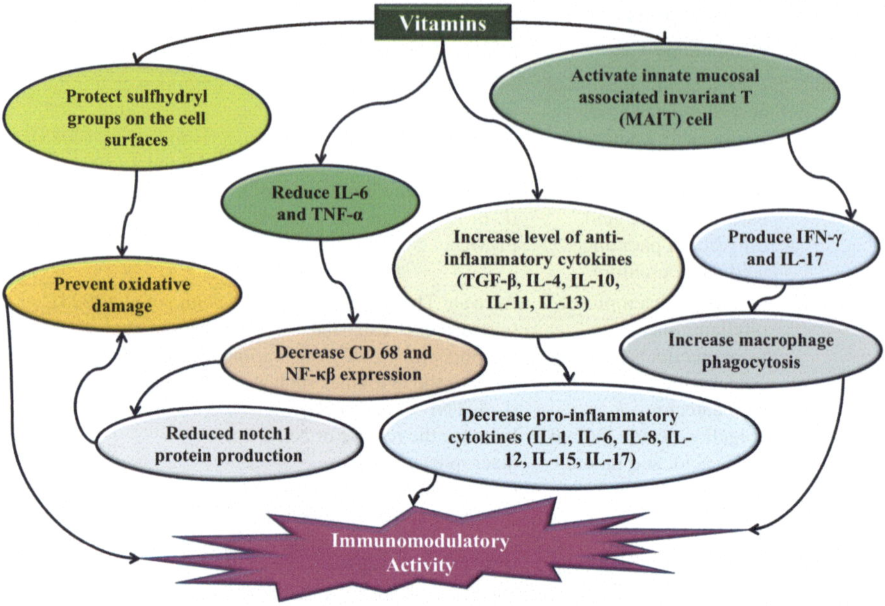

Fig. 15.2 Immunomodulatory mechanism of vitamins. (Mitra et al. 2022; Munteanu et al. 2023; Wimalawansa 2023)

15.3.2 Vitamin C

Vitamin C neutralizes reactive oxygen species and acts as a water-soluble electron donor, reducing agent, and important antioxidant. To act as an antioxidant and generate an intermediate free radical and semi-dehydroascorbic acid E, it transfers two electrons from the double-bond carbons at C-2 and C-3 (Arias et al. 2022). Vitamin C may have a part in boosting humoral and cellular immune responses as well as the efficiency of innate immunity. It is possible for vitamin C to reduce an overabundant immunological response. By blocking glyceraldehyde 3-phosphate dehydrogenase (GAPDH), vitamin C may stop immune cells from becoming too activated (Bae and Kim 2020). Research demonstrated that consuming insufficient amounts of micronutrients, such as vitamin C, reduces immunity against infection and heightens the consequences associated with illness (Calder 2020). The initial line of defense against external infections is the epithelial barrier, which is strengthened by vitamin C (Abobaker et al. 2020). Given that vitamin C appears to concentrate intracellularly in neutrophils, it is possible that vitamin C plays a part in preserving leukocytes' ability to function normally (Liugan and Carr 2019). Regular vitamin C supplementation may increase the innate immune response because it enhances natural killer cell activity, neutrophil chemotaxis, phagocytosis, and epithelial barrier integrity.

15.3.3 Vitamin D

Vitamins are the primary nutrients in food and play a crucial role in both the innate and acquired immune systems. Out of all the vitamins, vitamin D has a strong immune-boosting effect (Musavi et al. 2023). There are two types of Vitamin D: Vitamin D2 and Vitamin D3. Vitamin D2 is derived from ergosterol in yeast and mushrooms and is produced from vitamin D3 upon exposure to sunlight. Vitamin D3 is produced in the skin and is present in fish and fish oils. The macrophages at the infection site synthesize the active form 1, 25-dihydroxy vitamin D, which results in the production of antimicrobial peptides (Ismailova and White 2022; Ahsan et al. 2023). Immune system cells such as monocytes, neutrophils, and natural killer cells produce defensins and cathelicidins, which are antimicrobial peptides that protect the immune system. Accordingly, Vit-D has an effect on an increase in the antimicrobial peptide's expression (Ismailova and White 2024). Moreover, vitamin D regulates the synthesis of interferon (IFN), which boosts macrophages' ability to produce oxidative reactions against infections. Immune mediators between the innate and acquired immune systems are thought to be IFNs (Fig. 15.2). The strongest inducer of macrophage activation, IFN-c, also increases the formation of superoxide, boosts intracellular pathogen clearance, and increases the expression of genes that code for pro- and anti-inflammatory cytokines (Roy et al. 2023). Additionally, it converts Th1 to Th2 cells and suppresses IL-2 production, which prevents T cell growth. Th2 cells produce IL-4, IL-5, IL-10, and IL-13, whereas Th1 cells create pro-inflammatory cytokines like TNF-β and IFN-γ. According to Bae and Kim (2020), vitamin D enhances the body's immunity by inducing a shift from Th1 to Th2 phenotype, which results in a decrease in Th1 cytokines and an increase in Th2 cytokines.

15.3.4 Vitamin E

There are eight distinct forms of vitamin E, comprising α, β, γ, and δ for both tocopherol and tocotrienol isoforms. γ-tocopherol is the predominant form of vitamin E in foods, whereas α-tocopherol is the most prevalent form in serum and human tissues (Jiang 2022). Primary food sources of vitamin E include nuts, butter, olive oil, sunflower seeds, walnuts, spinach, and walnut oil. α-Tocopherol, a potent lipid-soluble antioxidant, can regulate host immunological responses by inhibiting free radicals. Vitamin E boosts helper T cell activity, which helps prevent oxidative reactions and is crucial in combating lipid peroxidation. It enhances immune function by protecting white blood cells and other components of the immune system, such as T cells and B cells. It enhances the body's immune system to combat diseases and illnesses (Stephen et al. 2023). Many proteins, enzymes, and immunological responses have been shown to benefit from vitamin E. This is attributed to its activation of protein phosphatase 2A (PP2A), which dephosphorylates and suppresses protein kinase C-α (PKC-α) (Shams et al. 2021). When PKC is inhibited, smooth muscle cells, neutrophils, monocytes, and macrophages proliferate less, and neutrophils and macrophages produce free radical species more quickly (Soares et al. 2023). Vitamin E also promotes the body's production of its antioxidant enzymes, such as catalase, glutathione peroxidase (GPX), and superoxide dismutase (SOD). Endogenous antioxidants in allergic disorders inhibit several mediators such as TNF-α, IL-1, IL-8, VCAM-1, NF-kβ, ROS, RNS, and TNF-α (Fig. 15.2) (Shams et al. 2021; Akram et al. 2023). It decreases the production of TNF-α and IL-6, two inflammatory cytokines. Vitamin E may be recommended as a pharmaceutical supplement in therapeutic settings to address and prevent many disorders, especially those associated with the immune system (Manosso et al. 2022).

15.3.5 Selenium

The immune system and selenium have a strong interaction. Selenium (Se) both increases and simultaneously alters the immunological response. The generation of ROS is essential for the innate immune system's ability to directly destroy microbes (Dalgaard et al. 2018). For phagocytes to lessen their oxidized products, selenium and selenioproteins are essential. Selenium also plays a role in adaptive immunity. It is known that this micronutrient improves IL2R signaling (Ala and Kheyri 2021). Selenium also stimulates human T-cell activation via T-cell receptors. This is due to increased expression of TXNRD1 and GPX1. Selenium, while it can boost T cell responsiveness when required, also keeps the immune system from activating uncontrollably in autoimmune illnesses. The latter may be accomplished through T reg cells, which proliferate following selenium administration (Barchielli et al. 2022). Selenium prevents autoimmunity and enhances the differentiation of mouse T cells. In mice with colitis brought on by dextran sodium sulfate (DSS), it boosted T reg cells and reduced the proportion of Th1 and Th17 cells (Ala and Kheyri 2021). Depending on the baseline selenium status, selenium supplementation is primarily thought to be immunostimulatory; when selenium levels raised from inadequate to adequate, immunostimulatory effects can be observed. However, the benefits of selenium supplementation to supra-nutritional levels on the immune system are less evident. In healthy mice, enough selenium (Se) levels are required to promote the production of selenoproteins, interferon (IFN)γ, and interleukin (IL)-6 (Fig. 15.3) (Razaghi et al. 2021). A human intervention trial showed that consuming selenium-fortified meals (200 μg per serving for 3 days) increases interleukin levels (IL-2, IL-4, IL-5, IL-13, and IL-22), indicating an activated Th2-type response (Bentley-Hewitt et al. 2014).

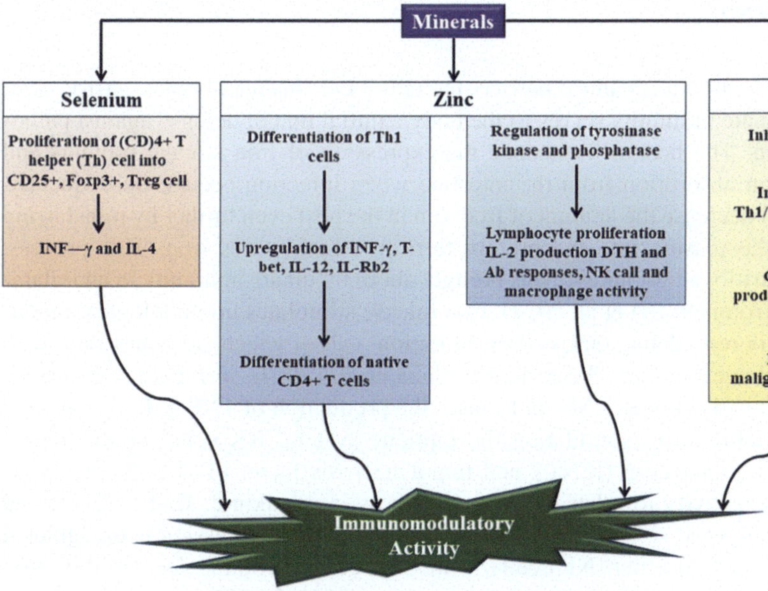

Fig. 15.3 Immunomodulatory mechanism of minerals. (Mitra et al. 2022; Toledano et al. 2023)

15.3.6 Zinc

Many people are only slightly low in zinc, a nutrient that is essential for the operation of about 300 proteins and enzymes and plays a key role in the control of the body's main genes. In the United States and Canada, adults over the age of 19 are advised to take 8 mg of zinc daily, while men should take 11 mg (McClung 2019; Godswill et al. 2020). Because zinc increases the immune system's antiviral activity by inducing leucocytes to produce interferon-α and interferon-γ, a zinc deficit may also compromise immunity. Blocking caspase 3, 6, and 9 causes cellular resistance to apoptosis and raises the Bcl-2/Bax ratio, which may help increase the number of T cells in the body (Marreiro et al. 2021). Since T cells are the primary source of most antigens, a zinc shortage impairs antigen production and prevents the body from producing antibodies in response to neoantigens (Fig. 15.3) (Maywald et al. 2017). Zinc alters the immune response by affecting inflammatory signaling for lymphocyte differentiation and interacting with serum proteins and signal transduction components that act as immuno-stimulants. Pathogen-sensing receptors, or Toll-like receptors, release pro-inflammatory cytokines when they detect pathogen-derived molecular structures and get activated. Zinc can either enhance or hinder these signaling pathways (Kumari et al. 2022).

15.3.7 Iron

By forming natural mechanical barriers including mucous membranes, gastric acid, and skin, innate immunity serves as the body's initial line of defense against pathogen invasion. The host will enhance the expression of iron storage proteins and decrease iron absorption from the intestine when infection occurs. Subsequently, the host will decrease the amount of free iron in the host even further by transferring iron from the plasma into the protein ferritin, which stores iron. This process, dubbed "nutritional immunity," was brought about by innate immunity in an inflammatory environment (Ni et al. 2022). Iron release stimulates immunological modulation, matrix remodeling, and cell proliferation—all of which are compatible with M2 macrophage function. According to Saha et al. (2017), iron excess in macrophages frequently activates M1 and causes the production of TNF α, IL-12p40, and CD163. In addition to stimulating the immune system, NK cells' production of cytokines like interferon (IFN)-γ and tumor necrosis factor (TNF) also increases NK cells' sensitivity to other cytokines, such as interleukin-2, IL-12, IL-15, and IL-18 (Samadi et al. 2023). Transferrin receptor (CD71) expression is upregulated by activated natural killer (NK) cells, while NK cell function is decreased by iron shortage (Loftus et al. 2018; Littwitz-Salomon et al. 2021). These all suggest that iron is essential for NK cell activation. Iron is a clonal growth promoter of lymphocyte subsets in the adaptive immune system (Ni et al. 2022). According to recent research, TfR1 (CD71) is necessary for serum iron for adaptive T-cell immunity (Frost et al. 2021). This emphasizes CD71's critical function in iron absorption. Without iron, T cell activation and proliferation during immune response processes (such as infection or tumor) will be impeded. Iron absorption through TfR1 (CD71) via the IL-2-dependent route triggers T cell activation (Liu et al. 2024)—tfR1 mutation results in decreased iron endocytosis and T cell dysfunction. Ultimately, this illness causes combined immunodeficiency sickness (Jabara et al. 2016). Th1 cell development and activity are inhibited by iron, an intrinsic component of adaptive immunity (Fig. 15.3).

15.3.8 Omega-3 Fatty Acid

Polyunsaturated fatty acids (PUFAs) are vital nutrients for the overall well-being of plants, animals, and humans. Essential polyunsaturated fatty acids (PUFAs) such as linolenic acid (LA), eicosapentaenoic acid (EPA), and docosahexaenoic acid (DHA) are commonly present in seafood, marine fish, and certain nuts and seeds (Kumar et al. 2022). Stearidonic acid (SDA) converts dietary linolenic acid (LA) into EPA and DHA. Omega-3 fatty acids refer to EPA and DHA acids. Cucchi et al. (2020) stated that fatty acids are crucial in critical biological processes for T cells, including their differentiation into diverse T cell subsets and the formation of immunological memory. Omega-3 polyunsaturated fatty acids can inhibit T-cell antigen presentation, activation, proliferation, and cytokine expression. Nevertheless, when consumed as dietary supplements, they might assist in managing systemic inflammation. N-3PUFA can reduce CD4+ T cell activation by affecting membrane raft-related signaling, enhance Treg differentiation by increasing peroxisome proliferator-activated receptor (PPAR) γ expression, and alter CD4+ T cell migration in a murine model by changing cytoskeletal dynamics and membrane microdomains (Camacho-Muñoz et al. 2022).

15.3.9 Bioactive Compounds

The immune response, a vital physiological mechanism in the human immune system, recognizes and eliminates alien, dangerous substances and organisms to preserve health. Natural products and their derivatives stimulate specific immune system components and have significant immunomodulatory effects; this indicates that they have tremendous promise as immunoregulatory agents (Wang et al. 2022). According to Stephen et al. (2023), phytochemicals are naturally occurring secondary metabolites that are not nutritional and may have defensive or disease-protective qualities. These phytochemicals served a variety of purposes in the reproductive and defensive processes of plants, such as hormone functions for growth and development, phytoalexins and allelochemicals for pathogen and herbivorous defense, and odor and color for pollination and protection (Banwo et al. 2021). Nuts, seeds, whole grains, fruits, and vegetables are rich sources of these. These compounds are commonly known as "secondary metabolites." Phytochemicals are glycosides, alkaloids, phenolics, saponins, and terpenoids. Secondary metabolites, also known as phytochemicals, are frequently generated in response to environmental stressors such as diet, infection, or climate change (Behl et al. 2021).

Glycosides are secondary plant metabolites with non-sugar components attached to a sugar moiety. These chemicals primarily activate the immune system (Khade et al. 2023). Zeng et al. (2020) studied the inhibitory impact of mangiferin on human epithelial ovarian cancer through experiments conducted in a controlled environment and within living organisms. MMP2 and MMP9 expressions were quantified via ELISA, western blotting, and immunohistochemistry. Zeng et al. (2020) demonstrated that mangiferin exhibited a dose-dependent impact on reducing the expression of MMP2 and MMP, thereby inhibiting the progression of human epithelial ovarian cancer.

Although in lower concentrations, secondary metabolites called alkaloids are present in many medicinal plants. Alkaloids are a primary source of improving immunological function and performance (Zandavar and Babazad 2023). Santos et al. have used the EAC model in vitro research to look into the immunomodulatory properties of the piperine analogue. Treatment with N-(p-nitrophenyl) acetamide piperine at doses of 6.25, 12.5, or 25 mg/kg reduced the expression of Th2-mediated IL-4 and IL-10 while increasing cell viability in Th1 cells that mediated TNF-α, IL-12, and IL-1β. According to data, the piperine analogue exhibits a notable capacity for immunomodulation (Santos et al. 2018)—Papaverine's powerful protection against brain ischemia-reperfusion injury through its modulation of many immunological pathways. Papaverine treatment decreased the expression of NF-κB, p38MAPK, RANKL, and IL-23, which in turn affected Th17 activity (Guan et al. 2020). A recent study utilized mice with colitis induced by dextran sodium sulfate to explore the immunomodulatory and anti-inflammatory effects of *Fumaria capreolata* L. (AFC) extract. The intestinal inflammation biomarkers, both macroscopic and microscopic, were considerably reduced upon treatment with AFC extract. Furthermore, AFC extract exhibited a delay in both pro- and anti-inflammatory colonic expressions, indicating its potential therapeutic benefit in the treatment of inflammatory bowel illness (Bribi et al. 2020).

Naturally occurring secondary metabolites, polyphenols, are present in various foods, including cereals, fruits, vegetables, and drinks. Due to their biological activity and use in the food sector, polyphenols have recently attracted a lot of attention (Iqbal et al. 2023). Several substances are now showing immunomodulatory effects via altering nitric oxide and eicosanoid synthesis and inhibiting the expression of proinflammatory genes and cytokines (Behl et al. 2021). Kalsum et al. studied the immunomodulatory effects of Propolis Trigona extract using a model of Prague Dawley rats. Propolis ethanolic extract treatment boosted nitric oxide generation, phagocytic index, and IgG antibody production, improving immunological responses (Kalsum et al. 2017). Park et al. studied the immunomodulatory impacts of water-soluble molokhia extract (WME) from *Corchorus olitorius* on a mouse model with immunosuppression induced by cyclophosphamide. Activity was evaluated by analyzing cytokine expression, NO production, and NK cell cytotoxicity. Research findings indicated that giving 50 and 100 mg/kg of WME notably boosted the number of white blood cells, thymus, and spleen sizes and acted as a potent immunostimulant (Park et al. 2018).

The naturally occurring glycoside class known as saponins is widely distributed throughout many plant parts, such as leaves, flowers, shoots, roots, tubers, and seeds (Oleszek and Oleszek 2020). Numerous studies, both *in vivo* and *in vitro*, have shown how plant-derived saponins can stimulate the immunogenicity of several vaccines. The immunomodulatory effects of saponins produced from *Astragalus brachycalyx* and *Astragalus karjaginii* on human whole blood were recently assessed by Aslanipour et al. The outcomes showed that the extracts from both plants enhance the induction of immunological responses and cytokine production (Aslanipour et al. 2017). Wu and associates assessed the protective effect of Ilexsaponin I, produced from Ilex pubescen, in LPS-induced RAW 264.7 cells in a different investigation. Data showed that treatment with 30 or 10 mM of ilexsaponin I inhibited the production of NO and PGE2 by downregulating the levels of iNOS and cyclooxygenase-2 [147]. By inhibiting IL-1β, Davidianoside F, saponin also demonstrates anti-inflammatory properties in human tumor cell lines (Zhou et al. 2017).

They include isoprene units, terpenoids, or secondary metabolites, which are naturally occurring chemicals sometimes referred to as isoprenoids. Higher plants frequently contain terpenoids, a broad class of phytochemicals (Tripathy et al.

2021a, b; Tripathy and Srivastav 2023a). According to Ludwiczuk et al. (2017), the main ways that terpenoids benefit the immune system are either enhancing T-cell response suppression or producing antibodies. Banerjee et al. studied the immunomodulatory effect of spergulin-A extracted from Glinus oppositifolius to manage visceral leishmaniasis (VL). At a 30 μg/mL concentration, spergulin-A exhibited significant activity against intracellular parasites. Spergulin-A usage in VL is supported by its increased pro- and anti-inflammatory effect (Banerjee et al. 2019). Mahmood et al. (2016) found that asiatic acid from Centella asiatica decreased the level of NO generated by LPS-stimulated RAW 264.7 macrophage cells in a dose-dependent manner at concentrations of 40 and 80 μmol/L. The Griess assay was utilized to examine this parameter. However, a contrasting effect was found when asiatic acid was administered at a higher quantity (160 μmol/L). As per another in vitro study, treatment of RAW 264.7 macrophage cells with asiatic acid at doses of 25, 50, and 100 μmol/L resulted in a dose-dependent reduction in LPS-induced prostaglandin E2 (PGE2), NO, interleukin-6 (IL-6), and interleukin-8 (IL-8). Furthermore, aseismic acid reduced p65 NF-κB phosphorylation in LPS-stimulated human gingival fibroblasts (HGFs) obtained from human normal gingival tissue explants. Asiatic acid efficiently inhibits the expression levels of IL-8 and IL-6 proteins generated by LPS in the gingival tissue of 7-week-old male Sprague-Dawley rats utilized as a model for periodontitis in the in vivo investigation. The levels of pro-inflammatory cytokines and protein expression were evaluated in this study using enzyme-linked immunosorbent assay (ELISA) and western blot, respectively (Hao et al. 2017). After exposing human corneal epithelial cells (HCECs) to lipopolysaccharide (LPS) for an in vitro study, Asiatic acid from the same plant at a concentration of 20 μmol/L also notably reduced the mRNA levels of IL-6, IL-8, IL-1β, and TNF-α (Chen et al. 2017).

15.4 Mechanism of Delivery into Body by Encapsulation

Customer demand for food products with health benefits has recently risen. This is due to an increasing awareness among individuals regarding the potential health impacts of consuming appropriate foods. Functional foods containing health-promoting bioactive compounds such as antioxidants, minerals, carotenoids, vitamins, and phytosterols are gaining popularity (Tripathy et al. 2023). Adding certain nutrients to food composition can be complicated due to the instability of certain nutrients during food processing and storage. Liu et al. (2022) stated that oxidation, triggered by external factors like light, temperature, oxygen, pH, etc., is responsible for the instability of these nutrients. These compounds, such as polyphenols, peptides, and phytochemicals, have an astringent taste and smell, restricting their application in food preparation (Mehta et al. 2022; Tripathy and Srivastav 2023b). The successful incorporation of health-promoting nutraceutical chemicals into functional meals is determined by their hydrophobicity and high sensitivity to light, heat, oxygen, and pH (Premathilaka et al. 2022).

The components of food consumption, particularly bioactive substances, have a positive impact on human metabolism and the function of the gastrointestinal tract (GIT); however, bioactive substances are known to be sensitive to a range of environmental factors, which leads to low bioavailability (Dahiya et al. 2023). Its high volatility, photodegradability, thermal instability, and limited bioavailability and bioaccessibility are further drawbacks. These restrict how many bioactives can be used (Hadidi et al. 2020). When food is processed, stored, and passes through the GIT, bioactives that are heat-and oxidative stress-sensitive—such as vitamin C or anthocyanins—degrade. Food products' quality and shelf life may suffer due to the oxidative destruction of phenolic substances, which can also produce free radicals (Shishir et al. 2018; Hadidi et al. 2023). Consequently, food storage and food passage through the GIT are made possible by encapsulation, which is suggested as a feasible method (Barajas-Álvarez et al. 2023; Rakshit et al. 2023). Furthermore, unwelcome scents resulting from bioactive components can also be concealed using bioactive encapsulation. For example, high doses of functional chemicals like polyphenols can impart a disagreeable bitter taste and astringency to the meal, while polyphenols that are encapsulated shield the customer from any unpleasant bitterness (Dahiya et al. 2023). Encapsulation in food science has concentrated chiefly on precise delivery to the human gastrointestinal tract. The primary objective has been facilitating the precise delivery of nutraceutical components to their destination. Coating materials improve the resilience of bioactive compounds to harsh conditions in the human gastrointestinal tract (Aguilar-Pérez et al. 2023). By including these healthy elements in different structures, they are protected from damage in challenging conditions such as food processing, exposure to stomach acid, or interaction with intestinal fluid. Moreover, this process improves the components' ability to dissolve and be absorbed by the body (Dima et al. 2020). Stomach acid or intestinal fluid can destroy bioactive molecules during meal digestion. Various kinds of bioactive compounds with distinct encapsulations are reported in Table 15.2. Each component can serve as a fundamental unit for constructing new configurations and enhancing features in food products. This could lead to extended shelf life and enhanced product quality (Tripathy et al. 2023).

15.5 Application of Encapsulated Bioactive Compounds in Food Fortification

Incorporating bioactive chemicals into food formulations is advantageous due to the global emphasis on developing health-promoting foods (Tripathy et al. 2023). As an illustration, several bioactive substances, such as vitamins, minerals, and phytochemicals, have been linked to immune-boosting qualities (Lee et al. 2024). Functional foods have attracted significant attention for incorporating bioactive compounds like vitamins, minerals, omega-3 fatty acids, carotenoids, antioxidants, phytosterols, and dietary fibers (Table 15.1). Due to their inherent instability during

Table 15.2 Various methods for encapsulating bioactive compounds with immunological properties

Encapsulation methods	Encapsulation matrix	Bioactive compound	Process parameter	Major findings	References
Electro-spraying/ Electrospinning	Zein/polyvinyl alcohol	Thymoquinone and resveratrol	Applied voltage: 25 kV Flow rate: 0.5 mL/h Distance between the syringe and the collector: 80 mm	Reduce the overall viable counts, peroxide values, TBARS values, and E. coli in fish fillet	Ahmadi et al. (2024)
	Zein	Vitamin B complex	Flow-rate: 0.15 mL/h Distance between the syringe and the collector: 3 cm	Preserve the vitamins by safeguarding them from external factors to keep their essential stability	Coelho et al. (2024)
	Milk protein isolates	Curcumin	Applied voltage: 30 kV Flow-rate: 0.01–10 mL/h	The curcumin nanoparticles exhibited over 70% heat resistance, enduring temperatures up to 120 °C for 10 min without degrading The antioxidant capacity in the aqueous solution was over 100% higher due to the suitable and undamaged coating	Rahbari et al. (2024)
	Poly (vinyl alcohol)/ Chitosan	Sage essential oil	Applied voltage: 20 kV Flow rate: 0.4 mL/h	The encapsulated nanosphere complex exhibited greater antifungal efficacy than the unbound essential oil	Erarslan et al. (2023)
	Chitosan	Oregano-olive oil	Applied voltage: 19–20 kV Flow rate: 0.3–0.4 mL/h Distance between the syringe and the collector: 9–13 cm	Antifungal and antimicrobial effects were increased with this technique	Vehapi et al. (2021)
	Polyvinyl alcohol	Phycocyanin	Applied voltage: 15–25 kV Flow rate: 50–200 μL/h Distance between the syringe and the collector: 12 cm	Ultrafine particles exhibited thermal stability against phycocyanin up to 216 °C while retaining their antioxidant properties	Schmatz et al. (2020)

(continued)

Table 15.2 (continued)

Encapsulation methods	Encapsulation matrix	Bioactive compound	Process parameter	Major findings	References
Nano-emulsion	Tween 80	Olive, flaxseed, and grapeseed oils	Timer: 5 min, amplitude: 60%, Pulser: 1 s ON/1 s OFF	Encapsulated group had significantly lower levels of malondialdehyde and nitric oxide compared to the other groups	Khalil et al. (2023)
	Lecithin, and grape seed oil	Thymol	Time: 15 min	Release experiments conducted in vitro demonstrated an immediate burst impact, followed by a continuous release	Saatkamp et al. (2023)
	Miglyol 812, soybean lecithin	Avocado peel extract	Instantaneous and spontaneous emulsion formation	The nano-formulation inhibited the movement of melanoma cells, indicating an antimetastatic impact	Cerda-Opazo et al. (2021)
	Glycerol, triton X-100/ Tween-80	Carotenoids	Vortexed and set in a water bath at 25 °C	The effectiveness of the delivery systems in encapsulating and stabilizing carotenoids for culinary applications	Chaari et al. (2018)
	Soy protein isolates	Anthocyanins	Sonicated at 50 °C for 1 h in ultrasound	Anthocyanins were efficiently shielded by W/O/W double emulsions from in vitro oral and gastric digestion and were released during the simulated small intestine phase	Xu et al. (2018)

Liposome	Phospholipid and cholesterol	Olive leaves and orange peels	Freeze-thawed: 5 times time: 15 min Pulsed mode: 3 min ON and 3 min OFF at an amplitude of 20% of full power	Encapsulating olive leaf extract in liposomes significantly increased its antibacterial effectiveness against S. aureus by tenfold	Prevete et al. (2024)
	Soy lecithin and stigmasterol	*Centella asiatica* leaf extract	A eliminates the organic solvent: 45 °C A homogeniser performed the dispersion of suspension.	The study found that liposomes effectively preserved the bioactivity of the leaf extract, making them suitable for creating enhanced food products with health advantages	Tripathy and Srivastav (2023b)
	Deoiled lecithin	Citrus essential oils	Sonication time: 5, 10, 20 and 30 min amplitudes: 25, 50 and 70% of full power Pulsed mode: on-time 5 s, off-time 2 s	The current approach is eco-friendly regarding a prospective food safety evaluation and could be appropriate for food-related uses	Ben-Fadhel et al. (2022)
	Soy lecithin	Naringenin and Naringin	Stirred: 4 h Passed through the microfluidizer: 3 times Pressure of 121.0 MPa to fabricate nanoliposomes	This will aid in creating innovative food-grade colloidal delivery systems to include naringenin or other lipophilic polyphenols into foods, supplements, or medications	Chen et al. (2021)
	Yolk lecithin, cholesterol, and Tween 80	Procyandins	Pressure of homogenizer: 30 MPa for 3 times	Liposomes provide a practical solution for preserving lychee pericarp procyanidins	Luo et al. (2020)
	Soy lecithin	Oleic palm oil	Mixing: 1000 rpm for 10 min at 35 °C Homogenization: 20,000 rpm for 2 min	Liposomes were destabilized in the gastrointestinal phase due to environmental factors such acidic pH and elevated ion concentration	Beltrán et al. (2019)

(continued)

Table 15.2 (continued)

Encapsulation methods	Encapsulation matrix	Bioactive compound	Process parameter	Major findings	References
Freeze drying	Whey protein isolates and modified tapioca starch	Chia seed oil	Freeze drying: −50 °C and 0.04 mbar for 24 h	The encapsulated chia seed oil with a balanced omega-3 to omega-6 ratio can be utilized as a functional component in foods to provide health advantages	Anand et al. (2024)
	Tween 80 and gum arabic	V D3	Freeze drying: −56 °C and 0.074 mbar for 48 h	Preserve the V-D3.	Bashir et al. (2024)
	Maltodextrin and gum acacia	*Pistacia terebinthus* L. fruit oil	Freeze drying: −45 °C for 2 days	α-pinene and linalool had exceptional retention rates, nearly reaching 100%	Yaman et al. (2023)
	Alginate and Maltodextrin	Canola oil	Freeze drying: −50 °C for 48 h	The oil's oxidation stability increased	Guerrero et al. (2022)
		Lycopene	Freeze drying: −42 °C for 48 h	The powder was utilized to functionalize dressing samples, enhancing antioxidant activity	Gheonea et al. (2021)
Spray drying	Tween 80 and gum arabic	V D3	Sample flow rate: 5 cm^3/min Drying air flow rate: 2.5 m^3/min Inlet air temperature: 120 °C Outlet temperature: 68 °C	Spray drying is more effective than freeze drying for encapsulating vitamin D3 because it has higher encapsulation efficiency and better release properties	Bashir et al. (2024)
	Maltodextrin and gum acacia	*Pistacia terebinthus* L. fruit oil	Sample flow rate: 4 mL/min Drying air flow rate: 30 m^3/min Inlet air temperature: 135 °C	Sublimation drying using an ultrasonic nozzle showed better retention of volatiles (73.19%) than freeze-drying	Yaman et al. (2023)
	Maltodextrin	Oregano polyphenols	Inlet air temperature: 145 °C	Spray drying is an excellent method for preserving and safeguarding the bioactive elements of oregano, making it suitable for use as a functional ingredient	Bernal-Millán et al. (2022)
	Gum acacia, and maltodextrin	V B12 and D3	Sample flow rate: 20 rpm Drying air flow rate: 1400 rpm Inlet air temperature: 140 °C Outlet temperature: 68 °C	SD provided an effective method to administer water-soluble vitamin B12 and fat-soluble vitamin D3 in a single product	Bajaj et al. (2021)

food processing and storage, including these components in food compositions is challenging (Hong et al. 2023). The fundamental reason for their instability is rapid deterioration due to oxidation or hydrolysis caused by environmental factors such as temperature, pH, light, oxygen, and moisture. These bioactive food compounds may interact adversely with other bioactive compounds in the food matrix (Banwo et al. 2021). Encapsulation technology is a practical method to safeguard bioactive substances in various industries such as food, pharmaceuticals, and cosmetics.

15.5.1 Bakery Products

These days, health-conscious consumers need goods that do more than just meet their nutritional needs; they also need to have qualities that promote health and general wellbeing. They anticipate a rise in the nutritional content of the common meals they eat (Bińkowska et al. 2024). Cereals and bakery products, such as bread and wheat, are common staple foods that can be enriched with various bioactive compounds (BACs) like probiotic bacteria, minerals, vitamins, ω-3 fatty acids, PUFAs, and curcumin. BACs are frequently utilized in the food industry to enhance the flavor, color, texture, shelf life, and nutritional composition of food, among other attributes. Nanoencapsulation technology is crucial for encapsulating bioactive food compounds in the food processing sector, enhancing their value (Tripathy et al. 2023). After conducting a thorough investigation, it has been ascertained that the functional food industry's overall market share is accounted for by around 20% by cereal and bakery items. Due to its large market presence, easy-to-use design, and smooth incorporation into food products, this specific classification is commonly acknowledged as the principal delivery method for BFCs (Tripathy et al. 2021a, b). The bakery products developed by incorporating encapsulated bioactive compounds are reported in Table 15.3.

In their, Pamunuwa et al. (2023) explored the use of liposomes to encapsulate α-tocopherol and curcumin, two substances with strong immune-boosting properties that can be used to fortify food. This study set out to examine the sensory and physicochemical characteristics, as well as the antioxidant capacity retention, of cookies after α-tocopherol or curcumin-encapsulated liposomes were added. Anticipatedly, liposomal encapsulation of α-tocopherol or curcumin reduced the reinforced cookies' decreased antioxidant capacity while they baked. In summary, liposomal encapsulation is an effective way to preserve the medicinal properties of α-tocopherol or curcumin in cookies while they bake (Pamunuwa et al. 2023). Additionally, da Cruz et al. (2023) examined the process of encapsulating essential lemongrass (*C. citratus*) oil via electrospinning into cassava starch fibers. The dough mixture was enriched with cassava starch containing C. citratus essential oil and utilized as Bread Fortification Compounds (BFCs) in the bread recipe. Although baked at a high temperature, the fibers effectively inhibited fungal spoilage by effectively releasing bioactive compounds derived from *Cymbopogon citratus* essential oil. Liposome technology has been used to encapsulate omega-3 fatty acids in soybean phospholipids for making bread. This technology helps mask the fish flavor while retaining the health advantages of omega-3 fatty acids. The temperature varied between 200 °C and 260 °C, while the baking duration ranged from 13 to 26 min. The enriched bread was found to have a higher concentration of nano-liposomal ω-3 fatty acids compared to the other samples through sensory evaluation. The inclusion of nano-liposomal ω-3 fatty acids did not have a negative effect on the texture or overall acceptance of the bread (Rasti et al. 2017; Ojagh and Hasani 2018). Increasing the amount of ω-3 fatty acids consumed through diet was the goal of this fortification. Bread and cakes, for example, are baked goods that are very susceptible to the growth of germs, yeasts, and mold. It follows that adding preservatives is essential to extending their shelf life. One practical method for prolonging the shelf life of baked goods is to incorporate fish oil or EO into different matrices.

Indian foods like chapatti and parantha are traditionally made on flatbreads. By including fruits, vegetables, and other cereal grains, parantha can be made into a functional food that is more bioactive and nutritionally balanced (Reshmi et al. 2017). As a result, paranthas were fortified with iron and zinc. It was discovered that these nutrients, regardless of the material used to package them, remained stable throughout time, improving the paranthas' nutritional profile (Kaul et al. 2022). Therefore, enhanced ready-to-bake frozen potato paranthas will be a useful technology to combat malnutrition while providing industrial applicability and convenience. Bryszewska et al. (2019) discovered that sourdough fermentation reduces iron bioavailability in bread samples fortified with encapsulated iron to 46%, compared to 85% in conventionally fermented bread. Furthermore, the iron transport efficiency is higher in conventionally fermented bread (6.1%) than in sourdough fermented bread (1.8%). Zhu et al. (2021) investigated the use of egg white and vitamin D microencapsulation for bread fortification. The authors used a new technique called ultrasonication, which employs soundwaves louder than 20 kHz. To be more precise, the authors sonicated an aqueous egg solution for 1 min at 160 W while layering vitamins on the solution's surface. Constantino and Garcia-Rojas (2022) found that after baking, vitamin D encapsulated in complexes formed by amaranth protein isolates and lactoferrin had a recovery rate of 88.65%, while the unencapsulated version had a recovery rate of 64.63%. Both studies by Zhu et al. (2021) and Constantino and Garcia-Rojas (2022) conducted in vitro gastrointestinal digestion of bread. They observed that the highest release of vitamin D from the microcapsules occurred during the intestinal phase. Costantino et al. (2022) reported that 22.98% of the vitamin was released from the encapsulated sample during the first 5 min, increasing to 82.96% at the 180-min point. Similarly, 67% of vitamin D encapsulated in an egg white microcapsule successfully reached the colon in its active form. Due to stomach breakdown, only 32% of the ingested vitamin was absorbed in the intestine from the free sample (Zhu et al. 2021). Bioactive food compounds (BFCs) can enhance the functionality of food products, especially bakery items, by incorporating anti-inflammatory, antioxidant, antifungal, antimicrobial, and immune-boosting properties. Encapsulation is essential for improving the durability and heat resistance of heat-sensitive BFCs in baked foods with BACs to enhance human body functionality.

15.5.2 Fruit and Vegetable-Based Products

Food fortification is the process of creating, producing, or developing specific food products with the intention of adding additional nutritional value or developing dietary supplements beyond the essential functions of the food (Dehnad et al. 2023). The growing acknowledgement and advocacy of functional foods and beverages by the World Health Organization is responsible for their growing global appeal. Additionally, customers are now more conscious of the significance of their nutritional intake and dietary composition. Fruit juices are a significant source of essential nutrients for the human diet due to their diverse bioactive compounds that promote human health and reduce the likelihood of diseases (Tripathy et al. 2021b). Natural fruit juices are inherently unstable and degrade rapidly, posing challenges in creating stable fresh beverages with a long shelf life. Fortifying fruit juices with nutrients is commonly practiced to enhance their shelf life and stability (Table 15.3) (Ephrem et al. 2018).

Ascorbic acid, another name for vitamin C, is a water-soluble vitamin that the human body is unable to produce. It is commonly recognized that vitamin C can prevent viral illnesses. An experiment conducted by Tao et al. (2017) found that fresh apple juice treated with an oil-in-water emulsion comprising steppogenin, butylated hydroxytoluene (BHT), ascorbic acid (vitamin C), and vitamin E, along with ethyl butyrate, tween 80, and PEG 400, showed significant improvement. The results of this investigation show that fresh apple juice's enzymatic browning has significantly decreased. This decrease was noted after 2 days at room temperature and 7 days at 4 °C (Tao et al. 2017). Furthermore, the anti-browning properties were enhanced by including 0.05% ascorbic acid. Co-encapsulating steppogenin in juice with BHT or vitamin E did not significantly enhance its anti-browning properties. Carvalho et al. (2020) utilized spray drying to produce pineapple chips containing iron microparticles and examined the impact of ethanol and ethanol combined with ultrasonic pretreatment on the drying process of pineapples. The combination of ethanol and ultrasound reduced drying time by 52%. A 7.5-min pretreatment with ethanol and ultrasonic considerably reduced the drying time and resulted in minimal residual ethanol in the sample. Genevois et al. (2017) delayed the textural change of pumpkin by fortifying it with both iron (Fe) and probiotics (*L. casei*). The final product contained 0.35 mg of iron per gram and had an average bioaccessibility of 54% following in vitro digestion. For 14 days, the probiotic level remained above 107 CFU g^{-1}, and the mineral insertion influenced the viability. Further scientific research is essential to investigate the process's bioavailability, consumer safety, and sustainability criteria. It is crucial to prioritize the creation of the product, ensure its stability during long-term storage, and conduct thorough evaluations of its sensory characteristics to ensure consumer approval.

Table 15.3 Food application of bioactive compounds containing immunological property

Class of food	Bioactive compounds	Encapsulation method	Coating material	Food product	Major findings	References
Baked food	Lemongrass essential oil	Electrospinning	Cassava starch fibers	Bread	The cassava starch-LEO fibers show potential as an antifungal agent to substitute synthetic additions in food and as active packaging for bread	da Cruz et al. (2023)
	Curcumin and α-tocopherol	Liposome	Soy lecithin	Cookies	Liposomal encapsulation is an effective technique for preserving the antioxidant properties of curcumin or α-tocopherol in cookies while baking	Pamunuwa et al. (2023)
	Clove essential oil	Emulsion-ionic gelation	Chitosan and Tween-80	Dry cakes	When chitosan was employed with CEO in a ratio exceeding 1:1, it effectively inhibited the growth of *E. coli* and *S. aureus*	Sharma et al. (2023)
	V D3	Emulsion	Amaranth protein isolate (API) and lactoferrin (LF)	Bread	Microcapsules of vitamin D3 produced as complexes with lactoferrin can be effectively used for fortifying food and baking	Constantino and Garcia-Rojas (2022)
	V A, D and E	Emulsion	Egg white solution	Bread	Vitamins enclosed in egg white shells might enhance the nutritional content of essential food items to address malnutrition in mothers and children	Zhu et al. (2021)
	Oregano and thyme essential oils	Nanoprecipitation	Zein	Bread	The nanocapsules exhibited strong temperature resistance during baking, safeguarding the bread from mold and yeast growth	da Rosa et al. (2020)

Fruit and vegetable based product	Palmyrah (*Borassus flabellifer* L) bioactive compounds	Spray drying	Maltodextrin and Gum Arabic	Functional beverage drink	FTIR analysis verified successful encapsulation and interactions within the final coating layer.	Sangeetha et al. (2023)
	Iron	Spray drying	Maltodextrin	Pineapple chips	The described technologies can be combined to produce iron-enriched pineapple chips with faster drying time and minimal ethanol residue	Carvalho et al. (2020)
	Hexanal and *trans*-2-hexenal of green pea	Nanoemulsion	Soy lecithin	Apple juice	Nano-emulsions have shown significant utility in the fruit juice industry, enhancing the quality of apple juice due to their organoleptic compatibility	Patrignani et al. (2020)
	Pomegranate flavedo phenolics	Freeze drying	Maltodextrin	Functional beverage	The enriched drink demonstrates a notable enhancement in phenolics, flavonoids, and antioxidant activity compared to the control sample	Thakur and Thakur (2020)

(continued)

Table 15.3 (continued)

Class of food	Bioactive compounds	Encapsulation method	Coating material	Food product	Major findings	References
Dairy products	Angelica essential oil	Emulsion	Tween 80	Dairy dessert	Samples enhanced with nanoemulsions, especially those with a higher concentration of AEO, had higher sensory ratings	Nanakali (2023)
	Zinc	Spray drying	Gum arabica and soy protein isolate	Milk	Zinc-fortified milk products can replace unfortified milk products to help treat zinc insufficiency in individuals of all ages	Mounika and Hemalatha (2023)
	Heracleum persicum essential oil	Emulsion followed by freeze drying	Chitosan and Tween 80	Yogurt	Utilizing HEO nanoparticles in yogurt shows potential for creating functional food products, including antioxidant-rich yogurt	Yousefi et al. (2023)
	Iron	Emulsion	Sodium chloride and Tween 80	Cream	Cream samples indicated that including encapsulated iron had less adverse effects compared to free iron. The sample prepared with iron-fortified emulsion exhibited acceptable PV and TBA values	Koohenjani and Lashkari (2022)
	Curcumin	Spray drying	Maltodextrin and whey protein isolate	Milk	Adding encapsulated curcumin to milk at a specific dose did not negatively impact its sensory characteristics compared to regular milk	Patel et al. (2022)
	V E	Emulsion	Goat milk powder (skimmed)	Yogurt-based beverage	Emulsification is a valuable technique for creating yogurt products enriched with fat-soluble vitamins	Raikos et al. (2021)
	Shrimp oil	Liposome	Lecithin	Skim milk	No significant changes in quality occurred during the preservation of skim milk enriched with shrimp oil	Gulzar et al. (2020)
	V B12	Emulsion	Polyglycerol polyricinoleate	Skim milk	No phase separation was detected in the emulsions after being optimized and stored at 37 °C for 7 days	Zaghian and Goli (2020)

Meat products	Clove EO	Metal organic frameworks	Beta cyclodextrin	Chinese bacon	The preservation benefits of microencapsulated clove essential oil on the production of Chinese bacon have improved	Wang et al. (2023)
	Oregano and thyme EO	Nanoemulsion	Chitosan, xanthan gum, and tween 80	Beef meat	Chitosan-based edible films with essential oils enhanced the quality and safety of chilled or fresh meat	Gaba et al. (2022)
	Fish oil	Nanoemulsion	Carrageenan and gum tragacanth	Chicken nuggets	Encapsulation of fish oil with tragacanth proved more efficient in maintaining the sensory properties	Pourashouri et al. (2020)
	Fish oil	Freeze drying	Soy protein isolate and inulin	Beef and Pork burger	Encapsulation of the fish oil preserved EPA and DHA during the cooking process	Rios-Mera et al. (2021)
	Tocopherol (Vitamin E)	Nanoemulsion	Canola oil, Tween 80	Fish sausage	Encapsulated tocopherol effectively delays lipid oxidation and enhances the quality of fish sausages without affecting their textural features during refrigerated storage	Feng et al. (2020)
	Cumin seed EOs	Nanoemulsion	Whey protein isolate, tween 80 and inulin	Beef hamburger	Using microencapsulated CSEO-loaded cellulose papers can prolong the shelf-life of beef hamburgers	Hemmatkhah et al. (2020)
	Pomegranate peel extract	Electrospinning	Chitosan and poly ethylene oxide	Beef meat	Nanofibers have shown outstanding physical, mechanical, and thermal capabilities, making them highly promising for use in food packaging	Surendhiran et al. (2020)

15.5.3 Dairy-Based Products

The food industry needs to work on upholding stringent quality standards while meeting consumer expectations for products considered pure and safe. This is particularly challenging due to increasing customer concerns and the economic sanctions imposed on synthetic and hazardous chemicals (Tripathy et al. 2023). These BFCs, according to Rao and Poonia (2023), have anti-inflammatory, antibacterial, and antioxidant properties that may improve the nutritional value of dairy products. Around the world, milk and dairy products are integral parts of many different dietary patterns (Verma et al. 2019a, b). The most common way to get a wide range of beneficial and health-promoting ingredients is through dairy products. Dairy products classified as functional have been fortified with extra nutrients or BFCs, which are known to have positive health effects (Table 15.3). Using nanoparticles has become a cutting-edge method of achieving these advantages. Immunologically active ingredients such as iron, probiotics, vitamin D, and omega-3 fatty acids have been enclosed in capsules to enhance the nutritional value of dairy products. Skim milk has been fortified with omega-3 fatty acids encapsulated in soy lecithin shrimp oil (Gulzar et al. 2020). There was significant bitterness when 100 ml of skim milk was enriched with 10 ml of shrimp oil nanoliposomes (SONL).

Siddique and Park (2019) found that cheddar cheese samples containing fetuses had increased gumminess and hardness ratings compared to regular cheese. Cheeses containing free and encapsulated iron have 72% and 74% iron recovery rates, respectively. Cheddar cheese containing both types of iron has sensory attributes that are considerably worse than control cheeses, as noted by Siddique and Park in 2019. The size of the microcapsules significantly affects the sensory characteristics and iron retention of Cheddar cheese. Smaller encapsulated FeSO4 resulted in improved sensory qualities and increased Fe retention in Cheddar cheese compared to bigger encapsulated FeSO4 (Arce and Ustunol 2018). In Santillán-Urquiza et al. 2017, Santillán-Urquiza, Méndez-Rojas, and Vélez-Ruiz studied how the size of iron particles affects the physical, chemical, and qualitative properties of yogurt when added as a supplement. Yogurts with nanosized Fe, zinc, and calcium phosphate showed reduced syneresis and improved digestion compared to yogurts with micro-sized minerals. Yogurts containing nano-sized minerals exhibited greater consistency and stiffness than those with micro-sized minerals. This enriched yogurt can help increase the body's iron levels and enhance our immune systems. Turkmen et al. (2019) found that dairy waste such as buttermilk whey, cheese whey, and milk or cheese whey permeate are promising for producing functional beverages.

15.5.4 Meat-Based Products

Animal-derived muscle tissues, such as meat and fish, are considered crucial sources of high-quality proteins containing essential amino acids necessary for biological functions. These muscle-building meals are rich in minerals, vitamin B complex, and several micronutrients (Tripathy et al. 2023). Protein-rich foods tend to spoil rapidly, reducing their quality and sustainability. Consequently, several researchers have developed innovative techniques to enhance the nutritional content of meat products or prolong their shelf life using natural preservatives (Banerjee et al. 2020; Tripathy et al. 2021a). These solutions involve using naturally derived preservatives or fortifying the food with additional nutrients (Table 15.3). Current study indicates that edible nanoparticles could extend the shelf life of meat products. Shavisi and Shahbazi (2022) demonstrated that applying chitosan/gum Arabic mats with anthocyanins on chicken fillets was a protective barrier, preserving the meat's quality. Furthermore, the nanofibers exhibited a color change from neutral white to pale khaki in the presence of ammonia vapor. These composite mats could serve as a reliable indicator of the quality of these products. This study emphasizes the importance of a detailed analysis of the color reaction to NH_3 vapor in volatile nitrogen compounds (Shavisi and Shahbazi 2022). Duan et al. (2023) have provided a theoretical explanation for the phenomenon of color modulation. When exposed to water, the gaseous NH3 molecule dissolves to form ammonium (NH_4^+) and hydroxide (OH^-) ions. The hydroxyl ion altered the electrospun nanofibers, creating an alkaline environment that transformed the phenolic hydroxyl group inside the curcumin structure into a phenolic oxygen anion (Duan et al. 2023).

Banerjee et al. (2020) demonstrated the frequent use of the encapsulation approach to enhance the fortification process. Consumer demand for incorporating essential oils (EOs) into meat-based food products to enhance safety has risen in recent years. Hasani et al. (2020) conducted a study investigating the impact of nonencapsulated lemon essential oils on the quality of fish burgers in the refrigerator. The scientists discovered that the quality characteristics of the fish burger were notably enhanced ($p < 0.05$) by using nanocapsules made from a mixture of chitosan and modified starch (1.5:8.5% w/V) with lemon essential oils. The burgers treated with lemon EO nanocapsules showed decreased total volatile base nitrogen (TVB-N) levels, thiobarbituric acid (TBA), and peroxide relative to the other samples after storage. People should include ω-3 polyunsaturated fatty acids (PUFAs) in their diet, particularly α-linolenic acid or its longer-chain forms like DHA (22:6 ω-3) and EPA (20:5 ω-3). (Abbasi et al. 2019). Pourashouri et al. (2020) increased the levels of ω-3 fatty acids in fish sausage by including fish oil. The previous paragraphs offered a comprehensive summary of the significant discoveries made by the researchers. Using nonencapsulated nutrients can be a strategic way to enhance a meat product's longevity and immune-boosting properties. Including encapsulated nutrients in the meat product has resulted in a notable reduction in microbe count, enhancing the safety of consuming meat. *Nanoencapsulation* is a crucial method that allows the incorporation of BACs into food products while minimizing oxidation.

15.6 Conclusions and Future Perspective

Extensive research has been conducted on phytochemicals due to their efficacy as alternative treatments for several chronic diseases. Food is the sole medication required for good health. Consuming the necessary amounts of immune-boosting foods helps maintain the body's immune system and metabolism. These immune-boosting foods are rich in vitamins, minerals, and antioxidants, which help fortify the body against sickness. This chapter elucidated the essential roles of innate and adaptive immunity in the human immune system for combating infectious cells. Research on clinical studies is lacking despite evidence from *in vitro* and *in vivo* studies demonstrating the efficacy of phytochemicals as medications in animal models and cell cultures while elucidating their mechanism of action. Shifting the research focus from laboratory testing to clinical trials is crucial to examining phytochemicals' impact on vital tissues such as the liver and immune system. The discussion also included the mechanisms of action of immune-enhancing nutrients such as dietary fiber, vitamin D, zinc, iron, and omega-3 fatty acids, as well as antioxidants including carotene, vitamin C, vitamin E, folate, selenium, and phytochemicals. In the end, these kinds of studies will help to understand the possible health benefits for customers. We should prioritize upholding a healthy lifestyle that includes regular exercise, enough sleep, stress management, and integrating immune-boosting foods into our meals. Additionally, it is imperative that you seek individualized counsel from medical specialists or registered dietitians, particularly if you have any underlying medical concerns or special dietary needs. We may actively boost our immune systems and advance our general well-being by realizing the importance of nutrition and choosing the foods we eat with awareness. Adopting a diet high in immune-boosting foods is a delightful and pleasurable approach to assisting our body's defense mechanisms and living a healthier and more resilient existence.

References

Abbasi F, Samadi F, Jafari SM, Ramezanpour S, Shams-Shargh M (2019) Production of omega-3 fatty acid-enriched broiler chicken meat by the application of nanoencapsultsed flaxseed oil prepared via ultrasonication. J Funct Foods 57:373–381

Abobaker A, Alzwi A, Alraied AHA (2020) Overview of the possible role of vitamin C in management of COVID-19. Pharmacol Rep 72(6):1517–1528

Aguilar-Pérez KM, Ruiz-Pulido G, Medina DI, Parra-Saldivar R, Iqbal HM (2023) Insight of nanotechnological processing for nano-fortified functional foods and nutraceutical—opportunities, challenges, and future scope in food for better health. Crit Rev Food Sci Nutr 63(20):4618–4635

Ahsan N, Imran M, Mohammed Y, Al Anouti F, Khan MI, Banerjee T, Haq A (2023) Mechanistic insight into the role of vitamin D and zinc in modulating immunity against COVID-19: a view from an immunological standpoint. Biol Trace Elem Res 201(12):5546–5560

Ahmadi SS, Aminzare M, Azar HH, Nikfarjam N, Roohinejad S, Greiner R, Tahergorabi R (2024) Application of electrospun zein/polyvinyl alcohol nanofibers incorporating thymoquinone and

electrosprayed resveratrol nanoparticles on the shelf life of fresh rainbow trout fillet and inoculated Escherichia coli O157: H7. Food Control 155:110089
Akram R, Muzaffar H, Anwar H, Malik SA, Sajid F, Iman T et al (2023) Role of endogenous and dietary antioxidants in brain disorders. In: The role of natural antioxidants in brain disorders. Springer International Publishing, Cham, pp 171–214
Ala M, Kheyri Z (2021) The rationale for selenium supplementation in inflammatory bowel disease: A mechanism-based point of view. Nutrition 85:111153
Alkhalaf MI (2021) Chemical composition, antioxidant, anti-inflammatory and cytotoxic effects of Chondrus crispus species of red algae collected from the Red Sea along the shores of Jeddah city. J King Saud University-Science 33(1):101210
Anand V, Ksh V, Kar A, Varghese E, Vasudev S, Kaur C (2024) Encapsulation efficiency and fatty acid analysis of chia seed oil microencapsulated by freeze-drying using combinations of wall material. Food Chem 430:136960
Arce A, Ustunol Z (2018) Effect of microencapsulated ferrous sulfate particle size on Cheddar cheese composition and quality. J Dairy Sci 101(8):6814–6822
Arias A, Feijoo G, Moreira MT (2022) Exploring the potential of antioxidants from fruits and vegetables and strategies for their recovery. Innovative Food Sci Emerg Technol 77:102974
Aslanipour B, Gülcemal D, Nalbantsoy A, Yusufoglu H, Bedir E (2017) Cycloartane-type glycosides from Astragalus brachycalyx FISCHER and their effects on cytokine release and hemolysis. Phytochem Lett 21:66–73
Avery JC, Hoffmann PR (2018) Selenium, selenoproteins, and immunity. Nutrients 10(9):1203
Bae M, Kim H (2020) The role of vitamin C, vitamin D, and selenium in immune system against COVID-19. Molecules 25(22):5346
Bajaj SR, Marathe SJ, Singhal RS (2021) Co-encapsulation of vitamins B12 and D3 using spray drying: wall material optimization, product characterization, and release kinetics. Food Chem 335:127642
Banerjee S, Mukherjee N, Gajbhiye RL, Mishra S, Jaisankar P, Datta S, Das Saha K (2019) Intracellular anti-leishmanial effect of Spergulin-A, a triterpenoid saponin of Glinus oppositifolius. Infect Drug Resist 12:2933–2942
Banerjee DK, Das AK, Banerjee R, Pateiro M, Nanda PK, Gadekar YP et al (2020) Application of enoki mushroom (Flammulina Velutipes) stem wastes as functional ingredients in goat meat nuggets. Food Secur 9(4):432
Banwo K, Olojede AO, Adesulu-Dahunsi AT, Verma DK, Thakur M, Tripathy S et al (2021) Functional importance of bioactive compounds of foods with potential health benefits: A review on recent trends. Food Biosci 43:101320
Barajas-Álvarez P, González-Ávila M, Espinosa-Andrews H (2023) Recent advances in probiotic encapsulation to improve viability under storage and gastrointestinal conditions and their impact on functional food formulation. Food Rev Intl 39(2):992–1013
Barakat H, Alkhurayji RI, Aljutaily T (2023) Immune-boosting potentiating properties of Brassica nigra hydroalcoholic extract in cyclophosphamide-induced immunosuppression in rats. Foods 12(19):3652
Barchielli G, Capperucci A, Tanini D (2022) The role of selenium in pathologies: an updated review. Antioxidants 11(2):251
Bashir I, Wani SM, Bhat AA, Khan AA, Hussain SZ, Ganai SA, Anjum N (2024) Effect of freeze drying and spray drying on physical properties, morphology and in vitro release kinetics of vitamin D3 nanoparticles. Powder Technol 432:119164
Behl T, Kumar K, Brisc C, Rus M, Nistor-Cseppento DC, Bustea C et al (2021) Exploring the multifocal role of phytochemicals as immunomodulators. Biomed Pharmacother 133:110959
Beltrán JD, Sandoval-Cuellar CE, Bauer K, Quintanilla-Carvajal MX (2019) In-vitro digestion of higholeic palm oil nanoliposomes prepared with unpurified soy lecithin: physical stability and nano-liposome digestibility. Colloids Surf A Physicochem Eng Asp 578:123603
Ben-Fadhel Y, Maherani B, Salmieri S, Lacroix M (2022) Preparation and characterization of natural extracts-loaded food grade nanoliposomes. Lwt 154:112781

Bentley-Hewitt KL, Chen RKY, Lill RE, Hedderley DI, Herath TD, Matich AJ, McKenzie MJ (2014) Consumption of selenium-enriched broccoli increases cytokine production in human peripheral blood mononuclear cells stimulated ex vivo, a preliminary human intervention study. Mol Nutr Food Res 58(12):2350–2357

Bernal-Millán MDJ, Gutiérrez-Grijalva EP, Contreras-Angulo L, Muy-Rangel MD, López-Martínez LX, Heredia JB (2022) Spray-dried microencapsulation of oregano (Lippia graveolens) polyphenols with Maltodextrin enhances their stability during in vitro digestion. J Chem 2022

Bińkowska W, Szpicer A, Wojtasik-Kalinowska I, Półtorak A (2024) Innovative methods of encapsulation and enrichment of cereal-based pasta products with biofunctional compounds. Appl Sci 14(4):1442

Bora G (2023) Antioxidant molecules and minerals in prevention of HIV and AIDS: a review. HIV Nurs 23(3):294–305

Bribi N, Rodríguez-Nogales A, Vezza T, Algieri F, Rodriguez-Cabezas ME, Garrido-Mesa J, Gálvez J (2020) Intestinal anti-inflammatory activity of the total alkaloid fraction from Fumaria capreolata in the DSS model of colitis in mice. Bioorg Med Chem Lett 30(18):127414

Bryszewska MA, Tomás-Cobos L, Gallego E, Villalba M, Rivera D, Saa DLT, Gianotti A (2019) In vitro bioaccessibility and bioavailability of iron from breads fortified with microencapsulated iron. LWT 99:431–437

Cai H, Chen L, Yin C, Liao Y, Meng X, Lu C et al (2020) The effect of micro-nutrients on malnutrition, immunity and therapeutic effect in patients with pulmonary tuberculosis: a systematic review and meta-analysis of randomised controlled trials. Tuberculosis 125:101994

Calder PC (2020) n-3 PUFA and inflammation: from membrane to nucleus and from bench to bedside. Proc Nutr Soc 79(4):404–416

Camacho-Muñoz D, Niven J, Kucuk S, Cucchi D, Certo M, Jones SW et al (2022) Omega-3 polyunsaturated fatty acids reverse the impact of western diets on regulatory T cell responses through averting ceramide-mediated pathways. Biochem Pharmacol 204:115211

Carvalho GR, Massarioli AP, Alvim ID, Augusto PED (2020) Iron-fortified pineapple chips produced using microencapsulation, ethanol, ultrasound and convective drying. Food Eng Rev 59:1–14

Cerda-Opazo P, Gotteland M, Oyarzun-Ampuero FA, Garcia L (2021) Design, development and evaluation of nanoemulsion containing avocado peel extract with anticancer potential: a novel biological active ingredient to enrich food. Food Hydrocoll 111:106370

Chaari M, Theochari I, Papadimitriou V, Xenakis A, Ammar E (2018) Encapsulation of carotenoids extracted from halophilic Archaea in oil-in-water (O/W) micro-and nano-emulsions. Colloids Surf B: Biointerfaces 161:219–227

Chen H, Hua XM, Ze BC, Wang B, Wei L (2017) The anti-inflammatory effects of asiatic acid in lipopolysaccharide-stimulated human corneal epithelial cells. Int J Ophthalmol 10(2):179–185

Chen M, Li R, Gao Y, Zheng Y, Liao L, Cao Y et al (2021) Encapsulation of hydrophobic and low-soluble polyphenols into nanoliposomes by pH-driven method: naringenin and naringin as model compounds. Foods 10(5):963

Coelho SC, Giron A, Rocha F, Estevinho BN (2024) Electrosprayed B-complex vitamins/zein microparticles for drug sustained release and antioxidant applications. J Chem Technol Biotechnol 99(1):217–226

Constantino ABT, Garcia-Rojas EE (2022) Vitamin D3 microcapsules formed by heteroprotein complexes obtained from amaranth protein isolates and lactoferrin: formation, characterization, and bread fortification. Food Hydrocoll 129:107636

Cooper EL, Ma MJ (2017) Understanding nutrition and immunity in disease management. J Tradit Complement Med 7(4):386–391

Cucchi D, Camacho-Muñoz D, Certo M, Pucino V, Nicolaou A, Mauro C (2020) Fatty acids-from energy substrates to key regulators of cell survival, proliferation and effector function. Cell Stress 4(1):9–23 da Cruz EP, Pires JB, dos Santos FN, Fonseca LM, Radünz M, Dal Magro J et al (2023) Encapsulation of lemongrass essential oil into cassava starch fibers for application as antifungal agents in bread. Food Hydrocoll 145:109105

Dahiya D, Terpou A, Dasenaki M, Nigam PS (2023) Current status and future prospects of bioactive molecules delivered through sustainable encapsulation techniques for food fortification. Sustain Food Technol. https://doi.org/10.1039/D3FB00015J

Dalgaard TS, Briens M, Engberg RM, Lauridsen C (2018) The influence of selenium and selenoproteins on immune responses of poultry and pigs. Anim Feed Sci Technol 238:73–83

da Rosa CG, de Melo APZ, Sganzerla WG, Machado MH, Nunes MR, Maciel MVDOB et al (2020) Application in situ of zein nanocapsules loaded with Origanum vulgare Linneus and Thymus vulgaris as a preservative in bread. Food Hydrocoll 99:105339

Dehnad D, Ghorani B, Emadzadeh B, Emadzadeh M, Assadpour E, Rajabzadeh G, Jafari SM (2023) Recent advances in iron encapsulation and its application in food fortification. Crit Rev Food Sci Nutr:1–17. https://doi.org/10.1080/10408398.2023.2256004

de Lima Nunes TA, Costa LH, De Sousa JMS, De Souza VMR, Rodrigues RRL, Val MDCA, da Franca Rodrigues KA (2021) Eugenia piauhiensis Vellaff. essential oil and γ-elemene its major constituent exhibit antileishmanial activity, promoting cell membrane damage and in vitro immunomodulation. Chem Biol Interact 339:109429

Díaz LD, Fernández-Ruiz V, Cámara M (2020) An international regulatory review of food health-related claims in functional food products labeling. J Funct Foods 68:103896

Dima C, Assadpour E, Dima S, Jafari SM (2020) Bioactive-loaded nanocarriers for functional foods: from designing to bioavailability. Curr Opin Food Sci 33:21–29

Domínguez-Andrés J, Dos Santos JC, Bekkering S, Mulder WJ, van der Meer JW, Riksen NP et al (2023) Trained immunity: adaptation within innate immune mechanisms. Physiol Rev 103(1):313–346

Duan M, Sun J, Huang Y, Jiang H, Hu Y, Pang J, Wu C (2023) Electrospun gelatin/chitosan nanofibers containing curcumin for multifunctional food packaging. Food Sci Human Wellness 12(2):614–621

Elalouf A, Elalouf H, Rosenfeld A (2023) Modulatory immune responses in fungal infection associated with organ transplant-advancements, management, and challenges. Front Immunol 14. https://doi.org/10.3389/fimmu.2023.1292625

Elmadfa I, Meyer AL (2019) The role of the status of selected micronutrients in shaping the immune function. Endocr Metab Immune Disord Drug Targets 19(8):1100–1115

Ephrem E, Najjar A, Charcosset C, Greige-Gerges H (2018) Encapsulation of natural active compounds, enzymes, and probiotics for fruit juice fortification, preservation, and processing: an overview. J Funct Foods 48:65–84

Erarslan A, Karakas CY, Bozkurt F, Sagdic O (2023) Enhanced antifungal activity of electrosprayed Poly(vinyl alcohol)/Chitosan nanospheres loaded with sage essential oil on the viability of Aspergillus niger and Botrytis cinerea. ChemistrySelect 8(21):e202300296

Feng X, Tjia JYY, Zhou Y, Liu Q, Fu C, Yang H (2020) Effects of tocopherol nanoemulsion addition on fish sausage properties and fatty acid oxidation. LWT 118:108737

Firmino JP, Fernández-Alacid L, Vallejos-Vidal E, Salomón R, Sanahuja I, Tort L et al (2021) Carvacrol, thymol, and garlic essential oil promote skin innate immunity in gilthead seabream (Sparus aurata) through the multifactorial modulation of the secretory pathway and enhancement of mucus protective capacity. Front Immunol 12:633621

Frost JN, Tan TK, Abbas M, Wideman SK, Bonadonna M, Stoffel NU et al (2021) Hepcidin-mediated hypoferremia disrupts immune responses to vaccination and infection. Med 2(2):164–179

Gaba ABM, Hassan MA, Abd EL-Tawab AA, Abdelmonem MA, Morsy MK (2022) Protective impact of chitosan film loaded oregano and thyme essential oil on the microbial profile and quality attributes of beef meat. Antibiotics 11(5):583

Gairola K, Gururani S, Dubey SK (2023) Polyphenols and its effect on the immune system. In: Nutraceuticals and functional foods in immunomodulators. Springer Nature Singapore, Singapore, pp 121–140

Genevois C, de Escalada Pla M, Flores S (2017) Novel strategies for fortifying vegetable matrices with iron and Lactobacillus casei simultaneously. LWT-Food Sci Technol 79:34–41

Ghafarifarsani H, Hoseinifar SH, Javahery S, Van Doan H (2022) Effects of dietary vitamin C, thyme essential oil, and quercetin on the immunological and antioxidant status of common carp (Cyprinus carpio). Aquaculture 553:738053

Ghahramanipour Z, Alipour S, Masoumi J, Rostamlou A, Hatami-Sadr A, Heris JA et al (2023) Regulation of dendritic cell functions by vitamins as promising therapeutic strategy for immune system disorders. Adv Biol 7:2300142

Gheonea I, Aprodu I, Cîrciumaru A, Râpeanu G, Bahrim GE, Stănciuc N (2021) Microencapsulation of lycopene from tomatoes peels by complex coacervation and freeze-drying: evidences on phytochemical profile, stability and food applications. J Food Eng 288:110166

Godswill AG, Somtochukwu IV, Ikechukwu AO, Kate EC (2020) Health benefits of micronutrients (vitamins and minerals) and their associated deficiency diseases: A systematic review. Int J Food Sci 3(1):1–32

Gray KJ, Gibbs JE (2022) Adaptive immunity, chronic inflammation and the clock. In: Seminars in immunopathology, vol 44, No. 2. Springer Berlin Heidelberg, Berlin/Heidelberg, pp 209–224

Grazul M, Kwiatkowski P, Hartman K, Kilanowicz A, Sienkiewicz M (2023) How to naturally support the immune system in inflammation—essential oils as immune boosters. Biomedicines 11(9):2381

Guan S, Liu Q, Gu H, Zhang YY, Wei PL, Qi YF et al (2020) Pluripotent anti-inflammatory immunomodulatory effects of papaverine against cerebral ischemic-reperfusion injury. J Pharmacol Sci 144(2):69–75

Guerrero I, Maldonado L, Marcia J, Aleman R, Moncada M, Fernández VM et al (2022) Freeze drying optimization of Canola Oil with Phytosterols using Alginate and Maltodextrin. Chem Eng Trans 93:139–144. https://doi.org/10.3303/CET2293024

Gulzar S, Benjakul S, Hozzein WN (2020) Impact of β-glucan on debittering, bioaccessibility and storage stability of skim milk fortified with shrimp oil nanoliposomes. Int J Food Sci Technol 55(5):2092–2103

Hadidi M, Khaksar FB, Pagan J, Ibarz A (2020) Application of Ultrasound-Ultrafiltration-Assisted alkaline isoelectric precipitation (UUAAIP) technique for producing alfalfa protein isolate for human consumption: optimization, comparison, physicochemical, and functional properties. Food Res Int 130:108907

Hadidi M, Tan C, Assadpour E, Kharazmi MS, Jafari SM (2023) Emerging plant proteins as nanocarriers of bioactive compounds. J Control Release, 355, 327–342. Polymer and lipid-based transporters. Trends Food Sci Technol 78:34–60

Hao C, Wu B, Hou Z, Xie Q, Liao T, Wang T, Ma D (2017) Asiatic acid inhibits LPS-induced inflammatory response in human gingival fibroblasts. Int Immunopharmacol 50:313–318. https://doi.org/10.1016/j.intimp.2017.07.005

Haque MA, Jantan I, Harikrishnan H, Ahmad W (2020) Standardized ethanol extract of Tinospora crispa upregulates pro-inflammatory mediators release in LPS-primed U937 human macrophages through stimulation of MAPK, NF-κB and PI3K-Akt signaling networks. BMC Complementary Med and Ther 20:1–13

Hasani S, Ojagh SM, Ghorbani M, Hasani M (2020) Nano-encapsulation of lemon essential oil approach to reducing the oxidation process in fish burger during refrigerated storage. J Food Biosci Technol 10(1):35–46

Hemmatkhah F, Zeynali F, Almasi H (2020) Encapsulated cumin seed essential oil-loaded active papers: characterization and evaluation of the effect on quality attributes of beef hamburger. Food Bioprocess Technol 13:533–547

Hooda P, Malik R, Bhatia S, Al-Harrasi A, Najmi A, Zoghebi K et al (2023) Phytoimmunomodulators: a review of natural modulators for complex immune system. Heliyon

Hong X, Zhao Q, Liu Y, Li J (2023) Recent advances on food-grade water-in-oil emulsions: instability mechanism, fabrication, characterization, application, and research trends. Crit Rev Food Sci Nutr 63(10):1406–1436

Huang Z, Liu Y, Qi G, Brand D, Zheng SG (2018) Role of vitamin A in the immune system. J Clin Med 7(9):258

Iqbal I, Wilairatana P, Saqib F, Nasir B, Wahid M, Latif MF et al (2023) Plant polyphenols and their potential benefits on cardiovascular health: a review. Molecules 28(17):6403

Ismailova A, White JH (2022) Vitamin D, infections and immunity. Rev Endocr Metab Disord 23:1–13

Ismailova A, White JH (2024) Vitamin D and antibacterial immunity. In: Feldman and pike's vitamin D. Academic, pp 995–1010

Jabara HH, Boyden SE, Chou J, Ramesh N, Massaad MJ, Benson H et al (2016) A missense mutation in TFRC, encoding transferrin receptor 1, causes combined immunodeficiency. Nat Genet 48(1):74–78

Jafari MM, Tabrizi ZA, Dayer MS, Kazemi-Sefat NA, Mohtashamifard M, Mohseni R et al (2023) Immune system roles in pathogenesis, prognosis, control, and treatment of Toxoplasma gondii infection. Int Immunopharmacol 124:110872

Jantan I, Haque MA, Ilangkovan M, Arshad L (2019) An insight into the modulatory effects and mechanisms of action of phyllanthus species and their bioactive metabolites on the immune system. Front Pharmacol 10:878

Jayawardena R, Sooriyaarachchi P, Chourdakis M, Jeewandara C, Ranasinghe P (2020) Enhancing immunity in viral infections, with special emphasis on COVID-19: a review. Diabetes Metab Syndr Clin Res Rev 14(4):367–382

Jiang Q (2022) Metabolism of natural forms of vitamin E and biological actions of vitamin E metabolites. Free Radic Biol Med 179:375–387

Kalsum N, Sulaeman A, Setiawan B, Wibawan IWT (2017) Preliminary studies of the immunomodulator effect of the propolis Trigona spp. extract in a mouse model. IOSR J Agric Vet Sci 10:75–80

Kaul S, Kaur K, Kaur P, Kaur J (2022) Development and shelf-life assessment of Ready-to-Bake frozen potato Paranthas fortified with iron and zinc. Int J Food Sci Technol 57(11):7214–7221

Khade OS, Sruthi K, Sonkar RM, Gade PS, Bhatt P (2023) Plant secondary metabolites: extraction, screening, analysis and their bioactivity. Int J Herbal Med. https://doi.org/10.22271/flora.2023.v11.i2a.855

Khalil WA, Hassan MA, Attia KA, El-Metwaly HA, El-Harairy MA, Sakr AM, Abdelnour SA (2023) Effect of olive, flaxseed, and grape seed nano-emulsion essential oils on semen buffalo freezability. Theriogenology 212:9–18

Koohenjani DK, Lashkari H (2022) Effects of double emulsion encapsulated iron on the properties of fortified cream. LWT 161:113296

Kumar N, Chandan NK, Gupta SK, Bhushan S, Patole PB (2022) Omega-3 fatty acids effectively modulate growth performance, immune response, and disease resistance in fish against multiple stresses. Aquaculture 547:737506

Kumari D, Garg S, Bhawrani P (2022) Zinc homeostasis in immunity and its association with preterm births. Scand J Immunol 95(4):e13142

Lange KW, Nakamura Y (2020) Food bioactives, micronutrients, immune function and COVID-19. J Food Bioact 2020(10):1–8

Lee MH, Do Kim H, Jang YJ (2024) Delivery systems designed to enhance stability and suitability of lipophilic bioactive compounds in food processing: a review. Food Chem 437:137910

Littwitz-Salomon E, Moreira D, Frost JN, Choi C, Liou KT, Ahern DK et al (2021) Metabolic requirements of NK cells during the acute response against retroviral infection. Nat Commun 12(1):5376

Liu L, Wang JR, Mei X (2022) Enhancing the stability of active pharmaceutical ingredients by the cocrystal strategy. CrystEngComm 24(11):2002–2022

Liu S, Lin Z, Qiao W, Chen B, Shen J (2024) Cross-talk between biometal ions and immune cells for bone repair. Eng Regen 5:375–408

Liugan M, Carr AC (2019) Vitamin C and neutrophil function: findings from randomized controlled trials. Nutrients 11(9):2102

Loftus RM, Assmann N, Kedia-Mehta N, O'Brien KL, Garcia A, Gillespie C et al (2018) Amino acid-dependent cMyc expression is essential for NK cell metabolic and functional responses in mice. Nat Commun 9(1):2341

Ludwiczuk A, Skalicka-Woźniak K, Georgiev MI (2017) Terpenoids. In: *Pharmacognosy*. Academic, pp 233–266

Luo M, Zhang R, Liu L, Chi J, Huang F, Dong L et al (2020) Preparation, stability and antioxidant capacity of nano liposomes loaded with procyandins from lychee pericarp. J Food Eng 284:110065

Maheshwari S, Kumar V, Bhadauria G, Mishra A (2022) Immunomodulatory potential of phytochemicals and other bioactive compounds of fruits: a review. Food Front 3(2):221–238

Mahmood A, Tiwari A, Şahin K, Küçük Ö, Ali S (2016) Triterpenoid saponin-rich fraction of Centella asiatica decreases IL-1ß andNF-? B, and augments tissue regeneration and excision wound repair. Turk J Biol 40(2):399–409

Manosso LM, Camargo A, Dafre AL, Rodrigues ALS (2022) Vitamin E for the management of major depressive disorder: possible role of the anti-inflammatory and antioxidant systems. Nutr Neurosci 25(6):1310–1324

Marreiro DDN, Cruz KJC, Oliveira AD, Morais JBS, Bjesa F, Melo SRDS et al (2021) Antiviral and immunological activity of zinc and possible role in COVID-19. Br J Nutr 127(8):1172–1179

Marshall JS, Warrington R, Watson W, Kim HL (2018) An introduction to immunology and immunopathology. Allergy, Asthma Clin Immunol 14(2):1–10

Masad RJ, Haneefa SM, Mohamed YA, Al-Sbiei A, Bashir G, Cabezudo MJF, Al-Ramadi BK (2021) The immunomodulatory effects of honey and associated flavonoids in cancer. Nutrients 13(4):1269

Matsuzaka Y, Yashiro R (2023) Unraveling the Immunopathogenesis of multiple sclerosis: the dynamic dance of Plasmablasts and pathogenic T cells. Biologics 3(3):232–252

Maywald M, Wessels I, Rink L (2017) Zinc signals and immunity. Int J Mol Sci 18(10):2222

McClung JP (2019) Iron, zinc, and physical performance. Biol Trace Elem Res 188:135–139

Mehta N, Kumar P, Verma AK, Umaraw P, Kumar Y, Malav OP et al (2022) Microencapsulation as a noble technique for the application of bioactive compounds in the food industry: a comprehensive review. Appl Sci 12(3):1424

Mishra S, Patel M (2020) Role of nutrition on immune system during COVID-19 pandemic. J Food Nutr Health 3(2):1–6

Mishra SB, Singh AK, Singh S, Sharma DR (2024) Consumption of Carotenoidrich vegetables and functions of human T-lymphocytes. In: Immune-boosting nutraceuticals for better human health. Apple Academic Press, pp 47–80

Mitra S, Paul S, Roy S, Sutradhar H, Bin Emran T, Nainu F et al (2022) Exploring the immune-boosting functions of vitamins and minerals as nutritional food bioactive compounds: A comprehensive review. Molecules 27(2):555

Mounika D, Hemalatha S (2023) Sensory evaluation of zinc microcapsule incorporated milk products

Munteanu C, Berindean I, Mihai M, Pop B, Popa M, Muntean L et al (2023) E, K, B5, B6, and B9 vitamins and their specific immunological effects evaluated by flow cytometry. Front Med 9:1089476

Musavi H, Abazari O, Barartabar Z, Kalaki-Jouybari F, Hemmati-Dinarvand M, Esmaeili P, Mahjoub S (2023) The benefits of vitamin D in the COVID-19 pandemic: biochemical and immunological mechanisms. Arch Physiol Biochem 129(2):354–362

Muthulakshmi M, Subramani PA, Michael RD (2016) Immunostimulatory effect of the aqueous leaf extract of Phyllanthus niruri on the specific and nonspecific immune responses of Oreochromis mossambicus. Iran J Vet Res 17(3):200–202

Nanakali NM (2023) Fabrication of nano-encapsulated angelica (Heracleum persicum) essential oil for enriching dairy dessert: physicochemical, rheological and sensorial properties. IET Nanobiotechnology 17(3):171–181

Ni S, Yuan Y, Kuang Y, Li X (2022) Iron metabolism and immune regulation. Front Immunol 13:816282

Ojagh SM, Hasani S (2018) Characteristics and oxidative stability of fish oil nano-liposomes and its application in functional bread. J Food Meas Charact 12:1084–1092

Olatunde A, Tijjani H, Ishola AA, Egbuna C, Hassan S, Akram M (2020) Carotenoids as functional bioactive compounds. In: Functional foods and nutraceuticals: bioactive components, formulations and innovations, pp 415–444

Oleszek M, Oleszek W (2020) Saponins in food. In: Handbook of dietary phytochemicals, pp 1–40

Pamunuwa GK, Prasadani M, Nanayakkara TU, Atapattu SN (2023) Effect of liposomal encapsulation of curcumin and α-tocopherol on sensory and physicochemical properties, and retention of antioxidant capacity of fortified cookies during baking. Food Chem Adv 3:100504

Pangrazzi L (2019) Boosting the immune system with antioxidants: where are we now? Biochemist 41(1):42–44

Parija SC (2023) Structure and function of immune system. In: Textbook of microbiology and immunology. Springer Nature Singapore, Singapore, pp 129–149

Park HY, Oh MJ, Kim Y, Choi I (2018) Immunomodulatory activities of Corchorus olitorius leaf extract: beneficial effects in macrophage and NK cell activation immunosuppressed mice. J Funct Foods 46:220–226

Park J, Wu Y, Li Q, Choi J, Ju H, Cai Y et al (2023) Nanomaterials for antigen-specific immune tolerance therapy. Drug Deliv Transl Res 13(7):1859–1881

Parker D (2018) CD80/CD86 signaling contributes to the proinflammatory response of Staphylococcus aureus in the airway. Cytokine 107:130–136

Patel SS, Pushpadass HA, Franklin MEE, Battula SN, Vellingiri P (2022) Microencapsulation of curcumin by spray drying: characterization and fortification of milk. J Food Sci Technol 59(4):1326–1340

Patrignani F, Siroli L, Braschi G, Lanciotti R (2020) Combined use of natural antimicrobial based nanoemulsions and ultra high pressure homogenization to increase safety and shelf-life of apple juice. Food Control 111:107051

Pecora F, Persico F, Argentiero A, Neglia C, Esposito S (2020) The role of micronutrients in support of the immune response against viral infections. Nutrients 12(10):3198

Pedroza-García KA, Careaga-Cárdenas G, Díaz-Galindo C, Quintanar JL, Hernández-Jasso I, Ramírez-Orozco RE (2023) Bioactive role of vitamins as a key modulator of oxidative stress, cellular damage and comorbidities associated with spinal cord injury (SCI). Nutr Neurosci 26(11):1120–1137

Phipps O, Brookes MJ, Al-Hassi HO (2021) Iron deficiency, immunology, and colorectal cancer. Nutr Rev 79(1):88–97

Pourashouri P, Shabanpour B, Kordjazi M, Jamshidi A (2020) Characteristic and shelf life of fish sausage: fortification with fish oil through emulsion and gelled emulsion incorporated with green tea extract. J Sci Food Agric 100(12):4474–4482

Pourashouri P, Shabanpour B, Heydari S, Raeisi S (2021) Encapsulation of fish oil by carrageenan and gum tragacanth as wall materials and its application to the enrichment of chicken nuggets. LWT 137:110334

Premathilaka R, Rashidinejad A, Golding M, Singh J (2022) Oral delivery of hydrophobic flavonoids and their incorporation into functional foods: opportunities and challenges. Food Hydrocoll 128:107567

Prevete G, Carvalho LG, del Carmen Razola-Diaz M, Verardo V, Mancini G, Fiore A, Mazzonna M (2024) Ultrasound assisted extraction and liposome encapsulation of olive leaves and orange peels: how to transform biomass waste into valuable resources with antimicrobial activity. Ultrason Sonochem 102:106765

Rahbari S, Tavakolipour H, Kalbasi-Ashtari A (2024) Application of electro-spraying technique and mathematical modelling for nanoencapsulation of curcumin. Heliyon

Raikos V, Pirie LP, Gürel S, Hayes HE (2021) Encapsulation of vitamin E in yogurt-based beverage emulsions: influence of bulk pasteurization and chilled storage on physicochemical stability and starter culture viability. Molecules 26(6):1504

Rakshit M, Tripathy S, Srivastav PP (2023) Encapsulation of natural polyphenols for food applications. In: Novel processing methods for plant-based health foods: extraction, encapsulation, and health benefits of bioactive compounds. Apple Academic Press

Rao V, Poonia A (2023) Bioactive compounds, nanoparticles synthesis, health benefits and potential utilization of edible flowers for the development of functional dairy products: a review. J Food Sci Technol:1–16. https://doi.org/10.1007/s13197-023-05853-6

Rasti B, Erfanian A, Selamat J (2017) Novel nanoliposomal encapsulated omega-3 fatty acids and their applications in food. Food Chem 230:690–696

Razaghi A, Poorebrahim M, Sarhan D, Björnstedt M (2021) Selenium stimulates the antitumour immunity: insights to future research. Eur J Cancer 155:256–267

Razizadeh MH, Zafarani A, Taghavi-Farahabadi M, Khorramdelazad H, Minaeian S, Mahmoudi M (2023) Natural killer cells and their exosomes in viral infections and related therapeutic approaches: where are we? Cell Commun Signal 21(1):261

Read SA, Obeid S, Ahlenstiel C, Ahlenstiel G (2019) The role of zinc in antiviral immunity. Adv Nutr 10(4):696–710

Renu K, Mukherjee AG, Gopalakrishnan AV, Wanjari UR, Kannampuzha S, Murali R et al (2023) Protective effects of macromolecular polyphenols, metal (zinc, selenium, and copper)-Polyphenol complexes, and pectin in different organs with an emphasis on arsenic poisoning: a review. Int J Biol Macromol 126715

Reshmi SK, Sudha ML, Shashirekha MN (2017) Starch digestibility and glycemic index of Paranthas supplemented with Citrus maxima (Burm.) Merr. fruit segments. J Food Sci Technol 54:4370–4377

Rios-Mera JD, Saldaña E, Patinho I, Selani MM, Contreras-Castillo CJ (2021) Enrichment of NaCl-reduced burger with long-chain polyunsaturated fatty acids: effects on physicochemical, technological, nutritional, and sensory characteristics. Meat Sci 177:108497

Roy A, Agnivesh PK, Sau S, Kumar S, Kalia NP (2023) Tweaking host immune responses for novel therapeutic approaches against Mycobacterium tuberculosis. Drug Discov Today 28:103693

Saatkamp RH, Sanches MP, Gambin JPD, Amaral BR, de Farias NS, Caon T et al (2023) Development of thymol nanoemulsions with potential application in oral infections. J Drug Delivery Sci Technol 87:104855

Saha S, Shalova IN, Biswas SK (2017) Metabolic regulation of macrophage phenotype and function. Immunol Rev 280(1):102–111

Samadi M, Kamrani A, Nasiri H, Shomali N, Heris JA, Shahabi P et al (2023) Cancer immunotherapy focusing on the role of interleukins; A comprehensive and updated study. Pathol Res Pract 249:154732

Sandner G, Heckmann M, Weghuber J (2020) Immunomodulatory activities of selected essential oils. Biomolecules 10(8):1139

Sangeetha S, Thuraisingam S, Jayawardane JAEC, Srivijeindran S (2023) A novel approach to the formulation of an encapsulated bioactive powder from palmyrah (Borassus flabellifer L) fruit pulp for nutraceutical applications. Food Chem Adv 3:100405

Santillán-Urquiza E, Méndez-Rojas MÁ, Vélez-Ruiz JF (2017) Fortification of yogurt with nano and micro sized calcium, iron and zinc, effect on the physicochemical and rheological properties. LWT 80:462–469

Santos J, Brito M, Ferreira R, Moura AP, Sousa T, Batista T et al (2018) Th1-biased immunomodulation and in vivo antitumor effect of a novel piperine analogue. Int J Mol Sci 19(9):2594

Schmatz DA, da Silveira Mastrantonio DJ, Costa JAV, de Morais MG (2020) Encapsulation of phycocyanin by electrospraying: a promising approach for the protection of sensitive compounds. Food Bioprod Process 119:206–215

Shams MH, Jafari R, Eskandari N, Masjedi M, Kheirandish F, Hakemi MG et al (2021) Anti-allergic effects of vitamin E in allergic diseases: an updated review. Int Immunopharmacol 90:107196

Sharma H, Ahuja A, Sharma B, Kulshreshtha A, Kadam A, Dutt D (2023) Vapor phase antimicrobial active packaging application of chitosan capsules containing clove essential oil for the preservation of dry cakes. Food Bioprocess Technol:1–11

Shavisi N, Shahbazi Y (2022) Chitosan-gum Arabic nanofiber mats encapsulated with pH-sensitive Rosa damascena anthocyanins for freshness monitoring of chicken fillets. Food Packag Shelf Life 32:100827

Shishir MRI, Xie L, Sun C, Zheng X, Chen W (2018) Advances in micro and nano-encapsulation of bioactive compounds using biopolymer and lipid-based transporters. Trends Food Sci Technol 78:34–60

Siddique A, Park YW (2019) Effect of iron fortification on microstructural, textural, and sensory characteristics of caprine milk Cheddar cheeses under different storage treatments. J Dairy Sci 102(4):2890–2902

Singh MR, Yadav K, Chaurasiya ND, Singh D (2022) Immune system and mechanism of immunomodulation. In: Plants and Phytomolecules for immunomodulation: recent trends and advances. Springer Nature Singapore, Singapore, pp 1–31

Singh DN, Bohra JS, Dubey TP, Shivahre PR, Singh RK, Singh T, Jaiswal DK (2023) Common foods for boosting human immunity: a review. Food Sci Nutr 11(11):6761–6774

Soares CLR, Wilairatana P, Silva LR, Moreira PS, Barbosa NMMV, da Silva PR et al (2023) Biochemical aspects of the inflammatory process: A narrative review. Biomed Pharmacother 168:115764

Stefanache A, Lungu II, Butnariu IA, Calin G, Gutu C, Marcu C et al (2023) Understanding how minerals contribute to optimal immune function. J Immunol Res 2023:1–26

Stephen J, Manoharan D, Radhakrishnan M (2023) Immune boosting functional components of natural foods and its health benefits. Food Product Process Nutr *5*(1):61

Suja SR, Krishnakumar NM, Bijukumar BS, Prakashkumar R (2024) Bioactive Flavonoids from natural sources: potential immune-boosters. In: Flavonoids as nutraceuticals, pp 143–166

Surendhiran D, Li C, Cui H, Lin L (2020) Fabrication of high stability active nanofibers encapsulated with pomegranate peel extract using chitosan/PEO for meat preservation. In: Food packaging and shelf life, vol 23, 100439

Sutoyo S, Ismono I, Mitarlis D (2018) Immunostimulant activity of flavonoid isolated from the acetone extract of silver fern (Pityrogramma calomelanos). Adv Eng Res 171:86–88

Tao J, Zhu Q, Qin F, Wang M, Chen J, Zheng ZP (2017) Preparation of steppogenin and ascorbic acid, vitamin E, butylated hydroxytoluene oil-in-water microemulsions: characterization, stability, and antibrowning effects for fresh apple juice. Food Chem 224:11–18

Thakur NS, Thakur A (2020) Microencapsulation of wild pomegranate flavedo phenolics by lyophilization: effect of maltodextrin concentration, structural morphology, functional properties, elemental composition and ingredient for development of functional beverage. LWT 133:110077

Toledano JM, Puche-Juarez M, Moreno-Fernandez J, Ochoa JJ, Diaz-Castro J (2023) Antioxidant and immune-related implications of minerals in COVID-19: A possibility for disease prevention and management. Antioxidants 12(5):1104

Trasino SE (2020) A role for retinoids in the treatment of COVID-19. Clin Exp Pharmacol Physiol 47(10):1765–1767

Tripathy S, Srivastav PP (2023a) Effect of dielectric barrier discharge (DBD) cold plasma-activated water pre-treatment on the drying properties, kinetic parameters, and physicochemical and functional properties of *Centella asiatica* leaves. Chemosphere 332:138901

Tripathy S, Srivastav PP (2023b) Encapsulation of *Centella asiatica* leaf extract in liposome: study on structural stability, degradation kinetics and fate of bioactive compounds during storage. Food Chemi Adv 2:100202

Tripathy S, Verma DK, Thakur M, Patel AR, Srivastav PP, Singh S et al (2021a) Encapsulated food products as a strategy to strengthen immunity against COVID-19. Front Nutr 8:673174

Tripathy S, Verma DK, Thakur M, Patel AR, Srivastav PP, Singh S et al (2021b) Curcumin extraction, isolation, quantification and its application in functional foods: A review with a focus on immune enhancement activities and COVID-19. Front Nutr 8:747956

Tripathy S, Verma DK, Thakur M, Chakravorty N, Singh S, Srivastav PP (2022) Recent trends in extraction, identification and quantification methods of Centella asiatica phytochemicals with potential applications in food industry and therapeutic relevance: A review. Food Biosci 49:101864

Tripathy S, Verma DK, Gupta AK, Srivastav PP, Patel AR, González MLC et al (2023) Nanoencapsulation of biofunctional components as a burgeoning nanotechnology-based approach for functional food development: a review. Biocatal Agric Biotechnol 53:102890

Turkmen N, Akal C, Özer B (2019) Probiotic dairy-based beverages: A review. J Funct Foods 53:62–75

Vehapi M, Yilmaz A, Özçimen D (2021) Fabrication of oregano-olive oil loaded PVA/chitosan nanoparticles via electrospraying method. J Nat Fibers 18(9):1359–1373

Verma DK, Thakur M (eds) (2021) Phytochemicals in food and health: perspectives for research and technological development. CRC Press

Verma DK, Goya MR, Suleria HAR (eds) (2019a) Nanotechnology and nanomaterial applications in food, health, and biomedical sciences. CRC Press

Verma DK, Patel AR, Prajapati JB, Thakur M, Abd Al-Manhel AJ, Srivastav PP (2019b) Starter culture and probiotic bacteria in dairy food products. In: Microbiology for food and health. Apple Academic Press, pp 3–49

Wang S, Liu Q, Zeng T, Zhan J, Zhao H, Ho CT et al (2022) Immunomodulatory effects and associated mechanisms of Momordica charantia and its phytochemicals. Food Funct 13:11986–11998

Wang Y, Du YT, Xue WY, Wang L, Li R, Jiang ZT et al (2023) Enhanced preservation effects of clove (Syzygium aromaticum) essential oil on the processing of Chinese bacon (preserved meat products) by beta cyclodextrin metal organic frameworks (β-CD-MOFs). Meat Sci 195:108998

Wimalawansa SJ (2023) Infections and autoimmunity—the immune system and vitamin D: A systematic review. Nutrients 15(17):3842

Xu W, Yang Y, Xue SJ, Shi J, Lim LT, Forney C et al (2018) Effect of in vitro digestion on water-in-oil-in-water emulsions containing anthocyanins from grape skin powder. Molecules 23(11):2808

Yamamoto E, Jørgensen TN (2019) Immunological effects of vitamin D and their relations to autoimmunity. J Autoimmun 100:7–16

Yaman DM, Koçak Yanık D, Elik Demir A, Uzun Karka H, Güçlü G, Selli S et al (2023) Effect of encapsulation techniques on aroma retention of Pistacia terebinthus L. Fruit Oil: spray drying, spray freeze drying, and freeze drying. Foods 12(17):3244

Yoshii K, Hosomi K, Sawane K, Kunisawa J (2019) Metabolism of dietary and microbial vitamin B family in the regulation of host immunity. Front Nutr 6:48

Yousefi M, Khanniri E, Sohrabvandi S, Khorshidian N, Mortazavian AM (2023) Encapsulation of Heracleum persicum essential oil in chitosan nanoparticles and its application in yogurt. Front Nutr 10:1130425

Zandavar H, Babazad MA (2023) Secondary metabolites: alkaloids and flavonoids in medicinal plants. In: Herbs and spices-new advances. IntechOpen

Zaghian N, Goli M (2020) Optimization of the production conditions of primary (W 1/O) and double (W 1/O/W 2) nano-emulsions containing vitamin B 12 in skim milk using ultrasound wave by response surface methodology. J Food Meas Charact 14:3216–3226

Zeng Z, Lin C, Wang S, Wang P, Xu W, Ma W et al (2020) Suppressive activities of mangiferin on human epithelial ovarian cancer. Phytomedicine 76:153267

Zhou M, Huang L, Li L, Wei Y, Shu J, Liu X, Huang H (2017) New furostanol saponins with anti-inflammatory and cytotoxic activities from the rhizomes of Smilax davidiana. Steroids 127:62–68

Zhu H, Mettu S, Cavalieri F, Ashokkumar M (2021) Ultrasonic microencapsulation of oil-soluble vitamins by hen egg white and green tea for fortification of food. Food Chem 353:129432

Zhu Y, Ma J, Shen R, Lin J, Li S, Lu X et al (2023) Screening for lipid nanoparticles that modulate the immune activity of helper T cells towards enhanced antitumour activity. Nat Biomed Eng 8:1–17

Chapter 16
Processing Effects on Food Bioactives and Functional Properties

Raquel Martins Martinez, Carolyne Pimentel Rosado, Monique de Barros Elias Campos, Carlos Eduardo de Faria Cardoso, Luana Oeby de Oliveira, Carolina de Oliveira Ramos Petra de Almeida, and Thuane Passos Barbosa Lima

Abstract Fresh foods typically undergo various levels of processing, such as fractionation, thermal and chemical processes. The need for these processes goes beyond the diversification of flavors, as they can contribute to the viability of consumption, conservation, storage, transportation and the development of new products. Domestic cooking and the technological processes used in the manufacture of foods and nutraceuticals affect nutritional, functional and biological properties of the different components present in the food matrix, and it is essential to better understand these transformations in order to optimize the use of nutritional and therapeutic resources. Current literature indicates that heat treatment, for example, is related to the degradation of polyphenols and anthocyanins, while carotenoids seem to be the most thermostable compounds. More recent methods such as freeze-drying, ultraviolet radiation, and ultrasound have been studied with promising

R. M. Martinez (✉)
Food and Nutrition Security Graduate Program, PPGSAN/UNIRIO – Federal University of the State of Rio de Janeiro, Rio de Janeiro, RJ, Brazil
e-mail: raquelm.martinez@hotmail.com

C. P. Rosado
Postdoctoral researcher in Food and Nutrition Graduate Program, PPGAN/UNIRIO - Federal University of the State of Rio de Janeiro, Rio de Janeiro, RJ, Brazil
e-mail: carolynerosado@gmail.com

M. de B. E. Campos · L. O. de Oliveira
Sciences Applied to Health Products Graduate Program, PPG-CAPS/UFF – Federal University Fluminense, Niteroi, RJ, Brazil
e-mail: moniquebarros.nutri@gmail.com; moniqueb@id.uff.br; luana.oeby@gmail.com; luanaoeby@id.uff.br

C. E. de F. Cardoso · C. de O. Ramos Petra de Almeida · T. P. B. Lima
Food and Nutrition Graduate Program, PPGAN/UNIRIO – Federal University of the State of Rio de Janeiro, Rio de Janeiro, RJ, Brazil
e-mail: caedufariac@gmail.com; caedufariac@edu.unirio.br; carolina.almeida@edu.unirio.br; thuanepassos@gmail.com

S. Amir Ashraf, M. Adnan (eds.), *Food Bioactives and Nutraceuticals*,
https://doi.org/10.1007/978-981-97-9723-3_16

effects on food preservation. However, the complexity of food matrices and the interaction between compounds still represents a major challenge in this field. Therefore, the aim of this chapter is to review the main aspects of the literature on the effects of different processing methods on the most well-known and studied classes of bioactive compounds and their properties.

Keywords Processing effects · Stability · Bioactive compounds · Carotenoids · Phenolic compounds

16.1 Introduction

Fresh ingredients have been traditionally sorted, cut, cooled and subjected to different cooking methods. Various levels of processing are often used, including the separation of parts (e.g., peel and pulp), fractionation, thermal and chemical processes. In food industry, the possibilities multiply as the need goes beyond the diversification of flavors to contribute to the viability of consumption, preservation, storage, transportation and the development of new products. Thanks to technological advances, new methods have made it possible to extend the shelf life of ingredients or even finished preparations by up to several months. Those methods have emerged to meet the new needs of the world's population, which has moved around the world, faced wars, crises and revolutionized ways of working, and social and cultural connections. Freezing and dehydration methods, canning, various types of packaging, and chemical additives were introduced, making large-scale industrial production possible (Ruiz-Rodriguez et al. 2008; Huebbe and Rimbach 2020).

Over the years, knowledge of food composition and nutritional aspects, as well as the effects of processing on physicochemical properties, has increased. Bioactive compounds have been the subject of studies into the various types of modification in an attempt to preserve their many beneficial properties for human health, such as antioxidant, antimutagenic and anti-obesity effects (Cilla et al. 2018; Arfaoui 2021). As summarized in Fig. 16.1, domestic cooking and the technological processes used in the food and nutraceutical industries can affect the nutritional, functional and biological properties of the different components present in the food matrix, and it is essential to better understand these transformations in order to optimize the use of nutritional and therapeutic resources (Ruiz-Rodriguez et al. 2008; Cilla et al. 2018; Arfaoui 2021). This chapter presents the most well-known and studied classes of bioactive compounds and the effects of processing on their respective properties.

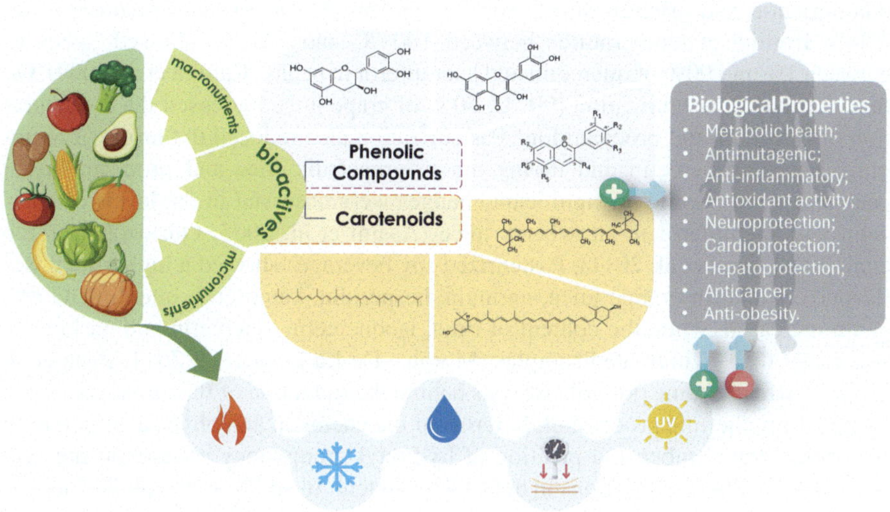

Fig. 16.1 Main classes of bioactive compounds studied in plant foods and their respective biological properties, enhanced or reduced by different processing methods

16.2 Phenolic Compounds

16.2.1 Flavonoids

Flavonoids are a class of polyphenols, with typical chemical structures containing three aromatic rings, including flavones, flavanols, flavanones, anthocyanidins, flavans, flavanonols and isoflavones. According to their physicochemical characteristics, each type of flavonoid can be affected by distinct variations during food processing (Fu et al. 2021). Heating is one of the oldest forms of food processing and several variations have been developed over time by the food industry. In addition to classic thermal processing, microwaves, pasteurization, blanching and other methods are often employed during food processing. However, studies show that flavonoids can be significantly affected by rising temperatures (Ahmed and Eun 2018). A loss of approximately 22% of total flavonoids was observed in foods cooked in water at a temperature of 50 °C for 90 s, while steaming for 40 min caused a 25% decrease in total flavonoid content (Viña and Chaves 2008; Huang et al. 2006; Zhang et al. 2010). Blanching different vegetables by immersion in water leads to a reduction in the total flavonoid content if carried out at over 100 °C for more than 4 min (Ahmed and Eun 2018; Ioannou et al. 2012).

Despite a tendency for flavonoids to degrade when exposed to high temperatures, studies show that the results can vary depending on the food matrix and methods (Ioannou et al. 2012). A 25% reduction in total flavonoid content was observed in mango juice after pasteurization (90 °C, 6 s) (Santhirasegaram et al. 2015). Rutin present in buckwheat proved to be more heat-stable than vitexin, isovitexin, homoorientin and orientin during roasting at 160 °C for 30 min (Zielinski et al. 2009). Boiling at temperatures between 100 °C and 121 °C showed losses of between 1% and 90% of quercetin and kaempferol in beans (Ranilla et al. 2009). On the other hand, pasteurization (90 °C, 60 s) of grape juice increased the concentration of catechins and procyanidins. Pasteurization also reduced the levels of naringin, narirutin, quercetin and naringenin in grapefruit juice and procyanidins in peach preserves, while no significant changes were observed in the levels of quercetin and kaempferol in strawberry juice (Asami et al. 2003; Odriozola-Serrano et al. 2008; Igual et al. 2011). Pasteurized soy beverages showed a higher total isoflavone content. In orange juice, naringin, hesperidin, kampferol, luteolin and apigenin increased, while the content of rutin, isoquercetin, quercetin and naringenin was lower than in untreated samples (Morales-De La Peña et al. 2011; Peña et al. 2017). There are some possible reasons behind the reduction of flavonoid content in foods during thermal processing. In terms of the chemical and physical structure of flavonoids, the number and position of hydroxyl groups may change during heat treatment, leading to variations in detectable content. In addition, flavonoids may be more degraded by increased time and temperature, especially when immersed in water, as they are water-soluble (Ahmed and Eun 2018).

Unexpectedly, microwaves seem to have the opposite effect of other thermal methods, increasing the levels of phenolic compounds and flavonoids. The concentrations of flavanones, flavones and flavanols increased in lemon powder after microwave treatment (Ahmed and Eun 2018). Polyphenols in two citrus pomace extracts has also been increased, from 1128 to 1317 μg/g DW and 620 to 854 μg/g DW, respectively, with increasing microwave power (0 to 500 W). In this study, the effects observed were enhanced by increasing time and power (Hayat et al. 2010). The consequences of this method reported by this and other authors can be explained by the fact that the processed vegetable samples contain a large amount of water, which effectively absorbs the microwave energy and increases the internal temperature, causing the cells to rupture and facilitating the extraction of these compounds (Khan et al. 2018). The use of non-thermal processing methods has recently increased, with particular benefits for food matrices containing temperature-sensitive bioactive ingredients such as flavonoids (Ahmed and Eun 2018). Phenolic compounds seem to be less affected by high pressure. Total phenolic compounds and antioxidant activity of mango and mandarin juice showed only a slight decrease after high pressure treatment, while pineapple juice showed an increase in total phenolic compounds (Fu et al. 2021). In onions, total flavonoid content increased after high pressure treatment. Treated carrots and spinach showed a higher total flavonoid content (Ahmed and Eun 2018). The positive correlation between total flavonoid content and high-pressure methods may be related to pressure-induced cell rupture, resulting in greater extraction of flavonoids. However, the increase or decrease in these levels may vary according to the type of flavonoid. Apple, orange and grape juices subjected to high hydrostatic pressure (250 Mpa, 10 min) showed a decrease in hesperidin, floridzin and epigallocatechin gallate in apple juice, while naringin, naringenin trisaccharide, luteolin rutinoside and quercetin trisaccharide were increased in orange juice. The concentration of proanthocyanidin was higher and that of epicatechin lower in grape juice (He et al. 2016).

Ultrasound appears to have even less effect on flavonoid levels. This method maintained the total flavonoid content in a peach drink compared to pasteurization and microwaves (Fu et al. 2021). Fresh pear juice treated with ultrasound (500 W, 25 kHz frequency, for 0, 15, 30, 45 and 60 min, at 25 °C) showed an increase in the total flavonoid content of the drink. In freshly squeezed orange juice treated with ultrasound (for 1, 10, 20 and 30 min at 24 kHz, at 43–45 ° C), there was a significant increase in total phenolic and flavonoid content (Guerrouj et al. 2016). The basic principle associated with the increase in phenolic and flavonoid content is the cavitation process in food components and the pressure exerted during this process, which results in the rupture of the outer cell envelope and the release of bound polyphenolic content from the food. Another reason for this increase is the addition of hydroxyl functional groups to aromatic compounds of polyphenols due to ultrasonic treatment (Khan et al. 2018). Irradiation has been increasingly used in the food industry and its effects seem to influence the modification of the enzymes responsible for the degradation of phenolic compounds, depending on the dose of irradiation (Khan et al. 2018). Treatment with ultraviolet radiation (2158 J/m^2) resulted in a significant increase in flavonoids and total phenolics in banana and guava. The longer the treatment time (10, 20 and 30 min), the higher the levels observed. In the same experiment, pineapple showed a significant increase only for total flavonoids content (Alothman et al. 2009). In apple juice, this method retained more flavonoids than pasteurization. Positive results were observed in treated onion slices, with a 35% increase in flavonols and quercetin. In tomato pulp, the concentrations of flavonoids and flavonols increased by 27% and 48%, respectively, after UV treatment (Morales-De La Peña et al. 2011). Despite promising studies on increasing the flavonoid content of foods, others have found no significant effects after irradiation. In grapes, UV treatment showed no effect on flavonoid or flavonol content, but could increase the resveratrol content. The same method also showed little effect on the flavonoids in green tea (Fu et al. 2021; Ahmed and Eun 2018).

Since flavonoids are strongly affected by high temperatures, exposing flavonoid-rich foods to shorter times and lower temperatures could prevent their degradation. If possible, non-thermal treatments seem to be the best option for flavonoids, especially ultrasound and ultraviolet radiation treatment (Fig. 16.2).

16.2.1.1 Quercetin

Five hydroxyl groups make up the chemical structure of quercetin, which has a flavonoid structure with two benzene rings linked to an oxygenated pyrene ring. Glycosidic derivatives of this structure are often formed by replacing one or more hydroxyl groups with sugar groups. Quercetin and its related derivatives show lipophilic behavior, with moderate solubility in ethanol (4.0 mg/ml, 37 °C), high solubility in dimethyl sulfoxide (150 mg/ml, 25 °C), but very low solubility in water (0.01 mg/ml, 25 °C). In general, its chemical stability can be affected by oxygen

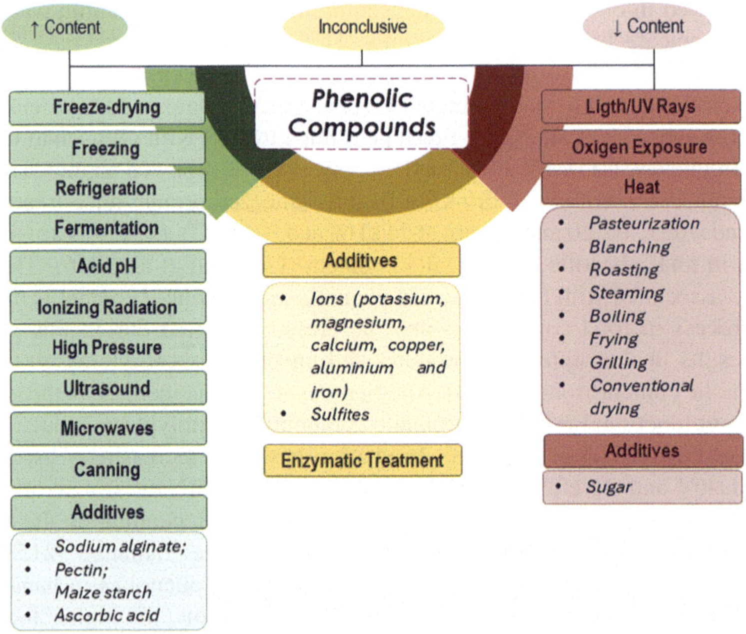

Fig. 16.2 Trends in the effects of processing on phenolic compounds

concentration, pH, temperature, concentrations of other antioxidants and the presence of metal ions. In alkaline conditions, the rate of quercetin degradation is significantly higher than in acidic and neutral environments (Lai and Wong 2022; Wang et al. 2016).

In relation to thermal processing, results can vary depending on the food matrix. In a study evaluating the heating of grape juice to produce a concentrated product, quercetin was the only polyphenol that did not decrease during the process, showing higher levels at the end. The authors suggest that this phenomenon can be explained by a possible glycosylation of the flavonoid, which does not allow a true quantification. Although quercetin may have better stability during thermal processing than other compounds, the phenomenon observed may generally affect its solubility, stability and bioaccessibility in foods. In this case, heating would have broken down the glycosylation, allowing the determination of free quercetin, just as it is likely to have affected the structure of the other degraded polyphenols on the food matrix. In some cases, the same thing happens when food is cooked and heat can lead to its degradation, modification and aggregation (García-Martínez et al. 2021; Palermo et al. 2014).

Low temperatures can also have an important influence on the content of flavonoids such as quercetin. However, these conditions may depend on the type of plant product, the kind and concentration of antioxidants present in the food, the parameters and technologies of the freezing process, the frozen storage temperature and the thawing conditions (Neri et al. 2020). Quercetin appears to be a more unstable compound than other flavonoids, especially in an organic environment, showing greater losses after storage at both ambient and refrigerated temperatures (Wang et al. 2016). For example, compared to fresh strawberries, frozen strawberries had significantly lower levels of quercetin, even though antioxidant activity tests generally showed no significant differences (Kamiloglu 2019). There is still no consensus on the effect of high-pressure processing on quercetin levels, which vary depending on the composition of the food in which the compound is found. After this treatment, extracts from kiwi fruit showed significantly higher quercetin levels, more than three times higher than untreated samples. Antioxidant capacity and total phenolic content were also increased (Błaszczak et al. 2021). However, 'Golden Delicious' apple extracts from different geographical origin, when subjected to high pressure processing, did not show significant variation in terms of catechin content, while quercetin content showed a tendency towards reduction (Fernández-Jalao et al. 2019).

Oxidation has been reported as the main cause of changes in quercetin during food production and storage. Due to the chemical structure of flavonoids and consequently their functional properties, it is expected that oxidative processes will directly affect these metabolites, which to some extent protect the food matrix due to their antioxidant nature (Lai and Wong 2022; Wang et al. 2016). According to Veiko et al. (2021), the apparent reaction rate constants (s^{-1}, 20 °C) and the activation energies (kJ/mol) of DPPH reduction were 0.34 ± 0.06 and 23.0 ± 2.5 in the case of quercetin, 0.09 ± 0.02 and 32.5 ± 2.5 in the case of catechin, respectively, with quercetin being a preferable reducing substance for the DPPH molecule than catechin.

Another important process is fermentation, as this type of food has been part of the human diet for a long time. In addition to the possibility of increasing shelf life and providing better flavors, fermented foods have relevant functional activity and can add benefits to consumers' health. In this process, several factors can affect the bioactive compounds in foods: food matrix, pH, temperature, fermentation time, cultivation environment and microorganism activity. Depending on the behavior of these factors, an increase in the content of phenolic compounds may be observed, which can be explained, for example, by the action of fermenting microorganisms, capable of degrading high molecular weight phenolic compounds into simple phenolics (Hur et al. 2014; Yang et al. 2023). When analyzing the effect of fermentation on the leaves of different grape varieties, very significant differences in the behavior of these compounds were observed. In general, the fermentation process caused a decrease in the antioxidant content, but a significant increase in the quercetin content was observed in all the varieties analyzed, reaching 445.91 mg/100 g (from 40.32 mg/100 g) in one of them (Al Juhaimi et al. 2019). The same behavior has been observed in other fermented foods, where the increase in quercetin content is associated with the degradation of other polyphenols (Yang et al. 2023).

16.2.1.2 Catechins

Catechins are classified as flavanols and contain two aromatic rings and several hydroxyl groups, on the basis of which they are divided into free catechins and esterified catechins. Because of the hydroxyl in the gallate group, epigallocatechin gallate and epigallocatechin are highly effective free radical scavengers compared to many other standard antioxidants such as ascorbic acid, tocopherol and Trolox. These compounds are usually colorless crystalline substances, often with a bitter, astringent taste. They are readily soluble in polar solvents such as water and methanol, but their solubility may depend on the extraction temperature, time and type of solvent. Catechins can be found in plant tissues bound to sugars or proteins, or they can form polymerized derivatives with variable solubility, and their chemical structures and interactions with other food components are not fully understood. Tea leaves are the best source of catechins among the many sources of plant materials (Bae et al. 2020; Gadkari and Balaraman 2015). Catechins are highly unstable to changes in pH, temperature, oxidation and light. They are also prone to epimerization, resulting in loss of activity (Bae et al. 2020; Gadkari and Balaraman 2015). A study evaluating the catechin content of dry green tea leaves and their infusions showed that the catechin content of both leaves and infusions varied, suggesting that thermal degradation and epimerization were the main causes of changes in tea catechins during processing (Donlao and Ogawa 2019). Conversely, powdered samples of various edible plants and their infusions evaluated after freeze-drying the raw material showed an increase in catechin content and antioxidant activity. In general, these studies report that the total polyphenolic content benefits from low temperatures, being higher in freeze-dried samples than in fresh ones, and remaining so for infusion months after processing (Dilucia et al. 2023; Wilhelm-Mouton et al. 2012).

Radiation, in turn, is also another process capable of affecting these flavonoids. Catechins are sensitive to heat and light, so UV-C irradiation alters the chemical structure of this compound after 20 h of exposure, opening the heterocyclic ring, also consequently altering its antioxidant action (Wilhelm-Mouton et al. 2012). On the other hand, quercetin which has 2,3-double bond, a 3-OH group and four additional aromatic hydroxy groups appear to need much less energy than other flavonoids to generate superoxide anion radical through UV irradiation and X-rays, which may indicate a more efficient antioxidant action. The lack of these specific components in the chemical structure of the metabolite may significantly increase such energy demand (Jiang et al. 2019).

For some foods matrix, enzymatic transformation plays an important fundamental role in the quality of the final product. When observing the impact of enzymatic treatment in the manufacturing processing of black tea, both the catechin content and total polyphenols were markedly affected by the enzymatic treatment methods. The content of total polyphenols, total catechins and individual catechins all decreased (Chiang et al. 2022). Quercetin, likewise, showed a slight tendency to degradation due to enzymatic processing, but with still satisfactory antioxidant performance (Atala et al. 2017). Fermentation can cause changes in the properties of catechins that are not fully understood. When analyzing the effect of fermentation on the leaves of different grape varieties for the catechin content, it has shown a really significant increase in most varieties (from 110.24 to 239.28 mg/100 g, for example), while some varieties have shown a large decrease (from 154.06 to 59.49 mg/100 g), not being possible to observe a pattern of behavior of this metabolite in this fermentation process (Al Juhaimi et al. 2019). However, some authors suggest that catechins can be cleaved during the fermentation process and that the change in pH could be responsible for altering the stability of these molecules, justifying the significant decrease in levels of these bioactives more often observed (Yang et al. 2023). Payne et al. (2010) found that fermentation of cocoa beans resulted in a significant decrease in catechin levels (>80%), while Tu et al. (2005) observed a decrease in catechins in green tea leaves after fermentation (29.6–47.6%).

16.2.1.3 Anthocyanins

Anthocyanins are glycoconjugates of flavonoids characterized by a C6-C3-C6 backbone, usually linked to one or more sugars. Their ability to acylate allows the formation of an even greater number of molecules of this class by linking them to aromatic acids. All these characteristics explain the existence of more than 700 different anthocyanins in nature, with a wide range of structural and physicochemical properties. However, only 6 of these compounds account for the majority of the quantities quantified in plants: pelargonidin, cyanidin, delphinidin, peonidin, petunidin and malvidin (Cheng et al. 2023; Zhang and Jing 2022; Enaru et al. 2021; Li et al. 2023). As anthocyanins are responsible for a wide range of colors in many plants, these compounds can be used as natural colors in the food industry. The main disadvantage is their extremely low stability, which is easily influenced by a wide range of parameters. Anthocyanins are soluble in water and polar solvents, with limited lipid solubility. Increasing methylation or decreasing hydroxylation can enhance their stability. It is also important to note that the cleavage of anthocyanins results in colorless compounds, while polymerization causes darkening and derivatization reactions produce several molecules of different colors. Monomeric anthocyanins generally have a low degree of stability, which is directly related to their structure and depends on the amount, type and nature of ligand interactions (Cheng et al. 2023; Zhang and Jing 2022; Enaru et al. 2021). Co-pigmentation is an important phenomenon in the stability of anthocyanins, as it is a unique property of this class of bioactives in which pigments and other colorless organic compounds or metal ions form molecular or complex associations that modify color intensity. The co-pigments can be flavonoids, alkaloids, amino acids, organic acids, nucleotides, polysaccharides, metals or another anthocyanin (Verma et al. 2023; Castañeda-Ovando et al. 2009).

By inducing different colors, pH variation significantly affects the photochemical degradation of anthocyanins. Reddish tones are observed at values below 7, turning purple at pH values between 7 and 8 and blue at values above 11. Stability is only observed under acidic conditions, with a pH equal to or lower than 3. Also, acylated anthocyanins can be more stable than their non-acylated counterparts.

Although the number of acyl groups in anthocyanins does not in all cases have a direct positive correlation with their stability, the conjugated double bond in anthocyanin generally gives the aromatic acyl group in the anthocyanin molecule a greater ability to absorb light energy and potentially release electrons (Li et al. 2023; Castañeda-Ovando et al. 2009). Overall, thermal processes above 50 °C result in partial or total degradation of anthocyanins. However, lower temperatures can also affect these compounds. Studies carried out on grape extracts have shown that heat treatment at 35 °C reduces the total anthocyanin content by more than 50% compared to the same extract treated at 25 °C. Therefore, degradation appears to be greater as the temperature increases and varies according to their molecular structure: hydroxylation of glycoside ligands decreases the thermal stability of anthocyanins, whereas methylation, glycosylation and acylation increase it. The thermal stability of anthocyanins is also influenced by acidity, being higher at pH values below 3 and lower at neutral values (Cheng et al. 2023; Li et al. 2023; Verma et al. 2023).

Different results can be observed for thermal processing depending on the food matrix. Studies indicate a lower degradation of anthocyanins in Brassicas such as red radish and red cabbage compared to other foods such as purple corn, purple carrots, sweet potatoes, grapes, blackcurrants and elderberries. The increased stability of Brassica anthocyanins may be due to intermolecular copigmentation, as it inhibits chalcone formation and prevents C-ring cleavage, important in the degradation process (Zhang and Jing 2022). The unsaturated chemical structure of anthocyanins facilitates their degradation by direct oxidative mechanisms or by the action of oxidizing enzymes, resulting in colorless or brown substances. Several studies have shown that high temperature and the presence of oxygen are the most damaging combination for the stability of anthocyanins, and it is possible to mitigate thermal degradation by removing oxygen (Cheng et al. 2023; Verma et al. 2023). On the other hand, cold temperatures have been suggested as a way to preserve anthocyanins. Freeze-drying, characterized by a low temperature dehydration process, ensures that the resulting compounds are immune to oxidation or chemical modification. In particular, this technique is used to improve the thermal and color stability of anthocyanins. However, lyophilization has its drawbacks, including its high cost and the length of the process (Verma et al. 2023).

In general, anthocyanins are unstable when exposed to ultraviolet rays, visible light or other types of ionic radiation. In addition to degradation being more pronounced in the presence of oxygen and under fluorescent lighting, anthocyanins with a C5 hydroxyl substitution may be even more susceptible than others. Therefore, packaging that can block light from the visible spectrum and ultraviolet rays can benefit products that want to preserve their properties (Cheng et al. 2023; Verma et al. 2023). The light stability of anthocyanins is also influenced by additives. Ascorbic acid usually eliminates peroxide radicals, exerting a protective influence, but with different effects between acylated and non-acylated anthocyanins (Li et al. 2023). Sugar and its degradation products can accelerate the degradation of anthocyanins, especially at low concentrations. Fructose, arabinose and lactose cause more degradation than glucose, sucrose or maltose (Cheng et al. 2023; Li et al. 2023). The use of sulfites is common in the preservation of fruit-based products, but these substances can cause a reduction in anthocyanin pigment. High concentrations of sulfites (above 10 g/kg) can cause irreversible degradation of anthocyanins, resulting in discoloration. Anthocyanins with two or more vicinal hydroxyl functionalities, mainly cyanidin, delphinidin and petunidin glycosides, can form complexes with divalent or trivalent cations such as Fe^{3+}, Al^{3+}, Mg^{2+}, Sn^{2+} or Cu^{2+} (Zhang and Jing 2022; Verma et al. 2023; Castañeda-Ovando et al. 2009). The addition of potassium ions had no significant effect on blueberries, while magnesium, calcium, copper and aluminum ions improved the color stability of blackberries. High concentrations of sodium, zinc and manganese ions increased the stability of blueberry pigments, whereas iron and lead ions decreased stability (Cheng et al. 2023). Sodium alginate, pectin and maize starch are indicated as additives capable of stabilizing anthocyanins. Studies on blackcurrants showed that starched pectin was even more effective than citrus pectin and apple pectin in retaining anthocyanins after storage (Verma et al. 2023). Anthocyanins are also subject to enzymatic degradation. β-Glucosidase is mentioned as one of the major enzymes affecting the stability of anthocyanins, producing anthocyanidins that are susceptible to further oxidation by polyphenol oxidase and/or peroxidases. Enzymatic action reduces the solubility of anthocyanins, reducing color intensity and forming colorless compounds (Verma et al. 2023).

16.2.2 Phenolic Acids

Phenolic acids make up one-third of the phenolic compounds and are divided into hydroxybenzoic acids and hydroxycinnamic acids based on their two different carbon structures and the position and number of hydroxyl groups on the aromatic ring. The concentration of these compounds in foods varies widely depending on processing methods, with thermal treatment being a prominent factor. They can also be used as functional additives in foods and have been used to retain color, retard microbial growth and inhibit lipid oxidation, thereby extending shelf life (Minatel et al. 2017; Rashmi and Negi 2020; Robbins 2003). Heat treatments introduces variability in phenolic acid concentrations, which is influenced by factors such as cooking time and cooling processes. High temperatures tend to degrade sensitive phenolic acids. Conversely, gentler techniques such as blanching or steam cooking may increase bioavailability, with studies suggesting potential increases of up to 25% (Arfaoui 2021; Khan et al. 2018; Lattanzio 2013). The wide variability in the food matrices in which phenolic acids are found adds complexity to the assessment of biochemical changes during cooking, revealing different effects on polyphenolic content and antioxidant activity. For example, a study of purple potatoes showed that vacuum cooking preserved 92% of the total polyphenolic content, while conventional cooking resulted in a significant reduction (Sun et al. 2021). In a study of peppers, cooking reduced capsaicinoids and increased phenolic acids, highlighting the complex biochemical changes during processing (Hamed et al. 2019).

Appropriately applied thermal treatment, including techniques such as far infrared radiation, can increase the bioavailability of phenolic acids in certain foods, releasing antioxidants in the process (Arfaoui 2021). Challenges remain as certain processing methods, such as beverage pasteurization, can result in the loss of phenolic compounds due to elevated temperatures. The variability in the impact of thermal processing on different foods is evident in studies of kale leaves and apricots, where blanching resulted in a 51% reduction in polyphenol content, while longer cooking resulted in a more substantial 73% reduction (Korus and Lisiewska 2011). Understanding these multifaceted dynamics is critical for optimizing cooking methods and preserving the bioactive components and functional properties of phenolic acids, providing valuable insights into the broader field of food science and nutrition. Although some studies show a loss of phenolic acid content during dehydration, there are opposite effects, including an increase in their antioxidant activity (Sun et al. 2021; Liu et al. 2019; Devic et al. 2010). Despite the variation in results observed in each study, in general, the more modern drying methods appear to be more effective at maintaining the amount of nutrients in dried fruits and vegetables at levels similar to those in fresh vegetables. While the conventional drying process uses heat, promoting moisture movement and heat transfer, lower temperatures and shorter processing times appear to be more beneficial for preserving the properties of bioactives such as polyphenolic acids, including sensory characteristics as color. In summary, controlled dehydration emerges as a strategic approach, preserving or concentrating phenolic acids and thereby enhancing the antioxidant activity of processed foods. This aligns with the growing demand for higher-quality dehydrated food products, driving the development of more efficient dehydration methods and operations (Khan et al. 2018; Minatel et al. 2017; Pereira et al. 2007; Aydin and Gocmen 2015).

Recent research has shown that high-pressure processing is effective in preserving phenolic acid content and minimizing degradation compared to traditional methods. It is estimated that this method is able to preserve an average of 85% of the phenolic compounds, as opposed to the degradation of approximately 20% observed with traditional methods. Even so, it is crucial to recognize that the specific effects of high pressure may vary based on the food matrix and applied pressure conditions. Despite the efficiency of using non-thermal techniques, such as high-pressure processing, to inactivate microbial loads with minimal impact on the nutritional and organoleptic properties of foods, it is important to note that the compounds responsible for antioxidant activity can still be affected during processing and storage (Khan et al. 2018; Pérez-Lamela et al. 2021; Salazar-Orbea et al. 2021). Promising effects of irradiation on phenolic content have been described, depending on the type of food and the intensity and duration of irradiation (Khan et al. 2018; Devic et al. 2010). da Quintão et al. (2020) observed the effects of irradiation on strawberries. Non-treated samples became unfit for consumption after 6 days, whereas samples treated with gamma radiation (0, 0.8, 1.6, and 2.4 kGy) maintained the physicochemical and microbiological qualities during 9 days of storage at 10°. As well as the color of the fruit, the phenolic compounds contents was preserved throughout the storage period. The increase in total phenolic compounds during gamma irradiation was attributed to the release of phenolic compounds from glycosidic components and the degradation of larger phenolic compounds into smaller ones (da Quintão et al. 2020; Sales and Resurreccion 2010). For the best application of this type of processing, tailored approaches are crucial, considering food matrix and desired outcomes.

Chemical or enzymatic pretreatments have been studied as an interesting tool for preserving bioactive compounds. For example, immersion in bisulfite solution removes oxygen through the SO_2 radical, thus avoiding molecular oxygen-induced degradation. Although sulfites are accepted as a safe additive, they can cause allergic effects in sensitive individuals and therefore caution is required in their use. Metabisulfite application before drying caused an increase in phenolic contents, bioaccessible phenolics, antioxidant capacities and phenolic acid levels of pumpkin flour (Aydin and Gocmen 2015). In turn, enzyme resources could release phenolic compounds, contributing to increased antioxidant activity. Enzymatic hydrolysis, for example, improves bioavailability, showing increases between 10% and 25% (Arfaoui 2021; Robbins 2003; Lattanzio 2013; Ribas-Agustí et al. 2018). Studies on the effects of fermentation on polyphenols suggest that fermented foods have higher levels of catechin, gallic acid and quercetin, while legumes have higher phenolic content and better antioxidant activity through this process. This traditional food processing method significantly affects phenolic acids and supports an overall increase in their content. Microbial activity during fermentation plays a critical role in breaking down the cell wall matrix of vegetables and complex phenolic compounds, thereby increasing their bioavailability and bioactivity (Arfaoui 2021; Rashmi and Negi 2020; Leonard et al. 2021; Bhanja Dey et al. 2016).

In summary, existing literature indicate that the best processing methods for greater retention of phenolic acids are pasteurization, canning, and freezing compared to frying, grilling, boiling, and steaming. Refrigerated storage is more efficient at retaining phenolic acids than room temperature. For these compounds, a combination of methods, such as refrigerated storage followed by blanching prior to freezing or the use of irradiation, appears to be the best alternative (Rashmi and Negi 2020; da Quintão et al. 2020).

16.3 Carotenoids

Carotenoids are the most abundant pigments in nature, giving bacteria, fungi, algae, plants and animals their yellow, orange, red and purple colors. These tetraterpenes consist of a polyene chain with conjugated double bonds and can be classified as carotenes or xanthophylls, depending on the presence of oxygen atoms. Among them, lycopene, β-carotene and lutein are the most common in food (Fig. 16.3) (Ngamwonglumlert et al. 2020; Hermanns et al. 2020).

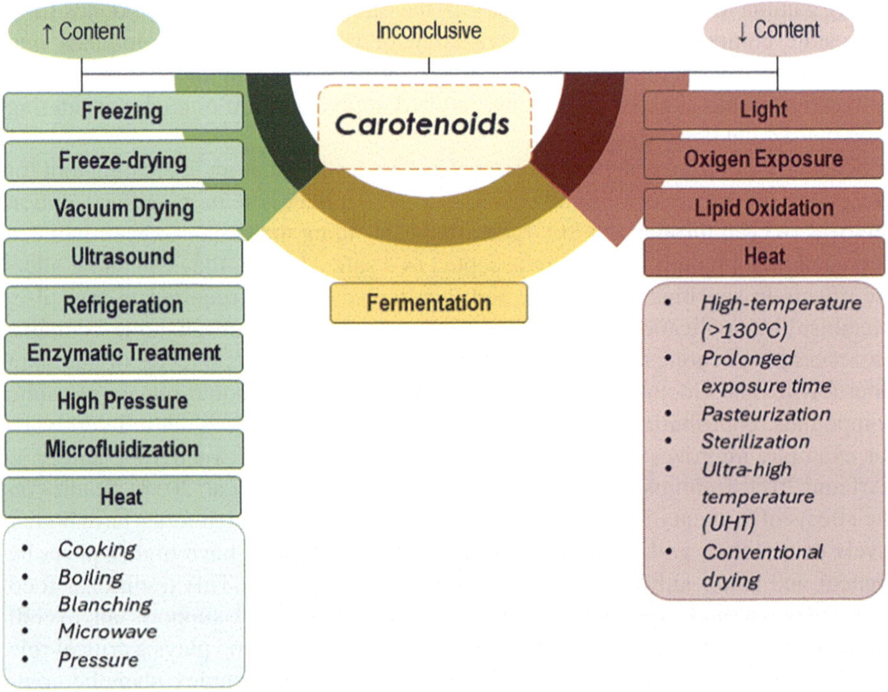

Fig. 16.3 Trends in the effects of processing on carotenoids

16.3.1 Lycopene

The bioavailability of lycopene appears to be related to its isomerism. Since heat has an effect on the modification of these isomeric forms, studies indicate that thermal methods may be able to increase the bioavailability of these bioactive molecules. In addition, heating could facilitate the release of lycopene from the cellular structures of foods, making it more accessible to the digestive tract. Many studies show that cooking tomatoes, one of the best-known sources of lycopene, significantly increases the amount of lycopene available for absorption compared to raw tomatoes, including products such as tomato extracts. However, the effect of thermal processing on lycopene may vary depending on the type of food and the method. Some studies suggest that certain types of processing, such as high-temperature industrial processing (up to 130 °C) or prolonged exposure time, may result in significant losses. Therefore, it is important to consider the specific processing technique when assessing the effect on lycopene concentration and bioavailability (Saini et al. 2020; Motilva et al. 2022; Khan et al. 2021; Martínez-Hernández et al. 2016). A study by Degwale et al. (2022) showed that dehydration by various methods, such as hot air dehydration, resulted in a reduction of lycopene in tomatoes. However, other dehydration methods, such as freeze-drying, were associated with less lycopene loss.

Although lycopene is often associated with heat processing, recent studies have investigated the effect of low-temperature storage on the stability of this carotenoid. A study conducted by Distefano et al. (2020) observed that lycopene maintained its stability during storage of tomatoes at lower temperatures, suggesting that cold conditions may be favorable for preserving the concentration of this antioxidant in foods. This finding has implications for the storage practices of lycopene-rich foods, highlighting the potential benefits of a cold environment in preserving the nutritional properties of these foods.

There is a significant lack of specific literature on the influence of pressure on lycopene stability, but some studies suggest that these methods may alter the structure of plant tissues, indirectly affecting lycopene availability. Researchers have studied biochemical changes in various compounds subjected to high hydrostatic pressure on foods, emphasizing the importance of understanding how this process can affect not only food quality and safety, but also the preservation of bioactive components such as lycopene, which play a critical role in promoting health (Wang et al. 2022; Carpentieri et al. 2023). Oxidation plays a crucial role in altering nutritional quality, including the potential degradation of bioactive compounds such as lycopene. Studies investigating the effects of oxidation on tomatoes have found that lipid oxidation, one of the most common oxidation processes in foods, can contribute to the degradation of lycopene, affecting its stability and bioavailability. As lipid oxidation is reported as major cause of lycopene degradation, these findings highlight the importance of proper storage strategies and the use of natural antioxidants to minimize oxidation and preserve the concentration of lycopene in foods (Xianquan et al. 2005; Liang et al. 2019). Processes such as improper storage, prolonged exposure to air, and the presence of catalytic metals can intensify the oxidation of lycopene. Exposure to ultraviolet (UV) radiation can affect the amount of lycopene in foods due to its ability to modulate metabolic processes and the biosynthesis of bioactive compounds. Increased lycopene synthesis is considered to be a plant defense strategy against the damaging effects of UV radiation. However, it is important to note that the response of lycopene to UV radiation can vary between different plant species and experimental conditions. Therefore, understanding the effects of UV radiation on the amount of lycopene in foods is essential to optimize cultivation and processing strategies that maximize the nutritional benefits of this carotenoid (Martínez-Hernández et al. 2016).

Fermentation is a food processing method that can influence lycopene levels in complex ways, as it involves microbial activity that can affect the structure and availability of bioactive compounds. Although specific literature on the effect of fermentation on lycopene is limited, some studies suggest the action of microorganisms during fermentation may lead to a change in the chemical environment of foods, potentially affecting the concentration of lycopene. A study of the effects of fermentation on lycopene-rich vegetables, such as tomatoes, showed changes in antioxidant composition. Results vary depending on the type of food and fermentation process used, so more in-depth research is needed to understand the dynamics of lycopene during fermentation and its potential impact on the nutritional quality of foods (Lu et al. 2020). Enzymatic action is capable of influencing the release and absorption of lycopene and modifying its concentration. Research on the effects of enzymatic processes, such as enzymatic hydrolysis, shows that enzymatic activity can contribute to the release of lycopene from its bound form, potentially increasing its bioavailability. These findings suggests that enzymatic methods may be promising strategies for improving the extraction and availability of lycopene in processed foods, thus providing optimized nutritional benefits (Sun et al. 2023).

16.3.2 *β-Carotene*

β-carotene, a secondary metabolite synthesized by plants, belongs to the group of non-oxidized carotenoids. With a polyene structure derived from an acyclic form, C40H56, it comprises a long chain of conjugated double bonds (Bogacz-Radomska and Harasym 2018). A common and stable natural pigment, β-carotene is characterized by low water solubility, a high melting point and limited oral bioavailability (Chen et al. 2021). As a primary precursor to vitamin A, β-carotene is found in orange and yellow fruits and green leafy vegetables, including carrots, squash and mangoes (Durante et al. 2014).

Cooking plays a crucial role in preserving and increasing the availability of β-carotene in foods. Studies show that cooking increases the β-carotene content in foods such as sweet corn compared to their raw state, and that boiling, which is preferred to frying, preserves the β-carotene content in different potato varieties (Mladenović-Drinić et al. 2021; Usman et al. 2022). De et al. (2023) compared the effect of three cooking methods (gas oven, pressure cooker and microwave) on carrot β-carotene. The microwave method produced the highest vitamin A equivalent. However, it was observed that chilling and subsequent reheating reduced the bioavailability of the compound. Singh et al. (2022) found a significant increase in β-carotene levels in carrots and spinach after various processing methods, including open pan boiling, blanching, pressure cooking and microwave cooking. Although β-carotene levels increased at high temperatures for short periods of time, continuous exposure to high temperature treatments could result in losses of up to 50%. Thus, studies show thermal processing could make β-carotene more accessible, outweighing potential losses due to isomerization or degradation. In addition, the inactivation or reduction of the activity of enzymes such as peroxidase and lipoxygenase involved in carotenoid degradation helps to preserve these compounds during cooking (Singh et al. 2022; Knockaert et al. 2015).

The effect of industrial thermal processing, including pasteurization and sterilization methods, on β-carotene preservation in foods has been demonstrated in recent studies. Both pasteurization and sterilization resulted in β-carotene losses in sweet potato-based baby purées. However, they showed higher bioaccessibility, giving better results in terms of vitamin A equivalents compared to homemade purees that had not undergone these processes (Dhuique-Mayer et al. 2018). On the other hand, ultra-high pressure (UHP) sterilization was the only technique that significantly preserved the β-carotene content in freshly squeezed lettuce juice compared to pasteurization and ultra-high temperature (UHT) sterilization, which significantly reduced β-carotene levels (Zhang et al. 2024). Solar and air drying are widely used methods for food preservation. However, these techniques can result in quality and nutritional losses due to prolonged drying times and/or high temperatures. In addition, the presence of oxygen during the process can lead to the oxidation of carotenoid pigments, including β-carotene (Ngamwonglumlert et al. 2020; Pott et al. 2003). When comparing different drying methods such as sun drying, convective drying or microwave drying in apricots, microwave-convective drying stands out as the best option for increasing β-carotene content, maintaining antioxidant activity and for its efficiency in terms of drying time. When comparing hot air-drying methods, freeze drying, vacuum drying and ultrasound-assisted vacuum drying in rose hips, freeze drying resulted in the highest β-carotene levels, followed by vacuum drying, ultrasound-assisted vacuum drying and hot air drying (Goztepe et al. 2022). Microfluidization, a type of non-thermal high-pressure homogenizer, has gained a prominent role in food processing due to advantages such as fast processing times, low temperature effects and minimal nutritional loss (Kavinila et al. 2023; Bhargava et al. 2021). The microfluidization process did not affect the β-carotene content of sea buckthorn juice, which remained constant at different treatment pressures. In addition, high-pressure microfluidization resulted in a significant increase in the β-carotene content of carrot juice. This improvement can be attributed to the intense shear and impact during the process, which ruptures cell membranes and releases carotenoids (Koley et al. 2020). Fermentation is a potential process for producing new products with functional properties. A study of alfalfa fermentation showed that although there was an initial loss of β-carotene in the first 3 days, the level gradually increased during the process and was positively correlated with *Lactobacillus* species such as *L. parabuchneri*, *L. kunkeei* and *L. kullabergensis*. However, factors such as high pH and low levels of propionic acid can reduce the retention of beta-carotene (Zong et al. 2023). Lactic fermentation using *L. plantarum* has been shown to be an effective method of preserving β-carotene, achieving a retention of 93.97% of β-carotene in sweet potatoes (Oloo et al. 2014). However, changes in β-carotene content during fermentation are highly variable, depending on the plant material and fermentation conditions (Mapelli-Brahm et al. 2020).

In the context of the mechanical extrusion process, a recent review article elucidated the conflicting results regarding the modification of β-carotene, highlighting the need for expanded investigation of the impact of extrusion on terpenes, given the paucity of research on this aspect (Mironeasa et al. 2023).

16.3.3 Lutein

Lutein is known for its yellow/orange color with a complex structure and long carbon chain that contribute to its biological and pharmacological properties (Tan et al. 2016; Sawa et al. 2020). This compound is unstable under certain processing conditions, including but not limited to heat, exposure to light and oxidation, which can affect its bioaccessibility and bioavailability (Yi et al. 2016). Specific information on the effects of different processing methods on lutein is still limited in the existing scientific literature. The high temperature used in pasteurization (86 °C) and the low storage temperature (6 °C) contributed to a significant reduction in the lutein content of ubu juice, with the same effect being observed in Valencia orange juice subjected to heat treatments in a variable range between 95 °C and 105 °C. Although some studies indicate that certain thermal methods can increase the bioavailability of lutein by breaking down the cell walls of the plant, favoring the release of the compound from the chromoplasts or even causing the isomerization of compounds such as zeaxanthin, others suggest that prolonged or intense heat can lead to its degradation (Gama and de Sylos 2007; Ribeiro et al. 2017; de Dutra et al. 2012). This can be observed in carrot pulp subjected to an extrusion process at different temperatures, in which there was a reduction in its content, but there was a significant increase in the *in vitro* bioavailability of the compound in the treatment subjected to a higher temperature (160 °C), suggesting that lutein has a relatively higher heat resistance, similar or even higher to the stability of other carotenoid compounds reported in dehydrated foods, as observed in pumpkins and spinach (Ortak et al. 2017; Song et al. 2018; Syamila et al. 2019; Provesi et al. 2011). The few studies of more modern processing methods, such as non-ionizing radiation, show significant differences in the effects on lutein compared to more conventional methods. An increase in lutein content has been observed in mustard leaves, while other matrices, such as broccoli, show a significant decrease (Zhang and Hamauzu 2004; Mehmood and Zeb 2023). A positive effect can also be observed with pressurization, where there is an inevitable increase in temperature due to the adiabatic heat of the technique, but the preservation of lutein content can be observed in food matrices such as broccoli, green beans and fresh cheese (McInerney et al. 2007; Sánchez et al. 2014; Gouvea et al. 2019). It is also noteworthy that lutein is susceptible to oxidative processes of enzymatic face and reactive oxygen species, due to the presence of double bonds in its molecular structure. In corn samples, there was an increase in the content of the molecule after thermal and mechanical processing (bleaching, drying and grinding), directly associated with the reduction in activity or inactivation of enzymes such as peroxidase and lipoxygenase, related to the degradation of carotenoids, suggesting that bleaching the corn before consumption can improve the extractability and bioavailability of the compound. Increasing its concentration and utilization by the body (Mamatha et al. 2012).

Changes in the chemical composition of foods affect the availability of these molecules and the enzymatic activity associated with their synthesis and degradation, as observed in fermentation processes, which can affect the levels of compounds such as lutein in complex ways by altering the chemical environment. Acidification caused by the metabolism of microorganisms can affect the concentration of these compounds, as observed in goat's milk yogurt, where the more acidic the environment, the greater the reduction in lutein concentration during storage (Barros et al. 2020). Since studies show rapid degradation of lutein and other carotenoids during storage, some precautions can be taken to reduce damage, such as control of exposure to oxygen, moisture, temperature and light (Morales-De La Peña et al. 2011; Provesi et al. 2011).

16.4 Conclusions

Most studies on the processing of bioactive compounds aimed at studying the stability of these molecules in different situations, in isolation or within a food matrix. The change in concentration of these substances is evaluated using various methods, often accompanied by the antioxidant activity of the samples. Although the interaction with different types of food matrices further increases the range of possibilities, it can be seen that there is a trend in the chemical behavior of each class of bioactive compounds, which guides the best choices for the use of these resources in the production of foods or dietary supplements. However, it is important to emphasize that the changes observed in the levels of each specific molecule may vary widely depending on its interaction with other substances, variations in each processing method, and its individual physicochemical characteristics. Due to the breadth of this topic, the literature on the influence of these modifications on the biological effects of bioactive compounds is still rather scarce and more studies are needed to better understand the effects of processing and its consequences on functional properties.

References

Ahmed M, Eun JB (2018) Flavonoids in fruits and vegetables after thermal and nonthermal processing: a review. Crit Rev Food Sci Nutr 58(18):3159–3188. https://doi.org/10.1080/10408398.2017.1353480

Al Juhaimi F, Uslu N, Özcan MM et al (2019) Effect of fermentation on antioxidant activity and phenolic compounds of the leaves of five grape varieties. J Food Process Preserv 43(7). https://doi.org/10.1111/jfpp.13979

Alothman M, Bhat R, Karim AA (2009) UV radiation-induced changes of antioxidant capacity of fresh-cut tropical fruits. Innov Food Sci Emerg Technol 10(4):512–516. https://doi.org/10.1016/j.ifset.2009.03.004

Arfaoui L (2021) Dietary plant polyphenols: effects of food processing on their content and bioavailability. Molecules 26(10). https://doi.org/10.3390/molecules26102959

Asami DK, Hong YJ, Barrett DM et al (2003) Processing-induced changes in total phenolics and procyanidins in clingstone peaches. J Sci Food Agric 83(1):56–63. https://doi.org/10.1002/jsfa.1275

Atala E, Fuentes J, Wehrhahn MJ et al (2017) Quercetin and related flavonoids conserve their antioxidant properties despite undergoing chemical or enzymatic oxidation. Food Chem 234:479–485. https://doi.org/10.1016/j.foodchem.2017.05.023

Aydin E, Gocmen D (2015) The influences of drying method and metabisulfite pre-treatment onthe color, functional properties and phenolic acids contents and bioaccessibility of pumpkin flour. LWT 60(1):385–392. https://doi.org/10.1016/j.lwt.2014.08.025

Bae J, Kim N, Shin Y et al (2020) Activity of catechins and their applications. Biomedical. Dermatology 4(1). https://doi.org/10.1186/s41702-020-0057-8

Barros RF, Torres FR, Silva PHF et al (2020) Lutein as a functional ingredient in sheep milk yogurt: development, characterization and extraction recovery. Food Sci Technol 40:683–690. https://doi.org/10.1590/fst.36019

Bhanja Dey T, Chakraborty S, Jain KK et al (2016) Antioxidant Phenolics and their microbial production by submerged and solid state fermentation process: a review. Trends Food Sci Technol 53(60–74):60–74. https://doi.org/10.1016/j.tifs.2016.04.007

Bhargava N, Mor RS, Kumar K et al (2021) Advances in application of ultrasound in food processing: a review. Ultrason Sonochem 70:105293. https://doi.org/10.1016/j.ultsonch.2020.105293

Błaszczak W, Latocha P, Jeż M et al (2021) The impact of high-pressure processing on the polyphenol profile and anti-glycaemic, anti-hypertensive and anti-cholinergic activities of extracts obtained from kiwiberry (Actinidia arguta) fruits. Food Chem 343. https://doi.org/10.1016/j.foodchem.2020.128421

Bogacz-Radomska L, Harasym J (2018) β-carotene—properties and production methods. Food Qual Saf 2(2):69–74. https://doi.org/10.1093/fqsafe/fyy004

Carpentieri S, Ferrari G, Donsì F (2023) High-pressure homogenization for enhanced bioactive recovery from tomato processing by-products and improved lycopene bioaccessibility during in vitro digestion. Antioxidants 12(10):1855. https://doi.org/10.3390/antiox12101855

Castañeda-Ovando A, de Pacheco-Hernández ML, Páez-Hernández ME et al (2009) Chemical studies of anthocyanins: a review. Food Chem 113(4):859–871. https://doi.org/10.1016/j.foodchem.2008.09.001

Chen Q-H, Wu B-K, Pan D et al (2021) Beta-carotene and its protective effect on gastric cancer. World J Clin Cases 9(23):6591–6607. https://doi.org/10.12998/wjcc.v9.i23.6591

Cheng Y, Liu J, Li L et al (2023) Advances in embedding techniques of anthocyanins: improving stability, bioactivity and bioavailability. Food Chem X 20:100983. https://doi.org/10.1016/j.fochx.2023.100983

Chiang SH, Yang KM, Wang SY et al (2022) Enzymatic treatment in black tea manufacturing processing: impact on bioactive compounds, quality, and bioactivities of black tea. LWT 163:113560. https://doi.org/10.1016/j.lwt.2022.113560

Cilla A, Bosch L, Barberá R et al (2018) Effect of processing on the bioaccessibility of bioactive compounds – a review focusing on carotenoids, minerals, ascorbic acid, tocopherols and polyphenols. J Food Compos Anal 68:3–15. https://doi.org/10.1016/j.jfca.2017.01.009

da Quintão ALC, Silva VDM, de Arquelau PBF et al (2020) Influence of gamma radiation treatment on the profile of phenolic compounds and on the quality parameters of strawberries cv. Albion during storage. Res Soc Dev 9(7):e991975147. https://doi.org/10.33448/rsd-v9i7.5147

de Dutra AS, Furtado AAL, Pacheco S et al (2012) Efeito do tratamento térmico na concentração de carotenóides, compostos fenólicos, ácido ascórbico e capacidade antioxidante do suco de tangerina murcote. Braz J Food Technol 15(3):198–207. https://doi.org/10.1590/s1981-67232012005000012

De S, Paul S, Manna A et al (2023) Phenolic phytochemicals for prevention and treatment of colorectal cancer: a critical evaluation of in vivo studies. Cancers (Basel) 15(3):10.3390/cancers15030993

Degwale A, Asrat F, Eniyew K et al (2022) Influence of dehydration temperature and time on physicochemical properties of tomato (Solanum lycopersicum L.) powder. Front Sustain Food Syst 6. https://doi.org/10.3389/fsufs.2022.839385

Devic E, Guyot S, Daudin J-D et al (2010) Effect of temperature and cultivar on polyphenol retention and mass transfer during osmotic dehydration of apples. J Agric Food Chem 58(1):606–614. https://doi.org/10.1021/jf903006g

Dhuique-Mayer C, Servent A, Messan C et al (2018) Bioaccessibility of biofortified sweet potato carotenoids in baby food: impact of manufacturing process. Front Nutr 5. https://doi.org/10.3389/fnut.2018.00098

Dilucia F, Rutigliano M, Libutti A et al (2023) Effect of a novel pretreatment before freeze-drying process on the antioxidant activity and polyphenol content of Malva sylvestris L., Calendula officinalis L., and Asparagus officinalis L. Infusions. Food Bioprocess Technol 16(10):2113–2125. https://doi.org/10.1007/s11947-023-03035-y

Distefano M, Arena E, Mauro RP et al (2020) Effects of genotype, storage temperature and time on quality and compositional traits of cherry tomato. Foods 9(12). https://doi.org/10.3390/foods9121729

Donlao N, Ogawa Y (2019) The influence of processing conditions on catechin, caffeine and chlorophyll contents of green tea (Camelia sinensis) leaves and infusions. LWT 116:108567. https://doi.org/10.1016/j.lwt.2019.108567

Durante M, Lenucci MS, D'Amico L et al (2014) Effect of drying and co-matrix addition on the yield and quality of supercritical CO_2 extracted pumpkin (Cucurbita moschata Duch.) oil. Food Chem 148:314–320. https://doi.org/10.1016/j.foodchem.2013.10.051

Enaru B, Drețcanu G, Pop TD et al (2021) Anthocyanins: factors affecting their stability and degradation. Antioxidants 10(12). https://doi.org/10.3390/antiox10121967

Fernández-Jalao I, Sánchez-Moreno C, De Ancos B (2019) Effect of high-pressure processing on flavonoids, hydroxycinnamic acids, dihydrochalcones and antioxidant activity of apple 'Golden Delicious' from different geographical origin. Innov Food Sci Emerg Technol 51:20–31. https://doi.org/10.1016/j.ifset.2018.06.002

Fu Y, Liu W, Soladoye OP (2021) Towards innovative food processing of flavonoid compounds: insights into stability and bioactivity. LWT 150:111968. https://doi.org/10.1016/j.lwt.2021.111968

Gadkari PV, Balaraman M (2015) Catechins: sources, extraction and encapsulation: a review. Food Bioprod Process 93:122–138. https://doi.org/10.1016/j.fbp.2013.12.004

Gama JJT, de Sylos CM (2007) Effect of thermal pasteurization and concentration on carotenoid composition of Brazilian Valencia orange juice. Food Chem 100(4):1686–1690. https://doi.org/10.1016/j.foodchem.2005.01.062

García-Martínez DJ, Arroyo-Hernández M, Posada-Ayala M et al (2021) The high content of quercetin and catechin in airen grape juice supports its application in functional food production. Foods 10(7). https://doi.org/10.3390/foods10071532

Gouvea FS, Walter EHM, Ferreira EHR et al (2019) Effects of carrot incorporation and high hydrostatic pressure processing on fresh cheese: Antilisterial activity, carotenoid degradation, and sensory characteristics. Food Sci Technol Int 25(7):597–607. https://doi.org/10.1177/1082013219843394

Goztepe B, Kayacan S, Bozkurt F et al (2022) Drying kinetics, total bioactive compounds, antioxidant activity, phenolic profile, lycopene and β-carotene content and color quality of rosehip dehydrated by different methods. LWT 153:112476. https://doi.org/10.1016/j.lwt.2021.112476

Guerrouj K, Sánchez-Rubio M, Taboada-Rodríguez A et al (2016) Sonication at mild temperatures enhances bioactive compounds and microbiological quality of orange juice. Food Bioprod Process 99:20–28. https://doi.org/10.1016/j.fbp.2016.03.007

Hamed M, Kalita D, Bartolo ME et al (2019) Capsaicinoids, polyphenols and antioxidant activities of Capsicum annuum: comparative study of the effect of ripening stage and cooking methods. Antioxidants 8(9):10.3390/antiox8090364

Hayat K, Zhang X, Farooq U et al (2010) Effect of microwave treatment on phenolic content and antioxidant activity of citrus mandarin pomace. Food Chem 123(2):423–429. https://doi.org/10.1016/j.foodchem.2010.04.060

He Z, Tao Y, Zeng M et al (2016) High pressure homogenization processing, thermal treatment and milk matrix affect in vitro bioaccessibility of phenolics in apple, grape and orange juice to different extents. Food Chem 200:107–116. https://doi.org/10.1016/j.foodchem.2016.01.045

Hermanns AS, Zhou X, Xu Q et al (2020) Carotenoid pigment accumulation in horticultural plants. Hortic Plant J 6(6):343–360. https://doi.org/10.1016/j.hpj.2020.10.002

Huang YC, Chang YH, Shao YY (2006) Effects of genotype and treatment on the antioxidant activity of sweet potato in Taiwan. Food Chem 98(3):529–538. https://doi.org/10.1016/j.foodchem.2005.05.083

Huebbe P, Rimbach G (2020) Historical reflection of food processing and the role of legumes as part of a healthy balanced diet. Foods 9(8). https://doi.org/10.3390/foods9081056

Hur SJ, Lee SY, Kim YC et al (2014) Effect of fermentation on the antioxidant activity in plant-based foods. Food Chem 160(346–356):346–356. https://doi.org/10.1016/j.foodchem.2014.03.112

Igual M, García-Martínez E, Camacho MM et al (2011) Changes in flavonoid content of grapefruit juice caused by thermal treatment and storage. Innov Food Sci Emerg Technol 12(2):153–162. https://doi.org/10.1016/j.ifset.2010.12.010

Ioannou I, Hafsa I, Hamdi S et al (2012) Review of the effects of food processing and formulation on flavonol and anthocyanin behaviour. J Food Eng 111(2):208–217. https://doi.org/10.1016/j.jfoodeng.2012.02.006

Jiang L, Yanase E, Mori T et al (2019) Relationship between flavonoid structure and reactive oxygen species generation upon ultraviolet and X-ray irradiation. J Photochem Photobiol A Chem 384. https://doi.org/10.1016/j.jphotochem.2019.112044

Kamiloglu S (2019) Effect of different freezing methods on the bioaccessibility of strawberry polyphenols. Int J Food Sci Technol 54(8):2652–2660. https://doi.org/10.1111/ijfs.14249

Kavinila S, Nimbkar S, Moses JA et al (2023) Emerging applications of microfluidization in the food industry. J Agric Food Res 12:100537. https://doi.org/10.1016/j.jafr.2023.100537

Khan MK, Ahmad K, Hassan S et al (2018) Effect of novel technologies on polyphenols during food processing. Innov Food Sci Emerg Technol 45(361–381):361–381. https://doi.org/10.1016/j.ifset.2017.12.006

Khan UM, Sevindik M, Zarrabi A et al (2021) Lycopene: food sources, biological activities, and human health benefits. Oxidative Med Cell Longev 2021. https://doi.org/10.1155/2021/2713511

Knockaert G, Lemmens L, Van Buggenhout S et al (2015) Changes in β-carotene during processing of carrots. In: Processing and impact on active components in food. Elsevier, pp 11–16. https://doi.org/10.1016/B978-0-12-404699-3.00002-0

Koley TK, Nishad J, Kaur C et al (2020) Effect of high-pressure microfluidization on nutritional quality of carrot (Daucus carota L.) juice. J Food Sci Technol 57(6):2159–2168. https://doi.org/10.1007/s13197-020-04251-6

Korus A, Lisiewska Z (2011) Effect of preliminary processing and method of preservation on the content of selected antioxidative compounds in kale (Brassica oleracea L. var. acephala) leaves. Food Chem 129(1):149–154. https://doi.org/10.1016/j.foodchem.2011.04.048

Lai WF, Wong WT (2022) Design and optimization of quercetin-based functional foods. Crit Rev Food Sci Nutr 62(26):7319–7335. https://doi.org/10.1080/10408398.2021.1913569

Lattanzio V (2013) Phenolic compounds: introduction. In: Ramawat KG, Mérillon J-M (eds) Natural products. Springer, Berlin/Heidelberg, pp 1543–1580. https://doi.org/10.1007/978-3-642-22144-6_57

Leonard W, Zhang P, Ying D et al (2021) Fermentation transforms the phenolic profiles and bioactivities of plant-based foods. Biotechnol Adv 49:107763. https://doi.org/10.1016/j.biotechadv.2021.107763

Li Q, Zhang F, Wang Z et al (2023) Advances in the preparation, stability, metabolism, and physiological roles of Anthocyanins: a review. Foods 12(21). https://doi.org/10.3390/foods12213969

Liang X, Ma C, Yan X et al (2019) Advances in research on bioactivity, metabolism, stability and delivery systems of lycopene. Trends Food Sci Technol 93. https://doi.org/10.1016/j.tifs.2019.08.019

Liu H, Liu J, Lv Z et al (2019) Effect of dehydration techniques on bioactive compounds in hawthorn slices and their correlations with antioxidant properties. J Food Sci Technol 56(5):2446–2457. https://doi.org/10.1007/s13197-019-03720-x

Lu Y, Mu K, McClements DJ et al (2020) Fermentation of tomato juice improves in vitro bioaccessibility of lycopene. J Funct Foods 71:104020. https://doi.org/10.1016/j.jff.2020.104020

Mamatha BS, Arunkumar R, Baskaran V (2012) Effect of processing on major carotenoid levels in corn (Zea mays) and selected vegetables: bioavailability of lutein and Zeaxanthin from processed corn in mice. Food Bioprocess Technol 5(4):1355–1363. https://doi.org/10.1007/s11947-010-0403-8

Mapelli-Brahm P, Barba FJ, Remize F et al (2020) The impact of fermentation processes on the production, retention and bioavailability of carotenoids: an overview. Trends Food Sci Technol 99:389–401. https://doi.org/10.1016/j.tifs.2020.03.013

Martínez-Hernández GB, Boluda-Aguilar M, Taboada-Rodríguez A et al (2016) Processing, packaging, and storage of tomato products: influence on the lycopene content. Food Eng Rev 8(1). https://doi.org/10.1007/s12393-015-9113-3

McInerney JK, Seccafien CA, Stewart CM et al (2007) Effects of high pressure processing on antioxidant activity, and total carotenoid content and availability, in vegetables. Innov Food Sci Emerg Technol 8(4):543–548. https://doi.org/10.1016/j.ifset.2007.04.005

Mehmood A, Zeb A (2023) Effects of different processing conditions on the carotenoid's composition, phenolic contents, and antioxidant activities of Brassica campestris leaves. Heliyon 9(11). https://doi.org/10.1016/j.heliyon.2023.e21191

Minatel IO, Borges CV, Ferreira MI et al (2017) Phenolic compounds: functional properties, impact of processing and bioavailability. In: Phenolic compounds – biological activity. InTech. https://doi.org/10.5772/66368

Mironeasa S, Coţovanu I, Mironeasa C et al (2023) A review of the changes produced by extrusion cooking on the bioactive compounds from vegetal sources. Antioxidants 12(7):1453. https://doi.org/10.3390/antiox12071453

Mladenović-Drinić S, Vukadinović J, Srdić J et al (2021) Effect of cooking on the content of carotenoids and tocopherols in sweet corn. Food Feed Res 48(2):119–129. https://doi.org/10.5937/ffr48-31960

Morales-De La Peña M, Salvia-Trujillo L, Rojas-Graü MA et al (2011) Changes on phenolic and carotenoid composition of high intensity pulsed electric field and thermally treated fruit juice-soymilk beverages during refrigerated storage. Food Chem 129(3):982–990. https://doi.org/10.1016/j.foodchem.2011.05.058

Motilva M-J, Costa BS, García MO et al (2022) Impact of industrial drying processes on phenolic compounds and lycopene in tomato by-products. SSRN Electron J. https://doi.org/10.2139/ssrn.4132193

Neri L, Faieta M, Di Mattia C et al (2020) Antioxidant activity in frozen plant foods: effect of Cryoprotectants, freezing process and frozen storage. Foods 9(12). https://doi.org/10.3390/foods9121886

Ngamwonglumlert L, Devahastin S, Chiewchan N et al (2020) Plant carotenoids evolution during cultivation, postharvest storage, and food processing: a review. Compr Rev Food Sci Food Saf 19(4):1561–1604. https://doi.org/10.1111/1541-4337.12564

Odriozola-Serrano I, Soliva-Fortuny R, Martín-Belloso O (2008) Phenolic acids, flavonoids, vitamin C and antioxidant capacity of strawberry juices processed by high-intensity pulsed electric fields or heat treatments. Eur Food Res Technol 228(2):239–248. https://doi.org/10.1007/s00217-008-0928-5

Oloo BO, Shitandi AA, Mahungu S et al (2014) Effects of lactic acid fermentation on the retention of Β-carotene content in Orange fleshed sweet potatoes. Int J Food Stud 3(1). https://doi.org/10.7455/ijfs/3.1.2014.a2

Ortak M, Caltinoglu C, Sensoy I et al (2017) Changes in functional properties and in vitro bioaccessibilities of β-carotene and lutein after extrusion processing. J Food Sci Technol 54(11):3543–3551. https://doi.org/10.1007/s13197-017-2812-4

Palermo M, Pellegrini N, Fogliano V (2014) The effect of cooking on the phytochemical content of vegetables. J Sci Food Agric 94(6):1057–1070. https://doi.org/10.1002/jsfa.6478

Payne MJ, Hurst WJ, Miller KB et al (2010) Impact of fermentation, drying, roasting, and dutch processing on epicatechin and catechin content of cacao beans and cocoa ingredients. J Agric Food Chem 58(19):10518–10527. https://doi.org/10.1021/jf102391q

Peña MM, Salvia-Trujillo L, Rojas-Graü MA et al (2017) Effects of high intensity pulsed electric fields or thermal pasteurization and refrigerated storage on antioxidant compounds of fruit juice-Milk beverages. Part I: phenolic acids and flavonoids. J Food Process Preserv 41(3). https://doi.org/10.1111/jfpp.12912

Pereira NR, Marsaioli A, Ahrné LM (2007) Effect of microwave power, air velocity and temperature on the final drying of osmotically dehydrated bananas. J Food Eng 81(1):79–87. https://doi.org/10.1016/j.jfoodeng.2006.09.025

Pérez-Lamela C, Franco I, Falqué E (2021) Impact of high-pressure processing on antioxidant activity during storage of fruits and fruit products: a review. Molecules 26(17):5265. https://doi.org/10.3390/molecules26175265

Pott I, Marx M, Neidhart S et al (2003) Quantitative determination of β-carotene stereoisomers in fresh, dried, and solar-dried mangoes (Mangifera indica L.). J Agric Food Chem 51(16):4527–4531. https://doi.org/10.1021/jf034084h

Provesi JG, Dias CO, Amante ER (2011) Changes in carotenoids during processing and storage of pumpkin puree. Food Chem 128(1):195–202. https://doi.org/10.1016/j.foodchem.2011.03.027

Ranilla LG, Genovese MI, Lajolo FM (2009) Effect of different cooking conditions on phenolic compounds and antioxidant capacity of some selected Brazilian bean (Phaseolus vulgaris L.) cultivars. J Agric Food Chem 57(13):5734–5742. https://doi.org/10.1021/jf900527v

Rashmi HB, Negi PS (2020) Phenolic acids from vegetables: a review on processing stability and health benefits. Food Res Int 136:109298. https://doi.org/10.1016/j.foodres.2020.109298

Ribas-Agustí A, Martín-Belloso O, Soliva-Fortuny R et al (2018) Food processing strategies to enhance phenolic compounds bioaccessibility and bioavailability in plant-based foods. Crit Rev Food Sci Nutr 58(15):2531–2548. https://doi.org/10.1080/10408398.2017.1331200

Ribeiro LO, Pontes SM, Ribeiro APO et al (2017) Avaliação do armazenamento a frio sobre os compostos bioativos e as características físico-químicas e microbiológicas do suco de umbu pasteurizado. Braz J Food Technol 20. https://doi.org/10.1590/1981-6723.9515

Robbins RJ (2003) Phenolic acids in foods: an overview of analytical methodology. J Agric Food Chem 51(10):2866–2887. https://doi.org/10.1021/jf026182t

Ruiz-Rodriguez A, Marín FR, Ocaña A et al (2008) Effect of domestic processing on bioactive compounds. Phytochem Rev 7(2):345–384. https://doi.org/10.1007/s11101-007-9073-1

Saini RK, Alaa AED, Roohinejad S et al (2020) Chemical stability of lycopene in processed products: a review of the effects of processing methods and modern preservation strategies. J Agric Food Chem 68(3). https://doi.org/10.1021/acs.jafc.9b06669

Salazar-Orbea GL, García-Villalba R, Tomás-Barberán FA et al (2021) High–pressure processing vs. thermal treatment: effect on the stability of polyphenols in strawberry and apple products. Foods 10(12):2919. https://doi.org/10.3390/foods10122919

Sales JM, Resurreccion AVA (2010) Phenolic profile, antioxidants, and sensory acceptance of bioactive-enhanced peanuts using ultrasound and UV. Food Chem 122(3):795–803. https://doi.org/10.1016/j.foodchem.2010.03.058

Sánchez C, Baranda AB, De Marañón IM (2014) The effect of high pressure and high temperature processing on carotenoids and chlorophylls content in some vegetables. Food Chem 163:37–45. https://doi.org/10.1016/j.foodchem.2014.04.041

Santhirasegaram V, Razali Z, George DS et al (2015) Comparison of UV-C treatment and thermal pasteurization on quality of Chokanan mango (Mangifera indica L.) juice. Food Bioprod Process 94:313–321. https://doi.org/10.1016/j.fbp.2014.03.011

Sawa M, Shunto T, Nishiyama I et al (2020) Effects of lutein supplementation in Japanese patients with unilateral age-related macular degeneration: the Sakai lutein study. Sci Rep 10(1). https://doi.org/10.1038/s41598-020-62483-0

Singh S, Verma L, Agarwal A et al (2022) Effect of different processing conditions on the carotene content of different vegetables. Ann Food Sci Technol 3(2):322–330

Song J, Wei Q, Wang X et al (2018) Degradation of carotenoids in dehydrated pumpkins as affected by different storage conditions. Food Res Int 107:130–136. https://doi.org/10.1016/j.foodres.2018.02.024

Sun Q, Du M, Navarre DA et al (2021) Effect of cooking methods on bioactivity of polyphenols in purple potatoes. Antioxidants 10(8):10.3390/antiox10081176

Sun R, Chen S, Chen X et al (2023) Enzymatic treatment to improve permeability and quality of cherry tomatoes for production of dried products. J Sci Food Agric 104:2718–2727. https://doi.org/10.1002/jsfa.13156

Syamila M, Gedi MA, Briars R et al (2019) Effect of temperature, oxygen and light on the degradation of β-carotene, lutein and α-tocopherol in spray-dried spinach juice powder during storage. Food Chem 284:188–197. https://doi.org/10.1016/j.foodchem.2019.01.055

Tan TB, Yussof NS, Abas F et al (2016) Forming a lutein nanodispersion via solvent displacement method: the effects of processing parameters and emulsifiers with different stabilizing mechanisms. Food Chem 194:416–423. https://doi.org/10.1016/j.foodchem.2015.08.045

Tu YY, Xia HL, Watanabe N (2005) Changes in Catechins during the fermentation of green tea. Appl Biochem Microbiol 41(6):574–577. https://doi.org/10.1007/s10438-005-0104-7

Usman A, Halliru K, Lawal U (2022) Effects of boiling and frying on nutritional and antioxidant properties of three sweet potato cultivars (Ipomoea Batatas L (Lam)) in North-Western Nigeria. Umaru Musa Yaradua University

Veiko AG, Lapshina EA, Zavodnik IB (2021) Comparative analysis of molecular properties and reactions with oxidants for quercetin, catechin, and naringenin. Mol Cell Biochem 476(12):4287–4299. https://doi.org/10.1007/s11010-021-04243-w

Verma D, Sharma N, Malhotra U (2023) Structural chemistry and stability of anthocyanins. Pharma Innov 12(7):1366–1373. https://doi.org/10.22271/tpi.2023.v12.i7p.21416

Viña SZ, Chaves AR (2008) Effect of heat treatment and refrigerated storage on antioxidant properties of pre-cut celery (Apium graveolens L.). Int J Food Sci Technol 43(1):44–51. https://doi.org/10.1111/j.1365-2621.2006.01380.x

Wang W, Sun C, Mao L et al (2016) The biological activities, chemical stability, metabolism and delivery Systems of Quercetin: a review. Trends Food Sci Technol 56:21–38. https://doi.org/10.1016/j.tifs.2016.07.004

Wang X, Chen F, Ma L et al (2022) Non-volatile and volatile metabolic profiling of tomato juice processed by high-hydrostatic-pressure and high-temperature short-time. Food Chem 371. https://doi.org/10.1016/j.foodchem.2021.131161

Wilhelm-Mouton A, Bonnet SL, Ding Y et al (2012) Photochemistry synthesis. Part 2: Enantiomerically pure polyhydroxy-1,1,3-triarylpropan-2-ols. J Photochem Photobiol A Chem 227(1):18–24. https://doi.org/10.1016/j.jphotochem.2011.10.019

Xianquan S, Shi J, Kakuda Y et al (2005) Stability of lycopene during food processing and storage. J Med Food 8(4). https://doi.org/10.1089/jmf.2005.8.413

Yang F, Chen C, Ni D et al (2023) Effects of fermentation on bioactivity and the composition of polyphenols contained in polyphenol-rich foods: a review. Foods 12(17). https://doi.org/10.3390/foods12173315

Yi J, Fan Y, Yokoyama W et al (2016) Characterization of milk proteins-lutein complexes and the impact on lutein chemical stability. Food Chem 200:91–97. https://doi.org/10.1016/j.foodchem.2016.01.035

Zhang D, Hamauzu Y (2004) Phenolics, ascorbic acid, carotenoids and antioxidant activity of broccoli and their changes during conventional and microwave cooking. Food Chem 88(4):503–509. https://doi.org/10.1016/j.foodchem.2004.01.065

Zhang N, Jing P (2022) Anthocyanins in Brassicaceae: composition, stability, bioavailability, and potential health benefits. Crit Rev Food Sci Nutr 62(8):2205–2220. https://doi.org/10.1080/10408398.2020.1852170

Zhang M, Chen H, Li J et al (2010) Antioxidant properties of tartary buckwheat extracts as affected by different thermal processing methods. LWT 43(1):181–185. https://doi.org/10.1016/j.lwt.2009.06.020

Zhang J, Cheng J, Li Z et al (2024) Effects of ultra-high pressure, thermal pasteurization, and ultra-high temperature sterilization on color and nutritional components of freshly-squeezed lettuce juice. Food Chem 435:137524. https://doi.org/10.1016/j.foodchem.2023.137524

Zielinski H, Michalska A, Amigo-Benavent M et al (2009) Changes in protein quality and antioxidant properties of buckwheat seeds and groats induced by roasting. J Agric Food Chem 57(11):4771–4776. https://doi.org/10.1021/jf900313e

Zong C, Xiao Y, Shao T et al (2023) Alfalfa as a vegetable source of β-carotene: the change mechanism of β-carotene during fermentation. Food Res Int 172:113104. https://doi.org/10.1016/j.foodres.2023.113104

Chapter 17
Fruits and Vegetables as a Source of Nutraceuticals: Twenty-First Century Perspective

Gadafi Iddrisu Balali, Derrick Opoku, Philip Asumang, Vera Afua Dela Gobe, Denis Dekugmen Yar, Abdulmanan Aziz, Zaid Abdulai, and Emmanuel Amihere Cobbinah

Abstract The emergence of nutraceuticals brought about tremendous changes in medicine and human nutrition. Nutraceuticals are gradually gaining ground in recent times, as they are not only a great source of nutrients but also an amazing source of health promotion and disease prevention. In this chapter of the book, we provide an insight into how vegetables and fruits are a reliable source of nutraceuticals. We aim to delve much into their diversity, action mechanisms, and importance. We begin the chapter by defining nutraceuticals and their necessity in the healthcare industry. For vegetables and fruits, we bring forth the presence of vitamins, minerals and phytochemicals, and the role they play in human health. Primarily, we are much more interested in nutraceutical diets and their role in human health such as heart

G. I. Balali (✉)
Department of Theoretical and Applied Biology, Kwame Nkrumah University of Science and Technology, Kumasi, Ghana

Department of Science, SDA College of Education, Agona, Ashanti, Ghana
e-mail: balaligadafi@sdacoe.edu.gh

D. Opoku
Department of Theoretical and Applied Biology, Kwame Nkrumah University of Science and Technology, Kumasi, Ghana

P. Asumang · E. A. Cobbinah
Department of Science, SDA College of Education, Agona, Ashanti, Ghana

V. A. Dela Gobe
Department of Food Science and Technology, Ho Technical University, Ho, Ghana
e-mail: vadgobe@htu.edu.gh

D. D. Yar
Department of Public Health, Akenten Appiah-Menka University of Skills Training and Entrepreneurial Development, Kumasi, Ghana

A. Aziz · Z. Abdulai
Department of Nursing, University of Sunderland, Sunderland, UK
e-mail: Bi17fc@student.sunderland.ac.uk

S. Amir Ashraf, M. Adnan (eds.), *Food Bioactives and Nutraceuticals*,
https://doi.org/10.1007/978-981-97-9723-3_17

health. Specifically, we looked at berries, tropical varieties, and vegetables including cruciferous vegetables and leafy greens. We also looked at extraction methods, regulations and regulatory bodies and ended up with discussions on research opportunities for novel compound discovery and advancement in nutrition.

Keywords Nutraceuticals · Fruits · Vegetables · Health benefits · Personalized nutrition

17.1 Introduction

Recently, the emergence of nutraceuticals has ignited scientists' minds to ponder their potential in preventing diseases and overwhelming health promotion (Reinarz 2023). In this chapter, we delve into fruits and vegetables as a significant source of nutraceuticals. In this chapter, we explore the importance of nutraceuticals and their origins, and provide insight into their relevance to our current health industry. Nutraceutical is a term derived from two different fields by the combination of the words "nutrition" and "pharmaceutical," indicating their paramount significance in the pursuit of good health. Mostly, compounds found in foods that usually provide health benefits aside from nutrition are known as nutraceuticals (Garza-Juárez et al. 2023). The nutraceutical revolution has transmogrified humans' perception of food by expanding their role from being solely nourishment to becoming an agent of improving human health (Marcus 2019). One captivating feature of this revolution lies in its methodology towards healthcare. Traditional medicine has habitually focused largely on treating illnesses that have already manifested (Egbuna et al. 2020). Notwithstanding, nutraceuticals prioritize prevention, this is by aiming to maintain health and reduce risks even before symptoms occur (Rosenberg et al. 2020). This metamorphosis aligns perfectly with the proverb that says "Prevention is better than cure." Additionally, the practice of nutraceuticals places paramountcy on harnessing health benefits found in compounds as a means of supporting good health. Right from the very beginning of life, humans have relied on the knowledge passed down over centuries to comprehend the benefits and qualities of herbs and foods. This jurisdiction combines the prospering forces of scientific understanding and wisdom creating a unique character for survival in the jungle of medicine and healing (Calado et al. 2018).

17.2 Fruits and Vegetables: History and Medicinal Values

Throughout the history and cultures, world have recognized the healing properties of fruits and vegetables incorporating them into their practices. Ancient civilizations such as the Egyptians, Greeks and Romans were well aware of the qualities found in these foods (Towns and Shackleton 2018). Hippocrates, often regarded as a pioneer in medicine famously emphasized the importance of maintaining a diet by stating, "Let your food be your medicine and your medicine be your food" (Jennings 2014). Traditional Chinese Medicine (TCM) has also embraced the concept of utilizing food as a form of medicine. In TCM fruits and vegetables are frequently incorporated into remedies to harmonize the body's energies and promote well-being. This holistic approach acknowledges that different foods possess energies that can influence the balance, between Yin and Yang within the body (Marshall et al. 2020). Throughout history cultures, around the world have understood the healing power of fruits and vegetables. In India, Ayurveda. A healing system deeply rooted in tradition. Emphasizes incorporating these foods into our diet to restore balance and enhance well-being. Ayurveda classifies foods based on their qualities (known as Guna). How they impact the body (Vata, Pitta and Kapha). This emphasizes how our dietary choices can affect our well-being. In North and South America indigenous communities share a connection, with the land they inhabit. They hold respect for fruits and vegetables like corn, beans and squash. Often referred to as the "Three Sisters." These cherished foods play a role in the healing traditions that have been passed down through generations (Delormier and Marquis 2019). Across Europe, monasteries and herbalists have long recognized the properties of fruits and vegetables. They cultivated gardens filled with these plants to harness their attributes. These gardens served as sources of knowledge and remedies for ailments during those times (Bloom 2018).

The use of fruits and vegetables as medicinal agents has transcended boundaries throughout history. This collective wisdom showcases how these natural resources have been highly praised for their ability to alleviate illnesses and promote well-being (Bommakanti et al. 2023). A testament to the timeless knowledge inherited from our ancestors. In today's world, this rich historical legacy seamlessly aligns with exploration as researchers uncover the bioactive compounds present in fruits and vegetables that contribute to their health-promoting qualities. The Nutraceutical Revolution combines wisdom with research highlighting the idea that the food we consume daily can greatly impact our overall well-being (Galanakis 2019).

17.3 Nutraceuticals; Definition and Importance

Although there is no accepted definition nutraceuticals can be described as compounds or substances, in food that offers health benefits beyond just nutrition. These compounds encompass vitamins, minerals, phytochemicals, antioxidants, probiotics and more (Alamgir and Alamgir 2018). What distinguishes nutraceuticals from nutrients is their ability to positively impact functions and contribute to the prevention or management of diseases. One noteworthy characteristic of nutraceuticals is their role as both food and medicine (Domínguez Díaz et al. 2020). While they are consumed as part of our diet their potential to enhance health has gained recognition. As a result, they have been integrated into healthcare practices (Gul et al. 2016). Nutraceuticals can be naturally found in a variety of foods such, as fruits, vegetables, whole grains, nuts and herbs (Verma and Mishra 2016). Furthermore, they are often incorporated into foods, supplements and beverages to provide targeted health effects through dosages. In today's fast-moving world, many people have to prioritize their health and take tremendous measures to attain good and quality health. Nutraceuticals have emerged as an important element of health-promoting and maintenance of good health (Egbuna and Dable-Tupas 2020). One important aspect is how nutraceuticals align seamlessly perfectly well with the concept of medicine which emphasizes the significance of preventing diseases before they occur. Incorporating phytonutrients into our diets can effectively help prevent the occurrence of diseases and promotion of good health (Gupta and Prakash 2015). A typical example is in the case of antioxidants such as vitamins C and E that play a role in combating stress, hence reducing the likelihood of developing conditions, like cancer or heart disease (Sharifi-Rad et al. 2020).

Nutraceuticals also usually work with other treatments to enhance their effectiveness; this is known as synergy. Some nutraceuticals are used along with some treatments to make them more successful or decrease any side effects (Kunnumakkara et al. 2017; Elkhalifa et al. 2021). Taking omega-3 fatty acids for instance, they are found in fish oil supplements and are often used with prescription drugs to lower inflammation and improve health. Moreover, it has been said that nutrition has become more popular as people have learned that their genes and health can affect what their body needs. The application of nutraceuticals provides a way to do this by recommending ingredients based on individual people's needs and their health goals (Freitas et al. 2012). In recent times, where many people are extremely busy and have to rely on fast-food joints. It is important to incorporate nutraceuticals in our diet to help us overcome vitamin deficiencies as a result of fast-food consumption (Sizer and Whitney 2022). Nutraceuticals can rightly fill the gaps in our diets, making sure we get all the vitamins, minerals, and other substances that we need for our bodies to stay healthy. Nutraceuticals are steadily gaining momentum in the global market, as reflected in the overwhelming market size estimates. A recent evaluation of the nutraceutical market revealed substantial growth, with values surpassing expectations (Fig. 17.1).

Another important aspect of nutraceuticals is their ability to make our lives better in addition to keeping us from getting ill. Nutraceuticals have been linked to benefits like better happiness, brain function, and increased energy (Rao and Singh 2022). For instance, some plants like the ginkgo biloba are said to improve memory and brain function (Sadeghinejad et al. 2019). More importantly, is how our research has helped us understand more about how nutraceuticals work. Evidence does not directly point to their applications in healthcare but promotes the persistent search

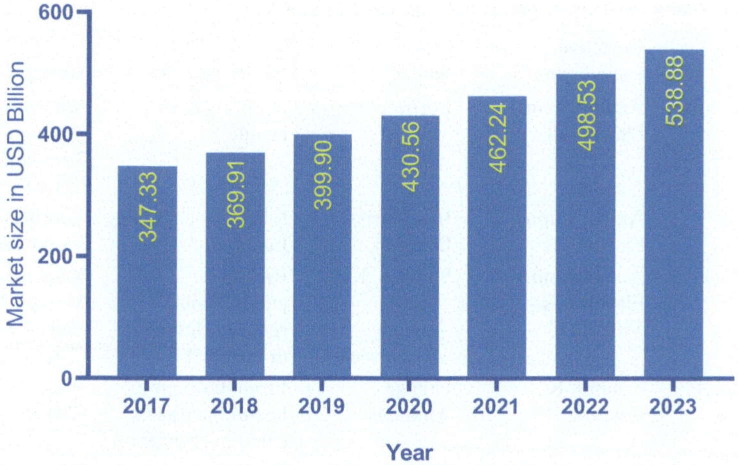

Fig. 17.1 Nutraceuticals' global market momentum. (Source: www.precedenceresearch.com)

for new chemicals and their usage. Nutraceuticals are a continuously changing field that combines food and health. Regarding their function in health, they improve health as a source of alternative treatments provide unique ways to deal with problems and improve the total quality of human health (Rathore et al. 2017). As we study much into nutraceuticals, they are becoming more important in shaping the future of human health. In the next parts of this chapter, we will delve much more into the diverse nutrients in vegetables and fruits and how they contribute to human health.

17.4 The Assortment of Nutraceuticals in Fruits and Vegetables

Vegetables and fruits are a great source of nutraceuticals. These substances contribute to good health by preventing the occurrence of diseases in humans (Santini et al. 2017). In this section, we delve into various nutraceuticals available in fruits and vegetables (Tables 17.1 and 17.2). Based on classification, we put them into phytochemicals and essential vitamins and minerals. Phytonutrients are naturally occurring compounds found in plants known as phytochemicals. Phytonutrients do not only provide fruits with their captivating flavours, vibrant colours and scents but also provide significant health benefits to human society (Agrawal et al. 2023). Is therefore important to delve into some of the phytochemicals mostly found in vegetables and fruits. In science, and specifically the field of pharmacy, compounds that act as natural defence systems against radicals are known as flavonoids. They are mostly found in diverse fruits such as berries and leafy greens (Kaurinovic and Vastag 2019). These vibrant compounds play vital roles in counteracting molecules

Table 17.1 Nutraceutical categories in fruits and vegetables

Type	Bioactive constituents	Source	Health benefits	Source(s)
Cardiovascular Health	Anthocyanins, Flavonoids	Berries, Grapes	Supports heart health	Kaurinovic and Vastag (2019), Rees et al. (2018)
Cognitive Health	Anthocyanins, Flavonoids	Blueberries, Broccoli	Enhances cognitive function	Kaurinovic and Vastag (2019)
Antioxidant Properties	Anthocyanins, Flavonoids, Vitamin C	Spinach, Kale	Rich in antioxidants, reduces chronic disease risk	Kaurinovic and Vastag (2019), Yadav et al. (2016)
Bone Health	Vitamin K, Calcium	Oranges, Almonds	Promotes bone health, facilitates calcium absorption	Wasilewski et al. (2019)
Immune System Support	Vitamin C, Zinc, Beta-Glucans	Citrus Fruits, Mushrooms	Strengthens immune system, aids in illness prevention	Basak and Gokhale (2022)
Anti-Inflammatory	Curcumin, Quercetin, Omega-3 Fatty Acids	Turmeric, Berries, Fatty Fish	Reduces inflammation, mitigates chronic diseases	Kumar et al. (2017), Gounden et al. (2018)
Digestive Health	Fiber, Probiotics	Whole Grains, Yogurt	Supports digestion, maintains gut microbiome balance	Dahiya and Nigam (2022), Anusha et al. (2018)
Vision Support	Lutein, Zeaxanthin, Vitamin A	Leafy Greens, Carrots, Eggs	Promotes eye health, reduces risk of age-related issues	Stice et al. (2018)
Joint Health	Curcumin	Turmeric	Supports joint function, reduces arthritis symptoms	Kumar et al. (2017), Daily et al. (2016)
Stress Management	Adaptogens (e.g., Ashwagandha, Rhodiola)	Ginseng, Ashwagandha	Helps the body adapt to stress, reduces stress levels	Serai (2021)

that can result in oxidative stress which is usually associated with the ageing process and diseases (Forni et al. 2019; Serai 2021). One captivating aspect of flavonoids is their ability to combat inflammation, which plays a role in various health issues. By calming the body's response flavonoids may help mitigate the risk of conditions, like heart disease, diabetes and arthritis (Shukla et al. 2019). Moreover, recent studies indicate that flavonoids may also support blood vessel function and reduce blood pressure thus contributing to health (Rees et al. 2018). The vibrant hues seen in fruits and vegetables are attributed to carotenoids. A group of pigments including beta carotene, lycopene and lutein (Maoka 2020). These compounds not

Table 17.2 Bioactive components and sources of nutraceuticals in fruits and vegetables

Type	Bioactive constituents	Source	Health benefits	Other important components	References
Flavonoids	Quercetin, Anthocyanins	Berries, Leafy Greens	Combats inflammation, reduces the risk of heart disease, diabetes, and arthritis	Supports blood vessel function, may reduce blood pressure	Kaurinovic and Vastag (2019)
Carotenoids	Beta Carotene, Lycopene, Lutein	Carrots, Sweet Potatoes, Tomatoes	Antioxidants, vision support, immune function, skin health	Beta-carotene is a precursor to Vitamin A	Maoka (2020), Zia-Ul-Haq (2021)
Glucosinolates	Isothiocyanates	Broccoli, Cabbage, Kale	Anti-cancer properties, inhibit cancer cell growth	Transform into isothiocyanates during enzymatic processes	Yadav and Dhankhar (2022)
Polyphenols	Resveratrol, Quercetin	Grapes, Red Wine, Onions	Antioxidant effects, anti-inflammatory, heart support	Resveratrol extends lifespan, Quercetin supports heart	Kaurinovic and Vastag (2019, Perez-Vizcaino and Fraga (2018)
Saponins	–	Legumes, Vegetables, Certain Grains	Regulates cholesterol levels, immune-boosting, anti-inflammatory	Create texture when mixed with water	Oakenfull and Sidhu (2023), Albuquerque et al. (2020)
Phytosterols	–	Nuts, Seeds, Whole Grains, Vegetable Oils	Lower LDL cholesterol levels, support heart health	Compete with cholesterol during absorption	Albuquerque et al. (2020), Poli et al. (2021)

(continued)

Table 17.2 (continued)

Type	Bioactive constituents	Source	Health benefits	Other important components	References
Allicin	–	Garlic, Onions	Protects against infections, maintains blood pressure, supports heart health	Associated with reduced risk of heart disease	Sachdeva et al. (2020), Aly et al. (2023)
Vitamin C	–	Oranges, Grapefruits, Strawberries, Bell Peppers	Immune-boosting, antioxidant, collagen formation	Aids in healing, supports healthy skin, tendons, vessels	Kaurinovic and Vastag (2019), Yadav et al. (2016)
Vitamin A	–	Carrots, Sweet Potatoes, Leafy Greens	Vision support, skin health, immune function	Retinol essential for vision and perception	Stice et al. (2018)
Vitamin K	–	Spinach, Kale	Blood clotting, bone and teeth health	Regulates calcium levels, supports bone density	Wasilewski et al. (2019)
Folate (Vitamin B9)	–	Leafy Vegetables, Avocados, Legumes	DNA synthesis, cell growth, crucial during pregnancy	Vital for blood cell production, brain health	Hanna et al. (2022)
Potassium	–	Potatoes, Bananas, Tomatoes	Blood pressure regulation, nerve signal transmission	Reduces risk of hypertension, heart muscle contraction	McLean and Wang (2021)
Magnesium	–	Nuts, Whole Grains, Leafy Greens	Muscle and nerve function, bone health, blood sugar regulation	Component of bone structure, regulates blood sugar	Mironenko and Eliseeva (2020)
Dietary Fiber	–	Fruits, Vegetables	Facilitates regular bowel movements, blood sugar regulation	Reduces inflammation, supports gut microbiome	Yusuf et al. (2022)

only captivate our eyes but also serve as antioxidants that safeguard our health. Beta carotene is often associated with carrots and sweet potatoes (Zia-Ul-Haq 2021). Acts as a precursor to vitamin A. A nutrient, for vision support, immune function maintenance and skin health. Lycopene, which can be found in tomatoes, watermelon and pink grapefruit has gained recognition, for its ability to reduce the risk of age-related macular degeneration and certain cancers, prostate cancer. Lutein, which is abundant in leafy greens and egg yolks acts as a protector for our eyes by shielding them from damage caused by ultraviolet (UV) light (Stice et al. 2018; Awadelkareem et al. 2022a).

Vegetables such as broccoli, cabbage, argula and kale are known for their health benefits due to the presence of glucosinolates. These sulfur-containing compounds have established themselves as agents in preventing cancer. When we consume vegetables the glucosinolates are transformed into compounds through enzymatic processes. Isothiocyanates, which are mainly found in vegetables such, as broccoli and cauliflower are versatile substances that possess strong anti-cancer properties. These natural compounds have garnered interest due to their ability to hinder the growth of cancer cells and initiate programmed cell death, also known as apoptosis (Yadav and Dhankhar 2022; Awadelkareem et al. 2022b). As a result, they serve as a defense against types of cancer like lung, breast and colon cancer. Moreover, they are thought to assist in detoxifying substances within the body serving as a barrier, against environmental toxins (Wagh et al. 2020). Polyphenols are widely present in fruits, vegetables and herbs. They are celebrated for their range of health benefits (Gorzynik-Debicka et al. 2018). Resveratrol found in grapes and red wine has attracted attention due to its role, in promoting health and extending lifespan (Perez-Vizcaino and Fraga 2018). Quercetin, which can be found in onions, apples and berries is highly valued for its ability to reduce inflammation and support a heart (Kumar et al. 2017). These compounds, with properties not only have powerful antioxidant effects but also demonstrate the ability to combat inflammation fight against microbes and protect the heart. It is commonly believed that these compounds play a vital role in the human diet and are associated with promoting life (Gorzynik-Debicka et al. 2018). Saponins are compounds that are commonly found in plant-based foods such as legumes, vegetables and certain grains. They acquired their name because of their ability to create a texture when mixed with water (Alcorta et al. 2021). One of the advantages of saponins is their potential to regulate cholesterol levels in our bodies. They achieve this by preventing the absorption of cholesterol thus reducing the amount that enters our bloodstream (Oakenfull and Sidhu 2023). This reduction in cholesterol can play a role in maintaining health since high levels of LDL (low-density lipoprotein) cholesterol increase the risk of heart disease. Additionally, saponins may also possess immune-boosting and inflammatory properties. Some studies suggest that they can stimulate our system and aid in defending against infections and diseases (Basak and Gokhale 2022). Phytosterols, also known as plant sterols or stanols are compounds in the cell membranes of plants. While they share similarities with cholesterol derived from animals their chemical composition is different (Moreau et al. 2018). Phytosterols are well known for their ability to lower LDL cholesterol levels in the bloodstream. They compete with cholesterol during absorption in our intestines effectively reducing its quantity (Zhang et al. 2022). Including foods in phytosterols like nuts, seeds, whole grains and certain vegetable oils can be an approach to managing cholesterol levels and promoting heart health (Albuquerque et al. 2020). Research indicates that consuming phytosterols is associated with a reduced risk of heart disease, a recognized benefit acknowledged by countries due to its impact on health (Poli et al. 2021). Incorporating foods that contain phytosterols, such as margarine can be beneficial for lowering cholesterol levels and supporting heart health. These compounds along with saponins have been proven to contribute to well-being by reducing cholesterol (Ogbe et al. 2015).

Now let's talk about allicin, a compound found in garlic and onions that has gained popularity due to its impact on health. Allicin is well known for its ability to protect against infections by inhibiting the growth of microorganisms and boosting our system's ability to fight off pathogens (Valenzuela-Gutiérrez et al. 2021). Acting as a guardian within our bodies' allicin plays a role during times when immune resilience is essential. Additionally, allicin has effects on our well-being by helping maintain blood pressure levels and supporting heart health. By promoting the relaxation of blood vessels allicin reduces stress on the heart and assists in managing blood pressure (Wang et al. 2022). Furthermore, one of the health benefits of allicin includes its potential to lower the risk of heart disease. Scientific research suggests that including foods with allicin in your diet may be associated with a reduced likelihood of developing heart disease. Allicin acts as a compound by influencing factors that affect heart health (Salehi et al. 2019). When it comes to nutrition, fruits and vegetables stand out as sources of vitamins and minerals. These natural food options, with their colors and diverse flavors, provide an abundance of nutrients that play roles in various bodily functions (Wallace et al. 2020). Let us embark, on a journey to explore the range of vitamins and minerals commonly found in these powerhouses.

Fruits like oranges, grapefruits, strawberries, bell peppers and some fungi such as mushrooms are packed with vitamin C, Zinc, and Beta-Glucans which are well known for their immune-boosting properties (Basak and Gokhale 2022). This water-soluble vitamin acts as an antioxidant that helps neutralize radicals that can compromise our defenses (Vinayagam and Xu 2015). Apart from supporting our system vitamin C also plays a role in collagen formation—a protein that is essential for maintaining healthy skin, tendons and blood vessels (Devaki and Raveendran 2017). It also aids in the healing process by promoting tissue growth. By incorporating fruits and vegetables into our diet to ensure an intake of vitamin C we can enhance our health and well-being (Khadim and Al-Fartusie 2021). Carrots, sweet potatoes and leafy greens are sources of vitamin A. This nutrient is widely recognized for its role in maintaining vision (Carazo et al. 2021). It supports the functioning of the retina assisting with both perception and color vision. Additionally, it contributes to skin health by keeping it supple and resilient while promoting cell production and preventing dryness or flakiness (Liu 2022). Vitamin A also plays a role in supporting our system as it acts as a defence against infections. Vitamin K can be found abundantly in greens such as spinach and kale. This essential vitamin has a role, in promoting wound healing and preventing excessive bleeding by aiding in blood clotting (Raseetha et al. 2022). Additionally, vitamin K helps to maintain the health of bones and teeth by regulating calcium levels ensuring that this mineral is deposited to enhance their strength and density. Including vegetables in vitamin K in your diet can contribute to promoting bone health (Wasilewski et al. 2019). Leafy vegetables, avocados and legumes are sources of folate or vitamin B9. This water-soluble nutrient is crucial for DNA synthesis and cell growth (Hanna et al. 2022). Its importance becomes more apparent during pregnancy as it supports the development of tubes. Folate is vital for the production of blood cells. Plays a role, in brain health and overall development making it an essential nutrient (Al Rahmad 2023).

Potassium is obtained from foods such as potatoes, bananas, and tomatoes and plays a vital role in the maintenance of blood pressure levels. Its ability to balance sodium effects helps to keep the blood pressure in a normal range (McLean and Wang 2021). Potassium consumption also reduces the risk of hypertension and the complications that come with it. Other functions include the transmission of nerve signals and the contraction of heart muscles (Gröber et al. 2015). Magnesium can be sourced from nuts, whole grains and leafy greens. Magnesium possesses great qualities in its ability to ensure muscle and nerve functions (Mironenko and Eliseeva 2020). It helps in the contraction and relaxation of muscles while influencing the transmission of nerve signals in the human body. It also functions to improve bone health as is a component of bone structure and also functions in the regulation of blood sugar levels (Ciosek et al. 2021). Although not officially classified as a vitamin or mineral, the dietary fibre found in fruits and vegetables deserves recognition for its role in promoting good health. Dietary fibre plays a part in maintaining health by facilitating regular bowel movements and preventing constipation (Yusuf et al. 2022). Additionally, dietary fibre aids in regulating blood sugar levels by slowing down the absorption of sugar and ensuring a release of energy. It also assists with weight management by creating a sense of fullness thereby reducing calorie intake (Farooqui and Farooqui 2015). Furthermore, embracing a high-fiber diet has been linked to a risk of diseases like heart disease, diabetes and certain types of cancer. This is achieved through the reduction of inflammation and the promotion of a gut microbiome (Bulsiewicz 2020).

17.5 Mechanisms of Action: Unraveling the Functional Complexities of Nutraceuticals

When it comes to nutraceuticals it is crucial to understand how these beneficial compounds function, in the body to fully grasp their impact, on overall health. This section delves into the workings of nutraceuticals shedding light on how they exert their effects. Additionally, it investigates the concepts of synergy and complementarity among these health-promoting compounds uncovering the hidden potential that arises when they come together. In the world of nutraceuticals, how these bioactive compounds bring about their health-enhancing benefits is as varied, as the compounds themselves. Let us take a journey through the way nutraceuticals work their magic in our bodies shedding light on their crucial roles in safeguarding and improving our health. A multitude of nutraceuticals, such as known vitamins like C and E along with polyphenols like resveratrol act as steadfast guardians in our body's ongoing battle against oxidative stress. Their role as antioxidants is vital as they diligently counteract radicals (Overby 2019). Those reactive molecules are notorious for causing damage within our cells. These free radicals, arising from external sources can inflict harm on cells and contribute to the development of chronic diseases. Conditions, like cancer, heart disease and even ageing itself bear the signature of these molecules (Aslani and Ghobadi 2016). Nutraceutical antioxidants possess the power to suppress these disturbances and emerge as defenders of our well-being. Chronic inflammation serves as a thread that weaves its presence into a range of health issues. Nutraceuticals, such as curcumin derived from the vibrant turmeric root and the omega-3 fatty acids abundant in fatty fish showcase their expertise in combating inflammation (Gounden et al. 2018). In the face of inflammation, these substances act as soothing agents calming the fires that pose a threat to our wellbeing. By reducing inflammation and managing its associated risks nutraceuticals provide relief from the effects of inflammation offering a path towards improved health (Zohn 2020).

Within the realm of nutraceuticals certain compounds like vitamin A take on the role of nurturing repair and regeneration. These unsung heroes play parts in maintaining tissues aiding in wound healing and supporting immune function (Abate 2017). Essentially acting as craftsmen within our body workshop, they diligently create tissue that ensures the seamless renewal of cellular structures. By supporting these functions nutraceuticals like vitamin A contribute to our body's lasting vitality (Mali et al. 2022). Specific compounds found abundantly in fruits and vegetables such as indole 3 carbinol present in vegetables, have a hand in regulating hormones. They have control, over the balancing act of hormones promoting equilibrium and harmony within the endocrine system (Werner-Gray 2020). In doing so, they contribute to reducing the risks of hormone-related cancers and help maintain a hormonal balance. Nutraceuticals, acting as guardians of harmony stand as protectors against imbalances and their potential health challenges (Eggers and de Blanco 2016). Nutraceuticals possess the ability to influence the symphony of gene expression within our cells. They can orchestrate the activation or deactivation of genes associated with disease risk (Sawikr et al. 2017). This intriguing phenomenon called epigenetics holds promise for improving health outcomes. By inducing changes, nutraceuticals can play a role in preventing cancer and achieving other health goals. They hold the potential to tune our destiny harmonizing our genetic melodies towards better health (Shanahan 2017). The gut often referred to as our "brain" houses a bustling community of microbes that profoundly affect our wellbeing. Within this ecosystem, prebiotics. A type of nutraceutical found in foods like garlic and onions. Take centre stage as nourishment, for our beneficial gut bacteria (Bauer et al. 2019). By supporting a community of microorganisms in our body, prebiotics have an impact not only on our digestion but also on our immune system and even our mental well-being. The nourishment of these microorganisms goes beyond the digestive system and affects every aspect of our overall health (Dahiya and Nigam 2022).

Our brain, which holds our abilities, deserves protection and certain natural substances found in delicious berries and soothing green tea play a crucial role in this regard. These substances, known as flavonoids act as guardians against decline (Darrell 2020). They help maintain function that lowers the risk of diseases and even has the potential to improve mood and mental well-being. By safeguarding our abilities these natural substances pave the way for a lifetime of vitality (Adan et al. 2019). Compounds like cinnamon and the important mineral chromium have proven their effectiveness in managing metabolic imbalances, particularly in individuals with diabetes or those at risk of developing this condition. These compounds are associated with control, over blood sugar levels offering hope to those striving for optimal metabolic health.

17.6 Synergy and How Nutraceuticals Work Together

When it comes to nutraceuticals, they do not usually work on their own in the body. Instead, they. Complement each other creating a network of health benefits that go beyond what individual compounds can achieve. Synergy happens when two or more nutraceuticals work together to create an effect that is greater than what they can do. For example, people often consume curcumin from turmeric and Piperine from pepper together (Pezzani et al. 2019). Piperine helps increase curcumin's effectiveness in reducing inflammation and providing antioxidant benefits. Complementarity refers to how different nutraceuticals target aspects of a health issue working together to address it comprehensively. For instance, when it comes to promoting heart health omega-3 fatty acids from fish oil reduce inflammation and support function while phytosterols from nuts and seeds help lower LDL cholesterol levels (Ansari 2021). Together they create a combined approach to heart well-being. Understanding the concept of synergy and complementarity among nutraceuticals emphasizes the importance of consuming a range of fruits, vegetables and other natural foods.

Whole foods provide a variety of nutraceuticals, with benefits and mechanisms of action (Leena et al. 2020). By incorporating a range of these foods into our meals, we can unlock the true potential of the gifts nature has provided us. Moreover, the way different nutraceuticals work together and complement each other suggests that a balanced diet is much more than nutrients. It's like a symphony of health-boosting compounds that harmonize to support our well-being (Khan et al. 2019). Take, for instance, the Mediterranean diet, which is widely known for its approach to health. It embraces a combination of fruits, vegetables, olive oil, nuts and seafood. All are packed with nutraceuticals that collaborate to promote longevity and vitality.

17.7 The Health Advantages of Rich Diets; Unleashing the Potential of Nature Abundance

Embracing diets, which are packed with a variety of fruits and vegetables, can bring about a wide range of benefits that positively affect every aspect of our wellbeing. These diets, characterized by their colors and diverse flavors provide an array of nutraceuticals, bioactive compounds sourced from food and that work wonders within our bodies (Gupta and Mishra 2021). In this section, we will delve into an in-depth exploration of the health advantages offered by rich diets from promoting cardiovascular health to enhancing cognitive function. Diets serve as strong guardians of our cardiovascular health. A combination of compounds found in fruits and vegetables orchestrates a symphony of benefits that support not the heart but the entire circulatory system (Pang et al. 2012). Firstly, potassium fruits and vegetables such as bananas, potatoes and tomatoes play a role in maintaining healthy blood pressure levels. Potassium counters the effects of sodium assisting in regulating blood pressure and reducing the risk of hypertension (Ramya and Patel 2019). Also, Nuts, seeds and whole grains are sources of phytosterols. Plant-based cholesterol counterparts aid in managing cholesterol levels. They can lower LDL (low-density lipoprotein) cholesterol levels, which promotes heart health and reduces the risk of atherosclerosis and related cardiovascular complications (Raj et al. 2023). Nutraceuticals, such as omega-3 fatty acids found in fish and polyphenols from fruits like berries and green tea contribute to reducing inflammation. This inflammation is a risk factor for heart disease. By soothing the process these compounds support the wellbeing of the heart (Ansari 2021). Vascular Health; Flavonoids, which are commonly found in fruits, berries and leafy greens promote health by improving endothelial function. This improvement ensures that blood flows smoothly through arteries reducing the risk of dysfunction and related cardiovascular disorders (Di Pietro et al. 2020). Vitamins C and E along with polyphenols like resveratrol, possess antioxidant properties that protect the system from oxidative stress. By neutralizing radicals within cells, these compounds help mitigate damage and reduce the risk of heart disease (Rana et al. 2022). A system is vital for good health. Diets rich in nutraceuticals play a role in fortifying the body's defenses against infections and illnesses. Citrus fruits, strawberries and bell peppers are packed with vitamin C, which is known to support the system (Alpert 2017). This essential nutrient not only helps strengthen function but also aids in collagen production and accelerates wound healing. Vitamin A, found in carrots, sweet potatoes and leafy greens is crucial, for maintaining a system. It plays a role in preserving the integrity of mucosal surfaces, which serve as our body's line of defense against harmful pathogens (Seow 2020). Some nutraceuticals such as onions and garlic possess great properties for the nourishment of gut bacteria. This intern not only supports digestion in humans but also plays a crucial role in boosting immune responses (Sachdeva et al. 2020). Additionally, allicins that come with great antimicrobial properties are sourced from onions and garlic. They are good at bolstering the body's defense against pathogens through the inhibition of the growth of such pathogens (Aly et al. 2023). Other compounds such as omega-3 fatty acids and curcumin of turmeric origin, possess great inflammatory properties that help to resolve chronic inflammation in humans (Neda et al. 2021). In all, this results in the system directing the body's energy to defend against pathogen invasion.

Most cancers are characterized by uncontrolled growth of body cells posing an important health risk to humans, well prevented by the consumption of diets that contain tremendous quantities of nutraceuticals (Alamgir and Alamgir 2018). Resveratrol polyphenols and antioxidants including vitamins E and C all found in nutraceuticals help in the reduction of oxidative stress within human cells. This is achieved through the neutralization of radicals. They also decrease the risk of DNA damage and mutations that can result in cancer (Simioni et al. 2018). Isothiocyanates are nutraceutical compounds found in broccoli and cauliflower that come with tremendous anticancer properties. They help in the detoxification of the body thereby reducing the risk of cancer initiation (Coscueta et al. 2022). Other nutraceuticals such as indole-3-carbino are mostly found in vegetables that function in hormonal metabolism. They also decrease the risk of hormone-related cancers including prostate and breast cancer by maintaining body balance (Vrânceanu et al. 2022). Nutraceuticals are capable of suppressing genes associated with the development of cancer due to their ability to influence gene expression. They do so by employing a process known as epigenetic effects that is capable of preventing cancer through signal regulation (Maiuolo et al. 2021). These processes are accomplished by nutraceuticals such as flavonoids that trigger apoptosis in cancer cells. These processes help in the distraction of damaged cells before they develop into tumors.

Human beings are the organisms with the highest intellect and rely mostly on this great assertion to carry out complex tasks involving cognitive function. It is important to know that fruits and vegetables contain some nutraceuticals that are capable of providing good brain health and enhancing cognitive function (Szot et al. 2022). Such nutraceuticals include flavonoids that are found in berries and green tea and are well known for their valuable contribution to the protection and functioning of the human brain. They are capable of reducing the risk of some health conditions including Parkinsons' disease and Alzheimer's and are also good at influencing mood (Avallone et al. 2019). Blood circulation in the human brain is an important component of maintaining good health, especially in the brain. Some nutraceuticals such as polyphenols sourced from cocoa and dark chocolate are chiefly responsible for improving blood circulation in the brain increasing brain functioning and decreasing the risk of cognitive decline (Grassi et al. 2016). Grapes-sourced vitamins E and C along with resveratrol shield the human brain from stress. This is achieved by minimizing the impact of radicals on brain cells which are important in contributing to cognitive function (Cui et al. 2020). Nutraceuticals such as omega-3 fatty acids are capable of affecting hormonal balances and neurotransmitter functionality in the human brain. As a result, they are good at influencing and contributing to stabilizing mood, and good and quality health (Nigam et al. 2018). Some nutraceuticals such as turmeric-sourced curcumin are good in the alleviation of inflammation in the human brain thereby supporting its resilience (Daily et al. 2016).

17.8 The Wonderful World of Nutraceutical-Rich Fruits and Vegetables

The astonishing diversity of fruits and vegetables reveals a captivating world occupied with the arrangement of colors compounds and flavors that support human life. These important natural gifts do not delight our taste buds only but also come with medicinally precious bioactive substances (Yahia et al. 2017) with a wide range of health benefits. The flowchart below illustrates the diverse nutraceutical contributions from different types of fruits and vegetables (Fig. 17.2).

From cardiovascular support in berries to cognitive enhancement in broccoli, each element plays a unique role in promoting overall health. It is important to explore the landscapes of nutraceutical-rich foods including berries, cruciferous vegetables, leafy greens and tropical fruits. For instance, berries with their colors and delightful tastes serve as ambassadors in the world of nutraceuticals. These small wonders are packed with antioxidants, vitamins and phytochemicals that offer a range of health advantages (Singh et al. 2018). Berries are abundant in antioxidants like anthocyanins, flavonoids and vitamin C. These compounds help

Fig. 17.2 Nutraceutical contributions from fruits and vegetables

neutralize radicals to protect cells from stress while reducing the risk of chronic diseases (Yadav et al. 2016). Incorporating berries into your diet has been associated with health. Their anti-inflammatory properties and ability to promote blood pressure and arterial function contribute to heart well-being (Arumugam et al. 2021). Berries and blueberries are renowned for their benefits. The presence of anthocyanins and flavonoids, in these fruits may enhance memory retention learning capabilities and overall brain function (De Amicis et al. 2022). Certain compounds found in berries, such as acid have demonstrated potential in inhibiting the growth of cancer cells and reducing the risk of types of cancer (George et al. 2017). Berries are a source of fiber and prebiotics which promote a healthy balance in the gut microbiome. A balanced gut microbiota is associated with health advantages, including improved digestion and strengthened immunity (Manoj et al. 2023).

Leafy greens, those gems from the plant kingdom are renowned for their concentration of nutrients and their valuable contributions to overall health. These nutritional powerhouses offer a range of vitamins, minerals and phytochemicals (Pinnock 2011). Leafy greens like spinach, kale and Swiss chard are abundant in vitamins and minerals such as vitamin K, folate and magnesium. These nutrients play roles in maintaining bones facilitating proper blood clotting processes and supporting cell growth (Reddy et al. 2019). Leafy greens contain carotenoids like lutein and zeaxanthin that promote eye health while reducing the risk of age-related macular degeneration (Kuo and McClements 2022). The high content of vitamin K in leafy greens contributes to bone health by facilitating calcium absorption and promoting bone mineralization (Wasilewski et al. 2019). The phytochemicals present in greens, such as flavonoids and glucosinolates, possess antiinflammatory properties that contribute to overall well-being (Oz and Kafkas 2017). Including greens in your diet can help reduce the risk of heart disease by lowering blood pressure and cholesterol levels (Alissa and Ferns 2017).

Cruciferous vegetables, such as broccoli, cauliflower Brussels sprouts and cabbage are fighters against cancer. They contain compounds called glucosinolates that have been shown to inhibit tumour growth and protect against types of cancer (Mitra et al. 2022). These veggies also support the body's natural detoxification processes due to their sulfur-containing compounds. This helps eliminate substances from our system (Fujioka et al. 2016). Moreover, cruciferous vegetables contribute to heart health by providing fibre, potassium and antioxidants. These nutrients help lower blood pressure reduce inflammation and maintain cholesterol levels (Rana et al. 2019). In addition to that, these veggies play a role in promoting bone health by supplying vitamin K and calcium. Both are essential for bones and preventing fractures (Webster et al. 2023). Furthermore, the fibre present in vegetables supports health by promoting regular bowel movements and maintaining a healthy gut microbiome (Makki et al. 2018). Moving on to fruits, they not only offer exotic flavours but also serve as sources of valuable nutraceuticals that can benefit our overall well-being. These fruits, often associated with locations offer a range of health benefits (Aguilera and Toledo 2022).

Tropical fruits, like oranges, papayas and guavas are packed with vitamin C. This essential nutrient boosts the system and aids in collagen formation (Sarkar et al.

2022). Helps heal wounds. The vibrant colors found in fruits, such as the beta carotene in mangoes and lycopene in watermelons indicate their high antioxidant content. These antioxidants fight against radicals and protect cells from damage (Deis et al. 2021). Pineapple and papaya contain enzymes like bromelain and papain respectively which aid digestion and reduce bloating (Anusha et al. 2018). Potassium-rich tropical fruits like bananas and kiwi fruits contribute to maintaining blood pressure levels and promote health (Butu and Rodino 2019). Tropical fruits have a water content along with vitamins that help keep you hydrated while promoting skin health (Kola–Fasanu OO 2018). Fruits and vegetables that are rich in nutrients have a range of flavours, textures and health-enhancing properties. Berries, are known for their antioxidant properties along with leafy greens that are packed with nutrients and vegetables that help fight against cancer and the exotic nutraceutical sources found in tropical fruits—all of these gifts from nature offer a variety of health benefits (Fernandes et al. 2017). By including these treasures in our diet we embark on a path, towards health savouring each delicious bite along the way.

17.9 Nutraceutical Extraction and Application

The exploration of nutraceuticals those health-enhancing substances found in fruits and vegetables goes beyond the fields and orchards. In this section, we dive into the world of extraction techniques and the art of incorporating these valuable compounds into functional foods. Our goal is to uncover nature's treasure trove and harness its benefits, for wellbeing. Extraction plays a role in revealing the true potential of nutraceuticals by isolating their bioactive compounds from natural sources. In the quest to unlock the nutraceuticals hidden within plant materials a wide range of extraction techniques has emerged. These methods can be compared to the crafted tools of an artisan each one carefully designed to target specific compounds and optimize extraction efficiency. Let us dive deeper into the intricacies of these extraction techniques. In this part, we discuss the various extraction techniques with the first being the usage of Solvents, this approach involves using solvents, like ethanol or hexane to dissolve and extract nutraceuticals from plant material. It is a technique that works well for compounds with polarities. Solvents act as messengers carrying away the cargo of bioactive compounds from within the plant matrix (Oreopoulou et al. 2019). Another technique is by Supercritical Fluid Extraction (SFE), in this extraction method fluids, typically, carbon dioxide (CO_2) take centre stage under precise conditions of high pressure and temperature (Bhusnure et al. 2015). SFE is renowned for its cleanliness and efficiency leaving no solvent residues behind. In its state CO_2 behaves both as a gas and a liquid, permeating plant structures and transporting nutraceuticals along with it only to be gently released during depressurization (Villacís-Chiriboga et al. 2021). Moreover, Ultrasound Assisted Extraction (UAE) is another technique and by harnessing the power of waves UAE generates cavitation bubbles within the solvent. When these bubbles collapse near the walls of plant cells they create a stress that helps release compounds (Khandare et al. 2021). It's, like a symphony of vibrations coaxing the compounds out from their places. Microwave Assisted Extraction (MAE) is another extraction technique were we utilize the power of microwave technology. Microwaves generate heat inside the plant material for a chef to carefully warm up a dish. This heat speeds up the extraction process (Bommakanti et al. 2023). Allows for the liberation of beneficial compounds. It's like artistry using microwaves as a tool. An additional extraction method is Enzyme Assisted Extraction, for this method, enzymes play a role as nature's craftsmen. We add enzymes to the plant material to delicately break down cell walls and facilitate the release of compounds. It's comparable to enlisting the skills of a master sculptor who skillfully carves out gems hidden within stone (Patil et al. 2021). There is also, Pressurized Liquid Extraction (PLE), this combines pressure and temperature with solvents in a manner that creates a perfectly coordinated dance between molecular forces. By controlling pressure and heat we persuade compounds to detach from plant structures (Dubazana 2022). It's, like a duet that ensures efficiency and precision. Interestingly, Solid Phase Microextraction (SPME) is preferred method of extraction when it comes to nutraceuticals. This technique uses a solid phase fibre that acts like an absorbent sponge capturing those compounds (Zheng et al. 2023). It's almost as if these volatile nutraceuticals willingly nestle themselves within the fibre ready to be analyzed once released. Another extraction method called Pulsed Electric Field (PEF) introduces bursts of voltage electric pulses into the equation. These pulses create pores, in the membranes of plant cells granting access to the treasures hidden within (Jin and Zhang 2020; Majid et al. 2023). Nutraceuticals that were once trapped within the plant's barriers can now be easily extracted. Like a master painter's brush strokes, each of these extraction techniques has its distinct qualities that make them suitable, for specific types of nutraceutical compounds and applications. Through their precision and innovative approaches, they allow us to unlock the health-enhancing potential found in nature's creations.

17.10 Contemplations; Exploring the Complex World of Nutraceuticals

An improved understanding of the different kinds of nutraceuticals available in diverse fruits and vegetables certainly prompts our senses on how complicated and mindboggling this whole field entails. Besides the nutritional and health benefits of nutraceuticals, there are also complex ethical issues associated with their acquisition and usage (Rucker and Rucker 2016). Such complicated issues must be resolved to improve the nutritional and health benefits we derive from nutraceuticals and also the health of our environment. Their acquisition and usage should not endanger the health of our environment or other animals and humans (Chopra et al. 2022). This part of the study provides insight into the controversial issues surrounding nutraceuticals and their considerations. Producing nutraceuticals involves encouraging the world to flourish pragmatically. In that, there should be a sustainable sourcing and utilization of these nutraceuticals without increasing the dangers to our natural environment (Nounou et al. 2018). It should be such that our vast biodiversity is preserved and with little or no harm. In this regard, it is important to place emphasis on various procurement methods and ensure that we do not endanger the environment in the name of nutraceutical acquisition (Chopra et al. 2022). We cannot promote the utilization of nutraceuticals without creating means and ways to obtain them, such as farming, extraction and consumption. It is, therefore, a necessity to make sure that there are guiding principles that are extended to regulate the farming, extraction and consumption of this health-improving aspect of fruits and vegetables (Ghafoor et al. 2017). Nutraceuticals hardly exhibit efficacy discretely; their effectiveness mostly amplifies when combined with compounds found in vegetables and fruits. The ability of nutraceuticals to exhibit synergistic effects makes them highly remarkable, especially in the fight against diseases and antimicrobial resistance (Martelli and Giacomini 2018). Scientists and nutritionists are now very interested in discovering products that synergize with nutraceuticals to maximize their benefits. A typical example is when someone consumes foods that contain great quantities of vitamin C and iron can influence the uptake of iron by the body while some phytochemicals also work hand in hand to combat stress (Gopi and Balakrishnan 2022). By uncovering and harnessing these interconnections, scientists and nutritionists can fully optimize the true value of diets that contain higher quantities of nutraceuticals.

Empowerment of consumers is crucial in harnessing the full utilization of nutraceuticals. This can be done through education where consumers will be made to know the potential benefits when they consume nutraceuticals (Peacock et al. 2019). When people are made to understand the nutritional value of some foods, they are capable of making their own choices. It is rather unfortunate that some individuals might resort to not taking some diets because of limited knowledge of the health benefits of such foods. Another empowerment area of consumers is the implementation of labelling and practices that help in providing accessible tools for people to gain an in-depth insight into the role of nutraceuticals in the prevention of diseases and promotion of health (Bagchi 2014).

17.11 Conclusion

Our study showed that fruits and vegetables are an important source of nutraceuticals and it can provide health benefits beyond human nutrition. Since, nutraceuticals are bioactive constituents derived from vegetables and fruits, which can support good and quality life style and hence it can improve disease prevention. Nutraceuticals act as a bridge connecting nutrition and medicine providing an approach, to quality health. Vegetables and Fruits contain various bioactive components like glucosinolates, antioxidants and polyphenols, which can provide novel benefits to enhance human health. Nutraceuticals wield their effects through diverse mechanisms including anti-inflammatory properties, hormone regulation, neuroprotection, antioxidant properties, cellular repair, and modulation of gene expression, blood sugar regulation and support for gut health. The multifaceted activities exhibited by nutraceuticals allow them to help in health maintenance and the prevention of diseases. Pragmatic consumption of fruits and vegetables based nutraceuticals will not only improve our health and strengthen our system but it can go a long way in preventing and managing cancer as well as improving cognitive function. Furthermore, funding rigorous state-of-the-art scientific studies to discover and deeply investigate the various functions of nutraceuticals will go a long way to harnessing their incorporation into modern medicine.

References

Abate L (2017) Healing health anxiety: the comprehensive guide. TEACH Services, Inc.

Adan RAH, van der Beek EM, Buitelaar JK, Cryan JF, Hebebrand J, Higgs S et al (2019) Nutritional psychiatry: towards improving mental health by what you eat. Eur Neuropsychopharmacol 29(12):1321–1332

Agrawal RS, Ranveer RC, Rathod NB, Nirmal NP (2023) Phytochemicals as bioactive ingredients for functional foods. Elsevier, Recent Frontiers of Phytochemicals, pp 95–108

Aguilera JM, Toledo T (2022) Wild berries and related wild small fruits as traditional healthy foods. Crit Rev Food Sci Nutr 64:1–15

Al Rahmad AH (2023) Scoping review: the role of micronutrients (Fe, Zn, Iodine, Retinol, Folate) during pregnancy. J Kesehatan Manarang 9(1):1–13

Alamgir ANM, Alamgir ANM (2018) Vitamins, nutraceuticals, food additives, enzymes, anesthetic aids, and cosmetics. In: Therapeutic use of medicinal plants and their extracts: volume 2: phytochemistry and bioactive compounds, pp 407–534

Albuquerque TG, Nunes MA, Bessada SMF, Costa HS, Oliveira MBPP (2020) Biologically active and health promoting food components of nuts, oilseeds, fruits, vegetables, cereals, and legumes. Elsevier, Chemical analysis of food, pp 609–656

Alcorta A, Porta A, Tárrega A, Alvarez MD, Vaquero MP (2021) Foods for plant-based diets: challenges and innovations. Food Secur 10(2):293

Alissa EM, Ferns GA (2017) Dietary fruits and vegetables and cardiovascular diseases risk. Crit Rev Food Sci Nutr 57(9):1950–1962

Alpert PT (2017) The role of vitamins and minerals on the immune system. Home Health Care Manag Pract 29(3):199–202

Aly SM, Elatta MA, Nasr AA, Fathi M (2023) Efficacy of garlic and cinnamon as an alternative to chemotherapeutic agents in controlling Saprolegnia infection in Nile tilapia. In: Aquaculture and fisheries

Ansari S (2021) Heart health and nutraceuticals. In: Preparation of phytopharmaceuticals for the management of disorders. Elsevier, pp 255–272

Anusha MB, Shivanna N, Kumar GP, Anilakumar KR (2018) Efficiency of selected food ingredients on protein efficiency ratio, glycemic index and in vitro digestive properties. J Food Sci Technol 55:1913–1921

Arumugam T, Sona CL, Maheswari MU (2021) Fruits and vegetables as superfoods: scope and demand. J Pharm Innov 10:119–129

Aslani BA, Ghobadi S (2016) Studies on oxidants and antioxidants with a brief glance at their relevance to the immune system. Life Sci 146:163–173

Avallone R, Vitale G, Bertolotti M (2019) Omega-3 fatty acids and neurodegenerative diseases: new evidence in clinical trials. Int J Mol Sci 20(17):4256

Awadelkareem AM, Al-Shammari E, Elkhalifa AEO, Adnan M, Siddiqui AJ, Snoussi M et al (2022a) Phytochemical and In Silico ADME/Tox analysis of eruca sativa extract with antioxidant, antibacterial and anticancer potential against Caco-2 and HCT-116 colorectal carcinoma cell lines. Molecules (Basel, Switzerland) 27(4):1409
Awadelkareem AM, Al-Shammari E, Elkhalifa AO, Adnan M, Siddiqui AJ, Patel M et al (2022b) Biosynthesized silver nanoparticles from Eruca sativa miller leaf extract exhibits antibacterial, antioxidant, anti-quorum-sensing, Antibiofilm, and anti-metastatic activities. Antibiotics 11(7):853
Bagchi D (2014) Nutraceutical and functional food regulations in the United States and around the world. Elsevier
Basak S, Gokhale J (2022) Immunity boosting nutraceuticals: current trends and challenges. J Food Biochem 46(3):e13902
Bauer KC, Rees T, Finlay BB (2019) The gut microbiota–brain axis expands neurologic function: a nervous rapport. BioEssays 41(10):1800268
Bhusnure OG, Gholve SB, Giram PS, Borsure VS, Jadhav PP, Satpute VV et al (2015) Importance of supercritical fluid extraction techniques in pharmaceutical industry: a review. Indo Am J Pharm Res 5:3785–3801
Bloom J (2018) Creating sanctuary: sacred garden spaces, plant-based medicine, and daily practices to achieve happiness and well-being. Timber Press
Bommakanti V, Puthenparambil Ajikumar A, Sivi CM, Prakash G, Mundanat AS, Ahmad F et al (2023) An overview of herbal nutraceuticals, their extraction, formulation, therapeutic effects and potential toxicity. Separations 10(3):177
Bulsiewicz W (2020) Fiber fueled: the plant-based gut health program for losing weight, restoring your health, and optimizing your microbiome. Penguin
Butu M, Rodino S (2019) Fruit and vegetable-based beverages—nutritional properties and health benefits. Elsevier, Natural beverages, pp 303–338
Calado R, Leal MC, Gaspar H, Santos S, Marques A, Nunes ML et al (2018) How to succeed in marketing marine natural products for nutraceutical, pharmaceutical and cosmeceutical markets. In: Grand challenges in marine biotechnology, pp 317–403
Carazo A, Macáková K, Matoušová K, Krčmová LK, Protti M, Mladěnka P (2021) Vitamin a update: forms, sources, kinetics, detection, function, deficiency, therapeutic use and toxicity. Nutrients 13(5):1703
Chopra AS, Lordan R, Horbańczuk OK, Atanasov AG, Chopra I, Horbańczuk JO et al (2022) The current use and evolving landscape of nutraceuticals. Pharmacol Res 175:106001
Ciosek Ż, Kot K, Kosik-Bogacka D, Łanocha-Arendarczyk N, Rotter I (2021) The effects of calcium, magnesium, phosphorus, fluoride, and lead on bone tissue. Biomol Ther 11(4):506
Coscueta ER, Sousa AS, Reis CA, Pintado MM (2022) Phenylethyl isothiocyanate: a bioactive agent for gastrointestinal health. Molecules 27(3):794
Cui X, Lin Q, Liang Y (2020) Plant-derived antioxidants protect the nervous system from aging by inhibiting oxidative stress. Front Aging Neurosci 12:209
Dahiya D, Nigam PS (2022) Probiotics, prebiotics, synbiotics, and fermented foods as potential biotics in nutrition improving health via microbiome-gut-brain axis. Fermentation 8(7):303
Daily JW, Yang M, Park S (2016) Efficacy of turmeric extracts and curcumin for alleviating the symptoms of joint arthritis: a systematic review and meta-analysis of randomized clinical trials. J Med Food 19(8):717–729
Darrell N (2020) Conversations with plants: the path back to nature. Aeon Books
De Amicis R, Mambrini SP, Pellizzari M, Foppiani A, Bertoli S, Battezzati A et al (2022) Systematic review on the potential effect of berry intake in the cognitive functions of healthy people. Nutrients 14(14):2977
Deis L, Quiroga AM, De Rosas MI (2021) Coloured compounds in fruits and vegetables and health. In: Psychiatry and neuroscience update: From epistemology to clinical psychiatry–Vol IV, pp 343–358

Delormier T, Marquis K (2019) Building healthy community relationships through food security and food sovereignty. Curr Dev Nutr 3(Supplement_2):25–31

Devaki SJ, Raveendran RL (2017) Vitamin C: sources, functions, sensing and analysis. In: Vitamin C. IntechOpen

Di Pietro N, Baldassarre MPA, Cichelli A, Pandolfi A, Formoso G, Pipino C (2020) Role of polyphenols and carotenoids in endothelial dysfunction: an overview from classic to innovative biomarkers. Oxidative Med Cell Longev 2020:1–19

Domínguez Díaz L, Fernández-Ruiz V, Cámara M (2020) The frontier between nutrition and pharma: the international regulatory framework of functional foods, food supplements and nutraceuticals. Crit Rev Food Sci Nutr 60(10):1738–1746

Dubazana B (2022) Analytical methodologies for extraction of Vitamins A, C and E in selected cosmetics. University of Johannesburg (South Africa)

Egbuna C, Dable-Tupas G (2020) Functional foods and nutraceuticals, vol 1. Springer Nature Switzerland AG, pp 1–632

Egbuna C, Mishra AP, Goyal MR (2020) Preparation of phytopharmaceuticals for the management of disorders: the development of nutraceuticals and traditional medicine. Academic

Eggers NA, de Blanco EJC (2016) Phytochemicals in nutrition and health. In: Nutrition: an approach to good health and disease management, p 201

Elkhalifa AEO, Alshammari E, Adnan M, Alcantara JC, Awadelkareem AM, Eltoum NE et al (2021) Okra (Abelmoschus Esculentus) as a potential dietary medicine with nutraceutical importance for sustainable health applications. Molecules (Basel, Switzerland) 26(3):696

Farooqui AA, Farooqui AA (2015) Importance and roles of fiber in the diet. In: High calorie diet and the human brain: metabolic consequences of long-term consumption, pp 193–218

Fernandes L, Casal S, Pereira JA, Saraiva JA, Ramalhosa E (2017) Edible flowers: a review of the nutritional, antioxidant, antimicrobial properties and effects on human health. J Food Compos Anal 60:38–50

Forni C, Facchiano F, Bartoli M, Pieretti S, Facchiano A, D'Arcangelo D et al (2019) Beneficial role of phytochemicals on oxidative stress and age-related diseases. Biomed Res Int 2019:8748253

Freitas AC, Rodrigues D, Rocha-Santos TAP, Gomes AMP, Duarte AC (2012) Marine biotechnology advances towards applications in new functional foods. Biotechnol Adv 30(6):1506–1515

Fujioka N, Fritz V, Upadhyaya P, Kassie F, Hecht SS (2016) Research on cruciferous vegetables, indole-3-carbinol, and cancer prevention: a tribute to lee W. Wattenberg Mol Nutr Food Res 60(6):1228–1238

Galanakis CM (2019) Nutraceuticals and natural product pharmaceuticals. Academic

Garza-Juárez A, Pérez-Carrillo E, Arredondo-Espinoza EU, Islas JF, Benítez-Chao DF, Escamilla-García E (2023) Nutraceuticals and their contribution to preventing noncommunicable diseases. Food Secur 12(17):3262

George VC, Dellaire G, Rupasinghe HPV (2017) Plant flavonoids in cancer chemoprevention: role in genome stability. J Nutr Biochem 45:1–14

Ghafoor K, Özcan MM, Al-Juhaımı F, Babıker EE, Sarker ZI, Ahmed IAM et al (2017) Nutritional composition, extraction, and utilization of wheat germ oil: a review. Eur J Lipid Sci Technol 119(7):1600160

Gopi S, Balakrishnan P (2022) Handbook of nutraceuticals and natural products. Wiley Online Library

Gorzynik-Debicka M, Przychodzen P, Cappello F, Kuban-Jankowska A, Marino Gammazza A, Knap N et al (2018) Potential health benefits of olive oil and plant polyphenols. Int J Mol Sci 19(3):686

Gounden V, Vashisht R, Jialal I (2018) Hypoalbuminemia. StatPearls Publishing

Grassi D, Ferri C, Desideri G (2016) Brain protection and cognitive function: cocoa flavonoids as nutraceuticals. Curr Pharm Des 22(2):145–151

Gröber U, Schmidt J, Kisters K (2015) Magnesium in prevention and therapy. Nutrients 7(9):8199–8226

Gul K, Singh AK, Jabeen R (2016) Nutraceuticals and functional foods: the foods for the future world. Crit Rev Food Sci Nutr 56(16):2617–2627

Gupta E, Mishra P (2021) Functional food with some health benefits, so called superfood: a review. Curr Nutr Food Sci 17(2):144–166

Gupta C, Prakash D (2015) Nutraceuticals for geriatrics. J Tradit Complement Med 5(1):5–14

Hanna M, Jaqua E, Nguyen V, Clay JB (2022) Vitamins: functions and uses in medicine. Perm J 26:89–97

Jennings HM (2014) Travelling people, travelling plants: an exploration into food-plant practice among Bengali women across transnational and generational landscapes. University of London, University College London

Jin TZ, Zhang HQ (2020) Pulsed electric fields for pasteurization: Food safety and shelf life. In: Food safety engineering, pp 553–577

Kaurinovic B, Vastag D (2019) Flavonoids and phenolic acids as potential natural antioxidants. IntechOpen, London

Khadim RM, Al-Fartusie FS (eds) (2021) Antioxidant vitamins and their effect on immune system. IOP Publishing

Khan SU, Khan MU, Riaz H, Valavoor S, Zhao D, Vaughan L et al (2019) Effects of nutritional supplements and dietary interventions on cardiovascular outcomes: an umbrella review and evidence map. Ann Intern Med 171(3):190–198

Khandare RD, Tomke PD, Rathod VK (2021) Kinetic modeling and process intensification of ultrasound-assisted extraction of d-limonene using citrus industry waste. Chem Eng Process Process Intensif 159:108181

Kola–Fasanu OO (2018) Biochemical analysis and antioxidant activity of Three different African tropical fruits (African Star Apple-Cherry, Avocado, and Lemon) Noted for their significant health benefits

Kumar R, Vijayalakshmi S, Nadanasabapathi S (2017) Health benefits of quercetin. Def Life Sci J 2(10):42–151

Kunnumakkara AB, Bordoloi D, Padmavathi G, Monisha J, Roy NK, Prasad S et al (2017) Curcumin, the golden nutraceutical: multitargeting for multiple chronic diseases. Br J Pharmacol 174(11):1325–1348

Kuo E, McClements DJ (2022) Food colloid-based delivery systems for tackling age-related macular degeneration by enhancing carotenoid bioavailability: a review. Food Hydrocoll Health 2:100093

Leena MM, Silvia MG, Vinitha K, Moses JA, Anandharamakrishnan C (2020) Synergistic potential of nutraceuticals: mechanisms and prospects for futuristic medicine. Food Funct 11(11):9317–9337

Liu J-K (2022) Natural products in cosmetics. Nat Prod Bioprospect 12(1):40

Maiuolo J, Gliozzi M, Carresi C, Musolino V, Oppedisano F, Scarano F et al (2021) Nutraceuticals and cancer: potential for natural polyphenols. Nutrients 13(11):3834

Majid I, Khan S, Aladel A, Dar AH, Adnan M, Khan MI et al (2023) Recent insights into green extraction techniques as efficient methods for the extraction of bioactive components and essential oils from foods. CyTA J Food 21(1):101–114

Makki K, Deehan EC, Walter J, Bäckhed F (2018) The impact of dietary fiber on gut microbiota in host health and disease. Cell Host Microbe 23(6):705–715

Mali S, Rathod S, Kale N, Shinde N (2022) Overview of nutraceuticals. Asian J Pharm Res 12(1):61–70

Manoj PM, Mohan JR, Khasherao BY, Shams R, Dash KK (2023) Fruit based probiotic functional beverages: a review. J Agric Food Res 14:100729

Maoka T (2020) Carotenoids as natural functional pigments. J Nat Med 74(1):1–16

Marcus JB (2019) Aging, nutrition and taste: nutrition, food science and culinary perspectives for aging tastefully. Academic

Marshall A, Dezuanni M, Burgess J, Thomas J, Wilson CK (2020) Australian farmers left behind in the digital economy–insights from the Australian digital inclusion index. J Rural Stud 80:195–210

Martelli G, Giacomini D (2018) Antibacterial and antioxidant activities for natural and synthetic dual-active compounds. Eur J Med Chem 158:91–105

McLean RM, Wang NX (2021) Potassium. In: Advances in food and nutrition research, vol 96. Elsevier, pp 89–121

Mironenko A, Eliseeva T (2020) Magnesium (Mg, Magnesium)–description, effect on the body, the best sources of magnesium. J Health Nutr Diet 4(14):60–72

Mitra S, Emran TB, Chandran D, Zidan BM, Das R, Mamada SS et al (2022) Cruciferous vegetables as a treasure of functional foods bioactive compounds: targeting p53 family in gastrointestinal tract and associated cancers. Front Nutr 9:951935

Moreau RA, Nyström L, Whitaker BD, Winkler-Moser JK, Baer DJ, Gebauer SK et al (2018) Phytosterols and their derivatives: structural diversity, distribution, metabolism, analysis, and health-promoting uses. Prog Lipid Res 70:35–61

Neda M-D, Beara I, Vojnović T, Helene CJ (2021) Bioactives for neuronal and immune functions. In: Food bioactives and health, pp 269–305

Nigam D, Yadav R, Tiwari U (2018) Omega-3 fatty acids and its role in human health. In: Functional food and human health, pp 173–198

Nounou MI, Ko Y, Helal NA, Boltz JF (2018) Adulteration and counterfeiting of online nutraceutical formulations in the United States: time for intervention? J Diet Suppl 15(5):789–804

Oakenfull D, Sidhu GS (2023) Saponins. In: Toxicants of plant origin. CRC Press, pp 97–142

Ogbe RJ, Ochalefu DO, Mafulul SG, Olaniru OB (2015) A review on dietary phytosterols: their occurrence, metabolism and health benefits. Asian J Plant Sci Res 5(4):10–21

Oreopoulou A, Tsimogiannis D, Oreopoulou V (2019) Extraction of polyphenols from aromatic and medicinal plants: an overview of the methods and the effect of extraction parameters. In: Polyphenols in plants, pp 243–259

Overby HB (2019) Exploring novel anti-obesity strategies: nutritional and pharmacological interventions to induce Browning and Brown Adipogenesis

Oz AT, Kafkas E (2017) Phytochemicals in fruits and vegetables. In: Waisundara V superfood and functional food London. IntechOpen, pp p175–p184

Pang G, Xie J, Chen Q, Hu Z (2012) How functional foods play critical roles in human health. Food Sci Human Wellness 1(1):26–60

Patil PD, Patil SP, Kelkar RK, Patil NP, Pise PV, Nadar SS (2021) Enzyme-assisted supercritical fluid extraction: an integral approach to extract bioactive compounds. Trends Food Sci Technol 116:357–369

Peacock M, Badea M, Bruno F, Timotijevic L, Laccisaglia M, Hodgkins C et al (2019) Herbal supplements in the print media: communicating benefits and risks. BMC Complement Altern Med 19:1–14

Perez-Vizcaino F, Fraga CG (2018) Research trends in flavonoids and health. Arch Biochem Biophys 646:107–112

Pezzani R, Salehi B, Vitalini S, Iriti M, Zuñiga FA, Sharifi-Rad J et al (2019) Synergistic effects of plant derivatives and conventional chemotherapeutic agents: an update on the cancer perspective. Medicina 55(4):110

Pinnock D (2011) Healing foods: prevent and treat common illnesses with fruits, vegetables, herbs, and more. Simon and Schuster

Poli A, Marangoni F, Corsini A, Manzato E, Marrocco W, Martini D et al (2021) Phytosterols, cholesterol control, and cardiovascular disease. Nutrients 13(8):2810

Raj R, Shams R, Pandey VK, Dash KK, Singh P, Bashir O (2023) Barley phytochemicals and health promoting benefits: a comprehensive review. J Agric Food Res:100677. https://doi.org/10.1016/j.jafr.2023.100677

Ramya V, Patel P (2019) Health benefits of vegetables. Int J Chem Stud 7(2):82–87

Rana N, Ghabru A, Vaidya D (2019) Defensive function of fruits and vegetables. J Pharmacogn Phytochem 8(3):1872–1877
Rana A, Samtiya M, Dhewa T, Mishra V, Aluko RE (2022) Health benefits of polyphenols: a concise review. J Food Biochem 46(10):e14264
Rao GP, Singh P (2022) Value addition and fortification in non-centrifugal sugar (jaggery): a potential source of functional and nutraceutical foods. Sugar Tech 24(2):387–396
Raseetha S, Aida FMNA, Roslan F, Vitamin K (2022) Handbook of food bioactive ingredients: properties and applications. Springer, pp 1–37
Rathore H, Prasad S, Sharma S (2017) Mushroom nutraceuticals for improved nutrition and better human health: a review. PharmaNutrition 5(2):35–46
Reddy BN, Naganathan L, Sunitha ANV, Jyothi RB, Chaitanya B, Sharma JVC (2019) A review on nutraceuticals. J Pharmacogn Phytochem 8(4):735–744
Rees A, Dodd GF, Spencer JPE (2018) The effects of flavonoids on cardiovascular health: a review of human intervention trials and implications for cerebrovascular function. Nutrients 10(12):1852
Reinarz J (2023) A cultural history of medicine in the age of Empire. Bloomsbury Publishing
Rosenberg A, Mangialasche F, Ngandu T, Solomon A, Kivipelto M (2020) Multidomain interventions to prevent cognitive impairment, Alzheimer's disease, and dementia: from FINGER to world-wide FINGERS. J Prev Alzheimers Dis 7:29–36
Rucker RB, Rucker MR (2016) Nutrition: ethical issues and challenges. Nutr Res 36(11):1183–1192
Sachdeva V, Roy A, Bharadvaja N (2020) Current prospects of nutraceuticals: a review. Curr Pharm Biotechnol 21(10):884–896
Sadeghinejad M, Soltani Z, Afzalpour ME, Khaksari M, Pourranjbar M (2019) What is the combined effect of intense intermittent exercise and Ginkgo biloba plant on the brain neurotrophic factors levels, and learning and memory in young rats? Pharmacol Rep 71(3):503–508
Salehi B, Zucca P, Orhan IE, Azzini E, Adetunji CO, Mohammed SA et al (2019) Allicin and health: a comprehensive review. Trends Food Sci Technol 86:502–516
Santini A, Tenore GC, Novellino E (2017) Nutraceuticals: a paradigm of proactive medicine. Eur J Pharm Sci 96:53–61
Sarkar T, Salauddin M, Roy A, Sharma N, Sharma A, Yadav S et al (2022) Minor tropical fruits as a potential source of bioactive and functional foods. Crit Rev Food Sci Nutr 63:1–45
Sawikr Y, Yarla NS, Peluso I, Kamal MA, Aliev G, Bishayee A (2017) Neuroinflammation in Alzheimer's disease: the preventive and therapeutic potential of polyphenolic nutraceuticals. Adv Protein Chem Struct Biol 108:33–57
Seow HF (2020) Nutrients and immune function. World Scientific
Serai M. Biologically active compounds like ashwagandha and ginseng work as stress relievers and alternatives to antidepressants. The Crow 2021:33
Shanahan C (2017) Deep nutrition: why your genes need traditional food. Flatiron Books
Sharifi-Rad M, Anil Kumar NV, Zucca P, Varoni EM, Dini L, Panzarini E et al (2020) Lifestyle, oxidative stress, and antioxidants: Back and forth in the pathophysiology of chronic diseases. Front Physiol 11:694
Shukla R, Pandey V, Vadnere GP, Lodhi S (2019) Role of flavonoids in management of inflammatory disorders. Elsevier, Bioactive food as dietary interventions for arthritis and related inflammatory diseases, pp 293–322
Simioni C, Zauli G, Martelli AM, Vitale M, Sacchetti G, Gonelli A et al (2018) Oxidative stress: role of physical exercise and antioxidant nutraceuticals in adulthood and aging. Oncotarget 9(24):17181
Singh K, Kumar S, Jangra A (2018) Berries healthiest foods on Earth. Kerala-Karshakan 5(10):18–21
Sizer F, Whitney E (2022) Nutrition: concepts & controversies. Cengage Learning
Stice CP, Xia H, Wang X-D (2018) Tomato lycopene prevention of alcoholic fatty liver disease and hepatocellular carcinoma development. Chronic Dis Trans Med 4(4):211–224

Szot M, Karpęcka-Gałka E, Dróżdż R, Frączek B (eds) (2022) Can nutrients and dietary supplements potentially improve cognitive performance also in esports? MDPI
Towns AM, Shackleton C (2018) Traditional, indigenous, or leafy? A definition, typology, and way forward for African vegetables. Econ Bot 72:461–477
Valenzuela-Gutiérrez R, Lago-Lestón A, Vargas-Albores F, Cicala F, Martínez-Porchas M (2021) Exploring the garlic (Allium sativum) properties for fish aquaculture. Fish Physiol Biochem 47(4):1179–1198
Verma G, Mishra MK (2016) A review on nutraceuticals: classification and its role in various diseases. Int J Pharm Ther 7(4):152–160
Villacís-Chiriboga J, Vera E, Van Camp J, Ruales J, Elst K (2021) Valorization of byproducts from tropical fruits: a review, part 2: applications, economic, and environmental aspects of biorefinery via supercritical fluid extraction. Compr Rev Food Sci Food Saf 20(3):2305–2331
Vinayagam R, Xu B (2015) Antidiabetic properties of dietary flavonoids: a cellular mechanism review. Nutr Metab 12:1–20
Vrânceanu M, Galimberti D, Banc R, Dragoş O, Cozma-Petruţ A, Hegheş S-C et al (2022) The anticancer potential of plant-derived nutraceuticals via the modulation of gene expression. Plan Theory 11(19):2524
Wagh NS, Pai SR, Sonkamble VV (2020) Phytochemicals in the prevention and cure of cancers. In: Plant-derived bioactives: chemistry and mode of action, pp 351–373
Wallace TC, Bailey RL, Blumberg JB, Burton-Freeman B, Chen CO, Crowe-White KM et al (2020) Fruits, vegetables, and health: a comprehensive narrative, umbrella review of the science and recommendations for enhanced public policy to improve intake. Crit Rev Food Sci Nutr 60(13):2174–2211
Wang Q, Lin P, Feng L, Ren Q, Xie X, Zhang B (2022) Ameliorative effect of allicin on vascular calcification via inhibiting endoplasmic reticulum stress. Vascular 30(5):999–1007
Wasilewski GB, Vervloet MG, Schurgers LJ (2019) The bone—vasculature axis: calcium supplementation and the role of vitamin K. Front Cardiovasc Med 6:6
Webster J, Dalla Via J, Langley C, Smith C, Sale C, Sim M (2023) Nutritional strategies to optimise musculoskeletal health for fall and fracture prevention: looking beyond calcium, vitamin D and protein. Bone Rep 19:101684
Werner-Gray L (2020) Anxiety-free with food: natural, science-backed strategies to relieve stress and support your mental health. Hay House, Inc
Yadav K, Dhankhar J (2022) Isothiocyanates-a review of their health benefits and potential food applications. Curr Res Nutr Food Sci J 10(2):476–502
Yadav A, Kumari R, Yadav A, Mishra JP, Srivatva S, Prabha S (2016) Antioxidants and its functions in human body-a review. Res Environ Life Sci 9(11):1328–1331
Yahia EM, Maldonado Celis ME, Svendsen M (2017) The contribution of fruit and vegetable consumption to human health. In: Fruit and vegetable phytochemicals: chemistry and human health, 2nd edn, pp 1–52
Yusuf K, Saha S, Umar S (2022) Health benefits of dietary fiber for the management of inflammatory bowel disease. Biomedicines 10(6):1242
Zhang R, Han Y, McClements DJ, Xu D, Chen S (2022) Production, characterization, delivery, and cholesterol-lowering mechanism of phytosterols: a review. J Agric Food Chem 70(8):2483–2494
Zheng J, Kuang Y, Zhou S, Gong X, Ouyang G (2023) Latest improvements and expanding applications of solid-phase microextraction. Anal Chem 95(1):218–237
Zia-Ul-Haq M (2021) Historical and introductory aspects of carotenoids. In: Carotenoids: structure and function in the human body, pp 1–42
Zohn R (2020) Break free from eczema: soothe chronic inflammation and itchy skin with at-home solutions and proven treatments for atopic dermatitis. Simon and Schuster

Chapter 18
Safety and Toxicity of Nutraceuticals: Concerns, Challenges and Global Regulatory Aspects

Pawan Prabhakar, Deepak Kumar Verma, and Mamoni Banerjee

Abstract The investigation of safety and toxicity concerns related to nutraceuticals is a prominent subject of study within the realms of nutrition and health. Nutraceuticals refer to bioactive compounds or food sources based products that provide additional health benefits with nutritional value. Vitamins, minerals, botanical extracts, nutritional supplements, and functional food products are potential constituents within this category. Consumers encounter challenges in determining the optimal dosage of nutraceuticals due to the frequent availability of diverse formulations. The safety of these products can be influenced by variations in quality, thus, it is crucial to ensure the uniformity and integrity of these substances. Although nutraceuticals are generally, considered safe when used as directed, excessive consumption can lead to toxicity. Owing to the wide range of nutraceutical products available, their potential physiological advantages, and the imperative to guarantee consumer well-being, regulatory concerns within the nutraceutical sector are intricate and multifaceted. These issues may vary across different countries and often evolve as fresh scientific research reveals new insights into the safety and efficacy concern. Manufacturers, distributors, and consumers of nutraceutical products should consider the regulatory framework in their respective regions. This chapter deals with safety and toxicity issues related to nutraceuticals and their constituents. It also explores the regulatory considerations of various agencies and outlook of Nutraceutical globally.

P. Prabhakar (✉)
Bio-Research Laboratory, Rajendra Mishra School of Engineering Entrepreneurship, Indian Institute of Technology Kharagpur, Kharagpur, West Bengal, India

School of Business, Woxsen University, Hyderabad, Telangana, India

D. K. Verma
Agricultural and Food Engineering Department, Indian Institute of Technology Kharagpur, Kharagpur, West Bengal, India

M. Banerjee
Bio-Research Laboratory, Rajendra Mishra School of Engineering Entrepreneurship, Indian Institute of Technology Kharagpur, Kharagpur, West Bengal, India
e-mail: mamoni@see.iitkgp.ac.in

S. Amir Ashraf, M. Adnan (eds.), *Food Bioactives and Nutraceuticals*,
https://doi.org/10.1007/978-981-97-9723-3_18

Keywords Nutraceuticals · Phytocompounds · Safety · Toxicity · Dietary supplements · Regulatory issues

18.1 Introduction

Nutraceuticals have gained significant attention due to their perceived safety and promise for providing nutritional and therapeutic benefits in the form of food. The burgeoning research in the field of nutrition is continuously advancing through the emergence of novel findings and cutting-edge technologies across the world (Nasri et al. 2014; Fleischhacker et al. 2020). Dr. Stephen DeFelice defined the term "nutraceuticals" in 1989 to refer to food or dietary components that offer health advantages, such as illness prevention and therapy. They are the compounds that possess both health-promoting and health-protective properties. They can be categorized into two groups: macronutrients and micronutrients (Blaze 2021). The macronutrients consist of carbohydrate, protein, lipid (fat), and crude fibre. Micronutrients encompass a variety of essential elements such as minerals, vitamins, antioxidants, and other compounds. Nutraceuticals are pharmacological formulations that contain food-derived phytochemicals as their active ingredients. These products can include individual nutrients, nutritional supplements, mainly specific diets, genetically modified foods, functional foods, herbal products, and processed foods including cereals, soups, and beverages (Puri et al. 2022; Prabhakar et al. 2023a). The issue of nutraceuticals is gaining popularity as people become more aware of the possible advantages of using dietary supplements alongside pharmacological therapy (Vishvakarma et al. 2023). Nutraceuticals possess the potential to enhance health, prolong the ageing process, mitigate chronic illnesses, augment life expectancy, and bolster the structure and functionality of the body. Additionally, they are employed in the prevention and management of mental health conditions and disorders (Daliu et al. 2019; Makkar et al. 2020). There is a significant rise in the utilisation of nutraceuticals for reproductive problems due to mounting clinical data supporting their therapeutic advantages. They are gaining popularity as an alternative to established medicines for reproductive diseases that have been ineffective due to various underlying conditions, nutritional inadequacies, and the limited response to traditional therapy (Nooshkam et al. 2020; Xiang et al. 2021). Several medically significant nutraceuticals, such as nutritional supplements, vitamins, and antioxidants, have demonstrated the ability to enhance reproductive performance and effectively address reproductive decline associated with ageing and infertility in humans (Silva et al. 2019; Horn and Yu 2020). Therefore, currently, emerging scientific research is progressively confirming the idea that diets, food products, and their components have a significant impact on reducing the likelihood of disease and addressing specific problems including high cholesterol, obesity, and low immunity (Liu et al. 2017; Verma et al. 2020; Barua et al. 2023). The bioactive compounds from plants have shown diverse pharmacological activities that suggest they could have a substantial effect in preventing chronic illnesses. Fruits, vegetables, and medicinal herbs are widely recognised for their many biological activities and antioxidant advantages. The bioactive compounds like phenolics have significant antioxidant activity present in plant as these bioactive molecules have the capability to reduce reactive oxygen species (Samtiya et al. 2021; Bautista-Hernández et al. 2021; Yuan et al. 2022). Regular consumption of synthetic drugs for the management of different ailments might lead to severe adverse reactions or organ dysfunction, whereas nutraceuticals derived from natural sources are considered to be safe (Prabhakar and Banerjee 2020a, 2020c). The daily dietary intake of many individuals lacks sufficient amounts of fibre, carbohydrates, proteins, as well as the appropriate quantities of vitamins and minerals required to sustain optimal health. The primary cause of the increased prevalence of chronic diseases such as cardiovascular diseases, cancer, diabetes which often result in mortality, is mostly attributed to a diet of inadequate quality and sedentary lifestyle (Bhattacharyya and Prabhakar 2015; Cena and Calder 2020). The mortality rates are rising each year mostly because people are not consuming enough fruits, nuts/seeds, whole grains, veggies, seafood, and minerals therefore, nutraceuticals are seen as means of enhancing health (Liu 2013). The treatment of diseases by modern therapeutics is prohibitively costly drugs derived from plants have the potential to be used for treating or preventing diseases, while food derived from plants can act as nutraceuticals by avoiding diseases and promoting health benefits (Sofowora et al. 2013; Elkhalifa et al. 2021).

The rapid rise of nutraceuticals can be attributed to heightened consumer awareness of escalating prescription expenses, the limited effectiveness of traditional medications, and the growing occurrence of negative side effects. The preference for natural products for the treatment of diseases has been reinforced by the lack of effective alternatives for long-term medical conditions, encounters with healthcare professionals without personal rapport or emotional involvement. The rapid expansion of the nutraceuticals market can be attributed to the growing trend of consumers incorporating high levels of nutrients, ideally derived from natural sources, into their regular meals. Nutraceutical products are readily accessible to customers through several distribution channels in the USA, European Union, South-east Asia and other regions (Martinez-Medina et al. 2021; Chopra et al. 2022). Specifically, these items can be acquired at pharmacies, herbalist's shops, supermarkets, or through online platforms. Although nutraceuticals are experiencing a rise in sales within the health and wellness product market, the European Union faces significant obstacles regarding plant-based nutraceuticals (Fehér et al. 2020. Both European Union and Russian customers frequently utilise conventional medicines derived from medicinal plants, Traditional Chinese Medicines (TCM), ayurvedic medicines, and products known as "dietary supplements" or "biologically active supplements to food." The utilisation of dietary supplements, which may consist of a combination of various herbal components and other natural items, is currently being examined by governmental agencies and legislators (Sammons et al. 2016; Prabhakar and Mamoni 2021). The main reason for this scrutiny are the inadequate research evidence regarding the effectiveness and safety of these nutraceuticals and food supplements, the inclusion of prohibited substances or potent conventional medicines in their formulations, and the absence of post-marketing surveillance for adverse reactions and interactions between herbs or herbs and drugs (Dwyer et al. 2018; Molin et al. 2019). These challenges pertain not only to claims made about the products, but also to concerns over their quality and safety. Regrettably, the belief that herbal dietary items are inherently natural and harmless at any dosage, combined with the absence of knowledge regarding the ingredients of these supplements, has resulted in individuals adopting harmful self-medication practices (Brown 2017; Banwo et al. 2021). The main challenge in discussing the regulations of dietary supplements is the lack of a consensus on the definition of the category of products commonly known as dietary supplements, natural health products (NHPs), complementary medicines, or food supplements in different countries. For example, a product that is classified as a dietary supplement and regulated as a food in the United States may be classed differently in another country (Nooshkam et al. 2020; Ng et al. 2022). It might be categorized as a dietary supplement, an adjunctive medicine, a prescribed medication, or perhaps a controlled drug. The scenario becomes further intricate when nations such as China or India, which already possess a regulatory framework for traditional medicine or phytomedicines encompassing raw botanicals, are taken into account (Dwyer et al. 2018; Liu et al. 2020). The improper use of dietary supplements and exposure to this substance can result in several harmful effects, including damage to the liver, lungs, cells, the potential to cause cancer, kidney damage, harm to blood cells, and toxicity to the heart. There is a lack of pharmacological and toxicological investigations for many nutraceuticals, resulting in the unavailability of toxicity and safety data (Bailey 2020). Nutraceuticals have made major profits by effectively achieving desirable therapeutic outcomes while minimising adverse effects in comparison to alternative treatment agents. Consuming plant-based meals, nuts, whole grains, cereals, and marine foods is essential for illness prevention and the promotion of optimal health. The scientific and regulatory challenges encountered in researching the safety, quality, and efficacies of dietary supplements are consistent worldwide, owing to the expanding global market for these products. This chapter presents a comprehensive examination of nutraceuticals, with a specific emphasis on their potential toxicity and the variables that influence their safety. It also discusses the many regulatory measures in place for assessing the safety of nutraceuticals.

18.2 Chemical Diversity and Classification of Nutraceuticals

The nutraceuticals have been divided into mainly three categories:

18.2.1 Dietary Supplements

Dietary supplements are specifically designed goods that are meant to enhance the diet by providing essential elements that may be lacking or insufficiently taken through food. They can be found in various forms, such as liquids, powders, capsules and pills. These supplements may include vitamins, minerals, amino acids, herbs, botanicals, enzymes, and other compounds. The ingredients comprise minerals, vitamins, carnitine, ginseng, Ginkgo, biloba, ashwagandha, Shilajit, St. John's wort, and others (Dwyer et al. 2018).

18.2.2 Medicinal Foods

Medicinal foods are defined as foods that offer additional health advantages beyond their fundamental nutritional value. They frequently consist of bioactive substances, such as antioxidants, vitamins, minerals, and phytochemicals, which have been associated with many health-enhancing characteristics. These foods can contribute to the prevention or management of specific diseases or ailments. Some potential interventions for various health conditions include the consumption of carbohydrates that are digested slowly, the use of lactoferrin to boost the immune system, the development of transgenic plants for oral immunisation against infectious diseases, and the creation of health bars that contain additional drugs (Mortazavian and Meybodi 2016).

18.2.3 Functional Foods

Functional foods are food products that possess the ability to exert beneficial effects on health beyond their fundamental nutritional value. Functional foods share similarities with medicinal foods as they both include bioactive components. However, functional foods are often ingested as part of a regular diet rather than for specific therapeutic purposes. Functional meals are frequently enhanced with extra nutrients or bioactive chemicals to offer health advantages. Foods like oats, bran, psyllium and lignin have been shown to have beneficial effects on heart disease and colon cancer and prebiotics such as oligofructose can help regulate intestinal flora (Ray et al. 2016; Banwo et al. 2021). On other hand Omega-3 milk can help prevent heart disease. The canola oil with reduced triglycerides can aid in reducing cholesterol levels. Stanols, such as Benecol, can help lower cholesterol absorption. Table salt fortified with iodine is also beneficial and there are other substances that may have similar effects as well (Bradberry and Hilleman 2013). Nutraceuticals can be broadly categorised according to their food origins, chemical composition, and pharmacological effects. Nutraceuticals can be obtained from either food sources, such as fortified foods and dietary supplements, or non-food sources, such as herbal extracts and marine products as shown in Table 18.1.

Table 18.1 Nutraceuticals classification based origin and source

Nutraceuticals	Origin	Source	Health benefits	References
Prebiotics and Probiotics	Microbes	Lactobacilli, Streptococci, Bifidobacteria, Oligofructose, Inulin, Galactooligosaccharide, Lactulose	Reduce lactose intolerance, help immune function, help prevent diarrhea	Kumar and Kumar (2015), Godwin (2018), Patra et al. (2021), Gupta et al. (2022), Ben Hsouna et al. (2023), Huda et al. (2024), Goel et al. (2024)
Carnitine, Glucosamine Bovine, Methylsulfonylmethane	Animals	Heart, skeletal muscles	Relief in chronic osteoarthritis, inflammation, promotes joint health, reduces cholesterol	
Curcumin, Lignin, Cellulose, Pectin, Lycopene, Flavonoids, Isoflavonoids, Antioxidant vitamins (vitamin C, vitamin E), Polyunsaturated fatty acids	Plants	Dietary fibres fruits, barley, citrus fruits, wheat germ oil, oats, tomatoes, soybeans, cotton seed oil, peanut oil turmeric, grapes	Protection from cardiovascular and heart diseases, diabetes, obesity, liver damage, renal failure	
Calcium, Chromium, Cobalt, Copper, Iron, Iodine, Magnesium Manganese, Phosphorous, Potassium, Zinc	Minerals	Aloe-vera, barley, black pepper, almonds, apples, apricots, beetroot, berries, buckwheat, caraway seeds, oysters, peanuts, peas, pine nuts, pork, poultry, prawns, shellfish, spirulina, sweet potato, tapioca, watermelon seeds, whole grains, cashew nuts, cherries, dates, leafy greens, legumes, oats, onions, potatoes, rye, green chili spices, spirulina, tomatoes, whole grains, milk, yogurt, cheese, creams, okra, dried beans, drumstick leaves, egg yolk, fennel, berries, grapes, guava, kiwi fruit, lamb, legumes, lemons, mango, millet, mushrooms, nuts, oats, pears, peas, beetroot, citrus fruits, egg yolk, fish liver oils, garlic, plums, potatoes, yogurt milk, meat, nuts, fish, beans, whole grains, avocado, potato, sweet potato, tomato, mushroom, pumpkin, radish, raisins, shellfish, spinach, spirulina, spring onions, strawberries, tea, walnuts, whole grains, yams brewer's yeast, broccoli, eggs, mushrooms, green beans, green leafy vegetables, broccoli, nuts, fish, organ meats, oysters, spinach, whole grains, halibut, pears, pineapples, prawns, sea salt, shellfish, spirulina, strawberries, turnip greens, apples, beef, caraway, cherries, chicory, cucumber, drumstick leaves, egg yolk, fennel, flaxseed, garlic, green tea, legumes, melon, milk, mulberries, mushrooms, oats	Bone strengthening, cardiac muscle functioning, erythropoiesis, fighting iron deficiency, enhancing memory, neural disorders, boosting immunity	

Nutraceuticals can also be categorized into six classes depending on their chemical composition and active functional groups. The categories encompass amino acid-based compounds, isoprenoid derivatives, phenolic compounds, carbohydrates and their derivatives, fatty acids and structural lipids, and minerals (Hoti et al. 2022). Nutraceuticals can further be classified based on their distinct health benefits, such as strengthening the immune system, improving cardiovascular health, enhancing cognitive function, and more. Furthermore, nutraceuticals can be categorised into two distinct groups: traditional nutraceuticals and non-traditional nutraceuticals. Traditional nutraceuticals are dietary ingredients derived from natural sources, containing various natural components that provide additional advantages beyond basic nutrition (Chanda et al. 2019). Some examples of beneficial compounds found in some foods are omega-3 fatty acids in salmon, lycopene in tomatoes, resveratrol in grapes, and saponin in soybeans. Non-traditional nutraceuticals refer to artificially produced food items that have been enriched with bioactive substances to increase human well-being (AlAli et al. 2021). The other two primary categories of nutraceuticals are fortified and recombinant. Fortified nutraceuticals are food products that have been fortified with additional nutrients by agricultural breeding or the addition of nutrients. Fortified nutraceuticals consist of orange juice that has been enriched with calcium, cereals that have been supplied with additional vitamins or minerals, flour that has been boosted with folic acid, and milk that has been fortified with cholecalciferol. Recombinant nutraceuticals are energy-providing meals that are produced using biotechnological methods (Lai et al. 2017; Verma et al. 2020; Mali et al. 2022).

18.3 Nutraceutical Development: Process and Procedure

The development of nutraceuticals is similar to the herbal drug manufacturing process which involves a systematic process that includes identifying plant species, selecting extraction techniques, purifying and identifying phytochemicals from crude extracts, and formulating the drug. This process also involves testing the drug's shelf life, efficacy, safety, and toxicity (Prabhakar and Mamoni 2021; Bommakanti et al. 2023). Every phase is equally crucial and should be approached with minimal complexities (Fig. 18.1). Nutraceuticals has also evolved from its traditional origins to become a highly specialised scientific sector. The efficacy and safety of the products are now substantiated by data, latest research, and state-of-the-art technical breakthroughs. Therefore, we need to thoroughly examine these processes, including the problems, obstacles, and potential areas for future study.

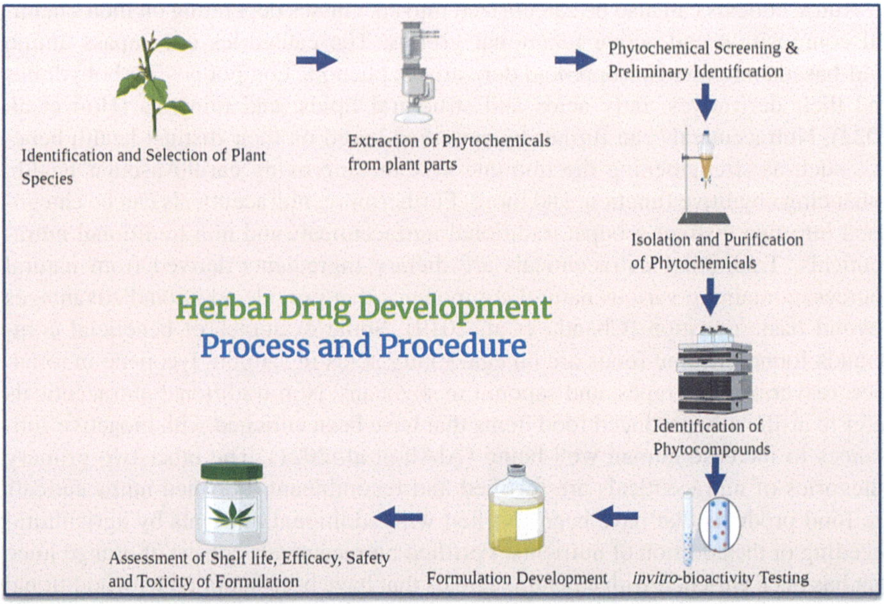

Fig. 18.1 Overview of herbal drug development from its source to final product development involving common process and procedure

18.4 Safety and Toxicity of Nutraceuticals

The antioxidant, anti-inflammatory, anti-proliferative, antimicrobial, and hypocholesterolemic effects are the most prominent, but the systemic effects of the many bioactive compounds found in plant nutraceuticals are the result of interactions with multiple targets in various organs. Nutraceuticals are typically considered safe at physiologic (nutritional) dosages according to toxicology research, but they can have harmful consequences at pharmacological measurements (AlFadhly et al. 2022). Therefore, it is not reasonable to assume that these substances pose no danger to human health or the environment. The toxic characteristics of nutraceuticals have not been explored to the same extent as those of medicines. There is little evidence to indicate that nutraceuticals and dietary supplements are hazardous. Most nutraceuticals available for purchase are generally safe for human use, although there are few cases where they may pose a risk due to their potential hazardous effects on human health (Kaur et al. 2021; Puri et al. 2022). Several researches have demonstrated that several commonly consumed nutraceuticals offer numerous health advantages with low harm when used in appropriate and manageable quantities. Anthocyanins, polyphenols, and catechins are often utilised nutraceuticals that are deemed safe for human consumption when administered under regulated conditions. Only a limited number of research have demonstrated the detrimental effects of these natural compound on human health (Prabhakar et al. 2018, 2023c; Verma et al. 2022). Calderon-Ospina et al. (2020) reported that the neurotropic vitamin B groups (B1, B6 and B12) are important for the optimal functioning of the neurological system. Deficiencies can lead to neurological illnesses such as peripheral neuropathy and mostly affect susceptible populations, such as the elderly, diabetics, and alcoholics. Neurological adverse effects resulting from vitamin B6 use are few and only manifest when large daily doses are taken or when the treatment duration is prolonged. The advantage of high-dose treatment with neurotropic B vitamins in conditions like peripheral neuropathy is that the benefit-risk ratio is regarded favourable, especially when following dosing recommendations and monitoring serum levels. Yao et al. (2019) reported that there is suspicion that certain by-products of epigallocatechin-3-gallate may not reduce, but instead increase oxidative stress and lead to liver damage. It is believed that the mitochondria could be a potential target for these effects. Gutiérrez-del-Río et al. (2018) reported that antioxidant, anti-inflammatory, antiproliferative, antimicrobial, and hypocholesterolemic effects are the most prominent, but the systemic effects of the many bioactive compounds found in plant nutraceuticals are the result of interactions with multiple targets in various organs. However, research on nutraceuticals has demonstrated that the advantages of their usage surpass the potential risks, and they are commonly authorised for human consumption when taken in appropriate quantities and dosages. Andres et al. (2018) reported that the flavonoid quercetin is commonly present in fruits and vegetables as a secondary metabolite, but in small quantities. Quercetin that has been separated from other substances is sold as a dietary supplement. It is commonly available as the free quercetin aglycone and is sometimes taken in daily amounts of up to 1000 mg/day, which is more than the amount typically consumed through regular diet. Out of the many human intervention trials that have been published, there have been very few reports of negative effects from taking quercetin supplements, and any reported effects were not severe. Currently, there is a lack of sufficient scientific data to assess the safety of using high dosages of quercetin supplements ($\geq$1000 mg) for a lengthy period (>12 weeks). Animal studies investigating the oral administration of quercetin have discovered significant safety concerns. These include the possibility of quercetin exacerbating kidney damage and promoting tumour growth, particularly in estrogen-dependent cancers.

Nevertheless, the improper utilisation and excessive consumption of these items can potentially endanger the well-being of individuals. The safety of these nutraceuticals for customers primarily relies on the specific type, duration, and dosage employed. The utilisation of some nutraceuticals, particularly when undergoing medication, might lead to drug-nutraceutical interactions that can have detrimental consequences on the body. Key attributes of nutraceuticals include their cost-effectiveness, extensive safety profiles in both humans and animals, tolerance, and widespread accessibility (Alongi and Anese 2021). Although generally harmless, a small number of them have been found to be contaminated by contamination with heavy metals, hazardous pesticides, medications with misuse potential, potentially toxic plants, fertilisers, and mycotoxins. Regrettably, there is still a lack of research on the safety of many nutraceuticals, resulting in a limited availability of safety data for these substances (Opuni et al. 2023). The pharmacokinetic characteristics of each medicine is of utmost significance for understanding its safety profile (toxico-kinetics), commencement of action, necessary dosage, and frequency of dosage. Additionally, the examination of nutraceutical safety involves considering the impact of interactions with other medications and nutrients/foods on factors such as efficacy, half-life, and therapeutic response (Bautista-Hernández et al. 2021; Prabhakar et al. 2023b).

Although the majority of nutraceuticals in the market are typically safe for human use, there are instances where some nutraceuticals might pose risks to human health. Studies have shown that regularly ingested nutraceuticals provide several health benefits with minimal risk when taken in suitable and controllable amounts (Ronis et al. 2018). Anthocyanins, polyphenols, and catechin are nutraceuticals that have extensive usage and are considered safe for restricted human consumption. Only a limited number of research have provided evidence on the detrimental effects of these chemicals on human health. Key attributes of nutraceuticals include their cost-effectiveness, extensive safety profiles in both people and animals, acceptability, and widespread availability (Chávez-González et al. 2020; Kaur et al. 2021). Although generally harmless, a limited number of nutraceuticals have been shown to be contaminated by contamination with heavy metals, hazardous pesticides, medications with misuse potential, potentially toxic plants, fertilisers, and mycotoxins. Regrettably, the safety characteristics of numerous nutraceuticals have not been well investigated, resulting in a lack of adequate safety data for these compounds. A comprehensive comprehension of the pharmacokinetic characteristics of each medicine is of utmost significance to grasp its safety profile, commencement of action, necessary dosage, and frequency of dosage (Gupta et al. 2022; AlAli et al. 2021). The evaluation of nutraceuticals safety includes the examination of their interactions with other medications and nutrients and foods. This assessment considers the impact of these interactions on safety, effectiveness, duration of action, and the resulting therapeutic response (Shrestha et al. 2021). Improving the accuracy and thoroughness of nutritional assessments in dietary supplements would be beneficial for regulatory agencies. This includes the use of bioactive marker chemicals to detect other substances, natural toxins, hazardous materials, and pesticides in both the components and end products (Vilas-Boas et al. 2021).

Nutraceuticals are generally categorised as dietary supplements in many countries, including the United States and European Union. As such, they are usually not subject to the safety regulations that are imposed on allopathic medications. In 1994, the United States Congress implemented a regulation called the "Dietary Supplements Health & Education Act," which included nutraceuticals in a roster of dietary items. Since these compounds are not categorised as medications, they can be sold without undertaking any evaluations of their safety and effectiveness (Gil et al. 2016). Nevertheless, in accordance with European Union laws, these herbal substances must provide scientific substantiation of their safety, effectiveness, and quality prior to obtaining a licence for public use. Evaluating the toxicity and therapeutic effects of nutraceuticals is challenging in comparison to products based on pure synthetic chemicals, as nutraceuticals typically consist of numerous compounds and are often a complicated mixture of phytocompounds (Helal et al. 2019; Prabhakar and Banerjee 2020a). Moreover, the chemical makeup of plants differs depending on the specific location where they are collected, the varying impacts of fertilisers and pesticides, and the influence of stress. The above mentioned issues, in addition to the lack of recognised techniques for extracting, identifying, and detecting the chemical composition, purity, potency, and safety of active medicinal substances, are the limitations of nutraceuticals. These variables lead to the discrepancies between batches and the incapacity to replicate therapeutic reactions (Kumar et al. 2019; Parihar and Sharma 2021). Furthermore, nutraceuticals are comprised of several compounds that may interact synergistically or have an antagonistic impact. Both of these influences can range from highly advantageous therapeutic outcomes to harmful effects and even a response that falls below the therapeutic threshold (Flore et al. 2023). Overall, there has been a significant rise in the use of nutraceuticals in both human and veterinary goods. As a result, the risks of toxicity linked to their usage are also heightened because there is a lack of extensive assessments about their safety and effectiveness in clinical trials. Moreover, it is crucial to regulate fluctuations and detect the existence of impurities such heavy metals, agricultural chemicals, and mycotoxins. Additionally, doing thorough pharmacokinetic and dynamic assessments is highly essential to enhance the utilisation of these significant alternative substances (Patel et al. 2018; Prabhakar and Banerjee 2020b; Prabhakar et al. 2023b).

To mitigate the dangers and toxicity associated with health and dietary supplements, it is imperative to provide patients or consumers with comprehensive education on the appropriate utilisation of these products. It is essential to emphasise that not all herbs and dosages are devoid of danger, and all consumers, including patients, should be educated about the possible risks and advantages associated with any such supplements (Dwyer et al. 2018; Puri et al. 2022). Together with this there is need to classify health and dietary supplements as therapeutic items to prevent their unnecessary and reckless consumption. Consuming nutraceuticals in excess of the required dosage can result in disasters. Therefore, customers should never think that nutraceuticals are exempt from this risk. It is recommended to get extensive guidance from medical professionals, particularly if the individuals are using allopathic medication (Bommakanti et al. 2023). If any adverse effects related to health and dietary supplements occur, users should be instructed to immediately stop using the product. In addition, it is imperative to educate customers about the importance of purchasing nutraceuticals and associated products exclusively from reliable and well-established suppliers. However, it is important to recognise that, in most cases, the cost of the items is not directly correlated with the quality of the products (Santini and Novellino 2018).

It is not advisable for those who are receiving nursing care or who are pregnant or breastfeeding to use herbal and dietary supplements, unless specifically instructed to do so by their healthcare providers. Furthermore, it may not be recommended for infants and children to utilise these supplements. Nutraceuticals have the potential to interact with prescription drugs, resulting in negative responses or decreased effectiveness. Prudence is essential, especially for persons with pre-existing medical issues or those who are taking many medications (Burkey and Holmes 2013). Herbal substances can get contaminated with poisons, heavy metals, or microbial infections during their production process or cultivation. Stringent quality control methods are required to reduce this risk. Overconsumption of nutraceuticals, surpassing the recommended dosages, can result in toxicity. Many vitamins, minerals, and herbal extracts can accumulate in the body over time and have adverse consequences. People who have allergies or sensitivities to specific components in nutraceuticals may encounter adverse responses (Wang et al. 2023; Prabhakar and Mamoni 2021). Accurate labelling of allergens and formulations that do not contain allergens can aid in the prevention of negative reactions. Ultimately, although nutraceuticals hold promise for improving health, it is crucial to have regulatory supervision, quality control protocols, clinical substantiation, and consumer education to guarantee their safety and mitigate any dangers of toxicity. Healthcare experts have a crucial role in providing guidance for the safe and appropriate use of healthcare based on an individual's specific health needs and circumstances (Fiocchi et al. 2021).

18.5 Global Regulatory for Safety and Toxicity of Nutraceuticals

Nutraceuticals refer to naturally occurring chemical compounds that possess therapeutic qualities, boosting health and preventing diseases. This classification of pharmaceuticals includes vitamins, minerals, herbal supplements, and specific animal-derived products. Furthermore, nutraceuticals encompass "functional foods," which are food products that claim a certain health advantage due to their contents. There is an increasing convergence between traditional food (including beverages) and food supplements, which include energy bars and teas or liquids. The extent of this overlap increases significantly when we consider functional foods and nutraceuticals (Santini and Novellino 2018; Kumar et al. 2022). The increasing popularity of nutraceuticals can be related to the rising consumer awareness of climbing prescription costs, the limited efficacy of conventional drugs, and the larger prevalence of adverse side effects. The inclination towards natural products in the prevention and treatment of diseases has been strengthened due to the absence of efficacious alternatives for chronic ailments, impersonal encounters with healthcare practitioners, the need for customised medications, a substantial population of elderly patients, and a renewed focus on preventive medicine (Ray et al. 2016; Thakkar et al. 2020). Hence, a significant proportion of the world's population, especially in developing nations such as India, is seeking alternative medical treatments and nutraceuticals (including herbal and botanical remedies) with high anticipation. The nutraceutical market is seeing fast growth, while the regulations governing nutraceuticals are constantly evolving without a definitive regulatory definition. Because of this lack of clarity, nutraceuticals are classified as food supplements rather than medicinal goods in several nations (Malve and Bhalerao 2023). Food supplements are commonly used in several countries to supplement the regular diet and ensure adequate intake of specific components, such as vitamins and minerals. The consumption of nutraceuticals exhibits significant variation across different countries, with some restricting its use solely to promote overall health and well-being, while others allow their usage for therapeutic objectives. The worldwide incidence of nutraceuticals is significantly high, although their regulation is comparatively little when compared to conventional pharmaceuticals (Puri et al. 2022).

The International Regulation Cooperation of Herbal Medicines (IRCH) is a global network of regulatory agencies that oversees the regulation of herbal medicines worldwide. This organisation convenes a meeting at regular intervals. The World Health Organisation provides significant support to this organisation. The objective is to deliberate on the regulation of herbal medicine to safeguard and advance public health. The most recent meeting, held in 2023, was organised by the WHO-IRCH Secretariat and had the participation of governmental agencies from 22 nations, including as Australia, India, Japan, Canada, and the United States (von Schoen-Angerer et al. 2023; ICRH 2024). In Australia, Nutraceuticals are classified as "Therapeutic Goods" and are subject to regulation under the Therapeutic Goods Act (TGA) of 1989. Within this classification, there exist pharmaceuticals, biologics, and medical devices. Fourteen complementary medicine is a regulatory phrase that encompasses herbs, vitamins, minerals, and other nutritional supplements within the broader category of medicine. Regulation of therapeutic goods is necessary to safeguard public health in light of the numerous dangers associated with these products. Australia has a risk-based strategy with a two-tiered structure to regulate all medications. Drugs with a low risk are eligible for inclusion in the Australian Register of Therapeutic Goods (ARTG), while drugs with a high risk are required to be registered on the ARTG. Medications that pose a significant risk are obligated to undergo assessment to determine their safety, effectiveness, and quality. Medications that are included in a list but have not been officially approved are exempt from this obligation (Blaze 2021). Food Standards Australia New Zealand (FSANZ) is an independent regulatory body created under the Food Standards Australia New Zealand Act 1991 (FSANZ Act). FSANZ is a component of the Health ministry of the Australian Government. FSANZ fulfils several roles in Australia, such as overseeing the nationwide coordination of food monitoring and food recall systems, conducting research, and aiding the Australian Department of Agriculture, Fisheries, and Forestry in the management of imported commodities. FSANZ is responsible with formulating food safety regulations that oversee the utilisation of ingredients, processing aids, colorings, additives, vitamins, and minerals. The Food Standards Code also governs the composition of certain food items, such as dairy, meat, and beverages. It also includes foods produced utilising new technologies, such as genetically modified foods. (Barker and Fleming 2023).

In India, nutraceuticals are largely categorised as functional foods, encompassing fortified food items and drinks, as well as dietary supplements, such as vitamins, minerals, protein supplements, and herbal products. Nutraceuticals, unlike pharmaceuticals, are frequently exempt from regulation under the Drugs and Cosmetic Act of 1940. As a result, several laws have been passed to oversee the control and management of nutraceuticals. The Prevention of Food Adulteration Act was established in 1954, and subsequently, the Fruit Product Order was implemented in 1955. In

1973, the Meat Food Products Order was put into effect, and in 1998, the Edible Oils Packaging (Regulation) Order was adopted. In addition, the regulations for Solvent Extracted Oil, de-oiled meal, and edible flour (control) were all implemented in 1967. The Milk and Milk Products Order was instituted in 1990, but the Essential Commodities Act, which specifically relates to food, was adopted in 1955. In order to consolidate the entire Act and streamline the process of obtaining licences, a single food regulatory authority was established in 1998 by the council of Prime Minister on trade and industry. This authority is responsible for recommending a unified set of rules for the food and agriculture industries, including common application forms, registration procedures, central licencing, and standards for safety, sanitation and hygiene. The Indian Food Safety and Standards Bill (FSSB) passed in 2005 and received formal approval from the President on August 23, 2006. The Food Safety and Standards Authority of India (FSSAI) was created by the Indian government in 2006. The primary goal of the Food Safety and Standards Act (FSSA) is to streamline and enhance the food processing sector by consolidating and implementing systematic and scientific food safety rules in the country. It also aims to shift from a regulated approach to a self-compliance strategy. The act had two main objectives: (i) to establish a comprehensive law on food and (ii) to promote the scientific progress of the food processing sector (Verma and Popli 2018; Bhupathiraju et al. 2019; Gulati et al. 2021; Ray et al. 2016).

In Canada, the regulation of food and medication goods is overseen by Health Canada, which has many directorates and divisions assigned to supervise specific types of products. Within the Canadian regulatory framework, nutraceuticals are categorized as either a food, a drug, or a NHP. The review primarily concentrates on products that are pertinent to the NHP category. NHPs, often known as complementary medicines or traditional remedies, are subject to regulation under the Food and Drugs Act and Regulations (Bagchi 2019; Umegaki 2021). The Chinese Food Safety Law was enforced in 2015 and a new centralised system was introduced by the National Medical Products Administration (NMPA). Their amended legislation has enhanced the mechanism for maintaining and documenting records and registrations. As per the Chinese Food Safety Law, natural goods are classified into two categories: "Supplements" or "Medicines." The TCM is highly prevalent in China and encompasses a wide range of goods that are classed as medicines. TCM includes goods that have been used by the Chinese population for a long time and have also influenced traditional medicine practices in nearby regions, such Kampo medicine in Japan. As a result, the structure for the creation, healthcare system, medical establishments, and regulatory procedures are well-established and effectively coordinated for TCM (Xiao et al. 2019; Liang et al. 2021).

Japan classifies nutraceuticals into two categories: "Foods in General" or "Food with Health Claims". The category "Food with Health Claims" can be divided into three distinct categories: (1) "Food with Nutrient Function Claims" (FNFC), which mostly consist of vitamins and minerals, (2) "Food for Specified Health Uses" (FOSHU), which serve other specific roles, and (3) "Foods with Function Claims" (FFC). The Ministry of Health, Labour, and Welfare serves as the primary governing authority for all pharmaceuticals and health supplements. The Ministry of Health,

Labour and Welfare (MHLW) enforces stringent rules to ensure the safety and quality of medications and dietary supplements. Japan pioneered the regulation of dietary supplements by introducing the dietary for FOSHU guidelines. The regulatory system was established in 1991 to authorise comments regarding the impact of food on the human body. Products that display the FOSHU emblem undergo rigorous clinical testing to verify their effectiveness and safety. These products are permitted to make health claims on their labels since they have been demonstrated to be effective and safe for consumption, as stated. Even if a health-promoting activity lacks scientific validation, it can still be permitted if the substance passes the safety requirements of FOSHU (Thakkar et al. 2020; Díaz et al. 2020). All drugs classified as medicine are subject to regulation under the pharmaceutical affairs act, which has been in effect since 1960 and was last amended in 2013. In 2015, the introduction of the "Foods with Function Claims" category aimed to classify dietary supplements with the purpose of promoting health and preventing diseases, distinct from FOSHU. Approval of FFC products requires submission to the Secretary-General of the Consumer Affairs Agency (CAA). In this category, the government does not assess the safety and effectiveness of functional claims. Businesses must comply with the Food Labelling Standards and "Guidelines on Notifications of Food with Function Claims" by submitting a premarket notice and properly labelling their product. The FFC system has successfully expedited the influx of new participants into the market, hence bolstering the expansion of items within the market (Blaze 2021). In European Union, items are categorised into two distinct groups: those that are classified as supplements and those that are classified as medicines. Within the classification of "Classified as Supplements," there are Food Supplements that are governed as food according to Directive 2002/46/EC (European Commission) (Prabhakar and Mamoni 2021). The European Union lacks a unified framework for pre-market authorization of food supplements. European Union Member States have the choice to request notification when a certain dietary supplement is introduced into their market. This enables the competent authority of the Member State to supervise its utilisation within their jurisdiction. The legislation within the European Union standardises the control of vitamins, minerals, and the chemicals utilised as their sources (Raposo et al. 2021). These regulations specifically apply to the manufacturing of dietary supplements. The relevant legislation was established in 2006 and updated in 2015. The European Commission has implemented standardised regulations to safeguard consumers from potential health hazards associated with ingredients other than vitamins and minerals. These regulations include a list of substances that are known or suspected to have negative effects on health, and their use is strictly regulated (Annex III of Regulation (EC) No 1925/2006). If a product intended for use in food supplements lacks a documented record of safe usage in the European Union before 1997, has undergone a new manufacturing method, or contains or is composed of engineered nanomaterials, it is categorised as a novel food. This prompts a solicitation for a safety evaluation by the European Food Safety Authority (EFSA) in accordance with Regulation (EC) No 2015/2283 on new foods substances. Legally, there is a big difference between "novel" and "conventional" foods; the new Foods Regulation states that "novel" foods must have official approval. Nutritional equality with substitute foods, correct labelling, and safety are the guiding principles of the rule. With the replacement of previous restrictions by Regulation (EU) 2015/2283 on January 1, 2018, the European Union market became more accessible to new and unique foods while still requiring high levels of safety (Djaoudene et al. 2023; Morán and Kilasoniya 2024).

In 1994, the United States Dietary Supplement Health and Education Act (DSHEA) created the regulatory structure for dietary supplements to be classified as food under the jurisdiction of the Food and Drug Administration (FDA). The legal definition, labelling standards, and process for reporting adverse events for dietary supplements were specified in the 1994 DSHEA. The DSHEA provides a clear definition of a dietary supplement as a product that is not related to tobacco and is designed to improve the diet. It must contain one or more of the following dietary ingredients: vitamins, minerals, herbs or other botanicals, amino acids, dietary substances used to increase daily intake, or combinations of these ingredients in the form of concentrates, metabolites, constituents, or extracts (Wallace 2015; Denham 2021; Komala et al. 2023). In contrast to the FDA's stringent requirements for proving safety and efficacy before approving pharmaceuticals, dietary supplements are largely subject to post-market surveillance and do not undergo any clearance process. The manufacturer bears the responsibility for ensuring the safety of nutraceuticals, and there is no requirement for FDA approval or product registration prior to sale. As per the FDA Modernization Act (1997), the FDA must be notified about any health claims or nutrient content claims on the label of a dietary supplement at least four months prior to its sale. These assertions might potentially get approval from the Academy of Science or another federal institution (Santini and Novellino 2018). Here, Table 18.2 provides a concise overview of important factors within each category, comparing them across different countries together with their regulatory authority for natural product supplements. Nevertheless, these assertions are typically subject to variation according on the jurisdiction.

Currently, several research works focussed on the utilisation of nutraceuticals due to its diverse health-enhancing and disease-preventing properties. However, the production of nutraceuticals faces various problems that are frequently overlooked due to a lack of authoritative regulation. The main obstacle in regulating dietary supplements is the absence of a worldwide agreement on the definition of this product category; there are no global standards in place to guarantee quality and integrity. In general, the Canadian restrictions are the most stringent, while the United States regulations are the most adaptable, with other countries having regulations that vary in between.

18.6 Future Perspectives, Challenges and Opportunities

Nutraceuticals, which are included in a well-balanced diet, provide the body with both energy and vital nutrients that are crucial for sustaining optimal health. Inadequate intake of vegetables, fruits, minerals, and vitamins is a significant

Table 18.2 Natural products are categorized as either supplements or medicines depending on the region

Regulatory agency	Country	Natural products as supplements	Natural products as Medicines	References
Food Standards Australia New Zealand (FSANZ)	Australia	Innovative Food Edible substances and derivatives derived from many sources such as plants and animals Food products and their derived substances obtained through manufacturing methods and techniques, as well as innovative technology		Blaze (2021), Barker and Fleming (2023)
Therapeutic Goods Administration (TGA)			Complementary medicines Homeopathy Minerals Nutritional supplements Microorganism (whole extracted) Vitamins	
Health Canada (HC)	Canada		Natural health products Herbal medicine Homeopathy Traditional medicine	Bagchi (2019)

(continued)

Table 18.2 (continued)

Regulatory agency	Country	Natural products as supplements	Natural products as Medicines	References
China Food and Drug Administration (CFDA)	China	Health foods	Traditional Chinese Medicine	Xiao et al. (2019), Liang et al. (2021)
European Medicines Agency (EMA) Herbal Medicinal	European Union		Herbal medicinal products Authorities at the national level of European Union member states	Raposo et al. (2021), Djaoudene et al. (2023), Morán and Kilasoniya (2024)
National competent authorities European Food Safety Authority (EFSA) if centralized procedures apply	European Union Member States	Ingredients (such as vitamins, minerals, and botanicals) that have a physiological or nutritional impact		
Food Safety and Standards Authority of India (FSSAI)	India	Nutritional supplements, which can take the form of vitamins, minerals, protein powders, herbal remedies, and fortified foods and drinks are all examples of functional foods		Verma and Popli (2018), Bhupathiraju et al. (2019), Gulati et al. (2021), Ray et al. (2016)
Ministry of Health, Labour and Welfare (MHLW) for Medicines Consumer Affairs Agency (CAA)	Japan	Health foods	Kampo medicine	Bagchi (2019), Umegaki (2021)

New Zealand Ministry for Primary Industries (MPI)	New Zealand	Consumption of foodstuffs that have been altered or supplemented with chemicals rendering them physiologically useful	Herbal remedies	Thakkar et al. (2020), Díaz et al. (2020), Blaze (2021), Barker and Fleming (2023)
New Zealand Medicines and Medical Devices Safety Authority (Medsafe)	New Zealand	Nutritional supplements	Herbal remedies	
Food Standards Australia New Zealand (FSANZ)	New Zealand	Innovative Food Edible substances and derivatives derived from many sources such as plants and animals Food products and their derived substances obtained through manufacturing methods and techniques, as well as innovative technology		
Food and Drug Administration (FDA)	USA	Dietary Supplement – Center for Food Safety and Applied Nutrition (CFASN)/FDA Herbs/Botanicals Vitamin Minerals Amino Acids A dietary supplement is a product intended for human consumption that is used to enhance the overall intake of nutrients in the diet. Concentrate, metabolite, constituent, extract or combination of the preceding substances.	Botanical drugs classified by Center for Drug Evaluation and Research (CDER)/FDA	Wallace (2015), Santini and Novellino (2018), Denham (2021), Komala et al. (2023)

(continued)

determinant in the onset of several illnesses. Incorporating antioxidants, dietary fibres, and reaching the necessary daily intake of proteins, vitamins, and minerals, together with participating in regular physical activity, can aid in the prevention and treatment of several illnesses. Unhealthy lifestyle habits, such as excessive consumption of unhealthy food and a lack of physical activity, have led to several illnesses caused by nutritional deficiencies (Nasri et al. 2014; Wallace 2015). Due to the notable adverse reactions and lack of efficacy of contemporary medications, individuals have been obliged to seek nutraceuticals as an alternate form of therapy. This phenomenon can be attributed to the promotion and extensive coverage of the advantages of nutraceuticals through advertising and media. Healthcare professionals, including general practitioners, nurses, chemists, and nutritionists, possess extensive knowledge of nutraceuticals and provide education to their patients or consumers (Galanakis 2019). Concerning the proper utilisation of these items nutraceuticals for serious disorders without medical supervision is not suitable, however their extended usage is considered safe and advantageous for preventing chronic diseases. Carnitine and flaxseed oil are primarily utilised in the treatment of cardiovascular illness, whereas antioxidants are primarily employed in the prevention of cancer in the context of nutraceuticals for serious disorders. Presently, a significant number of nutraceutical customers have the belief that dietary supplements may possess a higher level of safety compared to other synthetic chemicals. However, it is possible that this assumption is incorrect, and it is necessary to obtain a medical diagnostic to prescribe appropriate and effective conventional drugs for serious illnesses (Rizzo et al. 2023; Puri et al. 2022). Nutraceuticals used for therapeutic purposes are justified based on their ability to treat diseases caused by nutrient deficiencies. Definitive data has been documented indicating that the use of nutraceutical supplements enhances health and serves as a preventive measure against diseases. Nutraceutical supplementation does not necessitate diagnosis by a proficient practitioner and is expected to have beneficial effects on the overall body, rather than precisely addressing symptoms of a disease. Consumers of nutraceuticals may simply control their health without needing to communicate with their physicians. Extended use of nutraceuticals for self-medication may result in financial consequences for individuals, perhaps making them costlier compared to other medications, despite their advantages. (Lokhande 2018; Chanda et al. 2019).

The nutraceuticals industry is expanding rapidly as there is increasing interest in investigating new treatment possibilities utilising modern technology and scientific methods. While several nutraceuticals have been acknowledged for their beneficial impact on the immune system, it is imperative to carry out more rigorous clinical studies that provide conclusive proof. These trials are essential for conducting a comprehensive analysis of the long-term impacts and guaranteeing public safety. It is crucial to maximise the therapeutic capabilities and nutritional benefits of these chemicals separately, as nutraceuticals are still in a state of ambiguity. There is considerable uncertainty among individuals regarding whether they should utilise it as a medicinal treatment or as a fundamental dietary necessity (Verma et al. 2022; Tsiaka et al. 2022). Developing novel medicines incurs higher costs and has greater risks compared to existing treatments in the market. Consequently, many pharmaceutical companies are shifting their focus towards selling nutraceuticals and it is crucial to conduct a comprehensive investigation of the safety, bioactivity, and bioavailability of these natural nutraceutical components. Furthermore, there has been a significant increase in government sponsorship for clinical trials, along with a rise in funding for nutraceutical research (Nori et al. 2023). Despite the implementation of legislation aimed at enhancing the marketplace, several obstacles persist. Consequently, there might be significant discrepancies in the quality and safety of items, particularly those of herbal nature. Further endeavours and enhanced techniques are still required to evaluate the calibre of dietary supplements, particularly with regards to their purity, bioavailability, and safety (Blaze 2021). In addition, the use of developing technology and computational approaches can enhance the advantages and mitigate the safety issues associated with dietary supplements and herbal medications, which were previously limited to traditional drugs. Data analytics techniques can have a substantial impact on the creation of reliable quality assessment approaches to improve the evaluation of natural products (Thakkar et al. 2020; Vora et al. 2023).

18.7 Conclusion

Nutraceuticals are an emerging field of research that aims to create innovative health products with the potential to offer substantial benefits in terms of safety, efficacy, and cost-effectiveness. Globally, professionals have acknowledged the fact that sufficient nourishment and dietary supplements have the capacity to prevent and manage chronic diseases. A multitude of nutraceuticals have been derived from food sources, and substantial quantities are produced through the use of biotechnology and genetic engineering methods, leading to notable financial benefits in the pharmaceutical sector. Nonetheless, the capacity of nutraceuticals to prevent illnesses and uphold health mostly relies on the purity, efficacy, and safety of these products. Global regulatory markets lack consistency and agreement in classifying and regulating nutraceuticals. Due to the discrepancy between the increasing popularity of these goods and the limited knowledge about their possible benefits and downsides, it is essential for regulatory agencies globally to carefully consider their regulatory principles in order to fully comprehend their influence on public health. Although developing technologies have the potential to improve the alignment of regulatory frameworks for nutraceuticals, further discussion and study are required to effectively achieve this change.

Acknowledgement Authors are thankful to the Indian Institute of Technology Kharagpur (IIT Kharagpur), Kharagpur, India and Ministry of Education, Government of India for providing necessary research support for this work.

Conflict of Research Interests Authors declare no conflict of research interests.

References

AlAli M, Alqubaisy M, Aljaafari MN, AlAli AO, Baqais L, Molouki A et al (2021) Nutraceuticals: transformation of conventional foods into health promoters/disease preventers and safety considerations. Molecules 26(9):2540

AlFadhly NK, Alhelfi N, Altemimi AB, Verma DK, Cacciola F, Narayanankutty A (2022) Trends and technological advancements in the possible food applications of Spirulina and their health benefits: a review. Molecules 27(17):5584

Alongi M, Anese M (2021) Re-thinking functional food development through a holistic approach. J Funct Foods 81:104466

Andres S, Pevny S, Ziegenhagen R, Bakhiya N, Schäfer B, Hirsch-Ernst KI, Lampen A (2018) Safety aspects of the use of quercetin as a dietary supplement. Mol Nutr Food Res 62(1):1700447

Bagchi D (ed) (2019) Nutraceutical and functional food regulations in the United States and around the world. Academic Press

Bailey RL (2020) Current regulatory guidelines and resources to support research of dietary supplements in the United States. Crit Rev Food Sci Nutr 60(2):298–309

Banwo K, Olojede AO, Adesulu-Dahunsi AT, Verma DK, Thakur M, Tripathy S et al (2021) Functional importance of bioactive compounds of foods with potential health benefits: a review on recent trends. Food Biosci 43:101320

Barker J, Fleming F (2023) Regulations and guidance in Australia and New Zealand with implications for food factory design. In: Hygienic design of food factories. Woodhead Publishing, pp 155–169

Barua CC, Sharma D, Devi PV, Islam J, Bora B, Duarah R (2023) Nutraceuticals and bioactive components of herbal extract in the treatment and prevention of neurological disorders. In: Treatments, nutraceuticals, supplements, and herbal medicine in neurological disorders. Academic Press, pp 577–600

Bautista-Hernández I, Aguilar CN, Martínez-Ávila GC, Torres-León C, Ilina A, Flores-Gallegos AC et al (2021) Mexican oregano (Lippia graveolens Kunth) as source of bioactive compounds: a review. Molecules 26(17):5156

Ben Hsouna A, Sadaka C, Generalić Mekinić I, Garzoli S, Švarc-Gajić J, Rodrigues F et al (2023) The chemical variability, nutraceutical value, and food-industry and cosmetic applications of citrus plants: a critical review. Antioxidants 12(2):481

Bhattacharyya AS, Prabhakar P (2015) First order kinetics study of nanoparticle drug delivery. Pharmbit 2015:1–4

Bhupathiraju K, Krishnaraju AV, Sengupta K, Golakoti T, Akolkar SK, Datla P (2019) Regulations on nutraceuticals, functional foods, and dietary supplements in India. In: Nutraceutical and functional food regulations in the United States and around the world. Academic Press, pp 445–464

Blaze J (2021) A comparison of current regulatory frameworks for nutraceuticals in Australia, Canada, Japan, and the United States. Innov Pharm 12(2):8

Bommakanti V, Puthenparambil Ajikumar A, Sivi CM, Prakash G, Mundanat AS, Ahmad F et al (2023) An overview of herbal nutraceuticals, their extraction, formulation, therapeutic effects and potential toxicity. Separations 10(3):177

Bradberry JC, Hilleman DE (2013) Overview of omega-3 fatty acid therapies. Pharm Ther 38(11):681–691

Brown AC (2017) An overview of herb and dietary supplement efficacy, safety and government regulations in the United States with suggested improvements. Part 1 of 5 series. Food Chem Toxicol 107:449–471

Burkey BW, Holmes AP (2013) Evaluating medication use in pregnancy and lactation: what every pharmacist should know. J Pediatr Pharmacol Ther 18(3):247–258

Calderon-Ospina CA, Nava-Mesa MO, Paez-Hurtado AM (2020) Update on safety profiles of vitamins B1, B6, and B12: a narrative review. Ther Clin Risk Manag Volume 16:1275–1288

Cena H, Calder PC (2020) Defining a healthy diet: evidence for the role of contemporary dietary patterns in health and disease. Nutrients 12(2):334

Chanda S, Tiwari RK, Kumar A, Singh K (2019) Nutraceuticals inspiring the current therapy for lifestyle diseases. Adv Pharmacol Pharma Sci 2019

Chávez-González ML, Sepúlveda L, Verma DK, Luna-García HA, Rodríguez-Durán LV, Ilina A, Aguilar CN (2020) Conventional and emerging extraction processes of flavonoids. PRO 8(4):434

Chopra AS, Lordan R, Horbańczuk OK, Atanasov AG, Chopra I, Horbańczuk JO et al (2022) The current use and evolving landscape of nutraceuticals. Pharmacol Res 175:106001

Daliu P, Santini A, Novellino E (2019) From pharmaceuticals to nutraceuticals: bridging disease prevention and management. Expert Rev Clin Pharmacol 12(1):1–7

Denham BE (2021) Dietary supplements in the USA: problematic trends. Public Health Nutr 24(9):2771–2775

Díaz LD, Fernández-Ruiz V, Cámara M (2020) An international regulatory review of food health-related claims in functional food products labeling. J Funct Foods 68:103896

Djaoudene O, Romano A, Bradai YD, Zebiri F, Ouchene A, Yousfi Y et al (2023) A global overview of dietary supplements: regulation, market trends, usage during the COVID-19 pandemic, and health effects. Nutrients 15(15):3320

Dwyer JT, Coates PM, Smith MJ (2018) Dietary supplements: regulatory challenges and research resources. Nutrients 10(1):41

Elkhalifa AE, Alshammari E, Adnan M, Alcantara JC, Awadelkareem AM, Eltoum NE, Mehmood K, Panda BP, Ashraf SA (2021) Okra (Abelmoschus esculentus) as a potential dietary medicine with nutraceutical importance for sustainable health applications. Molecules 26(3):696

Fehér A, Gazdecki M, Véha M, Szakály M, Szakály Z (2020) A comprehensive review of the benefits of and the barriers to the switch to a plant-based diet. Sustain For 12(10)

Fiocchi A, Risso D, DunnGalvin A, Díaz SNG, Monaci L, Fierro V, Ansotegui IJ (2021) Food labeling issues for severe food allergic patients. World Allergy Organ J 14(10):100598

Fleischhacker SE, Woteki CE, Coates PM, Hubbard VS, Flaherty GE, Glickman DR et al (2020) Strengthening national nutrition research: rationale and options for a new coordinated federal research effort and authority. Am J Clin Nutr 112(3):721–769

Flore G, Deledda A, Lombardo M, Armani A, Velluzzi F (2023) Effects of functional and nutraceutical foods in the context of the Mediterranean diet in patients diagnosed with breast cancer. Antioxidants 12(10):1845

Galanakis CM (ed) (2019) Nutraceuticals and natural product pharmaceuticals. Academic Press

Gil F, Hernández AF, Martín-Domingo MC (2016) Toxic contamination of nutraceuticals and food ingredients. In: Nutraceuticals. Academic Press, pp 825–837

Godwin A (2018) Phytochemicals as nutraceuticals and pharmafoods. In: Phytochemistry. Apple Academic Press, pp 179–196

Goel K, Verma DK, Prabhakar P, Tripathy S, Dadrwal BK, Kumar P, Srivastav PP, Banerjee M (2024) In vitro and In silico evaluation of the antioxidant, anti-microbial and antihyperglycemic properties of giloy (Tinospora cordifolia L.) stem extract. Biocatal Agric Biotechnol 56:103059

Gulati K, Thokchom SK, Joshi JC, Ray A (2021) Regulatory guidelines for nutraceuticals in India: An overview. Forum Nutr:1273–1280

Gupta AK, Gurjar PS, Beer K, Pongener A, Ravi SC, Singh S et al (2022) A review on valorization of different byproducts of mango (Mangifera indica L.) for functional food and human health. Food Bioscience 48:101783

Gutiérrez-del-Río I, Fernández J, Lombó F (2018) Plant nutraceuticals as antimicrobial agents in food preservation: Terpenoids, polyphenols and thiols. Int J Antimicrob Agents 52(3):309–315

Helal NA, Eassa HA, Amer AM, Eltokhy MA, Edafiogho I, Nounou MI (2019) Nutraceuticals' novel formulations: the good, the bad, the unknown and patents involved. Recent Pat Drug Deliv Formul 13(2):105–156

Horn B, Yu W (2020) Nutritional influences on reproduction: a functional approach. In: Integrative and functional medical nutrition therapy: principles and practices, pp 533–561

Hoti G, Matencio A, Rubin Pedrazzo A, Cecone C, Appleton SL, Khazaei Monfared Y et al (2022) Nutraceutical concepts and dextrin-based delivery systems. Int J Mol Sci 23(8):4102

Huda H-S-A, Majid NBA, Chen Y, Adnan M, Ashraf SA, Roszko M, Bryła M, Kieliszek M, Sasidharan S (2024) Exploring the ancient roots and modern global brews of tea and herbal beverages: a comprehensive review of origins, types, health benefits, market dynamics, and future trends. Food Sci Nutr 00:1–18. https://doi.org/10.1002/fsn3.4346

IRCH (2024) International Regulatory Cooperation for Herbal Medicines. https://www.who.int/initiatives/international-regulatory-cooperation-for-herbal-medicines. Accessed 7 April 2024

Kaur K, Asthir B, Verma DK (2021) Biosynthesis, bioavailability, and metabolism of plant polyphenols: biological activities and their potential benefits in human health. In: Phytochemicals in food and health. Apple Academic Press, pp 231–255

Komala MG, Ong SG, Qadri MU, Elshafie LM, Pollock CA, Saad S (2023) Investigating the regulatory process, safety, efficacy and product transparency for nutraceuticals in the USA, Europe and Australia. Foods 12(2):427

Kumar K, Kumar S (2015) Role of nutraceuticals in health and disease prevention: a review. South Asian J Food Technol Environ 1(2):116–121

Kumar R, Kumar R, Prakash O (2019) The impact of chemical fertilizers on our environment and ecosystem Chapter 5

Kumar A, Rout RK, Rao PS, Prabhakar P (2022) Optimization of pulsed mode sonication and in silico molecular docking study for antioxidant properties of mandarin (Citrus reticulata Blanco) peels. J Food Process Eng 46(6):e14111

Lai WT, Khong NM, Lim SS, Hee YY, Sim BI, Lau KY, Lai OM (2017) A review: modified agricultural by-products for the development and fortification of food products and nutraceuticals. Trends Food Sci Technol 59:148–160

Liang Z, Lai Y, Li M, Shi J, Lei CI, Hu H, Ung COL (2021) Applying regulatory science in traditional chinese medicines for improving public safety and facilitating innovation in China: a scoping review and regulatory implications. Chin Med 16:1–16

Liu RH (2013) Health-promoting components of fruits and vegetables in the diet. Adv Nutr 4(3):384S–392S

Liu AG, Ford NA, Hu FB, Zelman KM, Mozaffarian D, Kris-Etherton PM (2017) A healthy approach to dietary fats: understanding the science and taking action to reduce consumer confusion. Nutr J 16:1–15

Liu X, Xia B, Hu LT, Ni ZJ, Thakur K, Wei ZJ (2020) Maillard conjugates and their potential in food and nutritional industries: a review. Food Front 1(4):382–397

Lokhande SS (2018) Role of nutraceuticals in various diseases: a comprehensive review. Asian J Pharma Res 8(4):236–240

Makkar R, Behl T, Bungau S, Zengin G, Mehta V, Kumar A et al (2020) Nutraceuticals in neurological disorders. Int J Mol Sci 21(12):4424

Mali S, Rathod S, Kale N, Shinde N (2022) Overview of nutraceuticals. Asian J Pharma Res 12(1):61–70

Malve H, Bhalerao P (2023) Past, present, and likely future of Nutraceuticals in India: evolving role of pharmaceutical physicians. J Pharm Bioallied Sci 15(2):68–74

Martinez-Medina GA, Chávez-González ML, Verma DK, Prado-Barragán LA, Martínez-Hernández JL, Flores-Gallegos AC et al (2021) Bio-funcional components in mushrooms, a health opportunity: Ergothionine and huitlacohe as recent trends. J Funct Foods 77:104326

Molin TRD, Leal GC, Müller LS, Muratt DT, Marcon GZ, Carvalho LMD, Viana C (2019) Regulatory framework for dietary supplements and the public health challenge. Rev Saude Publica 53:90

Morán J, Kilasoniya A (2024) Application of the "novel foods" regulation to botanicals in the European Union. Laws 13(1):10

Mortazavian AM, Meybodi NM (2016) Medicinal food products; a new approach from ordinary foods to medicine. Iranian Journal of Pharmaceutical Research: IJPR 15(1):1

Nasri H, Baradaran A, Shirzad H, Rafieian-Kopaei M (2014) New concepts in nutraceuticals as alternative for pharmaceuticals. Int J Prev Med 5(12):1487

Ng JY, Kim M, Suri A (2022) Exploration of facilitators and barriers to the regulatory frameworks of dietary and herbal supplements: a scoping review. J Pharma Pol Pract 15(1):55
Nooshkam M, Varidi M, Verma DK (2020) Functional and biological properties of Maillard conjugates and their potential application in medical and food: a review. Food Res Int 131:109003
Nori LP, Sowmya P, Madhusree T, Raju KV (2023) Nutraceuticals: regulatory landscape and market dynamics in the Indian scenario. J Young Pharm 15(4):595–603
Opuni KF, Kretchy JP, Agyabeng K, Boadu JA, Adanu T, Ankamah S et al (2023) Contamination of herbal medicinal products in low-and-middle-income countries: a systematic review. Heliyon 9:e19370
Parihar P, Sharma S (2021, June) Composting: a better alternative of chemical fertilizer. In: IOP Conference Series: Earth and Environmental Science, vol 795, No. 1. IOP Publishing, p 012038
Patel A, Thakkar P, Shah VDK, Srivastav P (2018) Chapter 7: Significance of food-grade microorganisms in biodegradation and biotransformation of harmful toxic compounds. In: Bioprocess technology in food and health: potential applications and emerging scope, pp 227–250
Patra S, Nayak R, Patro S, Pradhan B, Sahu B, Behera C et al (2021) Chemical diversity of dietary phytochemicals and their mode of chemoprevention. Biotechnol Rep 30:e00633
Prabhakar P, Banerjee M (2020a) Opportunities and challenges in current drug development process. J Crit Rev 7(18):3547–3563
Prabhakar P, Banerjee M (2020b) Nanotechnology in drug delivery system: challenges and opportunities. J Pharm Sci Res 12(4):492–498
Prabhakar P, Banerjee M (2020c) Antidiabetic phytochemicals: A comprehensive review on opportunities and challenges in targeted therapy for herbal drug development. Int J Pharma Res 12(2):1673–1696
Prabhakar P, Mamoni B (2021) Technical problems, regulatory and market challenges in bringing herbal drug into mainstream of modern medicinal practices. Res J Biotechnol 16:3
Prabhakar P, Banerjee M, Verma SK, Kandar P (2018) Rapid extraction and characterization of E. coli phospholipids and study of its potential application in liposomal drug delivery systems. Int J Eng Res Gen Sci 6(6):34–47
Prabhakar P, Mukherjee S, Kumar A, Kumar S, Verma DK, Dhara S et al (2023a) Optimization of MAE for Carica papaya phytochemicals, and its in silico, in vitro, and ex vivo evaluation: for functional food and drug applications. Food Biosci 54:102861
Prabhakar P, Mukherjee S, Kumar A, Kumar S, Verma DK, Dhara S et al (2023b) Optimization of microwave-assisted extraction (MAE) of key phenolic compounds from pigeon pea (Cajanus cajan L.), their characterization, and measurement of their anti-diabetic and cytotoxic potential. J Food Meas Charact 17(6):5697–5720
Prabhakar P, Pasayat AK, Kumar S, Banerjee M (2023c) An in silico evaluation of dipeptidyl peptidase-4 (DPP-4) inhibiting potential of Cajanus cajan, Coccinia grandis and Enhydra fluctuans. In: 2023 International Conference on Communication, Circuits, and Systems (IC3S). IEEE, pp 1–6
Puri V, Nagpal M, Singh I, Singh M, Dhingra GA, Huanbutta K et al (2022) A comprehensive review on nutraceuticals: therapy support and formulation challenges. Nutrients 14(21):4637
Raposo A, Saraiva A, Ramos F, Carrascosa C, Raheem D, Bárbara R, Silva H (2021) The role of food supplementation in microcirculation—a comprehensive review. Biology 10(7):616
Ray A, Joshi J, Gulati K (2016) Regulatory aspects of nutraceuticals: An Indian perspective. In: Nutraceuticals. Academic, pp 941–946
Rizzo M, Colletti A, Penson PE, Katsiki N, Mikhailidis DP, Toth PP et al (2023) Nutraceutical approaches to non-alcoholic fatty liver disease (NAFLD): a position paper from the International Lipid Expert Panel (ILEP). Pharmacol Res 189:106679
Ronis MJ, Pedersen KB, Watt J (2018) Adverse effects of nutraceuticals and dietary supplements. Annu Rev Pharmacol Toxicol 58:583–601
Sammons HM, Gubarev MI, Krepkova LV, Bortnikova VV, Corrick F, Job KM et al (2016) Herbal medicines: challenges in the modern world. Part 2. European Union and Russia. Exp Rev Clin Pharmacol 9(8):1117–1127

Samtiya M, Aluko RE, Dhewa T, Moreno-Rojas JM (2021) Potential health benefits of plant food-derived bioactive components: an overview. Food Secur 10(4):839
Santini A, Novellino E (2018) Nutraceuticals-shedding light on the grey area between pharmaceuticals and food. Expert Rev Clin Pharmacol 11(6):545–547
Shrestha R, Shrestha S, KC B, Shrestha S (2021) Evaluation of nutritional supplements prescribed, its associated cost and patients knowledge, attitude and practice towards nutraceuticals: a hospital based cross-sectional study in Kavrepalanchok, Nepal. Plos one 16(6):e0252538
Silva T, Jesus M, Cagigal C, Silva C (2019) Food with influence in the sexual and reproductive health. Curr Pharm Biotechnol 20(2):114–122
Sofowora A, Ogunbodede E, Onayade A (2013) The role and place of medicinal plants in the strategies for disease prevention. Afr J Tradit Complement Altern Med 10(5):210–229
Thakkar S, Anklam E, Xu A, Ulberth F, Li J, Li B et al (2020) Regulatory landscape of dietary supplements and herbal medicines from a global perspective. Regul Toxicol Pharmacol 114:104647
Tsiaka T, Kritsi E, Tsiantas K, Christodoulou P, Sinanoglou VJ, Zoumpoulakis P (2022) Design and development of novel nutraceuticals: current trends and methodologies. Forum Nutr 2(2):71–90
Umegaki K (2021) Regulatory aspects of nutraceuticals: Japanese perspective. In: Nutraceuticals. Academic Press, pp 1299–1307
Verma B, Popli H (2018) Regulations of nutraceuticals in India & US. Pharma Innov 7(7):811–816
Verma DK, Niamah AK, Patel AR, Thakur M, Sandhu KS, Chávez-González ML et al (2020) Chemistry and microbial sources of curdlan with potential application and safety regulations as prebiotic in food and health. Food Res Int 133:109136
Verma DK, Thakur M, Singh S, Tripathy S, Gupta AK, Baranwal D et al (2022) Bacteriocins as antimicrobial and preservative agents in food: biosynthesis, separation and application. Food Biosci 46:101594
Vilas-Boas AA, Pintado M, Oliveira AL (2021) Natural bioactive compounds from food waste: toxicity and safety concerns. Food Secur 10(7):1564
Vishvakarma P, Mandal S, Verma A (2023) A review on current aspects of nutraceuticals and dietary supplements. Int J Pharma Prof Res 14(1):78–91
von Schoen-Angerer T, Manchanda RK, Lloyd I, Wardle J, Szöke J, Benevides I et al (2023) Traditional, complementary and integrative healthcare: global stakeholder perspective on WHO's current and future strategy. BMJ Glob Health 8(12):e013150
Vora LK, Gholap AD, Jetha K, Thakur RRS, Solanki HK, Chavda VP (2023) Artificial intelligence in pharmaceutical technology and drug delivery design. Pharmaceutics 15(7):1916
Wallace TC (2015) Twenty years of the dietary supplement health and education act—how should dietary supplements be regulated? J Nutr 145(8):1683–1686
Wang H, Chen Y, Wang L, Liu Q, Yang S, Wang C (2023) Advancing herbal medicine: enhancing product quality and safety through robust quality control practices. Front Pharmacol 14:1265178
Xiang J, Liu F, Wang B, Chen L, Liu W, Tan S (2021) A literature review on maillard reaction based on milk proteins and carbohydrates in food and pharmaceutical products: advantages, disadvantages, and avoidance strategies. Food Secur 10(9):1998
Xiao X, Tang J, Mao Y, Li X, Wang J, Liu C et al (2019) Guidance for the clinical evaluation of traditional Chinese medicine-induced liver injuryIssued by China Food and Drug Administration. Acta Pharm Sin B 9(3):648–658
Yao HT, Li CC, Chang CH (2019) Epigallocatechin-3-gallate reduces hepatic oxidative stress and lowers CYP-mediated bioactivation and toxicity of acetaminophen in rats. Nutrients 11(8):1862
Yuan H, Luo Z, Ban Z, Reiter RJ, Ma Q, Liang Z et al (2022) Bioactive peptides of plant origin: distribution, functionality, and evidence of benefits in food and health. Food Funct 13(6):3133–3158

The manufacturer's authorised representative in the EU is Springer Nature Customer Service Centre GmbH, Europaplatz 3, 69115 Heidelberg, Germany. If you have any concerns regarding our products, please contact ProductSafety@springernature.com

Printed and bound by CPI Group (UK) Ltd, Croydon, CR0 4YY
01/10/2026
02225949-0018